14. $a_n = ar^{n-1}$ (nth term of a geometric sequence. Chapter 8)

15. $s_n = \dfrac{ar^n - a}{r - 1}, \quad r \neq 1$ (Sum of a geometric sequence. Chapter 8)

16. $I = Prt$ (Formula for simple interest. Chapter 8)

17. $P_n = P(1 + i)^n$ (Formula for compound interest. Chapter 8)

18. $A_m = P\left(1 + \dfrac{i}{q}\right)^m$ (Formula for compound interest. Chapter 8)

19. $E_q = \left[\left(1 + \dfrac{i}{q}\right)^q - 1\right] \cdot 100$ (Effective interest rate. Chapter 8)

20. $A_c = Pe^{i}$ (Continuous interest. Chapter 8)

21. $A = \dfrac{D}{j}[(1 + j)^{m+1} - (1 + j)]$ (Periodic deposits. Chapter 8)

22. $\overline{A} = \dfrac{D}{j}[(1 + j)^m - 1]$ (Amount for an ordinary annuity. Chapter 8)

23. $V = \dfrac{P}{j}[1 - (1 + j)^{-m}]$ (Present value of an annuity. Chapter 8)

24. $Z = \dfrac{X - \mu}{\sigma}$ (Conversion formula for X- to Z-numbers. Chapter 9)

FINITE MATHEMATICS FOR THE MANAGERIAL, SOCIAL, AND LIFE SCIENCES

FINITE MATHEMATICS FOR THE MANAGERIAL, SOCIAL, AND LIFE SCIENCES

F. Lane Hardy
DeKalb Community College

with the assistance of
Bevan K Youse
Emory University

West Publishing Company

St. Paul New York Los Angeles San Francisco

Copyediting: Beth Watson
Artwork: Bob Gallison
Interior design: Judith Fletcher Getman
Composition: York Graphic Services, Inc.
Production management: York Production Services
Cover art: Based on a design by Primo Angeli

Printed in the United States of America

Library of Congress Cataloging in Publication Data

Hardy, F. Lane (Flournoy Lane), 1928–
Finite mathematics for the managerial, social and life sciences.
Includes index.
1. Mathematics—1961– . I. Title.
QA39.2.H352 1984 510 83-23260
ISBN 0-314-77900-0

CONTENTS

PREFACE

The concerns which have motivated this new text mainly center around the following topics: exercises, applications, level of presentation, and depth of coverage.

EXERCISES

The philosophy which has been applied in constructing the exercises holds that the students who usually take a course based on this material have very little experience with mathematics and have not yet learned how to effectively study the subject. Therefore, I feel that it is necessary to present a large number of routine exercises which are closely related to the material of the text and examples. These drill exercises are carefully graded and gradually increased in difficulty until the point is reached at which the student is asked to do some independent thinking. I also strongly feel that it is very important to keep clearly in mind the objectives appropriate for each topic covered and to avoid complicating the learning process by asking the student to think of several unfamiliar or difficult topics at the same time. For example, students often have considerable difficulties mastering the simplex method. Frequently, the difficulties can be traced to the fact that the problems they are asked to work are so filled with complicated fractions, they cannot concentrate on the routine steps of the algorithm. It would appear to be appropriate to keep the arithmetic very simple until the complexities of the algorithm have been mastered.

APPLICATIONS

Applications are the basis for a book of this type and have been given careful consideration. We have provided applications that are honest, within the reach of the student, and have some basis in reality. However, they do not require a specialized knowledge of a discipline in order to master the idea. In addition, we have tried to avoid using the word *application* simply because a specialized vocabulary is used.

LEVEL OF PRESENTATION

A major objective of this course is to teach students to read and utilize mathematics. Therefore, the material has been carefully graded and we have attempted to provide students with sufficient opportunities to test their understanding as they read by including *practice problems*. The practice problems, along with the answers that follow immediately, should not only help but also encourage the student to read the text. In addition to the practice problems inserted after examples, exercises have also been keyed to the examples and problems have been labeled according to content when this has been appropriate.

DEPTH OF COVERAGE

The question of coverage is important as it relates to courses the student is likely to take after completing this one. For example, it is quite normal for a student to follow this course with one in statistics. It is therefore important that the concepts in statistics be covered in more than a superficial way. Such a student should be able to go into statistics knowing, for example, what Chebyshev's Theorem says. Another topic which is becoming increasingly important in finite mathematics is combinatorics. In this text, combinatorics has been covered in a separate chapter in order to emphasize its importance in its own right and not simply as a few rules one needs to do certain kinds of probability problems.

The only mathematical prerequisite for using this book is high-school algebra, and since many schools do not have a formal prerequisite for the material covered here, a review of elementary algebra has been included in Chapter 1.

It is a pleasure to acknowledge the help of Peter Marshall, Editor at West Publishing Company. His help, understanding, encouragement, and enthusiasm made this project possible; an author could not ask for a better editor than Peter. I also wish to thank Barbara Fuller for her excellent job as Production Editor. I am greatly indebted to Ms. Beth Watson for her expert copyediting of the manuscript; her many contributions have greatly improved the writing style. Special thanks go to Lorraine N. Ruff and Emma Hardin for their expert typing. Sincere appreciation is extended to my friend and coauthor Professor Bevan K Youse, without whose help this project would never have seen the light of day. I am especially grateful to the many reviewers for their criticisms, suggestions, and comments. Their names appear on the following page under the heading "Acknowledgments."

F. Lane Hardy
Doraville, Ga.
December 12, 1983

Directed graph for suggested chapter coverage

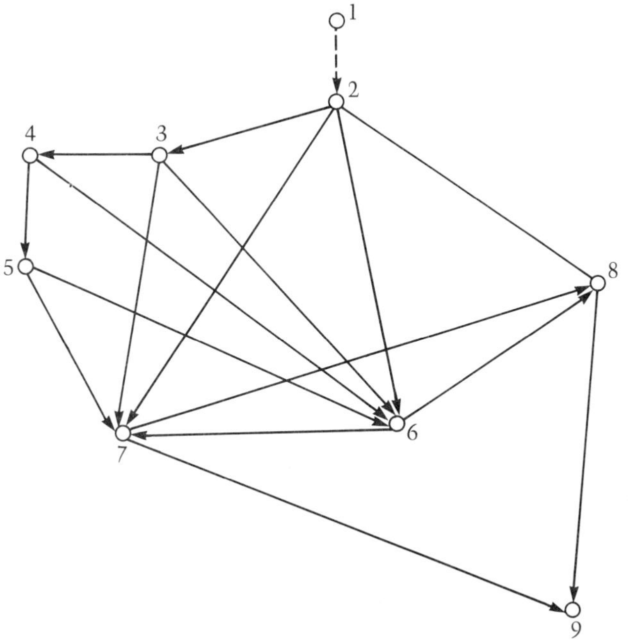

ACKNOWLEDGMENTS

Richard L. Andrews
University of Wisconsin, Oshkosh

Patricia Bannantine
Marquette University, Wisconsin

Carole Bernett
William Rainey Harper College, Illinois

Robert Bix
University of Michigan, Flint

Frederick J. Carter
St. Mary's University, Texas

Mary Jane Causey
University of Mississippi

Samuel Councilman
California State University, Long Beach

Carl Cuneo
Essex Community College, Maryland

Joe S. Evans
Middle Tennessee State

David Fonken
Austin Community College

William Roger Fuller
University of Portland

Herbert Gindler
San Diego State University

Joseph Howard
New Mexico Highlands University

D. Frank Hsu
Fordham University, New York

Alonzo F. Johnson
West Virginia University

Bruce A. Jensen
Portland State University

Robert A. Moreland
Texas Tech University

Rose Novey
Saginaw Valley State College, Michigan

Susan S. Nutter
Prestonburg Community College

Walter Roth
University of North Carolina, Charlotte

Wesley Sanders
Sam Houston State University

Bernard L. Schroeder
University of Wisconsin, Platteville

Linda B. Sherrell
Louisiana State University

John Spellman
Southwest Texas State University

Robin G. Symonds
Indiana University of Kokomo

CHAPTER

1

SETS, NUMBERS, AND COORDINATE SYSTEMS

1

1.1 INTRODUCTION

This chapter contains a brief review of some of the more commonly used basic concepts needed in this text. It is assumed that students have had a moderate exposure to most of the topics included. Therefore, the amount of time devoted to material here will depend on the needs of those students using other parts of the text. Instructors may wish to assign portions of this chapter as related topics are encountered in later chapters. ■

1.2 THE LANGUAGE OF SETS

When we speak of a set we may refer to it in one of two ways:

(a) We list or tabulate its members.

(b) We give a description or common property which characterizes the set.

braces, brackets, sets

To illustrate, we may say "the set whose members are 1, 2, 3" or "the set whose members are the first three counting numbers."* In each of these two cases it is convenient to use the **braces** or **bracket** notation. This means that the set is described by enclosing the members of the set between braces (also called "curly" brackets). The above set is indicated in this way:

$$\{1, 2, 3\}$$

or

$$\{x \mid x \text{ is one of the first three counting numbers}\}.$$

Both of these descriptions describe the same set. The second one should be read as "the set of all x such that x is one of the first three counting numbers." As another example of this notation, we may speak of the set consisting of Borman, Anders, and Lovell by saying

$$\{x \mid x \text{ is a human who orbited the moon in 1968}\}.$$

Consider the sets

$$\{6, 8, 10\} \qquad \text{and} \qquad \{4, 5, 6, 7, 8, 9, 10\}.$$

---

*The *natural numbers* or *counting numbers* are 1, 2, 3, 4, . . . , and the *whole numbers* are 0, 1, 2, 3 ,4, ...

It is evident that each member of the first set is a member of the second. For this reason we call the set {6, 8, 10} a *subset* of the second one.

subset

DEFINITION OF SUBSET Let A and B represent sets. Then A *is a* **subset** *of set* B if and only if each member of A is a member of B.

A common notation for subset is $\subseteq$, that is, we write "$A \subseteq B$" as a shorthand for the statement "A is a subset of B." For example, the statements below are true:

$$\{4, 6, 8\} \subseteq \{4, 5, 6, 7, 8, 9, 10\},$$
$$\{9, 1, 8, 2\} \subseteq \{y \mid y \text{ is a counting number}\}.$$

To indicate that a set A is *not* a subset of a set B, we write $A \not\subseteq B$.

A convenient way to represent a set S is to choose a closed figure such as the one illustrated in Figure 1.1 and let the points in the interior of the figure represent the elements of S. Using this device, we may indicate that $A \subseteq B$ by enclosing the figure for A within the figure for B (see Figure 1.2).

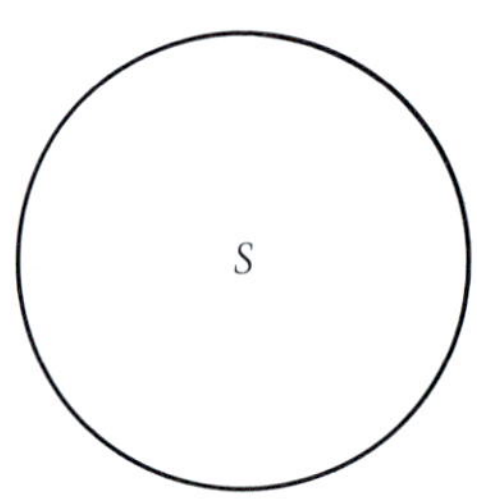

FIG. 1.1. Set S.

FIG. 1.2. $A \subseteq B$.

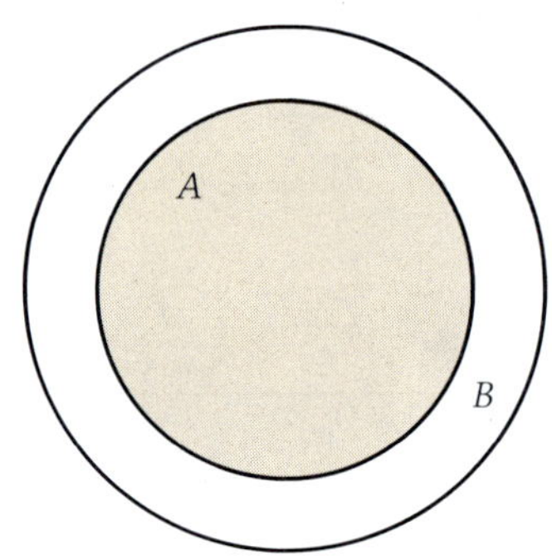

If an object a is a member of a set A, we use the symbol "$\in$" to indicate this fact and write: $a \in A$. We read the statement "$a \in A$" as "a is a member of A." Using this notation we may correctly make the following statements:

$$4 \in \{3, 4, 5\},$$
$$8 \in \{z \mid z \text{ is a counting number}\}.$$

With the set membership notation the definition of "subset" may be stated as follows:

$$A \subseteq B \quad \text{if and only if} \quad \text{for all } x, \quad \text{if } x \in A, \quad \text{then } x \in B.$$

It is convenient to employ the concept of a set which has no elements or members. This set is denoted by the symbol $\varnothing$ and called the **empty set** or **null set.** If X is any set, then no element of $\varnothing$ (since $\varnothing$ has no elements) fails to be a member of X. Therefore, it follows that $\varnothing$ is a subset of X (by the definition of subset). Hence, we have for all sets X,

empty set
null set

$$\varnothing \subseteq X.$$

set equality

DEFINITION OF SET EQUALITY If A and B are sets, then $A = B$ if and only if $A \subseteq B$ and $B \subseteq A$.

An immediate consequence of the definition is the fact that *every set is a subset of itself*, i.e., if A is any set, $A \subseteq A$.

Subsets

EXAMPLE 1 Determine all the subsets of the set $\{a, b, c\}$.

Solution: Since the set $\{a, b, c\}$ contains three elements, each subset will contain no members, 1 member, 2 members, or 3 members:

Subsets having 0 members: $\varnothing$

Subsets having 1 member: $\{a\}$, $\{b\}$, $\{c\}$.

Subsets having 2 members: $\{a, b\}$, $\{a, c\}$, $\{b, c\}$.

Subsets having 3 members: $\{a, b, c\}$.

PRACTICE PROBLEM Determine the subsets of the set $\{x, y\}$.

Answer: $\varnothing$, $\{x\}$, $\{y\}$, $\{x, y\}$. ■

In discussions involving sets it is sometimes convenient to choose some set which contains as subsets all other sets under consideration. The set so chosen is called the **universal set** and is often denoted by U. For example, the counting numbers

universal set

1, 2, 3, 4, . . . may be under discussion. In this case, it would be appropriate to take $U = \{1, 2, 3, 4, \ldots\}$. In another situation, plane geometry may be our concern. Then we could take U to be the set of all points in a plane.

Let us suppose that a set A is a subset of some universal set U, that is, $A \subseteq U$. We may then talk about the set of all those elements that are in U but not in A. This set is called the *complement* of A.

complement

DEFINITION OF COMPLEMENT The **complement** of a set A in U is the set of all elements in U that are not in A.

Since, in a particular discussion, it will be understood that all elements are in U, the definition of complement may be shortened to simply: *the set of elements not in A.* In Figure 1.3, the universal set U is represented by the points in the rectangle; and the complement of A, denoted by A', is represented by all the points of U *outside* the circle A.

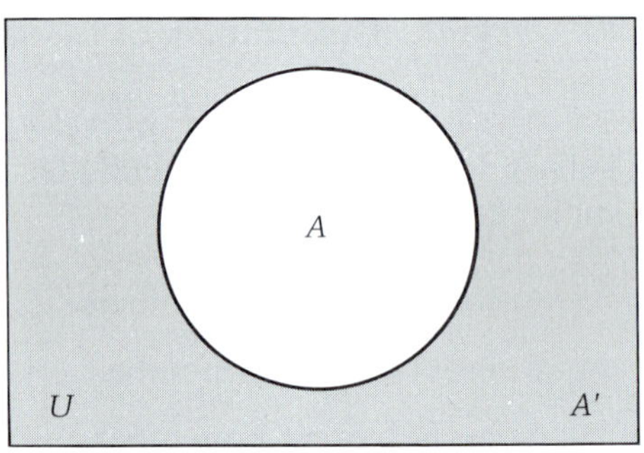

FIG. 1.3.

ILLUSTRATION Let $U = \{1, 2, 3, 4, 5, 6, 7, 8, 9, 10\}$. Then if

$$A = \{3, 5, 7, 9\},$$

we have

$$A' = \{1, 2, 4, 6, 8, 10\}.$$

(See Figure 1.4.)

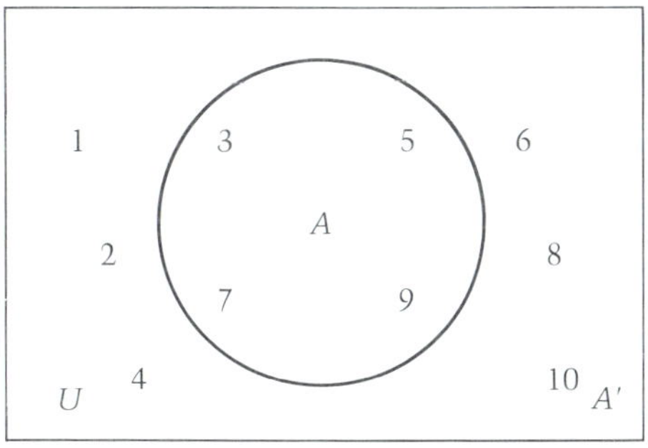

FIG. 1.4.

We have just seen that it is possible to associate with a given set A another set A'. Analogously, it is possible to determine, in many different ways, a set which is associated with two given sets A and B. Two of the most common ways of doing this lead to the *union* and *intersection* of two sets.

DEFINITION OF UNION AND INTERSECTION Let A and B be two sets. Then

union (a) the **union** of A, B is $\{y \mid y \in A \text{ or } y \in B\}$;

intersection (b) the **intersection** of A, B is $\{w \mid w \in A \text{ and } w \in B\}$.

Standard notations for union and intersection are $\cup$ and $\cap$, respectively. Using these symbols, we may write the definitions as follows:

$$A \cup B = \{y \mid y \in A \text{ or } y \in B\},$$

$$A \cap B = \{w \mid w \in A \text{ and } w \in B\}.$$

Venn diagram Figure 1.5 is an example of a **Venn diagram.** The rectangle represents the set U. In Figure 1.6 the points in the overlapping

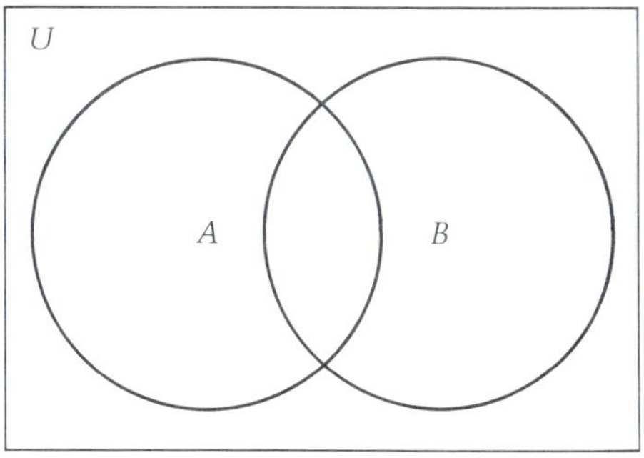

FIG. 1.5.

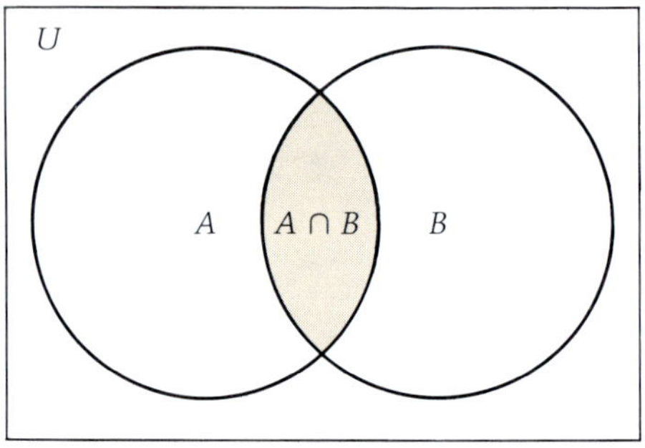

FIG. 1.6.

portion of A and B are in *both* circles and, hence, represent points of $A \cap B$. In Figure 1.7 the entire region of the sets A and B has been shaded and represents $A \cup B$; this is appropriate since a point in this region is either in the circle for A or for B or for both.

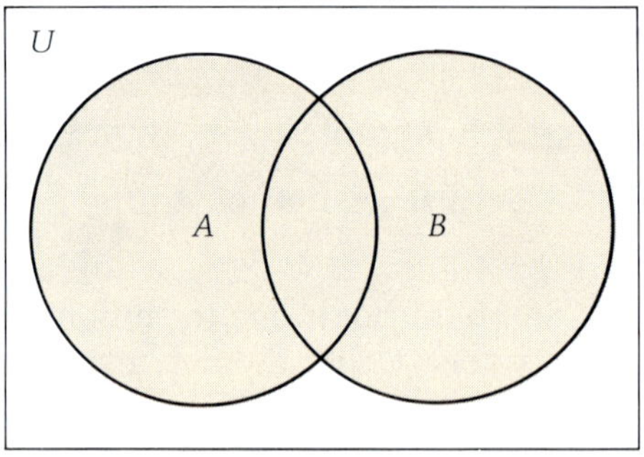

FIG. 1.7.

ILLUSTRATION For $A = \{2, 4, 5, 8, 9\}$ and $B = \{1, a, 2, t, 8\}$ we compute $A \cup B$ and $A \cap B$. In this case, the statement "$y \in A$ or $y \in B$" is true if and only if $y = 2, 4, 5, 8, 9, 1, a,$ or t. Therefore,

$$A \cup B = \{2, 4, 5, 8, 9, 1, a, t\}.$$

Similarly, the statement "$w \in A$ and $w \in B$" is true if and only if $w = 2$ or $w = 8$. Hence,

$$A \cap B = \{2, 8\}.$$

(See Figure 1.8.)

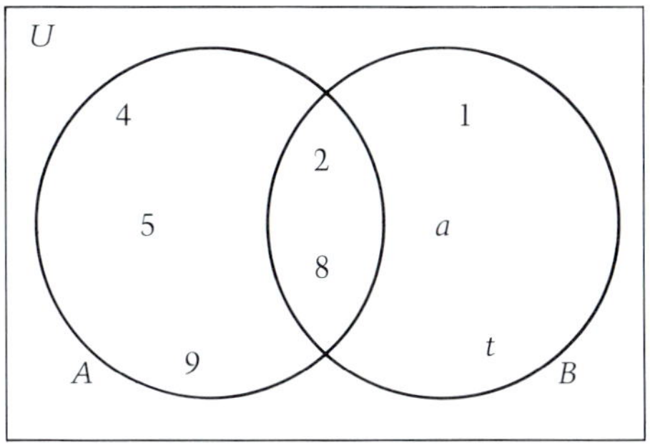

FIG. 1.8.

EXERCISES

Set A

In Exercises 1 through 12 list the members of each set.

1. $\{x \mid x = 1 \text{ or } x = 2\}$.
2. $\{x \mid x$ is a whole number less than 8$\}$.
3. $\{z \mid z$ is a counting number greater than 5 and less than 8$\}$.
4. $\{t \mid t$ is a letter of the word "music"$\}$.
5. $\{p \mid p \in \{1, 2, 3\}\}$.
6. $\{y \mid y$ is a whole number and $3y < 20\}$.
7. $\{w \mid w$ is a positive factor of 3$\}$.
8. $\{x \mid x - 3 = 8\}$.
9. $\{r \mid r$ is a whole number and is a factor of 24$\}$.
10. $\{z \mid (z - 1)(z - 2) = 0\}$.
11. $\{v \mid v \in \{2, 7\} \text{ and } v \in \{3, 9\}\}$.
12. $\{m \mid m$ is a number and is one of the first 5 positive multiples of 3$\}$.

Set B

In Exercises 13 through 20 use the brace/bracket notation to express each of the given sets. For example, both of the following are correct answers for Exercise 13:

$$\{2, 3, 4, 5, 6, 7\}, \qquad \{x \mid x \text{ is a whole number and } 1 < x < 8\}$$

13. The set of whole numbers greater than 1 and smaller than 8.
14. The set of natural numbers that are less than 15 and are divisible by 5.
15. The set of squares of natural numbers that are less than 25.
16. The set of natural numbers whose squares are less than 25.
17. The set of odd natural numbers that are not greater than 10.

18. The set whose members are elements of $\{2, 4, 6, 8\}$ or $\{1, 2, 3, 4\}$.
19. The set whose members are the elements of the set $\{1, 2, 3, 4, 5, \ldots, 30\}$ and are either multiples of 10 or perfect squares.
20. The set whose members are elements of the set $\{1, 2, 3, 4, 5, \ldots, 30\}$ and are divisible by either 3 or 5.

Set C

In Exercises 21 through 26, for each pair of sets A, B decide which of these possibilities is correct:

(a) $A \subseteq B$ (b) $B \subseteq A$ (c) $A = B$ (d) none of these

21. $A = \{x \mid x = 1 \text{ or } x = 2\}$, $B = \{x \mid x \text{ is a whole number}\}$
22. $A = \{y \mid y \text{ is a whole number}\}$, $B = \{w \mid w \text{ is an even whole number}\}$
23. $A = \{m \mid m \text{ is a triangle}\}$, $B = \{n \mid n \text{ is a triangle whose angle sum is 180 degrees}\}$
24. $A = \{x \mid x^2 = 1\}$, $B = \{1, -1\}$
25. $A = \{z \mid z \text{ is a positive factor of } 12\}$, $B = \{y \mid y \text{ is a positive factor of } 6\}$
26. $A = \{z \mid z \text{ is a positive factor of } 12\}$, $B = \{z \mid z \text{ is a positive factor of } 15\}$

Set D–Subsets

27. Let $S = \{a, b, c, d\}$.
    (a) List all subsets of S having 1 element.
    (b) List all subsets of S having 2 elements.
    (c) List all subsets of S having 3 elements.
28. Let $S = \{a, b, c, d, e\}$. List the subsets of S having (a) 1 element (b) 2 elements (c) 3 elements

Set E–Set Operations

In Exercises 29 through 36 let

$$W = \{1, 2, 3, 4, 5\},$$
$$X = \{2, 4, 6, 8, 10, 12\},$$
$$Y = \{1, 3, 5, 7, 9, 11\},$$

and determine each of the sets by listing its members.

29. $W \cup W$ 30. $X \cap X$ 31. $W \cup X$ 32. $W \cap X$
33. $W \cup Y$ 34. $W \cap Y$ 35. $X \cup Y$ 36. $X \cap Y$

For Exercises 37 through 42 let

$$U = \{1, 2, 3, 4, 5, 6, 7, 8, 9, 10, 11, 12\}$$

and determine A' in each case.

37. $A = \{1, 3, 5, 7, 9, 11\}$
38. $A = \{1, 5, 7, 9\}$
39. $A = \{2, 10\}$
40. $A = \varnothing$
41. $A = U$
42. $A = \{3, 6, 9, 12\}$

In Exercises 43 through 49, $X = \{1, 2\}$ and $Y = \{1, 2, 3, 4\}$. Decide in each case whether the statement is true.

43. $X \subseteq Y$
44. $X \subseteq (X \cup Y)$
45. $Y \subseteq (X \cup Y)$
46. $X \subseteq (X \cap Y)$
47. $Y \subseteq (X \cap Y)$
48. $(X \cap Y) \subseteq (X \cup Y)$
49. $(X \cup Y) \subseteq (X \cap Y)$

1.3 COUNTING ELEMENTS IN SETS

Problems frequently arise in finite mathematics which have to do with the number of elements in sets. In this section we briefly discuss one aspect of this kind of problem (Chapter 6 contains additional material related to this). More specifically, the following problem will be considered:

> Given two sets A and B. If we know the number of elements in A and B, how can the number of elements in $A \cup B$ be determined?

The notation $\#(A)$ will be used to denote the number of elements in set A. If $\#(A)$ is a whole number, that is, one of the numbers 0, 1, 2, 3, . . . , we say that A is a **finite** set.

finite set

Let us take a simple example to begin with. Consider the sets

$$A = \{1, 2, 3, 4, 5, 6\} \quad \text{and} \quad B = \{2, 4, 5, 8, 10\}.$$

Then $A \cup B = \{1, 2, 3, 4, 5, 6, 8, 10\}$ and

$$\#(A) = 6, \qquad \#(B) = 5, \qquad \#(A \cup B) = 8.$$

We might expect to find that the number of elements of A plus the number of elements of B is the number of elements of $A \cup B$. It is easy to see that this is not true since

$$\#(A) + \#(B) = 5 + 6 = 11,$$

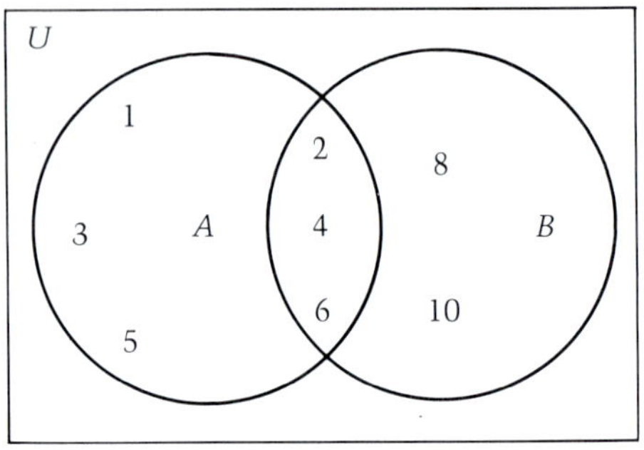

FIG. 1.9.

and

$$\#(A \cup B) = 8.$$

The Venn diagram given in Figure 1.9 shows why this is not true. If we take the sum $\#(A) + \#(B)$, the elements 2, 4, 6 are counted *twice*—once in A and once in B. Therefore, the number of elements in $A \cup B$, that is $\#(A \cup B)$, is $\#(A) + \#(B)$ *less* the number of elements *common* to A and B. Since this kind of reasoning may be applied to any two finite sets we have the following formula.

addition formula for sets

THE ADDITION FORMULA FOR SETS If A and B are finite sets, then

$$\#(A \cup B) = \#(A) + \#(B) - \#(A \cap B).$$

Number of students enrolled in language courses

EXAMPLE 1 If the Registrar's Office shows 115 students enrolled in French courses, 86 enrolled in German courses, and 28 students taking both French and German, how many are studying one or the other of these two languages?

Solution: Let F and G be the set of French and German students, respectively. Then we are given that $\#(F) = 115$, $\#(G) = 86$, and $\#(F \cap G) = 28$. The problem is to determine the number of students in F or G:

$$\#(F \cup G).$$

For this we may use the Addition Formula:

$$\#(F \cup G) = \#(F) + \#(G) - \#(F \cap G) = 115 + 86 - 28 = 173.$$

PRACTICE PROBLEM In Example 1 find the number of students taking at least one of the languages French or German if $F \cap G = \varnothing$.

Answer: 201. ■

A slightly more complicated problem occurs when we consider three sets A, B, C and try to determine $\#(A \cup B \cup C)$ in terms of $\#(A)$, $\#(B)$, $\#(C)$. Consider the Venn diagram in Figure 1.10 and the regions labeled 1, 2, 3, 4. We compare the numbers $\#(A \cup B \cup C)$ and $\#(A) + \#(B) + \#(C)$. The number $\#(A) + \#(B) + \#(C)$ counts the elements of regions 2, 3, 4 twice. Therefore, $\#(A) + \#(B) + \#(C)$ is possibly larger than $\#(A \cup B \cup C)$. Since we wish to count the elements of regions 2, 3, 4 only once, we may subtract $\#(A \cap B)$, $\#(A \cap C)$, and $\#(B \cap C)$ from $\#(A) + \#(B) + \#(C)$. This will result in the number

$$\#(A) + \#(B) + \#(C) - \#(A \cap B) - \#(A \cap C) - \#(B \cap C). \quad (1)$$

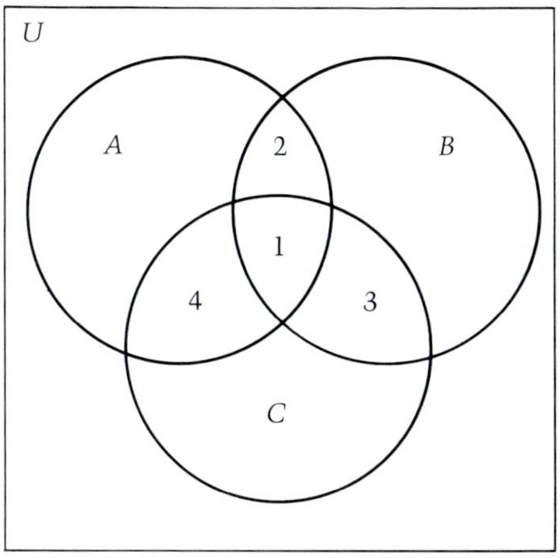

FIG. 1.10.

The number given in (1) is still not exactly right. In $\#(A) + \#(B) + \#(C)$ the elements of region 1 in Figure 1.10 have been *included* three times and in the expression $-\#(A \cap B) - \#(A \cap C) - \#(B \cap C)$ they have been *excluded* three times. Therefore, (1) does not count the elements of region 1 at all! Thus, to make sure that the elements in region 1 are counted, we add

$\#(A \cap B \cap C)$ to (1). This will result in the following formula.

$$\#(A \cup B \cup C) = \#(A) + \#(B) + \#(C) - \#(A \cap B) - \#(A \cap C) - \#(B \cap C) + \#(A \cap B \cap C). \quad (2)$$

Formula (2) and the Addition Formula are two instances of what is called the *inclusion-exclusion principle.*

Market survey

EXAMPLE 2 The Nix-All Drug Company makes aspirin, mouthwash, and nose drops. After making a survey of 1500 consumers, the company reports the results listed in the accompanying table.

Product	Number of Consumers Using Product
Aspirin	1356
Mouthwash	1232
Nose drops	853
Aspirin and mouthwash	911
Aspirin and nose drops	815
Mouthwash and nose drops	618
All three products	405

Show that the data given in this survey cannot be correct.

Solution: Let us use the following notation:

A: set of persons who use aspirin,

M: set of persons who use mouthwash,

N: set of persons who use nose drops.

Note that the total number of persons in the survey who use at least one of these three products cannot exceed the number of persons surveyed. Another way to state this is to say that $\#(A \cup M \cup N)$ cannot be greater than 1500, the total number of surveyed consumers. The given information may be restated using the notation which we have selected:

$$\#(A) = 1356, \qquad \#(A \cap M) = 911,$$
$$\#(M) = 1232, \qquad \#(A \cap N) = 815,$$
$$\#(N) = 853, \qquad \#(M \cap N) = 618,$$
$$\#(A \cap M \cap N) = 405.$$

Substituting this information into formula (2), we get

$$\begin{aligned}\#(A \cup M \cup N) &= \\ &\#(A) + \#(M) + \#(N) - \#(A \cap M) - \#(A \cap N) - \#(M \cap N) \\ &\qquad + \#(A \cap M \cap N) \\ &= 1356 + 1232 + 853 - 911 - 815 - 618 + 405 \\ &= 1502.\end{aligned}$$

This result shows that the information reported cannot possibly be correct since it implies that more persons use at least one of the three products than the total number in the survey. The reported information is therefore inconsistent.

PRACTICE PROBLEM Assuming the information in Example 2 is correct for nose drops and mouthwash, how many of the 1500 persons surveyed use either nose drops or mouthwash?

Answer: 1467. ■

Problems of this type can sometimes be solved easily by using a Venn diagram as the following example shows.

Library use

EXAMPLE 3 The school library reported that on May 2 a total of 731 students used the library facilities. It was also reported that on that date particular facilities were used as shown in the accompanying table.

Library Facility	Number of Users of Facility
Card catalog	256
Reference collection	183
Browsing room	318
Catalog and Reference	95
Catalog and Browsing	152
Reference and Browsing	103
All three facilities	67

On May 2 how many of the 731 students used none of the three facilities given in the table above?

Solution: We use the following notation and draw a Venn diagram as shown in Figure 1.11.

> C: the set of students using the card catalog,
>
> R: the set of students using the reference collection,
>
> B: the set of students using the browsing room.

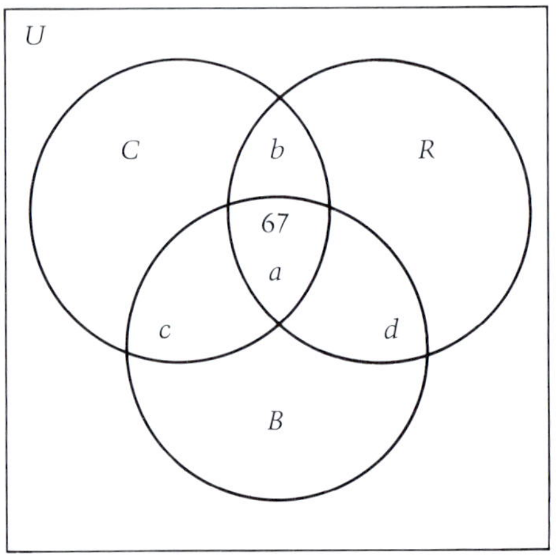

FIG. 1.11.

We see that $\#(C \cap R \cap B) = 67$ and this number may be placed in region a on the diagram. Region a and region b together make up $C \cap R$, and since we are given that $\#(C \cap R) = 95$, it follows that in region b we must have $95 - 67 = 28$. In a similar way we have $152 - 67 = 85$ in region c and $103 - 67 = 36$ in region d. Now the Venn diagram appears as in Figure 1.12. At this point we have already accounted for $67 + 28 + 85 = 180$ elements in C, and since $\#(C) = 256$, there must be an additional $256 - 180 = 76$ elements in C outside of regions a, b, c. Similarly, we find that in regions a, b, and d, there are $67 + 28 + 36 = 131$ elements, and in set R there are $183 - 131 = 52$ elements outside of regions a, b, and d. In regions a, c, and d, there are $67 + 85 + 36 = 188$ elements, and in set B there are $318 - 188 = 130$ elements outside of a, c, and d. This information is recorded in Figure 1.13, and from the diagram, by simply adding the numbers in all the regions, it is possible to determine the number of students who used at least one of the three facilities:

$$76 + 85 + 67 + 28 + 52 + 36 + 130 = 474.$$

FIG. 1.12.

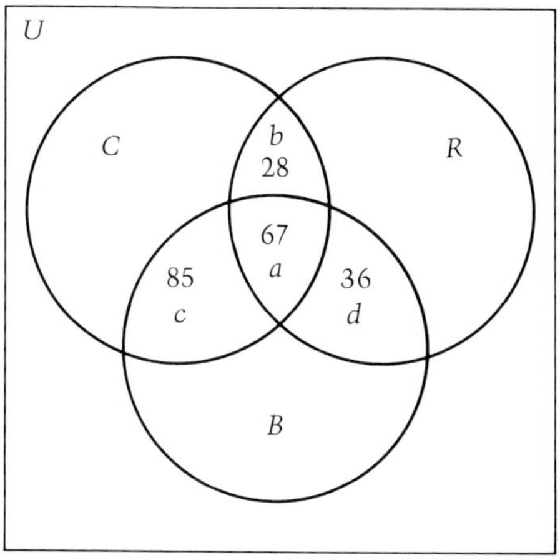

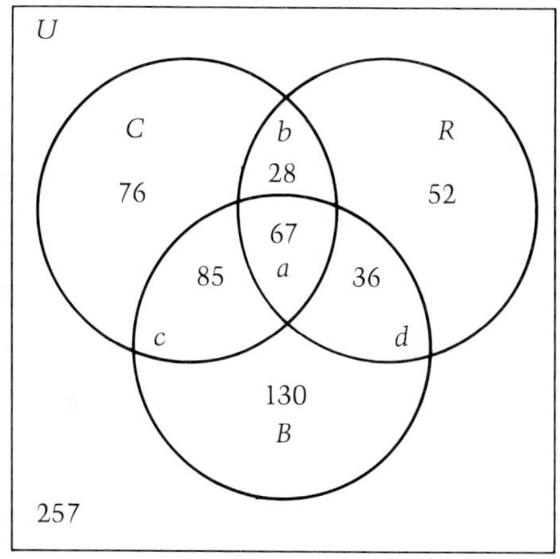

FIG. 1.13.

Since 731 students visited the library on the day in question, there are 731 elements in the set U. Inside the three circles, there are 474 elements, and hence, outside there are $731 - 474 = 257$ elements. Therefore, on the second of May, 257 of the 731 students used none of the three facilities.

PRACTICE PROBLEM The library of Example 3 reported the following information on May 3 about the use of facilities. Use Venn diagrams to determine the number of students who used none of the facilities on this date. Total number of students in library on May 3 was 763.

Library Facility	Number of Users of Facility
Card catalog	258
Reference collection	188
Browsing room	318
Catalog and Reference	87
Catalog and Browsing	154
Reference and Browsing	112
All three facilities	76

Answer: 276. ■

EXERCISES

Set A–Addition Formula

In Exercises 1 through 8 use the Addition Formula to solve the problems.

1. If $\#(A) = 5$, $\#(B) = 6$, and $\#(A \cap B) = 3$, find $\#(A \cup B)$.
2. If $\#(A) = 24$, $\#(B) = 31$, and $\#(A \cap B) = 11$, find $\#(A \cup B)$.
3. If $\#(A) = 32$, $\#(A \cap B) = 11$, and $\#(A \cup B) = 40$, find $\#(B)$.
4. If $\#(A \cap B) = 20$, $\#(A \cup B) = 60$, and $\#(B) = 25$, find $\#(A)$.
5. If $\#(A) = 35$, $\#(B) = 18$, and $\#(A \cup B) = 50$, find $\#(A \cap B)$.
6. Given $\#(A) = 25$, $\#(B) = 30$, $\#(C) = 36$, $\#(A \cap B) = 3$, $\#(A \cap C) = 5$, $\#(B \cap C) = 10$, and $\#(A \cap B \cap C) = 1$. Find $\#(A \cup B \cup C)$.
7. Given $\#(A) = 41$, $\#(B) = 30$, $\#(A \cap B) = 3$, $\#(A \cap C) = 5$, $\#(B \cap C) = 10$, $\#(A \cap B \cap C) = 2$, and $\#(A \cup B \cup C) = 70$. Find $\#(C)$.
8. Given $\#(A) = 39$, $\#(B) = 30$, $\#(C) = 15$, $\#(A \cap B) = 5$, $\#(A \cap C) = 6$, $\#(B \cap C) = 7$, and $\#(A \cup B \cup C) = 70$. Find $\#(A \cap B \cap C)$.

In Exercises 9 and 10 test for consistency:

9. $\#(A \cup B \cup C) = 70$, $\#(A) = 39$, $\#(B) = 37$, $\#(C) = 24$, $\#(A \cap B) = 24$, $\#(A \cap C) = 15$, $\#(B \cap C) = 20$, $\#(A \cap B \cap C) = 14$.
10. $\#(A \cup B \cup C) = 70$, $\#(A) = 39$, $\#(B) = 37$, $\#(C) = 24$, $\#(A \cap B) = 24$, $\#(A \cap C) = 15$, $\#(B \cap C) = 20$, $\#(A \cap B \cap C) = 29$.

Set B

Class composition

11. In the senior class at Mill University, 84 students are Latin majors, 75 are Greek majors, and 23 have a double major in Greek and Latin. How many members of the class are majoring in either Latin or Greek?

Divisibility

12. How many of the numbers from 1 to 24 (inclusive) are divisible by
(a) 6? (b) 8? (c) 6 or 8?

Divisibility

13. How many of the numbers from 1 to 244 (inclusive) are divisible by
(a) 6? (b) 8? (c) 6 or 8?

Divisibility

14. How many of the numbers from 1 to 700 (inclusive) are divisible by
(a) 2? (b) 5? (c) 7?
(d) 2 and 5? (e) 2 and 7? (f) 5 and 7?
(g) 2, 5, and 7? (h) 2 or 5 or 7?

Market survey

15. In a market survey the Backlog Cereal Company asked 800 customers about their tastes in cereal. As a result of the survey the company reported the data provided in the accompanying table.

Kind of Cereal	Number in Survey Who Liked Cereal
Sugar-coated	395
Cereal with raisins	354
Cereal with marshmallows	186
Sugar-coated with raisins	101
Sugar-coated with marshmallows	94
Cereal with raisins and marshmallows	72
Sugar-coated with raisins and marshmallows	13

On the basis of this information, how many of the 800 consumers liked none of these three varieties of cereal?

Medical data

16. A large metropolitan hospital compiled information over a year's period on patients admitted for treatment of cancer. It was found that these patients often had maladies other than cancer, such as heart disease, diabetes, and kidney problems. The accompanying table gives data on 597 patients. From this information determine the number of patients having at least one of the three ailments and also the number having none of these.

Ailment	Number of Patients Having Ailment
Heart disease	162
Diabetes	103
Kidney problems	94
Heart disease and diabetes	47
Heart disease and kidney problems	32
Diabetes and kidney problems	73
All three ailments	18

1.4 THE REAL NUMBERS

The **real number system** is composed of the following sets of numbers:

whole numbers

Whole numbers: 0, 1, 2, 3, 4, 5, . . .

integers

Integers: . . . , −5, −4, −3, −2, −1, 0, 1, 2, 3, 4, 5, . . .

rational numbers
repeating decimals

Rationals: Ratios of integers such as $\frac{1}{2}$, $-\frac{3}{4}$, $\frac{101}{23}$; these can be expressed as finite or infinite repeating decimals.

irrational numbers
nonrepeating decimals

Irrationals: Real numbers which are not rational such as $\sqrt{2}$, $\sqrt{131}$, π; these can be expressed as infinite nonrepeating decimals.

Rational numbers according to the description just given are ratios a/b, where a and b are any integers except that b cannot be 0.

The usual decimal notation for real numbers will be used frequently in this text. The decimal 2.387, for example, is a short way of writing the rational number

$$\frac{2387}{1000},$$

and $\frac{1}{3}$ as a decimal is 0.333. . . or $0.\overline{3}$. The "bar" over 3 in the second notation means that 3 repeats indefinitely. The number $0.\overline{3}$ is an example of an infinite repeating decimal. Another example is $-45.003\overline{72}$. In this example, 72 is the infinitely repeating part. The number 2.387 can be written as $2.387\overline{0}$ but it is customary not to indicate the infinite string of 0s. For this reason, 2.387 is called a **terminating decimal** in contrast to the nonterminating $0.\overline{3}$.

terminating decimal

In dealing with rational numbers we must sometimes convert from one representation to another, i.e., convert from a fraction to a decimal and vice versa. If we are given a terminating decimal such as 2.387 and are asked to express this as a fraction, the solution is easy:

$$2.\underset{\substack{\uparrow\\ 1}}{3}\;\underset{\substack{\uparrow\\ 2}}{8}\;\underset{\substack{\uparrow\\ 3}}{7} = \frac{2\;3\;8\;7}{1\;\underset{\substack{\uparrow\\ 1}}{0}\;\underset{\substack{\uparrow\\ 2}}{0}\;\underset{\substack{\uparrow\\ 3}}{0}}$$

The rule of thumb in cases like this is to count the smallest number of places the decimal point must be moved to produce a whole number. In the above example, it must be moved three places to the right. We then write the resulting whole number divided by the appropriate number, where the appropriate number is 1 followed by a number of 0s equal to the number of places the point was moved. Another example is

$$0.00315 = \frac{315}{100000}.$$

If the rational number is not terminating, such as $2.\overline{387}$, then the solution is not quite as easy. We are given the infinite repeating decimal

$$2.\overline{387} = 2.387387387\ldots$$

and we wish to find a fraction a/b which is equal to $2.\overline{387}$. The technique is similar to that used with terminating decimals; the basic idea again is to move the decimal point. To see how this works let

$$x = 2.387387387\ldots.$$

Now we want to move the point like this:

$$x = 2.387\ 387\ 387\ 387\ldots. \tag{1}$$

To accomplish this multiply by 1000:

$$1000x = 2\,387.387\ 387\ 387\ldots$$

Next, write Eq. (1) under the above equation and subtract:

$$\begin{aligned} 1000x &= 2\,387.387\,387\,387\ldots \\ x &= 2.387\,387\,387\ldots \\ \hline 999x &= 2\,385.000\,000\,000\ldots \end{aligned}$$

Dividing by 999 in the last equation produces the answer

$$x = \frac{2385}{999} = \frac{265}{111}.$$

Decimal conversion

EXAMPLE 1 Convert the following decimal numbers to fractions:

(a) 825.95507. (b) $825.\overline{25}$.

Solution

(a)

$$825.95507 = \frac{82{,}595{,}507}{100{,}000}.$$

(b) Let $x = 825.\overline{25}$ and multiply this by 100:

$$\begin{aligned} 100x &= 82{,}525.\overline{25} \\ x &= 825.\overline{25} \\ \hline 99x &= 81{,}700.00 \end{aligned}$$

Then

$$x = \frac{81{,}700}{99}.$$

PRACTICE PROBLEM Convert the following decimal numbers to fractions:

(a) 17,386.05. (b) $17{,}386.\overline{05}$.

Answers: (a) $\frac{1738605}{100}$. (b) $\frac{1721219}{99}$. ■

The problem of converting fractions to decimals presents no unusual difficulties. For any given fraction we only have to carry out the indicated division. For example, to convert the fraction $\frac{106}{5}$ to a decimal, simply divide 5 into 106 in the usual way or use a hand-held calculator. In either case the answer obtained is 21.2.

Fraction conversion

EXAMPLE 2 Convert the following fractions to decimal notation.

(a) $\frac{35}{8}$. (b) $\frac{1}{11}$.

Solution

(a) By division,

$$\frac{35}{8} = 4.375.$$

(b) $\frac{1}{11} = 0.\overline{09}$.

PRACTICE PROBLEM Convert (a) $\frac{258}{25}$ and (b) $\frac{1}{15}$ to decimals.

Answers: (a) 10.32. (b) $0.0\overline{6}$. ■

Although the decimal notation is most often used in computations, it has important theoretical uses as well. For example, we may distinguish between rational and irrational numbers through this notation. Each decimal is either terminating or nonterminating, and each nonterminating decimal is either repeating or nonrepeating. The rational numbers are those decimals which are either terminating or repeating (see the diagram below).

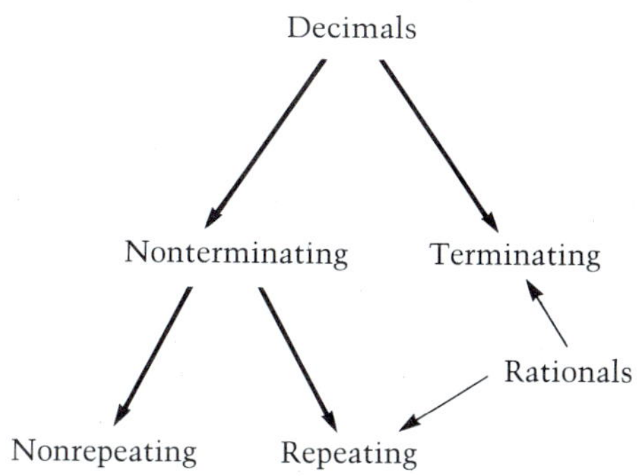

Frequent use will be made of the **real number line** in this text. It is a geometric way of representing the real numbers. Some of the basic features will be covered at this point. Choose any line and a positive direction on the line to the right:

Now choose any point to represent 0 and any point to the right of 0 to represent 1:

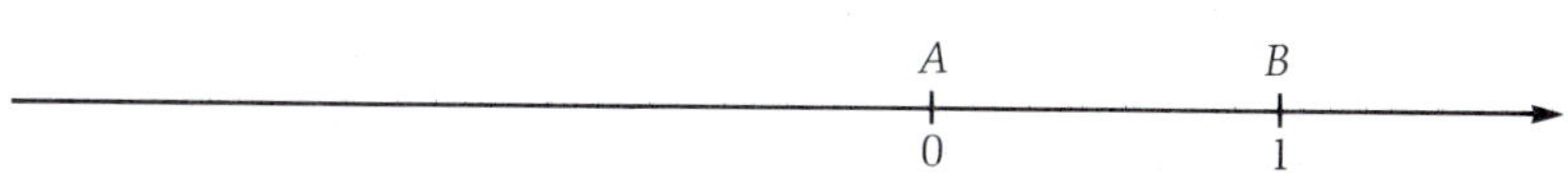

The unit length from A to B is now repeated indefinitely on the line:

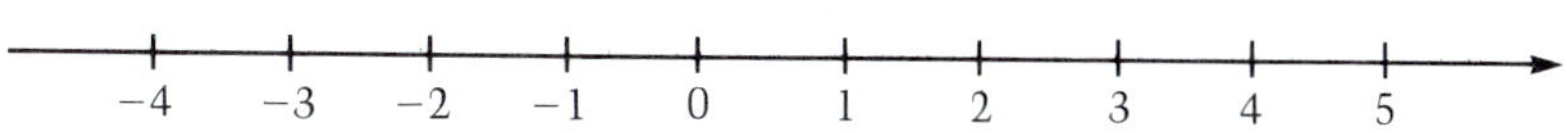

At this stage we have a representation of all the integers on the line. The rational numbers may now be added to the line by making appropriate divisions of the unit segments. For example, the number 12.5 is located as the midpoint of the segment from 12 to 13:

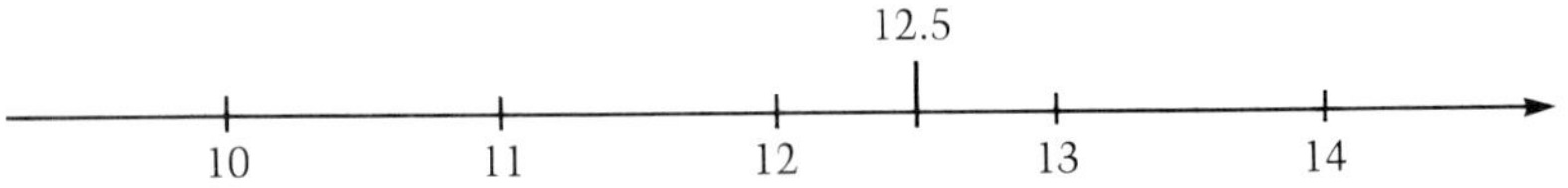

The number -2.25 is located on the line between -2 and -3:

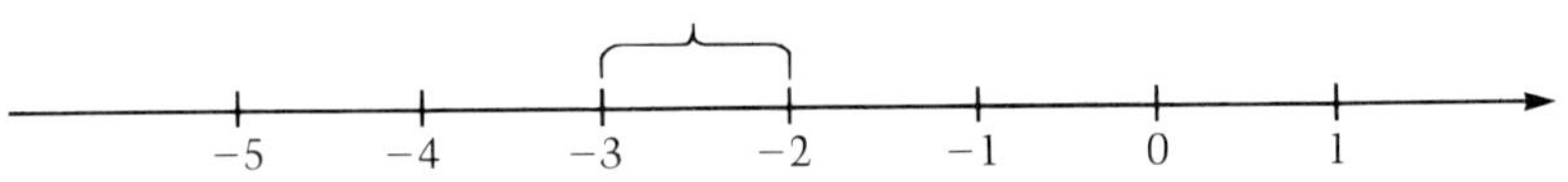

The precise point is found by dividing the segment from -3 to -2 into 100 equal segments:

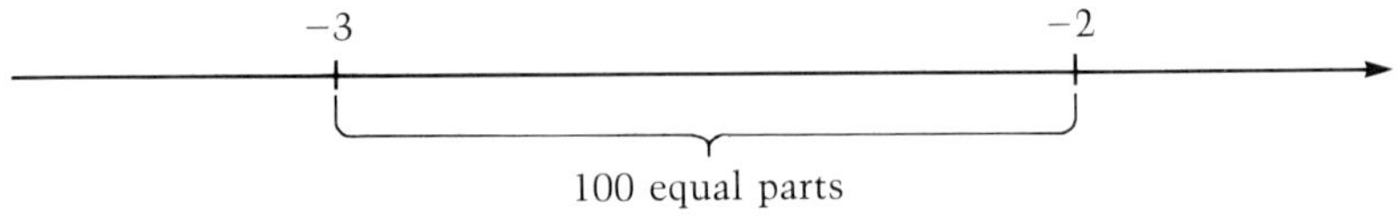

Once this has been done, we count from -2 for 25 points in the direction of -3. The point arrived at in this way is the representation of -2.25:

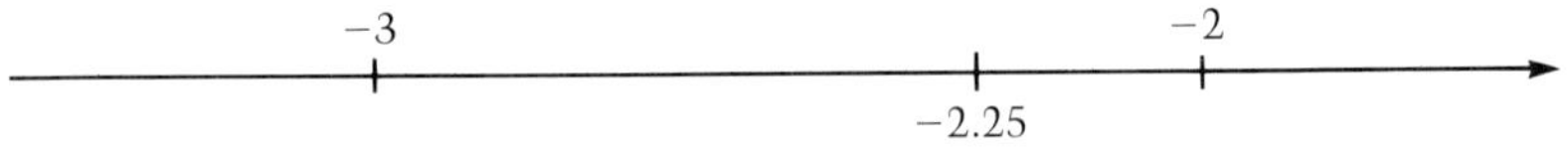

Numbers such as $\sqrt{2}$ may be located on the number line by finding a decimal approximation for the number. For example,

$$\sqrt{2} \doteq 1.41421356.$$

This approximation was obtained by using a hand-held calculator and is accurate to 8 places. Now we may use any of the numbers

1.4
1.41
1.414
1.4142
1.41421

1.414214
1.4142136
1.41421356

to approximate the position of $\sqrt{2}$ on the line. The one we choose will depend on how accurate we wish to be.

Number line plot

EXAMPLE 3 Plot the numbers below on a number line.

(a) 5.7. (b) -4.8. (c) $-\sqrt{6}$.

Solution

(a) 5.7 lies between 5 and 6:

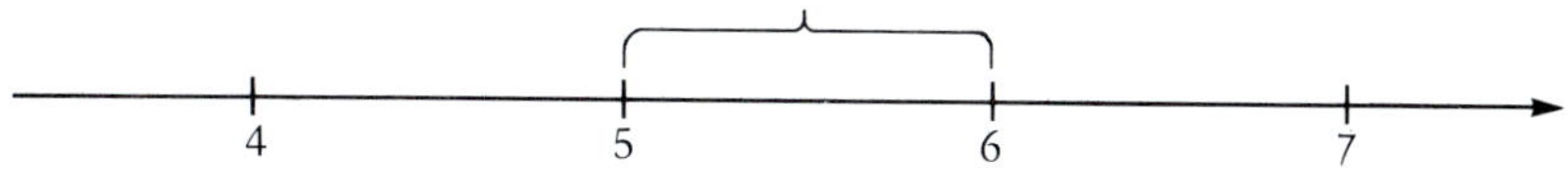

Divide the segment from 5 to 6 into ten equal parts:

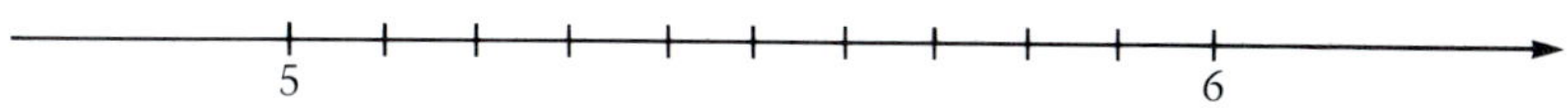

From 5 we count 7 points in the direction of 6. The point arrived at then represents the number 5.7:

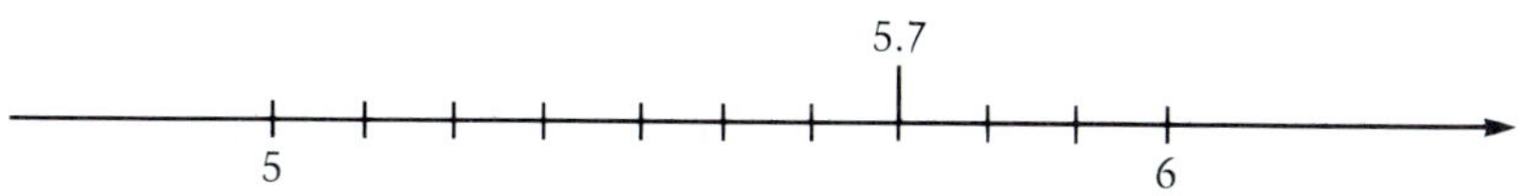

(b) The number -4.8 lies between -4 and -5:

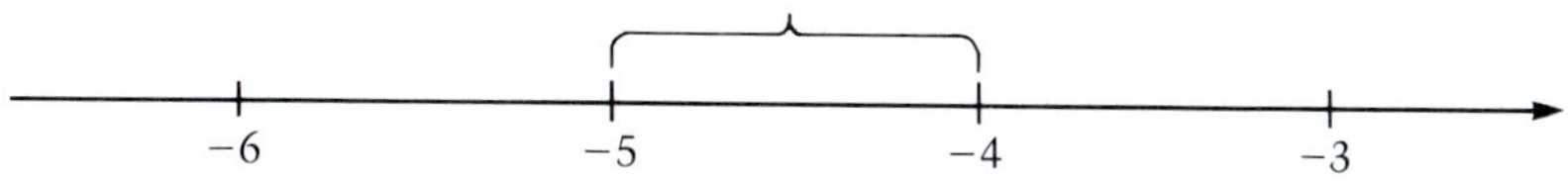

Divide the segment into ten equal parts:

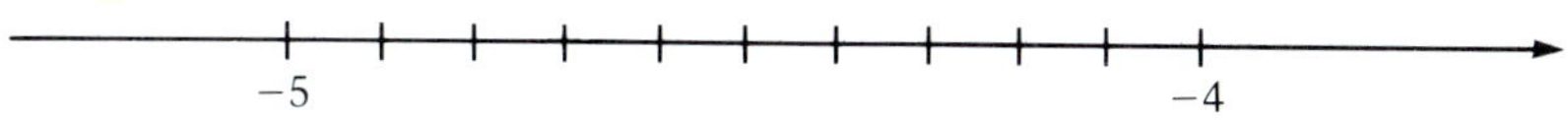

Now count from -4 toward -5 eight points. This brings you to the correct representation of -4.8:

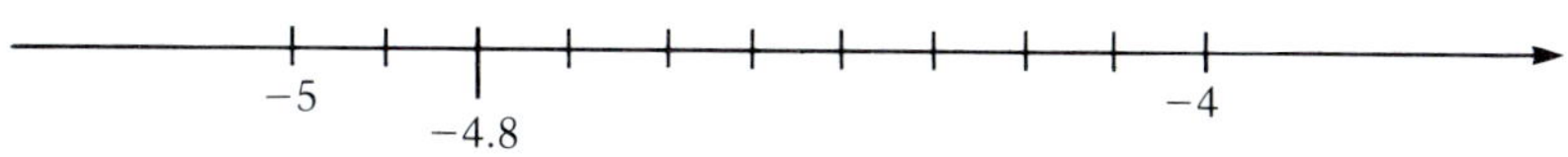

(c) We can only approximate the position of $-\sqrt{6}$. To do this we take a rational approximation of $\sqrt{6}$ using a hand-held calculator:

$$\sqrt{6} \doteq 2.44948974.$$

We will use one decimal place for this approximation. Therefore, the point to find on the number line is -2.4 and this lies between -2 and -3:

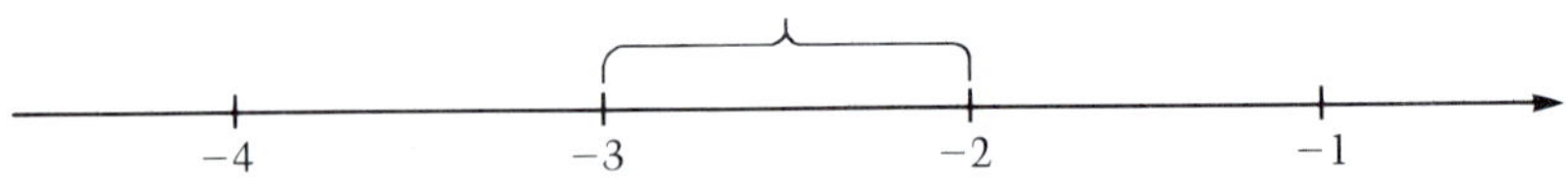

Dividing the segment from -2 to -3 into ten equal parts produces the following representation:

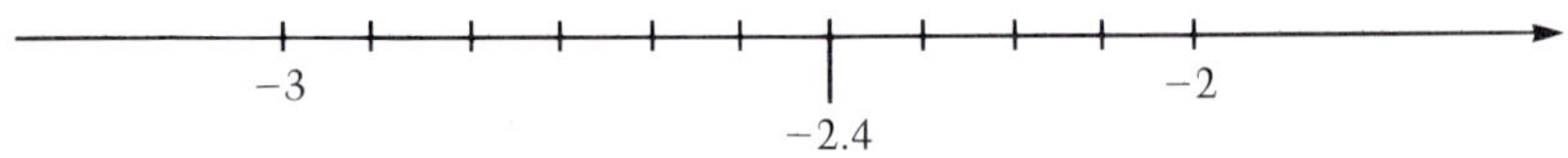

EXERCISES

Set A–Converting Decimals to Fractions

In Exercises 1 through 18, express each decimal as a fraction in lowest terms.

1. 1.1 2. 3.1 3. 6.2 4. 9.4
5. $6.\overline{8}$ 6. $7.\overline{4}$ 7. 2.81 8. 0.91
9. $8.\overline{37}$ 10. $8.\overline{24}$ 11. $5.\overline{34}$ 12. $47.\overline{64}$
13. $-2.\overline{452}$ 14. $0.\overline{766}$ 15. $-80.7\overline{52}$
16. $98.2\overline{67}$ 17. $5.6\overline{71}$ 18. $0.009\overline{29}$

Set B–Converting Fractions to Decimals

In Exercises 19 through 44, express each fraction as a decimal.

19. $\frac{3}{2}$ 20. $\frac{5}{2}$ 21. $\frac{2}{5}$ 22. $\frac{3}{5}$
23. $\frac{7}{10}$ 24. $\frac{3}{10}$ 25. $\frac{11}{10}$ 26. $\frac{13}{10}$
27. $\frac{2}{25}$ 28. $\frac{1}{50}$ 29. $\frac{3}{100}$ 30. $\frac{7}{100}$
31. $\frac{19}{100}$ 32. $\frac{21}{100}$ 33. $-\frac{5}{8}$ 34. $-\frac{2}{3}$
35. $\frac{7}{6}$ 36. $\frac{17}{12}$ 37. $-\frac{1}{6}$ 38. $\frac{1}{12}$
39. $-\frac{4}{15}$ 40. $-\frac{145}{16}$ 41. $\frac{1}{9}$ 42. $\frac{1}{18}$
43. $-\frac{1}{27}$ 44. $\frac{100}{7}$

Set C–Locating Numbers on the Number Line

In Exercises 45 through 58, find the point on a number line which represents the given numbers. For irrational numbers approximate the location using the indicated decimal approximation.

45. 2.6
46. −3.5
47. $\frac{2}{3}$
48. 21.4
49. −2.75
50. 3.25
51. $\frac{3}{8}$
52. $-\frac{33}{8}$
53. $\sqrt{10}$ (1 decimal place)
54. $\sqrt{20}$ (1 decimal place)
55. $-\sqrt{24}$ (2 decimal places)
56. $-\sqrt{37}$ (1 decimal place)
57. $\sqrt{41}$ (2 decimal places)
58. $\sqrt{26}$ (2 decimal places)

1.5 RECTANGULAR COORDINATE SYSTEMS

rectangular coordinate system

A *rectangular coordinate system* is constructed by forming two number lines which are perpendicular to each other at their origins. This is illustrated in Figure 1.14 below. The horizontal line is the *x-axis* and the vertical the *y-axis.* Each point P in the plane may be specified relative to a coordinate system as shown in Figure 1.14. Through P we draw lines parallel to the two axes: these lines intersect the axes in points as shown. (See Figure 1.14.) The number a on the x-axis is called the **x-coordinate** of P, and b on the y-axis is called the **y-coordinate** of P. To indicate the coordinates of P, we write $P(a, b)$ or (a, b). If we are given coordinates (a, b) and are asked to find the corresponding point in the plane, the process just described is followed in reverse. We sometimes say "plot the point (a, b)." Such a phrase means that we should find the point having coordinates (a, b).

x-, y-coordinates

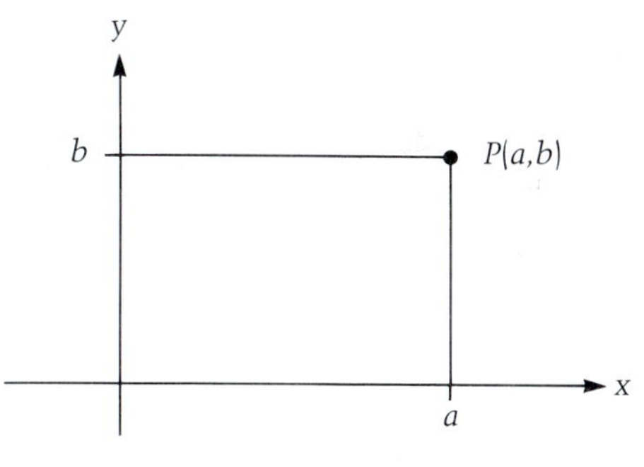

FIG. 1.14.

Plotting points

EXAMPLE 1 Plot the points (1, 1), (5, 3), (−2, 4), (−3, −1).

Solution: These points are plotted in Figure 1.15.

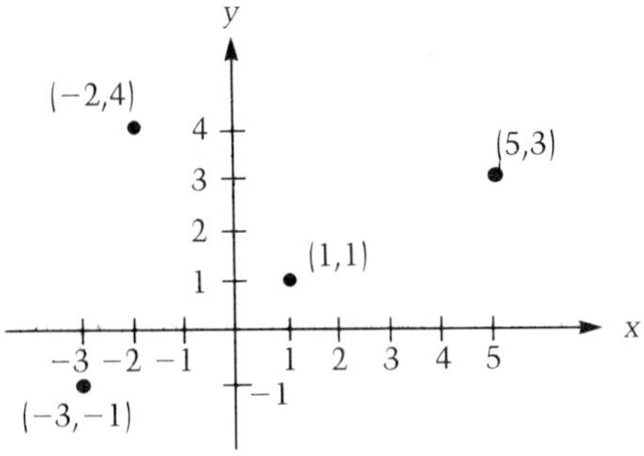

FIG. 1.15.

PRACTICE PROBLEM Plot the points (2, 3), (3, −2).

Answer:

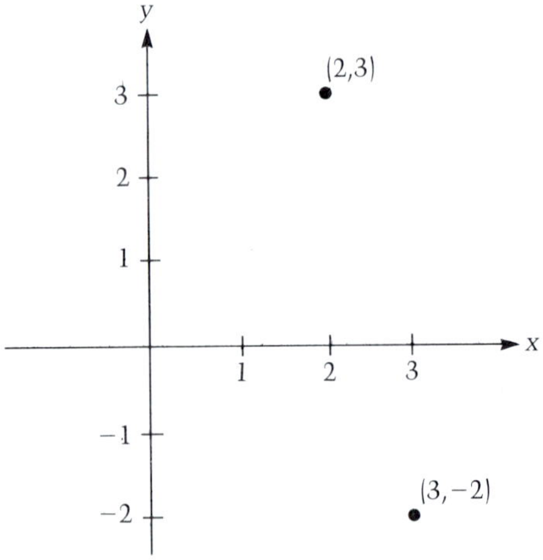

If two points $P(a, b)$ and $Q(c, d)$ are given, we may join P and Q with a straight line to obtain the *line segment* PQ. (See Figure 1.16.) It is important for us to know the midpoint of PQ and also the length. We denote the length of PQ by $\overline{PQ}$.

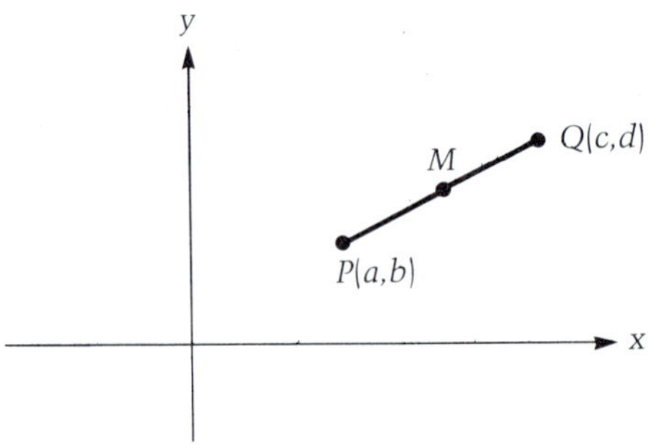

FIG. 1.16.

If $M(x, y)$ is the midpoint of the segment joining $P(a, b)$ and $Q(c, d)$, then

midpoint formula

$$x = \frac{a + c}{2} \qquad \text{and} \qquad y = \frac{b + d}{2}.$$

Finding midpoints

EXAMPLE 2 Find the midpoint of the segment joining $P(1, 1)$, $Q(5, -3)$.

Solution: If $M(x, y)$ is the midpoint, then

$$x = \frac{(1 + 5)}{2} = \frac{6}{2} = 3$$

and

$$y = \frac{(1 + (-3))}{2} = -\frac{2}{2} = -1.$$

Therefore, the midpoint is $M(3, -1)$.

PRACTICE PROBLEM Find the midpoint of $(-2, 4)$ and $(-4, 6)$.

Answer: $(-3, 5)$. ■

The formula which gives us the length of the line segment PQ is illustrated in Figure 1.17. In the figure, we have completed the right triangle PQR with short sides u, v and hypotenuse d. By the Pythagorean Theorem we know that

$$d^2 = u^2 + v^2 \qquad \text{or} \qquad d = \sqrt{u^2 + v^2}$$

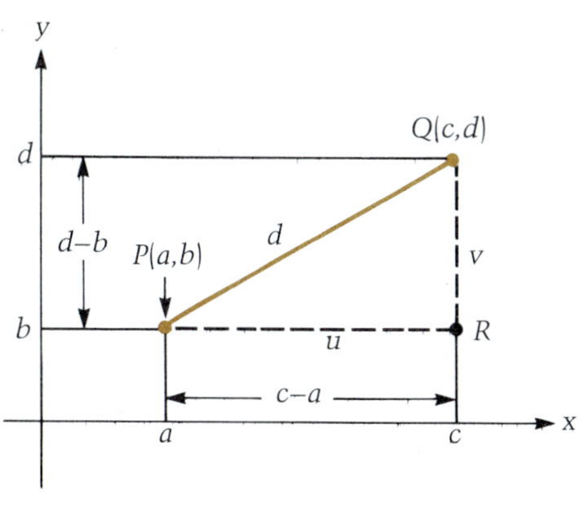

FIG. 1.17.

(d must be positive since it is a length). But from the figure we see that

$$u = c - a \quad \text{and} \quad v = d - b.$$

Therefore,

$$d = \sqrt{(c - a)^2 + (d - b)^2}.$$

This expression is called the *distance formula.*

distance formula

DISTANCE FORMULA The **distance** between any two points $P(a, b)$ and $Q(c, d)$ is given by the formula

$$d = \sqrt{(c - a)^2 + (d - b)^2}.$$

Using the distance formula

EXAMPLE 3 Find the distance between the points (3, 4) and (7, 8).

Solution: Using the distance formula we have

$$\begin{aligned} d &= \sqrt{(7 - 3)^2 + (8 - 4)^2} \\ &= \sqrt{4^2 + 4^2} \\ &= \sqrt{2 \cdot 4^2} \\ &= 4\sqrt{2}. \end{aligned}$$

PRACTICE PROBLEM Find the distance between the points (−3, 4) and (0, 8).

Answer: 5. ■

We next review the concept of slope of a line. The slope of a line conveys information about the steepness of the line. The following formula gives the desired information.

slope formula

SLOPE FORMULA If $P(a, b)$ and $Q(c, d)$ are points and $a \neq c$, then the **slope of the segment** PQ is

$$\frac{b - d}{a - c}.$$

The slope of a line can be positive, negative, or 0. The three possibilities are related to movement in the positive direction along the x-axis and are illustrated in Figure 1.18. For points $P(a, b)$, $Q(c, d)$, if $a = c$, then the slope formula does not apply and the line on P and Q is vertical. Therefore, slope is not defined for vertical lines.

FIG. 1.18.

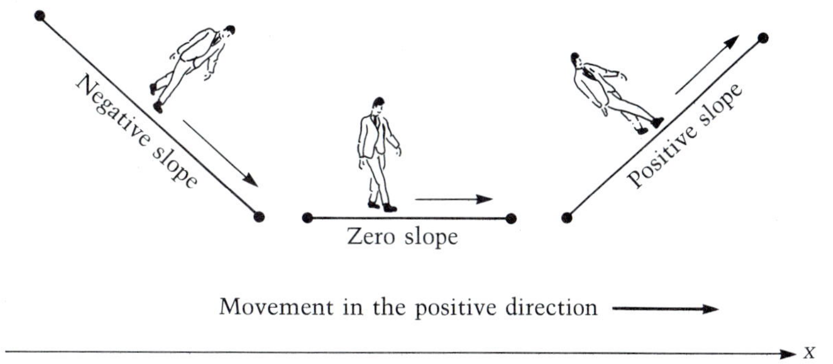

We often use the notation m for slope. The slope formula may be expressed in terms of m as

$$m = \frac{b - d}{a - c}.$$

The way in which the slope is related to steepness is illustrated in Figure 1.19.

Computing slopes

EXAMPLE 4 Compute the slopes of the lines on the following points

(a) $(-1, -1)$, $(2, 2)$. (b) $(-5, 2)$, $(0, 1)$. (c) $(1, 0)$, $(1, 5)$

Solution

(a)
$$\frac{-1 - 2}{-1 - 2} = \frac{-3}{-3} = 1.$$

Therefore, the slope is 1 and the line is rising.

(b)
$$\frac{2 - 1}{-5 - 0} = \frac{1}{-5} = -\frac{1}{5}.$$

The slope is $-\frac{1}{5}$ and the line is falling.

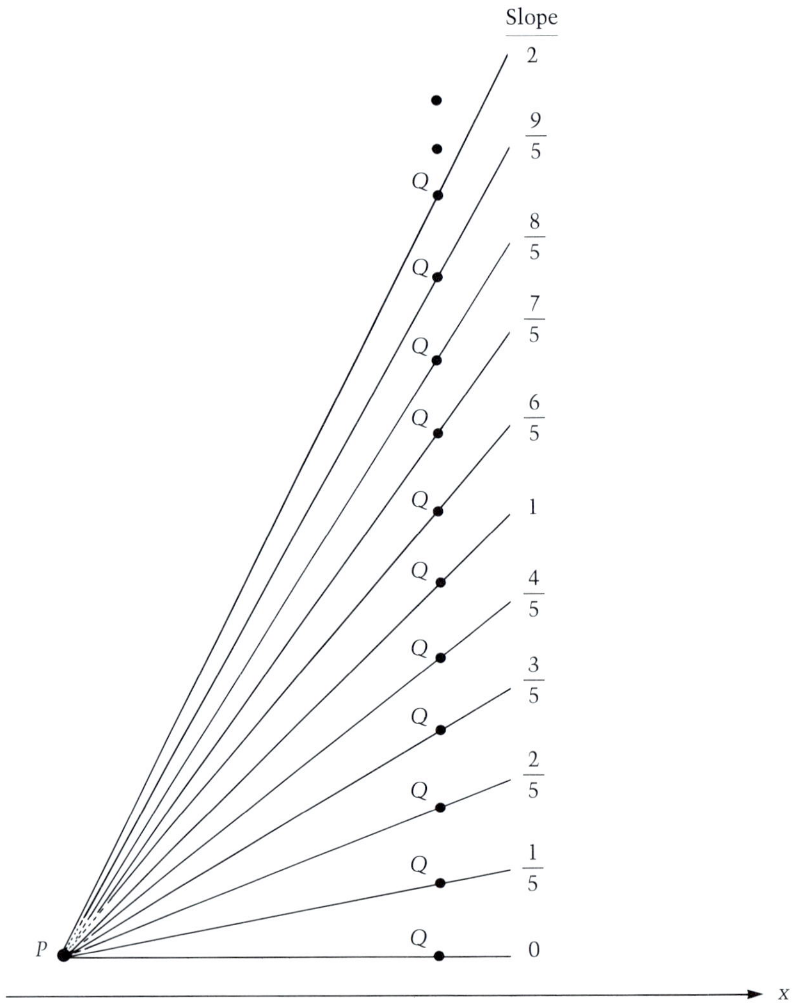

FIG. 1.19.

(c) Since the x-coordinates are the same, the line on (1, 0) and (1, 5) is vertical and has no slope.

PRACTICE PROBLEM Find the slope of the line on the points (6, −1) and (−3, 2).

Answer: $-\frac{1}{3}$. ■

Comparing steepness

EXAMPLE 5 Given the lines PQ and RS. If PQ has slope −2 and RS has slope −3.5, which line is steeper?

Solution: Since both PQ and RS have negative slopes, RS is steeper since -3.5 is smaller than -2.

PRACTICE PROBLEM If PQ has slope $\sqrt{2}$ and RS has slope 1, which line is steeper?

Answer: PQ. ■

Coordinate systems are especially useful in the study of equations. For example, suppose we are given the equation

$$x + y = 1. \tag{1}$$

We may draw a "picture" of this equation by plotting all those points (x, y) which satisfy the equation (a point having coordinates (a, b) is said to *satisfy* the equation if the substitution $x = a$ and $y = b$ results in the equation being true). That is, we plot all points which are solutions of the equation. The resulting points of the coordinate system are called the *graph* of the equation. To determine the graph of Eq. (1), we may substitute any number for x and then solve for y. In fact, it is easier to write Eq. (1) as

$$y = -x + 1. \tag{2}$$

Equation (2) tells us exactly how each second coordinate is related to the first: "Each second coordinate is obtained from the first by increasing the negative of the first by 1." The graph of (2) may be constructed from a table such as that given below.

x	$y = -x + 1$	Point (x, y) to plot
-3	4	$(-3, 4)$
-2	3	$(-2, 3)$
-1	2	$(-1, 2)$
0	1	$(0, 1)$
1	0	$(1, 0)$
2	-1	$(2, -1)$
3	-2	$(3, -2)$

linear equations

The points in the table have been plotted in Figure 1.20. Equations (1) and (2) are **linear equations** because the degree of the variables is equal to 1. Note also that the graph in Figure 1.20

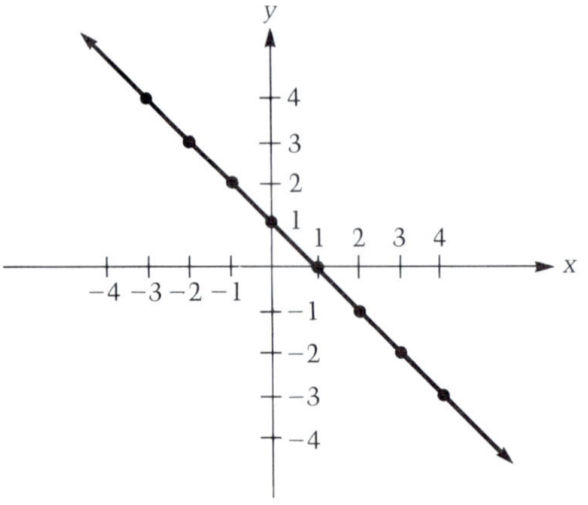

FIG. 1.20.

appears to be a straight line. This is no accident since *every linear equation has a graph which is a straight line.* This fact makes it especially easy to graph linear equations: find two points on the graph and connect them with a line. The resulting line is the graph of the equation.

Graphing linear equations

EXAMPLE 6 Graph the linear equation

$$2x + 3y = 6.$$

Solution: Choose $x = 0$ in the equation:

$$\begin{aligned} 2(0) + 3y &= 6, \\ 3y &= 6, \\ y &= 2. \end{aligned}$$

Therefore, (0, 2) is on the graph. Now choose $y = 0$:

$$\begin{aligned} 2x + 3(0) &= 6, \\ 2x &= 6, \\ x &= 3. \end{aligned}$$

Then the point (3, 0) is on the graph. Now plot the points (0, 2) and (3, 0) and connect them with a line. This gives the graph shown in Figure 1.21.

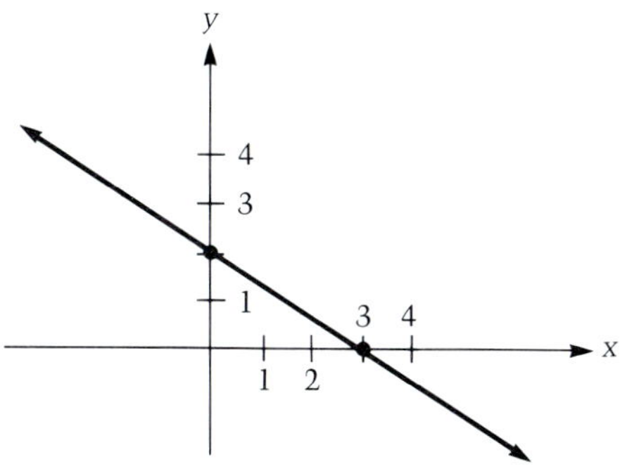

FIG. 1.21.

PRACTICE PROBLEM Graph the equation $3x - 4y = 12$.

Answer:

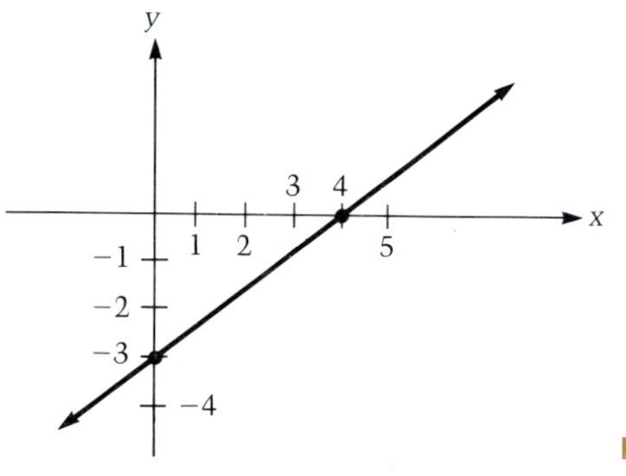

Two kinds of linear equations should be singled out. They are equations of the following types:

$$x = c \qquad [\text{or } x + (0)y = c], \tag{3}$$

$$y = d \qquad [\text{or } (0)x + y = d]. \tag{4}$$

In these equations, c and d are constants. Equation (3) has no y-term (its coefficient is 0) and Eq. (4) has no x-term. Since the coefficient of y is 0 in Eq. (3), y may have any value whatsoever. Therefore, to graph Eq. (3), we plot all those points (x, y), where $x = c$ and y takes all possible values. This results in a vertical line. (See Figure 1.22.) Equation (4) has a horizontal graph as shown in Figure 1.22.

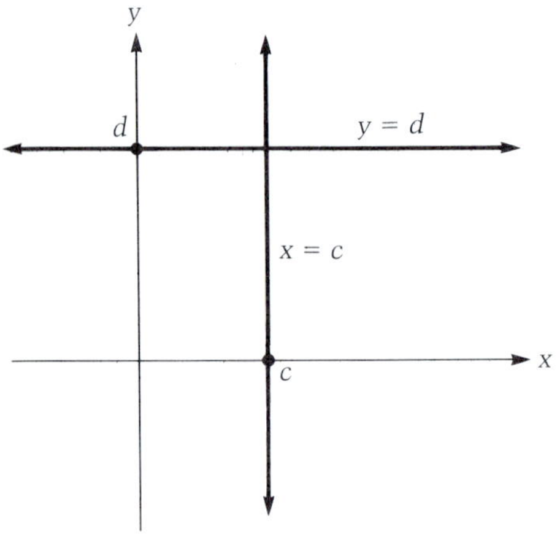

FIG. 1.22.

Describing lines

EXAMPLE 7 Give a verbal description of the graph of the following equations. Include the slope of the line in your description.

(a) $y = 5$. (b) $x = 3$. (c) $x = -4$. (d) $y = -3$.

Solution

(a) The graph of $y = 5$ is a line parallel to and 5 units above the x-axis. The line has slope 0.

(b) The graph of $x = 3$ is a line parallel to and 3 units to the right of the y-axis. The line has no slope.

(c) The graph of $x = -4$ is a line parallel to and 4 units to the left of the y-axis. The line has no slope.

(d) The graph of $y = -3$ is a line parallel to and 3 units below the x-axis. The line has slope 0.

PRACTICE PROBLEM Give a verbal description of the graph of the following equations. Include the slope of the line in your description.

(a) $x = 10$. (b) $y = -13$.

Answer: (a) A line parallel to and 10 units to the right of the y-axis. The line has no slope. (b) A line parallel to and 13 units below the x-axis. The line has slope 0. ■

EXERCISES

Set A–Plotting Points

In Exercises 1 through 10, construct a coordinate system and plot the given points.

1. $(9, -5)$
2. $(-8, -9)$
3. $(-6, 2)$
4. $(-6, -5)$
5. $(-3, 8)$
6. $(7, 1)$
7. $(-8, 4)$
8. $(0, -8)$
9. $(-5, -6)$
10. $(1, -1)$

Set B–Midpoints

In Exercises 11 through 20, find the coordinates of the midpoints of the segment which connects the given points.

11. $(-8, -7), (-2, 0)$
12. $(-4, 4), (9, -3)$
13. $(5, -4), (3, 0)$
14. $(1, -8), (-3, 1)$
15. $(-5, -9), (-5, -1)$
16. $(-8, 1), (5, 0)$
17. $(-5, 5), (-1, -1)$
18. $(6, -4), (8, -4)$
19. $(9, -7), (9, 0)$
20. $(-9, 7), (-8, -2)$

Set C–The Distance Formula

In Exercises 21 through 30, find the distance between the given points.

21. $(-1, 4), (-6, 9)$
22. $(3, -9), (8, 5)$
23. $(3, -4), (4, -3)$
24. $(0, 2), (0, -8)$
25. $(8, 9), (-4, 2)$
26. $(-8, 2), (8, 7)$
27. $(2, 1), (9, 9)$
28. $(0, -4), (7, -4)$
29. $(-6, -5), (9, 0)$
30. $(-9, -8), (0, 4)$

Set D–Computing Slopes

In Exercises 31 through 40, compute the slope of the line on the given points.

31. $(6, 6), (-9, 7)$
32. $(-6, 8), (-3, 8)$
33. $(9, -5), (-4, 7)$
34. $(0, -6), (3, -2)$
35. $(5, -2), (-7, 8)$
36. $(4, -9), (3, 6)$
37. $(5, 5), (6, -3)$
38. $(0, -8), (-8, -7)$
39. $(9, -1), (-2, 0)$
40. $(3, 2), (1, -2)$

Set E–Comparing Steepness

In Exercises 41 through 50, assume that the first given number is the slope of segment PQ and that the second is the slope of segment RS. Determine which segment is steeper.

41. 2, 1	42. 1, 8	43. −6, −9	44. −4, −4
45. −2, −8	46. 4, 2	47. −2, −8	48. −8, −1
49. 2, 2	50. −9, −3		

Set F–Graphing Linear Equations

In Exercises 51 through 65, construct the graph of each given equation.

51. $5x - 3y = 15$	52. $-x + 7y = 7$	53. $7x - 4y = 28$
54. $5x + 7y = 35$	55. $x = -2$	56. $3x - 7y = 21$
57. $-8x + 8y = -8$	58. $3x - 3y = -9$	59. $-6x + 9y = -1$
60. $-9x - 8y = 9$	61. $-7x - 7y = 0$	62. $4y = -4$
63. $5x + 2y = -7$	64. $5y = 4$	65. $5x - y = 0$

1.6 SUMMARY AND REVIEW EXERCISES

REVIEW OF CONCEPTS

A. Sets

1. *Equality of Sets.* If A and B are two sets, then $A = B$ if and only if $A \subseteq B$ and $B \subseteq A$.
2. *Complement of a Set.* If A is a set and U is a universal set, then A' (the complement of A in U) is the set of all elements in U not in A.
3. *Intersection of Sets.* If A and B are sets, then
$$A \cap B = \{x \mid x \in A \text{ and } x \in B\}$$
4. *Union of Sets.* If A and B are sets, then
$$A \cup B = \{x \mid x \in A \text{ or } x \in B\}.$$
5. *Subset of a Set.* If A and B are sets, then $A \subseteq B$ if and only if each element of A is also an element of B.
6. *Addition Formula for Sets.* If A and B are finite sets, then
$$\#(A \cup B) = \#(A) + \#(B) - \#(A \cap B).$$

B. Real Numbers

1. *Whole Numbers.* The whole numbers are the numbers 0, 1, 2, 3, 4, 5, ...

2. *Integers.* The integers are the numbers

$$\ldots, -5, -4, -3, -2, -1, 0, 1, 2, 3, 4, 5, \ldots.$$

3. *Rational Numbers.* The rational numbers are those real numbers which can be expressed as a/b, where a and b are integers and b is not 0.

4. *Irrational Numbers.* The irrational numbers are those real numbers which are not rational.

5. *Real Numbers.* The set of real numbers consists of all the decimal numbers.

C. Coordinate Systems

1. *Linear Equation.* A linear equation in two variables is an equation of the form

$$Ax + By = C,$$

where A, B, C are real numbers and not both A and B are 0.

2. *Distance Formula.* The distance between two points (a, b) and (c, d) is given by the formula

$$\sqrt{(a - c)^2 + (b - d)^2}.$$

3. *Midpoint Formula.* If (m, n) is the midpoint of the line segment joining the two points (a, b) and (c, d), then

$$m = \frac{a + c}{2} \quad \text{and} \quad n = \frac{b + d}{2}.$$

4. *Slope Formula.* If (a, b) and (c, d) are two points and a is not equal to c, then the slope m of the line on these points is

$$m = \frac{b - d}{a - c}.$$

REVIEW EXERCISES

1. List the members of the set
   (a) $\{x \mid x \in \{1, 2, 3\} \text{ or } x \in \{3, 6, 9\}\}$
   (b) $\{y \mid y = 2x + 1 \text{ and } x \in \{0, 1, -1\}\}$

2. List all the three-element subsets of the set $\{a, b, c, d, e\}$.

3. If W is the set of whole numbers and E is the set of even integers less than 10, list the elements of the set $W \cap E$.

4. If the universal set U is the set of whole numbers and P is the set of positive even integers, describe the set P'.

5. (a) Given $\#(A) = 7$, $\#(B) = 9$, $\#(A \cap B) = 2$. Find $\#(A \cup B)$.
   (b) Given $\#(A) = 49$, $\#(B) = 34$, $\#(C) = 12$, $\#(A \cap B) = 18$, $\#(A \cap C) = 6$, $\#(B \cap C) = 4$, $\#(A \cap B \cap C) = 3$. Find $\#(A \cup B \cup C)$.

6. Convert to a decimal:
   (a) $\frac{14}{25}$ (b) $\frac{38}{15}$

7. Convert to a fraction:
   (a) 2.15 (b) $52.\overline{81}$

8. Use a decimal approximation (2 places) to locate $\sqrt{35}$ on a number line.

9. Plot the points $(-3, -4)$, $(8, 9)$, $(7, 0)$, $(5, -1)$ on a coordinate system.

10. Find the midpoint of the segment connecting the points
    (a) $(-7, 7)$ and $(1, 7)$ (b) $(-9, 1)$ and $(5, 0)$

11. Find the distance between the points
    (a) $(-7, 7)$ and $(1, 7)$ (b) $(-9, 1)$ and $(5, 0)$

12. Find the slope of the line on the points
    (a) $(-7, 7)$ and $(1, 7)$ (b) $(-9, 1)$ and $(5, 0)$

13. Graph the following:
    (a) $x + y = 4$ (b) $y = x + 1$ (c) $x = 4$
    (d) $y = -5$ (e) $3x + 5y = 15$

CHAPTER

2

FUNCTIONS AND LINEAR MODELS

2

2.1 INTRODUCTION

The goals of this chapter are twofold: (1) to discuss some concepts and techniques from elementary algebra which are useful in constructing mathematical models of practical problems, and (2) to provide an opportunity for the student to practice constructing models for such problems and, in turn, obtain their solutions. Some of the problems we shall learn how to solve are:

1. A manufacturer has two television models to sell. The profit on each of the Superior models is \$200 and the profit on each of the Standard models is \$125. Since the Standard model sells at a lower price, its sales exceed the sales of the Superior model by 30%. If in a three-month period the profit from the sale of both models is \$87,000, how many units of each model have been sold? (Exercise 18, Section 2.2)
2. Alice and Bob each take 18 hours to prepare the monthly statements for their company. If Carol can do the same job in 15 hours, how long would it take all three of them working together to do the job? (Exercise 31, Section 2.2.)
3. Two trucks must make deliveries over a route totaling 240 miles. The pickup truck gets 12 miles per gallon of fuel and the panel truck gets 10 miles per gallon. If the total fuel consumption for the deliveries cannot exceed 22 gallons of fuel, what is the minimum number of miles that the pickup truck must travel? (Exercise 23, Section 2.3.)
4. A woman estimates that 30% of her gross income will go to taxes. If she wants to save each year 10% of her after-tax income and needs \$20,000 annually for living expenses, what should be her minimum gross income? (Example 5, Section 2.3.)
5. The mayor of a city determines that 3 subdivisions are located relative to City Hall as follows.

   Subdivision A: 3 miles east and 1 mile north
   Subdivision B: 6 miles east and 8 miles north
   Subdivision C: 9 miles east and 3 miles north

   Suppose a fire station is to be built equidistant from each of the three subdivisions. What should be its location in reference to City Hall? (Example 3, Section 2.5.) ■

2.2 LINEAR EQUATIONS IN ONE VARIABLE

linear equation

In this section we shall consider a variety of applied problems whose solutions are obtained from what are called linear equations. An equation of the form $ax + b = 0$ where a and b are constants and $a \neq 0$ is called a **linear equation** (or *first-degree equation*) in the *variable* x. For example,

$$3x + 5 = 0$$

is a linear equation in x. Since the equation

$$7x + 3 = 11x - 5$$

may be written in the equivalent form

$$-4x + 8 = 0$$

solution of linear equation

by using the standard operations on equalities, it is also an example of a linear equation in one variable. The **solution** of the linear equation $ax + b = 0$ is $x = -b/a$; it is the real number which, if substituted for x in the equation, produces a true statement. The solution of $-4x + 8 = 0$ is $x = -8/(-4) = 2$; thus 2 is the solution of the original equation $7x + 3 = 11x - 5$.

Linear equation

EXAMPLE 1 Find the solution of the equation

$$\frac{3x - 1}{4} + \frac{6x + 5}{3} = 8.$$

Solution: Multiplying each side of the given equation by the common denominator 12 yields

$$\begin{aligned} 3(3x - 1) + 4(6x + 5) &= 96, \\ 9x - 3 + 24x + 20 &= 96, \\ 33x &= 79, \\ x &= \frac{79}{33}. \end{aligned}$$

Let us now consider several practical problems whose solutions can be found by solving linear equations constructed as models to describe the given situations.

Unit cost of merchandise

EXAMPLE 2 A large department store receives a shipment of 135 inexpensive radios containing two makes. One make costs \$20 apiece, and the other one costs \$35 apiece. If the total bill for the shipment is \$3600, how many \$20 radios are in the shipment?

Solution: Let x equal the number of \$20 radios in the shipment. Then $135 - x$ equals the number of \$35 radios in the shipment. Since for each of the models

$$\text{total cost} = (\text{cost per unit})(\text{number of units}),$$

the total cost of the \$20 radios is $\$20x$ and the total cost of the \$35 radios is $\$35(135 - x)$. Therefore the total cost in dollars of the shipment is

$$20x + 35(135 - x) = 3600.$$

Solving for x, we obtain

$$20x + 4725 - 35x = 3600,$$
$$-15x = -1125,$$
$$x = 75.$$

Therefore, the shipment contains 75 of the \$20 radios.

PRACTICE PROBLEM A large department store receives a shipment of 150 radios containing two makes. One make costs \$30 apiece and the other one costs \$42 apiece. If the total bill for the shipment is \$5196, how many of the \$30 radios are in the shipment?

Answer: 92 radios. ■

Hourly wage

EXAMPLE 3 A college student obtained a summer job paying a fixed amount per hour with "time and one-half" for each hour worked over 40 hours in a given week. If the student's gross income for working 52 hours in one week was \$243.60, what was the student's base salary per hour?

Solution: Let x equal the student's hourly wage. For the first 40 hours of work the student earned

$$40x \text{ dollars.}$$

Since the student's hourly overtime wage is $1.5x$ dollars, for the 12 hours overtime the student earned

$$12(1.5)x \qquad \text{or} \qquad 18x \text{ dollars.}$$

Therefore,

$$\begin{aligned} 40x + 18x &= \$243.60, \\ 58x &= \$243.60, \\ x &= \$4.20. \end{aligned}$$

Fuel consumption

EXAMPLE 4 A store uses a van and a pickup truck for deliveries. While making deliveries the van averages 10.5 miles per gallon of gasoline and the pickup averages 8.5 miles per gallon. On a given day of deliveries the two vehicles traveled 465 miles and used a total of 50 gallons of gasoline. How far did the van travel? How many gallons of gasoline did the van use?

Solution: First recall that

$$(\text{miles per gallon})(\text{gallons used}) = \text{distance traveled.}$$

Therefore,

$$\text{gallons used} = \frac{\text{distance traveled}}{\text{miles per gallon}}.$$

Let x equal the distance the van traveled. Then $465 - x$ is the distance the pickup truck traveled. Since the van used $x/10.5$ gallons of gasoline and the truck used $(465 - x)/8.5$ gallons,

$$\frac{x}{10.5} + \frac{465 - x}{8.5} = 50.$$

Multiplying each side of the preceding equation by the common denominator $(10.5)(8.5)$ yields

$$\begin{aligned} 8.5x + (10.5)(465 - x) &= 50(10.5)(8.5), \\ 8.5x + 4882.5 - 10.5x &= 4462.5, \\ -2.0x &= -420, \\ x &= 210. \end{aligned}$$

The number of gallons of gasoline used by the van to drive the 210 miles is

$$\frac{210}{10.5} = 20 \text{ gallons.}$$

PRACTICE PROBLEM A store uses a van and a pickup truck for deliveries. While making deliveries the van averages 12 miles per gallon of gasoline and the pickup truck averages 9 miles per gallon. On a given day of deliveries the two vehicles traveled 699 miles and used a total of 65 gallons of gasoline. How far did the van travel? How many gallons of gasoline did the van use?

Answer: 456 miles, 38 gallons. ■

Average of test grades

EXAMPLE 5 A student achieves the following scores on five tests in a mathematics course: 88, 70, 94, 92, 96. If the test grades count three-fourths toward the final grade and if the final examination counts one-fourth, what grade on the final examination will give the student an average of 90?

Solution: The average grade on the first five tests is

$$\frac{88 + 70 + 94 + 92 + 96}{5} = \frac{440}{5} = 88.$$

If x is the grade on the final examination, then

$$\frac{3}{4}(88) + \left(\frac{1}{4}\right)x = 90.$$

Multiplying each side of the preceding equation by 4 yields

$$264 + x = 360,$$
$$x = 96.$$

Travel time for truck deliveries

EXAMPLE 6 Town A and Town C are 130 miles apart and Town B is on the highway between them. Suppose a truck can average 50 mph traveling from Town A to Town B and 40 mph traveling from Town B to Town C. If the total time required to drive from Town A to Town C is 3 hours, how long does it take the truck to get from Town A to Town B? How far is it from Town A to Town B? (See Figure 2.1.)

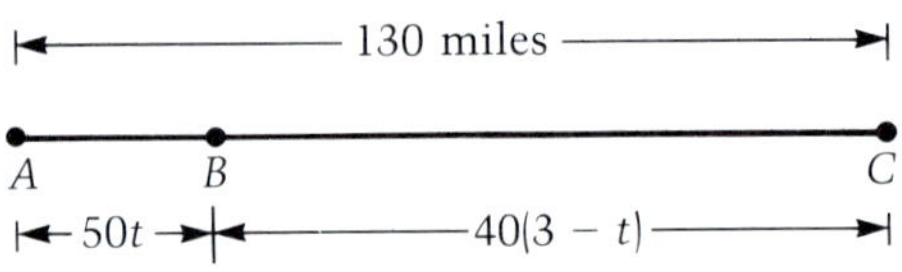

FIG. 2.1.

Solution: First recall that

> distance = (rate)(time).

That is, if d is the distance traveled in miles, r is the rate in miles per hour, and t is the time in hours, then

> $d = rt.$

Let t be the time in hours required to go from A to B. Then

$3 - t$ is the time required to go from B to C.

Consequently,

$50t$ is the distance between A and B, and

$40(3 - t)$ is the distance between B and C.

Since the total distance from A to C is 130 miles,

$$\begin{aligned} 50t + 40(3 - t) &= 130, \\ 50t + 120 - 40t &= 130, \\ 10t &= 10, \\ t &= 1. \end{aligned}$$

Therefore, it requires 1 hour to go from A to B; the distance between Towns A and B is (50 mph)(1 hour) = 50 mi.

PRACTICE PROBLEM Town A and Town C are 218 miles apart and Town B is on the highway between them. Suppose a truck can average 60 mph traveling from A to B and 48 mph traveling from B to C. If the total time required to drive from A to C is 4 hours, how long does it take the truck to get from A to B? How far is it from A to B?

Answer: 2 hours and 10 minutes; 130 miles. ■

EXERCISES

Set A–Solutions to Linear Equations

Find the solution to each equation in Exercises 1 through 16.

1. $3x + 17 = 28$
2. $7 + 3x = -11$
3. $5x - 12 = 20$
4. $35 - 3x = -7$
5. $-8x = \frac{2}{5} - 10x$
6. $12x + \frac{1}{4} = 22x$
7. $\frac{2}{3}x - \frac{5}{9} = \frac{7}{6}$
8. $\left(\frac{3}{4}\right)x - \frac{11}{6} = \left(\frac{5}{2}\right)x + \frac{5}{24}$
9. $0.03x - 5 = 1.12$
10. $-0.06x + 4 = 1.84$
11. $1.12x + 4 = 1.16x - 2$
12. $1.1x - 4.2 = 5.1 - 4.3x$
13. $\dfrac{x - 1}{3} + \dfrac{x + 1}{2} = 1$
14. $\dfrac{x - 1}{5} - \dfrac{x + 1}{6} = 2$
15. $\dfrac{2x + 3}{4} - \dfrac{3x + 1}{2} = 6$
16. $\dfrac{3x - 6}{2} - \dfrac{6x - 3}{3} = 2$

Set B–Applications Involving Linear Equations

Unit costs for automobiles

17. In a three-month period a car dealer receives two different models in shipments totaling 60 cars. The Slowback model costs the dealer \$5600 and the Fastback model costs the

dealer \$6200. If the dealer's total cost for the 60 cars was \$349,800, how many cars of each model did the dealer receive?

Unit sales for television sets

18. A manufacturer has two television models to sell. The profit on each of the Superior models is \$200 and the profit on each of the Standard models is \$125. Since the Standard model sells at a lower price, its sales exceed the sales of the Superior model by 30%. If in a three-month period the profit from the sale of both models is \$87,000, how many of each model have been sold?

Hourly wage

19. A college student obtained a summer job paying a fixed amount per hour and "time and one-half" for each hour worked over 40 hours in a given week. For 50 hours worked in one week the student's gross income was \$198. What was the student's hourly wage?

Hourly wage

20. A college student obtained a summer job paying a fixed amount per hour and "time and one-half" for each hour worked over 40 hours in a given week. For 48 hours worked in one week the student's gross income was \$176.80. What was the student's hourly wage?

Fuel consumption for delivery vehicle

21. A company uses a van and a pickup truck for deliveries. In making deliveries the van averages 11.5 miles per gallon of gasoline and the pickup truck averages 8 miles per gallon. On a given day of deliveries the two vehicles travel a total of 381 miles and use 38 gallons of gasoline. How far did the truck travel? How many gallons of gasoline did the truck use?

Final grade average

22. A student makes the following grades on four tests in a mathematics course: 82, 76, 89, 99. If the average test grade counts two-thirds toward the final grade and if the final exam counts one-third, what grade on the final exam will give the student a final grade average of 90?

Final grade average

23. A student makes the following grades on five tests in a mathematics course: 82, 76, 94, 90, 88. If the average test grade counts three-fifths toward the final grade and if the final exam counts two-fifths, what grade on the final exam will give the student a final grade average of 90? A final grade average of 80?

Distance traveled by delivery vehicle

24. Town A and Town C are 120 miles apart and Town B is on the highway between them. Suppose a truck can average 55 mph traveling from Town A to Town B and 45 mph traveling from Town B to Town C. If the total time required to go from A to C is $2\frac{1}{2}$ hours, how long does it take the truck to travel from A to B? How far is it from A to B?

Distance traveled by delivery vehicle

25. A delivery truck averages 40 mph traveling from Town A to Town B and 50 mph traveling from Town B to Town C. If it takes $3\frac{1}{2}$ hours to cover the 150-mile distance between A and C, how far is it from A to B?

Child's savings

26. A child's piggy bank contains 300 coins made up of dimes and nickels. If the total amount in the bank is \$26.50, how many dimes are in the bank?

Revenue from parking meter

27. The total amount collected from a parking meter containing 425 coins made up of dimes and quarters is \$59.75. How many quarters were in the meter?

Division of an estate

28. A man's will provides that his \$110,000 estate be divided in the following way. A university is to receive \$5000 for a scholarship fund, and from the remaining amount his wife is to receive 4 times as much as each of their 3 children who are to receive equal amounts. How much does each child receive? How much does his wife receive?

Average speed

29. A driver averages 40 mph going from home to work and then averages 60 mph on the return trip. What should be the driver's average speed in order to make a round trip home-office-home in the same amount of time?

Increased personnel for a task

30. Don and Carol can each construct a given number of boxes for a department store in 16 hours. If Tim can do the same job in 12 hours, how long would it take all three of them working together to do the job? [*Hint:* In x hours, the fractional part of the job Don does is $x/16$.]

Increased personnel for a task

31. Alice and Bob each take 18 hours to prepare the monthly statements for their company. If Carol can do the same job in 15 hours, how long would it take all three of them working together to do the job? [*Hint:* In x hours, the fractional part of the job Alice does is $x/18$.]

2.3 LINEAR INEQUALITIES IN ONE VARIABLE

inequality properties

In this section our main goal is to consider problems whose solutions can be obtained from linear inequalities in one variable. Although operations on inequalities are very similar to those on equalities, there are some significant differences. For real numbers a, b, and c, the basic **inequality properties** are as follows.

PROPERTY 1 If $a < b$, then

$$a + c < b + c \quad \text{and} \quad a - c < b - c.$$

(The same number can be added or subtracted from each side of an inequality without changing the direction of the inequality.) For example, since $4 < 7$, $4 + 8 < 7 + 8$ and $4 - 8 < 7 - 8$.

PROPERTY 2 If $a < b$ and $c > 0$, then

$$ac < bc \quad \text{and} \quad \frac{a}{c} < \frac{b}{c}.$$

(Each side of an inequality can be multiplied or divided by the same positive number without changing the direction of the inequality.) For example, since $3 < 8$ and $2 > 0$, $(3)(2) < (8)(2)$.

PROPERTY 3 If $a < b$ and $c < 0$, then

$$ac > bc \quad \text{and} \quad \frac{a}{c} > \frac{b}{c}.$$

(The direction of an inequality is reversed if each side is multiplied or divided by a negative number.) For example, since $8 < 11$ and $-5 < 0$, $-40 > -55$.

PROPERTY 4 If $a < b$ and $c < d$, then

$$a + c < b + d.$$

(Inequalities can be added provided the direction of the inequality sign for each is the same.) For example, since $-2 < 7$ and $5 < 8$, $-2 + 5 < 7 + 8$.

linear inequality
solution set

A **linear inequality** in one variable can be written in the form $ax + b < 0$ where $a \neq 0$. Its **solution set** consists of all real numbers x satisfying $x < -b/a$. Other forms of linear inequalities can be changed to the standard form by the preceding inequality properties. To find the solution set of the linear inequality $3x + 7 < 5x - 3$, we proceed as follows.

$$
\begin{aligned}
3x + 7 &< 5x - 3, \\
3x - 5x &< -3 - 7, && \text{(Property 1)} \\
-2x &< -10, \\
x &> 5. && \text{(Property 3)}
\end{aligned}
$$

The graph of this solution set is the set of all points on the real number line whose coordinates make the inequality a true statement. (See Figure 2.2.) *Note:* The "hole," or open dot, at 5 indicates that it is not in the solution set.

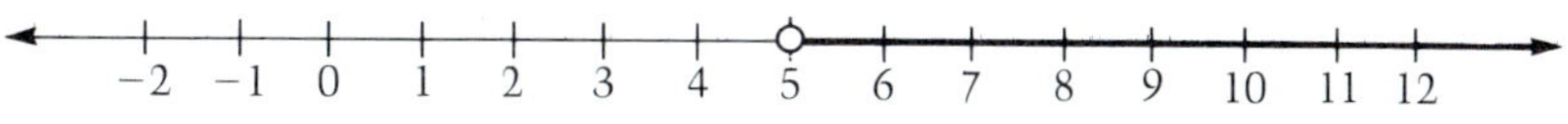

FIG. 2.2.

Linear inequality

EXAMPLE 1 Find the solution set of $5x + 7 < 8x + 11$.

Solution

$$
\begin{aligned}
5x + 7 &< 8x + 11, \\
5x - 8x &< 11 - 7, && \text{(Property 1)} \\
-3x &< 4, \\
x &> -\frac{4}{3}. && \text{(Property 3)}
\end{aligned}
$$

Let us now look at some practical problems whose solutions are obtained from linear inequalities.

Maximum unit cost

EXAMPLE 2 Assume that the combined cost of one wristwatch and one pocket watch is \$100. If it is known that 5 wristwatches and 3 pocket watches may be bought for \$424 or less, what is the maximum cost of one wristwatch?

Solution: Let x equal the cost of a wristwatch. Then

$100 - x$ is the cost of one pocket watch,

$5x$ is the cost of 5 wristwatches, and

$3(100 - x)$ is the cost of 3 pocket watches.

Since the combined cost cannot exceed \$424,

$$\begin{aligned} 5x + 3(100 - x) &\le 424, \\ 5x + 300 - 3x &\le 424, \\ 2x &\le 124, && \text{(Property 1)} \\ x &\le 62. && \text{(Property 2)} \end{aligned}$$

The maximum cost of a wristwatch is \$62.

Maximum delivery distance

EXAMPLE 3 Two trucks must make deliveries totaling 500 miles. Truck A gets 15 miles per gallon and Truck B gets 10 miles per gallon. If the total fuel consumption for the deliveries cannot exceed 40 gallons, what is the minimum number of miles that Truck A must travel?

Solution: Let x be the number of miles Truck A travels. Then

$500 - x$ is the number of miles Truck B travels,

$\frac{x}{15}$ is the number of gallons of fuel Truck A uses, and

$\frac{500 - x}{10}$ is the number of gallons of fuel Truck B uses.

Since the total fuel consumption cannot exceed 40 gallons,

$$\frac{x}{15} + \frac{500 - x}{10} \le 40.$$

Multiplying each side of the preceding inequality by 30, we obtain

$$2x + 1500 - 3x \le 1200.$$

Consequently,

$$\begin{aligned} -x &\le -300, && \text{(Property 1)} \\ x &\ge 300. && \text{(Property 3)} \end{aligned}$$

Therefore, Truck A must make at least 300 miles of deliveries.

PRACTICE PROBLEM Two trucks must make deliveries totaling 600 miles. Truck A gets 15 miles per gallon and Truck B gets 12 miles per gallon. If the total fuel consumption for the deliveries is not to exceed 42 gallons, what is the minimum number of miles that Truck A must travel?

Answer: 480 miles. ■

Minimum unit cost

EXAMPLE 4 Suppose the total cost of one reel mower and one rotary mower is \$372. If the total cost of 3 reel mowers and 7 rotary mowers is not more than \$2204, what is the minimum cost of a reel mower?

Solution: Let x be the cost of one reel mower. Then

$372 - x$ is the cost of a rotary mower,

$3x$ is the total cost of 3 reel mowers, and

$7(372 - x)$ is the total cost of 7 rotary mowers.

Since the total cost of the 10 mowers cannot exceed \$2204,

$$\begin{aligned} 3x + 7(372 - x) &\leq 2204 \\ 3x + 2604 - 7x &\leq 2204 \\ -4x &\leq -400 \\ x &\geq 100. \end{aligned}$$

Consequently, a reel mower costs at least \$100.

PRACTICE PROBLEM In Example 4, what is the maximum cost of a rotary mower?

Answer: \$272. ■

Minimum salary requirement

EXAMPLE 5 A woman estimates that 30% of her gross income will go to taxes. If she wants to save each year 10% of her after-tax income and needs \$20,000 annually for living expenses, what should be her minimum gross income?

Solution: Let x represent her gross income in dollars. Since 30% of her gross income goes to taxes,

$$x - 0.30x = 0.70x \text{ dollars}$$

is her income after taxes. Since \$20,000 is needed for living expenses and since

$$0.10(0.70x) = 0.07x$$

is needed for savings, she wants her net income to be at least

$$20{,}000 + 0.07x.$$

Hence, her salary x must be such that

$$0.70x \geq 20{,}000 + 0.07x,$$
$$0.63x \geq 20{,}000,$$
$$x \geq \$31{,}746.03$$

PRACTICE PROBLEM A man estimates that 35% of his gross income will go to taxes. If he needs a total of \$18,000 annually for living expenses and if he wants to deposit 15% of his after-tax income into a savings account each year, what should be his minimum gross income?

Answer: \$32,579.19. ■

Minimum sales to make a profit

EXAMPLE 6 The *fixed costs* (costs independent of the level of production) for publishing a certain textbook are \$74,250. If each book costs \$18 to produce and sells for \$27, how many textbooks must be sold before a profit is made?

Solution: Since

total cost

$$\text{total cost} = (\text{variable costs}) + (\text{fixed costs}),$$

the total cost in dollars for producing x textbooks is

$$18x + 74{,}250.$$

Since

revenue

$$\text{revenue} = (\text{selling price per unit})(\text{number of units sold}),$$

the revenue from selling x textbooks is $27x$ dollars. A profit results when revenue exceeds costs; that is, when

$$27x > 18x + 74{,}250,$$
$$9x > 74{,}250,$$
$$x > 8250.$$

A profit results from selling more than 8250 textbooks.

EXERCISES

Set A–Solutions to Linear Inequalities

In Exercises 1 through 20, find the solution set for the given inequalities.

1. $3x - 5 < 7x + 13$
2. $8x - 23 < 12x + 32$
3. $7 - 8x \geq 5x + 13$
4. $6 - 4x > 7x - 15$
5. $-6x + 6 \leq -1 - 5x$
6. $-8 - 5x \geq -4x + 9$
7. $7(4x + 2) < 2x + 5$
8. $5 - 9x > -6(4 + 9x)$
9. $3(1 + 8x) \leq 3x + 3(3 + 4x)$
10. $3(-3x - 7) \geq 5(-6 - 5x)$
11. $\dfrac{5x - 7}{4} < \dfrac{3x - 2}{5}$
12. $\dfrac{2x + 5}{7} > \dfrac{2 - 3x}{4}$
13. $\dfrac{2}{3}x - \dfrac{4}{7} \leq \dfrac{1}{6}x + \dfrac{5}{42}$
14. $\dfrac{3}{4}x + \dfrac{7}{2} < \dfrac{11}{6}x - \dfrac{5}{3}$
15. $\dfrac{3}{4}x + \dfrac{5}{2} > \dfrac{2}{3}x - \dfrac{5}{6}$
16. $\dfrac{4}{5}x + \dfrac{3}{5} \geq \dfrac{5}{9}x + \dfrac{7}{3}$
17. $0.3x + 1.7 < 0.25x + 3.75$
18. $0.06x + 1.4 > 0.04x - 2.8$
19. $1.25x + 8.1 > 0.75x - 9.3$
20. $-1.5x - 9.3 \leq -1.7x + 5.7$

Set B–Applications Involving Linear Inequalities

Maximum unit cost

21. Assume that the combined cost of a 19-inch television set and a 21-inch television set is \$912 and that a shipment containing seven 19-inch sets and four 21-inch sets can be purchased for \$4620 or less. What is the maximum cost per 19-inch set?

Minimum unit cost

22. Assume that the combined cost of a 19-inch television set and a 21-inch television set is \$858. Suppose a shipment containing nine 21-inch sets and six 19-inch sets can be purchased for \$6642 or less. What is the minimum cost per 19-inch set?

Minimum delivery distance

23. Two trucks must make deliveries over a route totaling 240 miles. The pickup truck gets 12 miles per gallon and the panel truck gets 10 miles per gallon. If the total fuel consumption for the deliveries cannot exceed 22 gallons, what is the minimum number of miles that the pickup truck must travel?

Maximum delivery distance

24. Two trucks must make deliveries totaling 300 miles. The pickup truck gets 14 miles per gallon and the panel truck gets 12 miles per gallon. If the total fuel consumption for the deliveries cannot exceed 23 gallons, what is the maximum number of miles that the panel truck can travel?

Maximum and minimum unit costs

25. Suppose the total cost of a reel mower and a rotary mower is $576.
   (a) If the total cost of 5 reel mowers and 9 rotary mowers is no more than $4336, what is the minimum cost of a reel mower?
   (b) What is the maximum cost of a rotary mower?

Minimum salary requirement

26. A man estimates that 35% of his gross income will go to taxes. If he needs a total of $30,000 annually for living expenses and if he wants to save each year 6% of his after-tax income, what should be his minimum gross income?

Minimum sales to make a profit

27. The fixed costs for a company producing metal storage containers are $150,000. Each storage unit sells for $12 and costs $9 to produce. How many units must be sold for the company to make a profit?

Minimum sales to make a profit

28. The fixed costs for publishing a certain textbook are $60,000. If each book costs $16 to produce and sells for $21.50, how many books must be sold for the company to make a profit?

2.4 LINEAR EQUATIONS IN TWO VARIABLES

In Chapter 1, we found that for two points $P(x_1, y_1)$ and $Q(x_2, y_2)$ in the rectangular coordinate plane, the coordinates of the **midpoint** of the segment PQ (see Figure 2.3.) is given by

$$\left(\frac{x_1 + x_2}{2}, \frac{y_1 + y_2}{2}\right),$$

and the **length** of the segment PQ is given by

$$d = \sqrt{(x_2 - x_1)^2 + (y_2 - y_1)^2}.$$

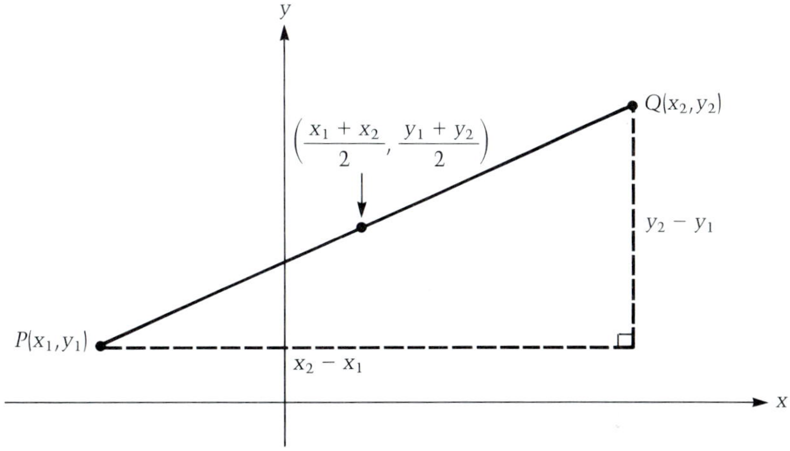

FIG. 2.3. $|PQ| = \sqrt{(x_2 - x_1)^2 + (y_2 - y_1)^2}$.
Midpoint: $\left(\frac{x_1 + x_2}{2}, \frac{y_1 + y_2}{2}\right)$.

PRACTICE PROBLEM For the points $P(2, -7)$ and $Q(5, 9)$, find the distance between them and the midpoint of the segment PQ.

Answer: $\sqrt{265}$ and $(\frac{7}{2}, 1)$. ■

Location of postal pickup stations

EXAMPLE 1

(a) In a given city, two postal pickup stations are located from the central post office A as follows.

Pickup station B: 3 miles east and 4 miles north
Pickup station C: 9 miles east and 6 miles north

Suppose a third pickup station P is to be located midway (on a line) between these two stations. What would be its location relative to the central post office?

(b) How far would the third pickup station be from the central post office?

(c) How far would it be from each of the other stations?

Solution

(a) Note that the points B and C in Figure 2.4 can be used to represent the locations of the first two pickup stations, provided the central post office is located at the origin. Since

$$\left(\frac{3+9}{2}, \frac{4+6}{2}\right) = (6, 5)$$

is the midpoint of the segment with B and C as endpoints, the third pickup station P should be located 6 miles east and 5 miles north of the central post office.

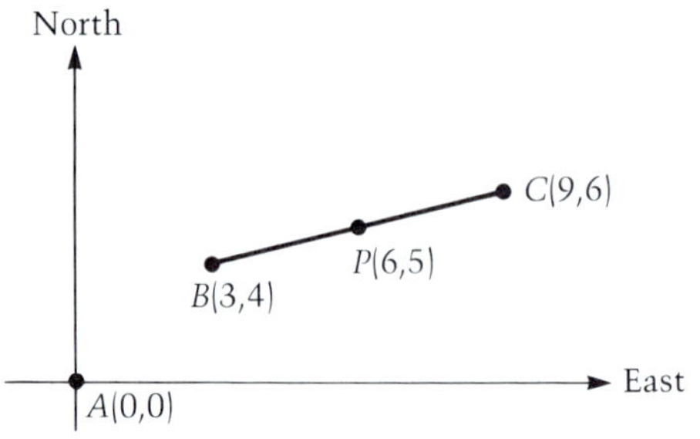

FIG. 2.4.

(b) The distance d between $P(6, 5)$ and the central post office $A(0, 0)$ is

$$d = \sqrt{(6-0)^2 + (5-0)^2} = \sqrt{61} \doteq 7.81 \text{ miles.}$$

(c) The distance D between $P(6, 5)$ and $B(3, 4)$ (which is the same as between P and C) is

$$D = \sqrt{(6-3)^2 + (5-4)^2} = \sqrt{10} \doteq 3.16 \text{ miles.}$$

---

slope

If (x_1, y_1) and (x_2, y_2) are any two points for which $x_1 \neq x_2$ (that is, if the line containing the two points is not parallel to the y-axis), then the **slope** of the line containing (x_1, y_1) and (x_2, y_2), denoted by m, is defined by

$$m = \frac{y_2 - y_1}{x_2 - x_1}.$$

Essentially, the slope is the ratio of the change in y to the change in x between the two points. Lines with positive slope rise to the right, lines with negative slope fall to the right, and lines with zero slope are horizontal (parallel to the x-axis). If $x_1 = x_2$ then the two points are on a vertical line (a line parallel to the y-axis); slope is not defined for such lines. (See Figure 2.5.)

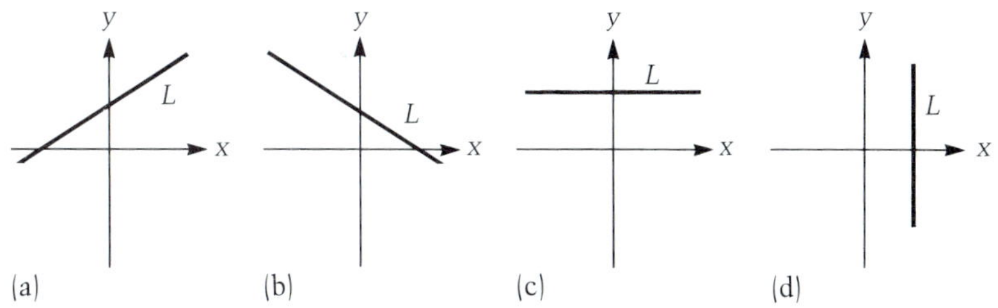

FIG. 2.5. (a) $m > 0$. (b) $m < 0$. (c) $m = 0$. (d) m undefined.

For two points (x_1, y_1) and (x_2, y_2) with $x_1 \neq x_2$, note that

$$\frac{y_2 - y_1}{x_2 - x_1} = \frac{-(y_2 - y_1)}{-(x_2 - x_1)} = \frac{y_1 - y_2}{x_1 - x_2}.$$

This means, for example, that to find the slope of the line containing (2, 3) and (7, 5) either point can be called (x_1, y_1).

Slope of a line

EXAMPLE 2 Find the slope of the line containing the points (2, 5) and (7, 9).

Solution: $m = (9 - 5)/(7 - 2) = \frac{4}{5}$.

PRACTICE PROBLEM Find the slope of the line containing the points (2, −7) and (5, 9).

Answer: $m = \frac{16}{3}$. ■

parallel

perpendicular

It is not difficult to prove that two nonvertical lines are **parallel** if and only if they have the same slope. It is also true, though not as easy to prove, that two nonvertical lines are **perpendicular** if and only if the slope of one line is the negative reciprocal of the slope of the other. (See Figure 2.6.)

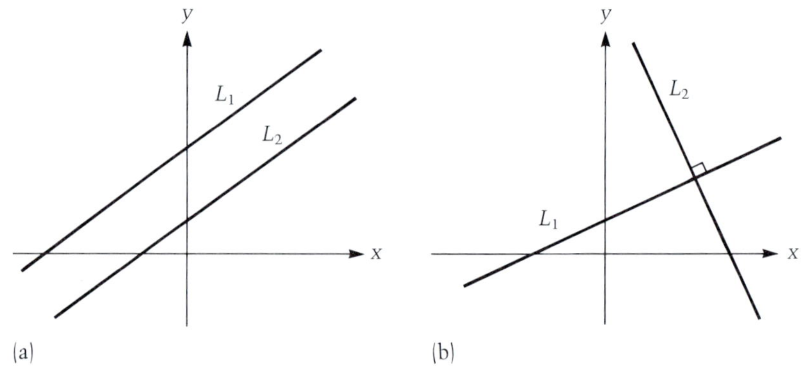

FIG. 2.6. (a) Parallel lines, $m_1 = m_2$. (b) Perpendicular lines, $m_1 = -1/m_2$.

Locations of subdivisions

EXAMPLE 3 The mayor of a city determines that 3 subdivisions can be located relative to City Hall as follows.

Subdivision A: 3 miles east and 1 mile north
Subdivision B: 6 miles east and 8 miles north
Subdivision C: 9 miles east and 3 miles north

(a) Construct a coordinate system and exhibit the locations of the subdivisions. Put City Hall at the origin of the coordinate system.

(b) Determine the locations of the midpoints between A and B, B and C, and A and C.

(c) Construct the perpendicular bisectors of each segment AB, BC, and AC. (The perpendicular bisectors of the sides of a triangle all meet in one point. Later we will determine the coordinates of their point of intersection.)

Solution

(a) See Figure 2.7.

(b) The midpoint P of AC is $(6, 2)$. The midpoint Q of BC is $(\frac{15}{2}, \frac{11}{2})$. The midpoint R of AB is $(\frac{9}{2}, \frac{9}{2})$.

(c) Since the slope of the line containing A and C is $\frac{1}{3}$, the slope of the perpendicular bisector is -3. Hence, if we move 1 unit to the left of P and 3 units up, the resulting point $(5, 5)$ will be another point on the perpendicular bisector of AC. The other perpendicular bisectors can be constructed in a similar manner.

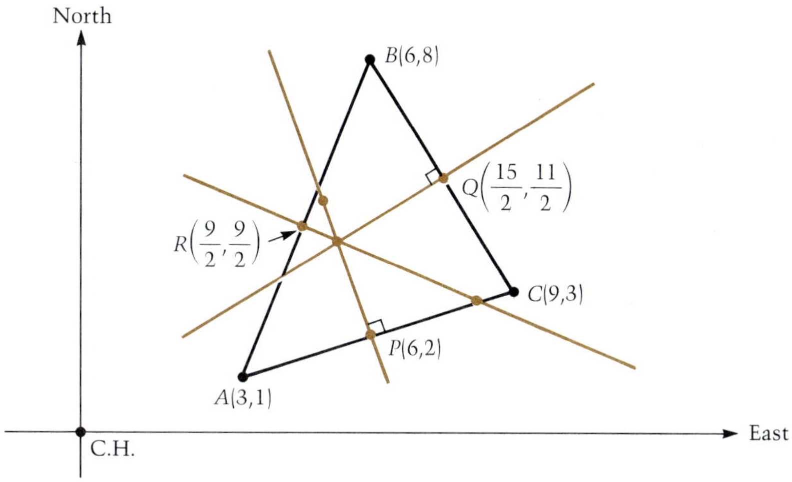

FIG. 2.7.

---

equation of a line

Let (x_1, y_1) and (x_2, y_2) be any two points in the rectangular coordinate plane such that $x_1 \neq x_2$ and let S be the set of points on the line containing the two points. An equation defining S is called an *equation of the line* and it can be found as follows. A point (x, y) different from (x_1, y_1) is on the line (in the set S) if and only if the slopes

$$\frac{y - y_1}{x - x_1} \quad \text{and} \quad \frac{y_2 - y_1}{x_2 - x_1}$$

are equal. (See Figure 2.8.) Therefore the point (x, y) is on the line if and only if

$$\frac{y - y_1}{x - x_1} = \frac{y_2 - y_1}{x_2 - x_1}$$

or

$$y - y_1 = \frac{y_2 - y_1}{x_2 - x_1}(x - x_1).$$

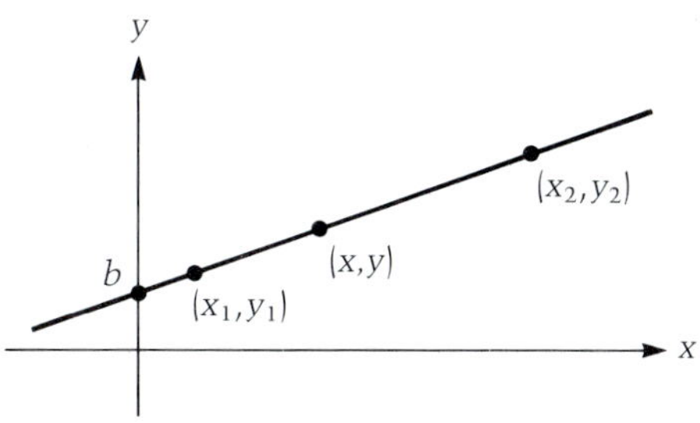

FIG. 2.8.
$y - y_1 = m(x - x_1)$
or $y = mx + b$.

two-point form equation

The preceding equation is called the *two-point form* for an equation of the line. Since m represents the slope of the line,

$$y - y_1 = m(x - x_1)$$

point-slope form equation

is also an equation of the line; it is called the *point-slope form.* In the point-slope form, if $x_1 = 0$ and $y_1 = b$, then the equation reduces to $y - b = m(x - 0)$, $y - b = mx$, or

slope-intercept form equation

$$y = mx + b.$$

Since $y = mx + b$ is an equation of a line with slope m crossing the y-axis at b, it is called the *slope-intercept form.*

Equation of a line

EXAMPLE 4 Find an equation of the line containing (3, 5) and (7, 8).

Solution: The slope of the line is

$$\frac{8-5}{7-3} = \frac{3}{4}.$$

Using this slope and the point (3, 5), we obtain

$$y - 5 = \left(\frac{3}{4}\right)(x - 3).$$

Simplifying the preceding equation yields $3x - 4y = -11$. (See Figure 2.9.) *Note:* If the point (7, 8) is used in the formula, we obtain

$$y - 8 = \left(\frac{3}{4}\right)(x - 7)$$

or

$$3x - 4y = -11,$$

which, of course, is the same as the previous result.

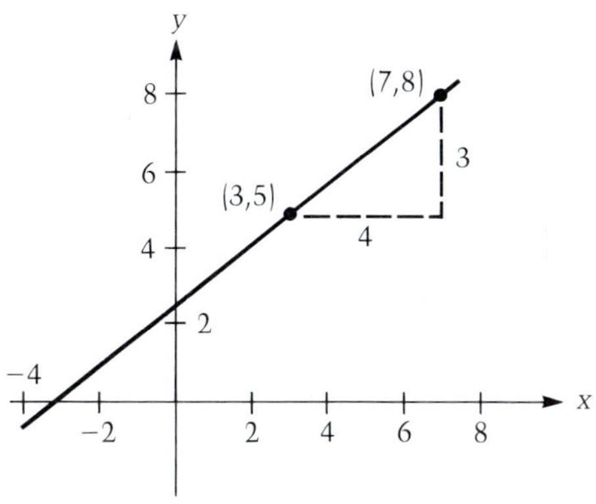

FIG. 2.9. $3x - 4y = -11$.

Equation of a line

EXAMPLE 5 Determine an equation for each of the following lines in the coordinate plane: (a) the line containing (3, −7) having slope $-\frac{8}{5}$, (b) the line having 3 as slope and 9 as y-intercept, and (c) the line parallel to the x-axis containing (5, 3).

Solution

(a) $y - (-7) = -(\frac{8}{5})(x - 3)$. Such an equation is often written in the form $Ax + By = C$; it is obtained as follows.

$$5y + 35 = -8x + 24,$$
$$8x + 5y = -11,$$

(b) $y = 3x + 9$.

(c) Since the line has slope 0 and y-intercept 3, its equation is $y = 3$.

PRACTICE PROBLEM Determine an equation for each of the following lines in the coordinate plane: (a) the line containing (−4, 2) and having slope $\frac{3}{7}$, (b) the line having slope −2 and y-intercept 5, and (c) the line parallel to the y-axis containing (−3, 9).

Answer: (a) $3x - 7y = -26$. (b) $y = -2x + 5$. (c) $x = -3$. ■

Any equation that can be written in the form

linear equation in two variables

$$Ax + By = C$$

is called a **linear equation in two variables.** Any straight line in the coordinate plane can be represented by a linear equation in two variables. Furthermore, any linear equation in two variables has a straight line as its graph. Provided $B \neq 0$, we can solve $Ax + By = C$ for y and obtain

$$y = -\frac{A}{B}x + \frac{C}{B}.$$

Since the preceding equation is in the slope-intercept form, we see that $Ax + By = C$ is a line with slope $-A/B$ and y-intercept C/B. (See Figure 2.10.)

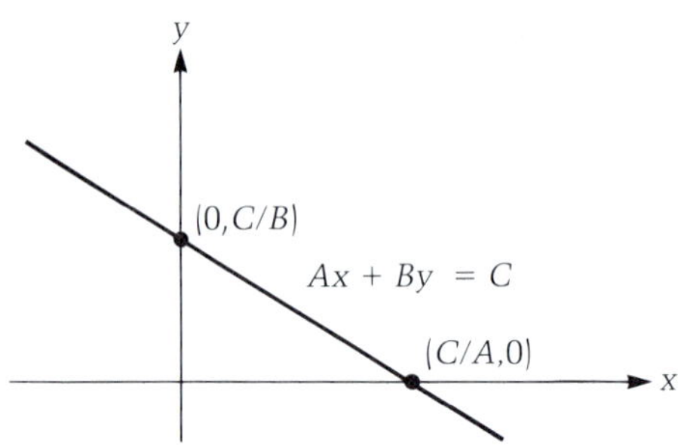

FIG. 2.10. $y = -\frac{A}{B}x + \frac{C}{B}$ and $m = -\frac{A}{B}$.

Graph of a line

EXAMPLE 6 Graph the equation $3x + 5y = 15$. That is, graph the set of points S defined by

$$S = \{(x, y) \mid 3x + 5y = 15\}$$

(read, "the set of all points (x, y) such that $3x + 5y = 15$).

Solution: (See Figure 2.11.) Since the graph (in two dimensions) of a linear equation in x and y is a straight line, two points whose

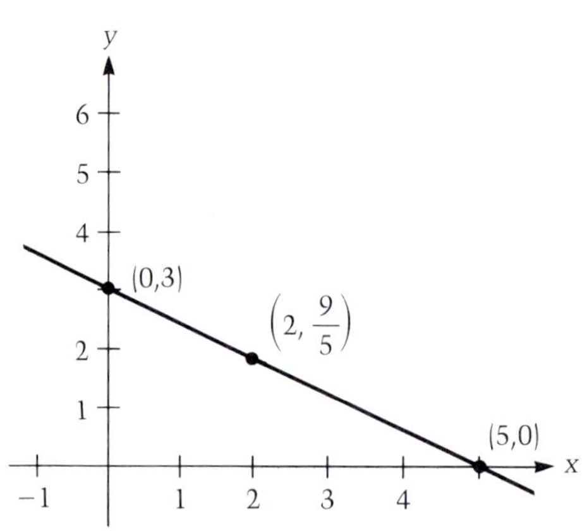

FIG. 2.11. $3x + 5y = 15$.

coordinates satisfy the equation are enough to determine the graph. However, plotting an extra point provides a check on computations. Usually, the x-intercept (where $y = 0$) and the y-intercept (where $x = 0$) are two of the points plotted.

The fact that the line $Ax + By = C$, $B \neq 0$, has slope $-A/B$ is quite useful in determining an equation of a line. For example, consider the line containing the point (1, 5) with slope $-\frac{2}{3}$. Since $-\frac{2}{3} = -A/B$, we can let $A = 2$ and $B = 3$. Its equation has the form $2x + 3y = C$ where C is a constant to be determined from the fact that (1, 5) must satisfy the equation. Since $2(1) + 3(5) = 17$, (1, 5) satisfies the equation if and only if $2x + 3y = 17$. Consequently, an equation of the line is $2x + 3y = 17$. (It is also true that the equation could have the form $-2x - 3y = C$, but in this case $C = -17$ and the equivalent equation $-2x -3y - 17$ is obtained. In fact, A and B could be any constants such that $-A/B = -\frac{2}{3}$.)

Equation of a line

EXAMPLE 7 Find an equation of the line with slope $\frac{7}{5}$ containing the point (2, 3).

Solution: An equation of the line has the form $7x - 5y = C$. (*Note:* This is equivalent to $7x + (-5)y = C$ where $A = 7$ and $B = -5$, and $-A/B = \frac{7}{5}$.) For (2, 3) to satisfy the equation, $7(2) - 5(3) = C$ and $C = -1$. Hence an equation of the line is $7x - 5y = -1$.

PRACTICE PROBLEM Find an equation of the line with slope $\frac{9}{4}$ containing the point (1, 2).

Answer: $9x - 4y = 1$. ■

Equation of perpendicular bisector

EXAMPLE 8 Find an equation of the perpendicular bisector of the line segment with $P(-1, 4)$ and $Q(7, 10)$ as endpoints.

Solution: The midpoint of the line segment with (−1, 4) and (7, 10) as endpoints is (3, 7). The slope of the line containing (−1, 4) and (7, 10) is $\frac{3}{4}$, and therefore, the slope of the line perpendicular to it is $-\frac{4}{3}$. An equation of the perpendicular bisector of the segment is

$$y - 7 = -\left(\frac{4}{3}\right)(x - 3) \qquad \text{or} \qquad 4x + 3y = 33.$$

PRACTICE PROBLEM Find an equation of the perpendicular bisector of the line segment with $(-3, 6)$ and $(5, 10)$ as endpoints.

Answer: $2x + y = 10$. ■

EXERCISES

Set A–Midpoint of a Line Segment

In Exercises 1 through 6, for each segment PQ find its length and the coordinates of the midpoint.

1. $P(3, 7)$ and $Q(8, 10)$
2. $P(-3, 6)$ and $Q(7, 12)$
3. $P(-7, 3)$ and $Q(5, -4)$
4. $P(-8, 3)$ and $Q(5, -5)$
5. $P(-\frac{5}{2}, 7)$ and $Q(\frac{7}{2}, -5)$
6. $P(\frac{3}{4}, -\frac{1}{2})$ and $Q(\frac{19}{4}, \frac{11}{2})$

Set B–Slope of a Line

In Exercises 7 through 12, graph the line containing the two given points and find its slope.

7. $(-4, 7)$ and $(8, 13)$
8. $(-3, 8)$ and $(5, 2)$
9. $(-3, -1)$ and $(6, -3)$
10. $(-5, -5)$ and $(7, 7)$
11. $(-\frac{5}{2}, 3)$ and $(\frac{9}{2}, 7)$
12. $(-\frac{3}{4}, -\frac{3}{2})$ and $(\frac{17}{4}, \frac{7}{2})$

Set C–Equation of a Line

In Exercises 13 through 22, find an equation of the line containing the given point and having the given slope.

13. $(-7, -7)$, $m = 0$
14. $(-2, -7)$, $m = \frac{5}{7}$
15. $(2, -8)$, m undefined
16. $(1, 2)$, $m = 4$
17. $(-2, 2)$, $m = -2$
18. $(8, -5)$, $m = -\frac{7}{3}$
19. $(7, -2)$, $m = \frac{4}{3}$
20. $(3, -3)$, $m = -\frac{3}{2}$
21. $(2, -6)$, $m = \frac{11}{4}$
22. $(\frac{2}{3}, \frac{3}{5})$, $m = 0$

In Exercises 23 through 28, find an equation of the line containing the two given points.

23. $(2, 6)$ and $(-7, 9)$
24. $(-1, 3)$ and $(2, -9)$
25. $(-3, 5)$ and $(-3, 8)$
26. $(-7, 3)$ and $(8, 1)$
27. $(0, 6)$ and $(7, -4)$
28. $(0, -2)$ and $(6, 0)$
29. Find an equation of the line containing $(-1, 4)$ and parallel to $3x - 5y = 11$.
30. Find an equation of the line containing $(-\frac{2}{3}, \frac{3}{4})$ and parallel to $4x + 7y = 15$.
31. Find an equation of the perpendicular bisector of the line seg-

ment with $(-4, 2)$ and $(2, 5)$ as endpoints. Sketch a graph of the segment and the line in the same coordinate plane.

32. Find an equation of the perpendicular bisector of the line segment with $(-2, -4)$ and $(7, 6)$ as endpoints. Sketch a graph of the segment and the line in the same coordinate plane.

Set D–Applications for a Rectangular Coordinate System

Locations of subdivisions

33. The mayor of a city determines that three subdivisions can be located from City Hall as follows.

Subdivision A: 3 miles east and 2 miles north
Subdivision B: 7 miles east and 10 miles north
Subdivision C: 11 miles east and 3 miles south

(a) Construct a coordinate system and exhibit the locations of the subdivisions. Put City Hall at the origin of the coordinate system.
(b) Determine the locations of the midpoints between A and B, B and C, and A and C.
(c) Construct the perpendicular bisector of each segment AB, BC, and AC.

Locations of subdivisions

34. The mayor of a city determines that three subdivisions can be located relative to City Hall as follows.

Subdivision A: 5 miles west and 4 miles north
Subdivision B: 3 miles east and 5 miles south
Subdivision C: 10 miles east and 2 miles north

(a) Construct a coordinate system and exhibit the locations of the subdivisions. Put City Hall at the origin of the coordinate system.
(b) Determine the locations of the midpoints between A and B, B and C, and A and C.
(c) Construct the perpendicular bisectors of each segment AB, BC, and AC.

Distances of warehouses

35. A company has two warehouses A and B and plans to open a new Warehouse C. Warehouse A is 40 miles west and 60 miles north of company headquarters, Warehouse B is 100 miles east and 40 miles north of company headquarters, and Warehouse C will be 20 miles east and 140 miles north of company headquarters. Which of Warehouses A and B will be closer to C?

Distances of warehouses

36. (a) Suppose the company described in Exercise 35 decides to open a fourth warehouse, D, midway (on a line) between

Warehouses A and B. Give its location from company headquarters.

(b) Is Warehouse D closer to Warehouse A or Warehouse C?

(c) How far is Warehouse D from company headquarters?

2.5 LINEAR SYSTEMS AND THEIR APPLICATIONS

A pair of linear equations of the form

$$Ax + By = E, \quad (1)$$

$$Cx + Dy = F \quad (2)$$

linear system in two variables

solution of linear system

is called a **linear system in two variables.** (*Note:* It is assumed for Eq. (1) that not both A and B are 0 and for Eq. (2) that not both C and D are 0.) As we know, the solution set for the linear equation $Ax + By = E$ (or $Cx + Dy = F$) contains infinitely many ordered pairs of real numbers, the ordered pairs associated with the points on the graph of the equation. The **solution set** of the given linear system is the set containing exactly those ordered pairs of real numbers in the solution sets of both equations; that is, the solution set is the intersection of the solution sets for Eqs. (1) and (2). In this section, we learn how to solve a linear system in two variables by three methods: (1) the method of elimination, (2) the method of substitution, and (3) determinants. We shall also learn how linear systems can be used to solve practical problems. (A discussion of linear systems with more than two variables is given in Chapter 3.)

Intersection of lines

EXAMPLE 1 Find the point of intersection of the lines $2x - 3y = 3$ and $4x + 5y = 17$. (See Figure 2.12.)

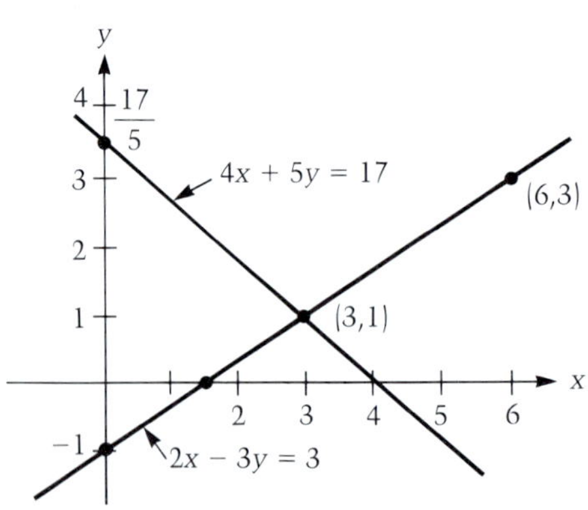

FIG. 2.12.

Solution: The point of intersection (x, y) is on both lines and, hence, must be in the solution set for each equation in the following linear system.

$$2x - 3y = 3,$$
$$4x + 5y = 17.$$

If numbers x and y exist which make both equalities true statements, then they also satisfy the system obtained by multiplying the first equation by 2. (This is done to make the coefficients of x the same in each equation.) That is, they satisfy

$$4x - 6y = 6,$$
$$4x + 5y = 17.$$

Subtracting the second equation from the first, we find that

$$-11y = -11,$$
$$y = 1.$$

Substituting $y = 1$ in either of the original equations, we get $x = 3$. The point of intersection of the two lines is $(3, 1)$.

---

elimination technique

The preceding system of equations was solved by what is called the **elimination technique.** Basically, one or both of the equations are multiplied by constants selected to make either the coefficients of x equal or the coefficients of y equal. Then one variable is eliminated by subtracting one equation from the other, and the resulting equation can be solved by methods learned in Section 2.3.

Ticket sales

EXAMPLE 2 A theater with 500 seats takes in \$1347.50 for a given performance that is sold out. If a child's ticket costs \$1.75 and an adult's ticket costs \$3.00, how many children and how many adults attended the performance?

Solution: Let x be the number of children and let y be the number of adults attending the performance. Then

$$x + y = 500. \qquad (1)$$

The children's tickets bring in \$1.75x and the adults' tickets bring in \$3.00y; hence

$$1.75x + 3.00y = 1347.50. \qquad (2)$$

Solving Eq. (1) for y and substituting in Eq. (2), we get

$$\begin{aligned} 1.75x + 3.00(500 - x) &= 1347.50, \\ 1.75x + 1500 - 3x &= 1347.50, \\ -1.25x &= -152.50, \\ x &= 122. \end{aligned}$$

Therefore 122 children attended the performance and $500 - 122 = 378$ adults attended.

---

substitution technique

In the preceding example, the solution to the linear system was obtained by what is called the **substitution technique.** In this procedure, one equation is solved for one of the variables, and then this expression is substituted into the other equation resulting in a linear equation in one variable. The resulting equation is solved, and this solution is substituted into either of the original equations to obtain the solution to the system.

PRACTICE PROBLEM The total sales from tickets to a concert came to \$98,875 for 12,000 persons attending. If only two types of tickets were available, one selling for \$6.50 and the other selling for \$9.00, how many tickets of each type were sold?

Answer: 3650 of the \$6.50 tickets and 8350 of the \$9.00 tickets. ■

Location of fire station

EXAMPLE 3 Consider the information given in Example 3, Section 2.4, and suppose a fire station is to be built equidistant from each of the three subdivisions. What should be its location in reference to City Hall?

Solution: Note in Figure 2.13 that the three subdivisions are located at the points $A(3, 1)$, $B(6, 8)$, and $C(9, 3)$. The midpoint P of AC is $(6, 2)$ and the midpoint Q of BC is $(\frac{15}{2}, \frac{11}{2})$. The slope of the line containing A and C is $\frac{1}{3}$, so the slope of the perpendicular bisector of AC is -3. The equation of the perpendicular bisector of AC is

$$3x + y = 20. \qquad (1)$$

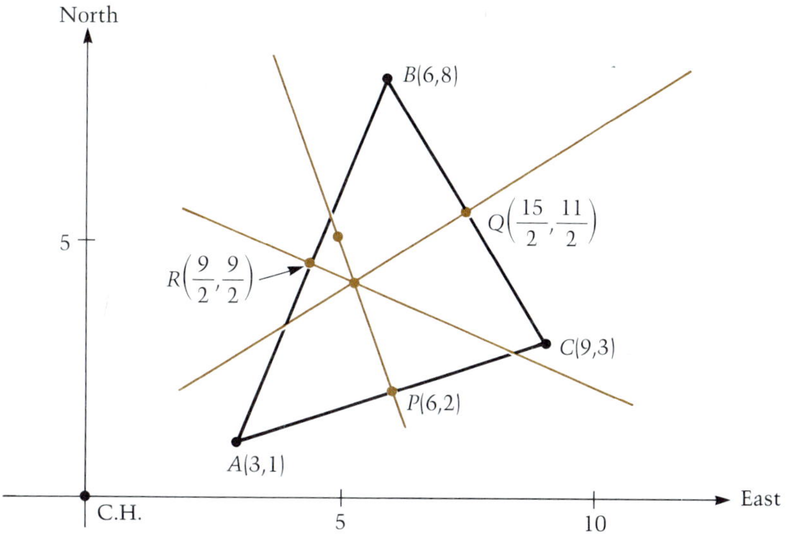

FIG. 2.13.

Similarly, the slope of the line containing B and C is $-\frac{5}{3}$, so the slope of the perpendicular bisector of BC is $\frac{3}{5}$. The equation of the perpendicular bisector of BC is

$$3x - 5y = -5. \tag{2}$$

Since the perpendicular bisectors intersect in a point equidistant from A, B, and C, we find the intersection of the lines given by Eqs. (1) and (2). Subtracting Eq. (2) from Eq. (1), we obtain

$$6y = 25,$$

$$y = \frac{25}{6}.$$

Substituting this value for y in Eq. (1), we get

$$3x + \frac{25}{6} = 20,$$

$$3x = \frac{95}{6},$$

$$x = \frac{95}{18}.$$

Therefore, the fire station should be located $\frac{95}{18} \doteq 5.28$ miles east and $\frac{25}{6} \doteq 4.17$ miles north of City Hall.

Since two lines in a plane can

(1) intersect in a point,

(2) be parallel, or

(3) be coincident,

the solution set of a linear system can

(1) consist of exactly one pair of real numbers,

(2) be empty, or

(3) contain infinitely many ordered pairs of real numbers, respectively.

For the equations $Ax + By = E$ and $Cx + Dy = F$, if the lines they represent are distinct and not parallel, then the lines will intersect in one point and the solution set will contain exactly one pair of real numbers. If neither line is vertical, then the slopes of the lines are $-A/B$ and $-C/D$. Furthermore, since the lines are not parallel,

$$-\frac{A}{B} \neq -\frac{C}{D},$$

$$AD \neq CB,$$

and

$$AD - CB \neq 0.$$

If one of the lines is vertical and if the two lines are not parallel, it is still true that $AD - CB \neq 0$. Consequently, the linear system

$$Ax + By = E,$$

$$Cx + Dy = F$$

has a unique solution provided $AD - CB \neq 0$. (If $AD - CB = 0$, the lines are either parallel or coincident.)

If we solve the preceding general linear system by either the elimination or substitution techniques, we get

solution to general linear system

$$x = \frac{FD - BE}{AD - CB} \quad \text{and} \quad y = \frac{AF - CE}{AD - CB}.$$

determinant of order two

Let us now discuss a useful tool not only for remembering the preceding solution but also for solving linear systems. We denote the number $AD - CB$ by what is called a **determinant of order two.** By definition

$$\begin{vmatrix} A & B \\ C & D \end{vmatrix} = AD - CB.$$

The **value** of any such determinant is the product of the numbers in the upper left and lower right minus the product of the numbers in the lower left and upper right. For our purposes, a determinant of order two is just a square array of numbers that is useful in expressing (remembering) the solution set of the linear system

$$Ax + By = E,$$
$$Cx + Dy = F$$

where $AD - CB \neq 0$. The solution set is given in determinant notation as follows.

$$x = \frac{\begin{vmatrix} E & B \\ F & D \end{vmatrix}}{\begin{vmatrix} A & B \\ C & D \end{vmatrix}} \quad \text{and} \quad y = \frac{\begin{vmatrix} A & E \\ C & F \end{vmatrix}}{\begin{vmatrix} A & B \\ C & D \end{vmatrix}}.$$

Cramer's rule

Note that the determinant in the numerator of the solution for x can be obtained from the determinant in the denominator by replacing the *coefficients of* x by the constants E and F. Similarly, the determinant in the numerator of the solution for y can be obtained from the determinant in the denominator by replacing the *coefficients of* y by the constants E and F. (This is known as **Cramer's rule.**)

Solving a linear system by determinants

EXAMPLE 4 Find the solution set of the following system by using determinants:

$$3x - 2y = 1,$$
$$2x + 7y = 4.$$

Solution: Since $AD - CB = (3)(7) - (2)(-2) = 25 \neq 0$, the system has a unique solution. It is

$$x = \frac{\begin{vmatrix} 1 & -2 \\ 4 & 7 \end{vmatrix}}{25} \quad \text{and} \quad y = \frac{\begin{vmatrix} 3 & 1 \\ 2 & 4 \end{vmatrix}}{25},$$

$$x = \frac{15}{25} = \frac{3}{5} \quad \text{and} \quad y = \frac{10}{25} = \frac{2}{5}.$$

In other words, the two lines represented by the linear equations intersect in the point $(\frac{3}{5}, \frac{2}{5})$.

For the linear system $Ax + By = E$ and $Cx + Dy = F$, if $AD - CB = 0$, then the lines have the same slope. If the lines have the same slope and are distinct, they do not intersect and there is no solution to the system. If $AD - CB = 0$ and the lines are not distinct, then one equation is a constant multiple of the other and there are infinitely many ordered pairs in the solution set, namely, the ordered pairs associated with the points on the line.

PRACTICE PROBLEM Find the solution set of the following linear system by using determinants: $5x - 2y = 10$ and $2x + 3y = 6$.

Answer: $x = \frac{42}{19}$ and $y = \frac{10}{19}$. ■

Stock portfolio

EXAMPLE 5 An investor bought two stocks, one selling at \$30 a share and the other selling at \$12 a share, for a net total of \$18,900. The dividend on the higher priced stock is \$2.60 per share, and the dividend on the lower priced stock is \$1.12 per share. If the investor expects to receive a total of \$1694 in dividends from the two stocks in a given year, how many shares of each stock did the investor buy?

Solution: Let x be the number of shares bought of the higher priced stock and let y be the number of shares bought of the lower priced stock. Therefore, the cost of the higher priced stock was $30x$ dollars, and the cost of the lower priced stock was $12y$ dollars. Consequently,

$$30x + 12y = 18{,}900 \qquad \text{(Total investments)}.$$

Similarly, we conclude from the information on dividends that

$$2.60x + 1.12y = 1694 \qquad \text{(Total dividends)}.$$

Dividing each side of the first equation by the common factor 6 and multiplying each side of the second equation by 100 to get rid of decimals, we get the equivalent system

$$\begin{aligned} 5x + 2y &= 3150, \\ 260x + 112y &= 169{,}400. \end{aligned}$$

Using determinants to solve the linear system, we find

$$x = \frac{\begin{vmatrix} 3150 & 2 \\ 169{,}400 & 112 \end{vmatrix}}{\begin{vmatrix} 5 & 2 \\ 260 & 112 \end{vmatrix}} = \frac{352{,}800 - 338{,}800}{560 - 520} = \frac{14{,}000}{40} = 350.$$

Since the determinant of the coefficients is equal to 40, we have

$$y = \frac{\begin{vmatrix} 5 & 3150 \\ 260 & 169{,}400 \end{vmatrix}}{40} = \frac{847{,}000 - 819{,}000}{40} = \frac{28{,}000}{40} = 700.$$

Thus the investor bought 350 shares of the higher priced stock and 700 shares of the lower priced stock.

Gasohol mixture

EXAMPLE 6 A gasoline supplier has two large gasohol tanks, one containing 8% alcohol and the other containing 14% alcohol. If 1000 gallons of gasohol with 10% alcohol is needed, how many gallons should be taken from each tank in order to provide the proper mixture?

Solution: Let x be the number of gallons required of the 8% mixture and let y be the number of gallons required of the 14% mixture. Since there is to be a total of 1000 gallons,

$$x + y = 1000.$$

Now, $0.08x$ is the amount of alcohol in x gallons of the first mixture and $0.14y$ is the amount of alcohol in y gallons of the second mixture. Since there is to be

$$0.10(1000) = 100$$

gallons of alcohol in the 1000 gallons,

$$0.08x + 0.14y = 100.$$

The problem therefore results in the following system:

$$x + y = 1000, \tag{1}$$

$$8x + 14y = 10{,}000. \tag{2}$$

The system can be solved

(1) by determinants,

(2) by multiplying the first equation by 8 and then eliminating x, or

(3) by substitution.

Let us use the third technique. From Eq. (1), we get

$$y = 1000 - x.$$

From Eq. (2) and the preceding expression for y, we obtain

$$\begin{aligned} 8x + 14(1000 - x) &= 10{,}000, \\ 8x + 14{,}000 - 14x &= 10{,}000, \\ -6x &= -4{,}000, \\ x &= 666\frac{2}{3} \text{ gallons.} \end{aligned}$$

Therefore, $y = 1000 - x = 333\frac{1}{3}$ gallons.

Size of a shipment

EXAMPLE 7 A retailer pays \$178.50 for a box containing Gidgets and Widgets. He sells the Gidgets for \$3.50 apiece and the Widgets for \$7.80 apiece and receives a total of \$547 for the sale of the shipment. If the Gidgets cost \$1.05 apiece and the Widgets cost \$2.66 apiece, how many of each were in the shipment?

Solution: Let x be the number of Gidgets in the shipment and let y be the number of Widgets in the shipment. From the given information, we obtain the following linear system:

$$3.50x + 7.80y = 547,$$

$$1.05x + 2.66y = 178.50.$$

The preceding linear system is equivalent to

$$35x + 78y = 5470,$$
$$105x + 266y = 17{,}850.$$

By using determinants, we find the solutions to be

$$x = \frac{\begin{vmatrix} 5470 & 78 \\ 17{,}850 & 266 \end{vmatrix}}{\begin{vmatrix} 35 & 78 \\ 105 & 266 \end{vmatrix}} = \frac{1{,}455{,}020 - 1{,}392{,}300}{9310 - 8190} = \frac{62{,}720}{1120} = 56,$$

$$y = \frac{\begin{vmatrix} 35 & 5470 \\ 105 & 17{,}850 \end{vmatrix}}{1120} = \frac{624{,}750 - 574{,}350}{1120} = \frac{50{,}400}{1120} = 45.$$

Therefore, there were 56 Gidgets and 45 Widgets in the shipment.

EXERCISES

Set A–Solving Linear Systems

For each linear system in Exercises 1 through 10, find the solution set.

1. $3x - 5y = 21,$
   $4x + 7y = -13$
2. $2x + 4y = 14,$
   $7x - 2y = 17$
3. $2x + 7y = 8,$
   $4x - 9y = -7$
4. $4x - 3y = 2,$
   $x + 3y = 3$
5. $5x - 7y = -12,$
   $3x + 4y = 8$
6. $7x + 8y = 10,$
   $2x - 3y = 12$
7. $(\frac{3}{4})x + (\frac{2}{3})y = -2,$
   $(\frac{1}{2})x - (\frac{5}{6})y = 14$
8. $(\frac{5}{8})x - (\frac{3}{4})y = \frac{1}{4},$
   $(\frac{3}{4})x + (\frac{3}{2})y = \frac{3}{8}$
9. $1.5x - 3.2y = 6,$
   $4.5x + 2.8y = 2.5$
10. $2.4x + 3.2y = 12.4,$
    $1.5x - 5.4y = -1.5$

Set B–Applications Involving Linear Systems

In Exercises 11 through 25, use linear systems to solve the problems.

Size of shipment

11. A large department store ordered a shipment of two different types of shirts, one costing \$6 apiece and another costing \$9 apiece. The total cost of the shipment was \$7680. A second

order was placed later for the same number of shirts of each type. By this time the cost of the less expensive shirt had gone up to $6.25 and the cost of the more expensive shirt had gone up to $9.50. If the second shipment cost $8050, how many shirts of each type were in each shipment?

Size of shipment

12. A store ordered a shipment of two different types of sweaters, one costing $12 each and another costing $20 each. The total cost of the shipment was $768. A second order was placed later for the same number of sweaters of each type. By this time the cost of the less expensive sweater had gone up to $13 and the cost of the more expensive sweater had increased to $22. If the cost of the second shipment was $842, how many sweaters of each type were in each shipment?

Location of fire station

13. Consider the information given in Exercise 33, Section 2.4, and suppose a fire station is to be built equidistant from each of the three subdivisions. What should be its location in reference to City Hall?

Location of fire station

14. Consider the information given in Exercise 34, Section 2.4, and suppose a fire station is to be built equidistant from each of the three subdivisions. What should be its location in reference to City Hall?

Stock portfolio

15. An investor bought two stocks, one selling at $20 a share and the other selling at $15 a share, for a net total of $16,500. The dividend on the higher priced stock is $1.60 per share and the dividend on the lower priced stock is $1.00 per share. If the investor expects to receive a total of $1300 in dividends from the two stocks in a given year, how many shares of each stock did the investor buy?

Stock portfolio

16. An investor bought two stocks, one selling at $18.75 a share and the other selling at $10.50 a share, for a net total of $9150. The dividend on the higher priced stock is $1.50 per share, and the dividend on the lower priced stock is $1.00 per share. If the investor expects to receive a total of $800 in dividends from the two stocks in a given year, how many shares of each stock did the investor buy?

Gasohol mixture

17. A gasoline supplier has two large gasohol tanks, one containing 9% alcohol and the other containing 13% alcohol. If 2000 gallons of gasohol with 10% alcohol are needed, how many gallons should be taken from each tank in order to provide the proper mixture?

Gasohol mixture

18. A gasoline supplier has two large gasohol tanks, one containing 8% alcohol and the other containing 14% alcohol. If 3000 gallons of gasohol with 10% alcohol are needed, how many

gallons should be taken from each tank in order to provide the proper mixture?

Size of shipment

19. A retailer pays $111.50 for a box containing Chewey and Crispy candy bars. He sells the Chewey bars for 75 cents each and the Crispy bars for 35 cents each and takes in a total of $171.25 for the sale of the shipment. If the Chewey bars cost 50 cents apiece and the Crispy bars cost 22 cents apiece, how many of each were in the shipment?

Size of shipment

20. A retailer pays $115 for a box containing nut and caramel candy bars. He sells the nut bars for 65 cents apiece and the caramel bars for 40 cents apiece and receives a total of $175.25 for the sale of the shipment. If the nut bars cost 45 cents apiece and the caramel bars cost 25 cents apiece, how many of each were in the shipment?

Acid solution mixture

21. The Arbor Chemical Company wants to produce 500 gallons of an 18% acid solution by mixing 25% acid solution with a 14% acid solution. How many gallons of each solution are needed?

Acid solution mixture

22. The Arbor Chemical Company wants to obtain 800 gallons of an 18% acid solution by mixing a 20% acid solution with a 14% acid solution. How many gallons of each solution are needed?

Stock portfolio

23. A person has $85,000 invested in stocks and bonds. If the return per year from stocks is 12% and that from bonds is 8% and if these investments yield $8680 in one year, what amount is invested in stocks?

2.6 FUNCTIONS AND THEIR GRAPHS

One extremely important mathematical concept is that of a function. In this section we shall not only learn what a function is but also investigate some of the basic ideas associated with functions which make them exceptionally useful in applied problems.

relation **graph** **domain** **range**

As we know, a set of ordered pairs of real numbers can be paired with a set of points in the rectangular coordinate plane. Such a set is called a **relation,** and the set of corresponding points in the plane is called the **graph** of the relation. If $f = \{(x, y)\}$ is such a relation, then its **domain** is the set of all first elements x in the ordered pairs and its **range** is the set of all second elements y. All of the following are examples of relations; their graphs are given in Figures 2.14 through 2.17.

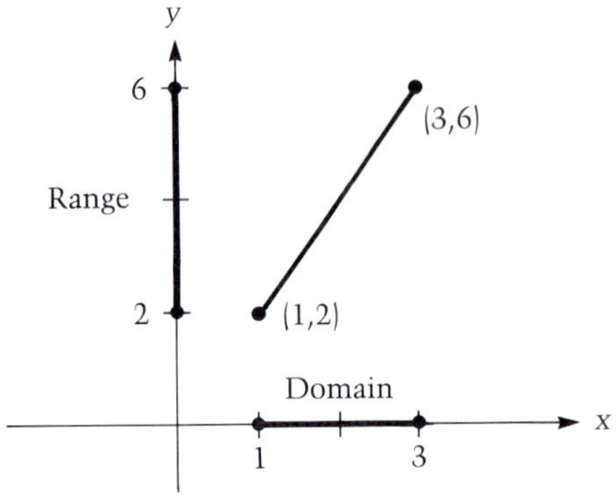

FIG. 2.14.

FIG. 2.15.

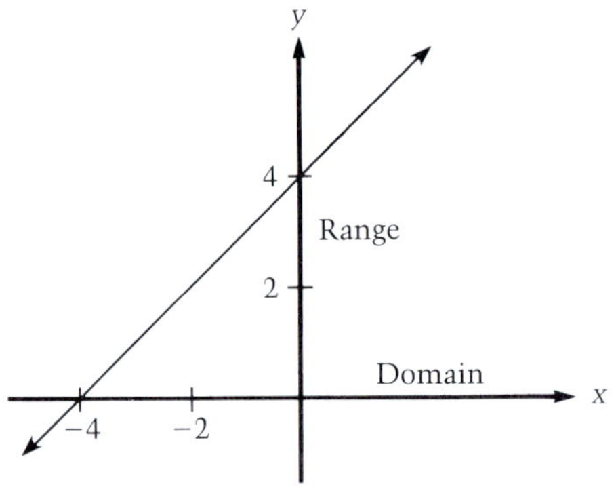

$f_1 = \{(x, y) \mid y = 2x \text{ and } 1 \leq x \leq 3\}$
Domain of f_1: $[1, 3]$
Range of f_1: $[2, 6]$
Graph of f_1: Figure 2.14

$f_2 = \{(x, y) \mid y = x + 4, x \text{ a real number}\}$
Domain of f_2: set of real numbers
Range of f_2: set of real numbers
Graph of f_2: Figure 2.15

$f_3 = \{(x, y) \mid y = x^2, x \text{ a real number}\}$
Domain of f_3: set of real numbers
Range of f_3: set of nonnegative reals
Graph of f_3: Figure 2.16

$f_4 = \{(x, y) \mid y^2 = x, x \geq 0\}$
Domain of f_4: set of all nonnegative reals
Range of f_4: set of all real numbers
Graph of f_4: Figure 2.17

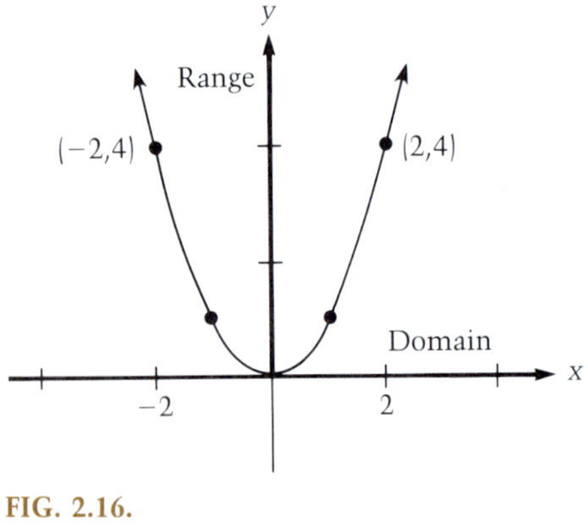

FIG. 2.16.

FIG. 2.17.

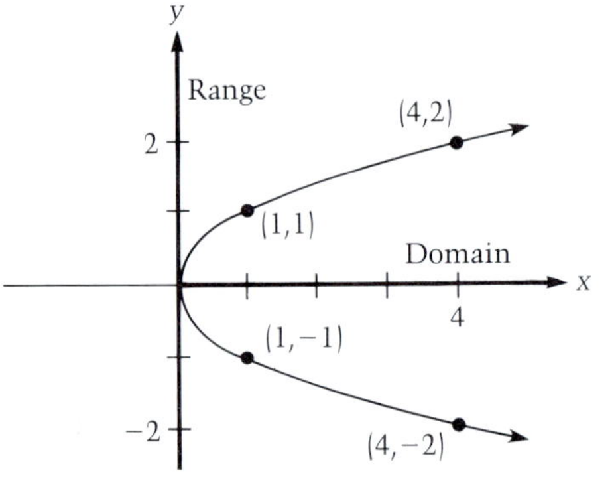

function

A special class of relations which are exceptionally useful in building mathematical models to solve practical problems are called functions. A **function** (of one variable) is a relation where no two ordered pairs have the same first element. Graphically, a relation is a function provided no line parallel to the y-axis (range axis) intersects its graph in more than one point. For example, the relation defined by $y = x^2$ where $x \geq 0$ is a function. (See Figure 2.18a.) However, the relation defined by $y^2 = x$ where $x \geq 0$ is not a function since, for example, both ordered pairs (4, 2) and (4, −2) are elements in the relation. (See Figure 2.18b.)

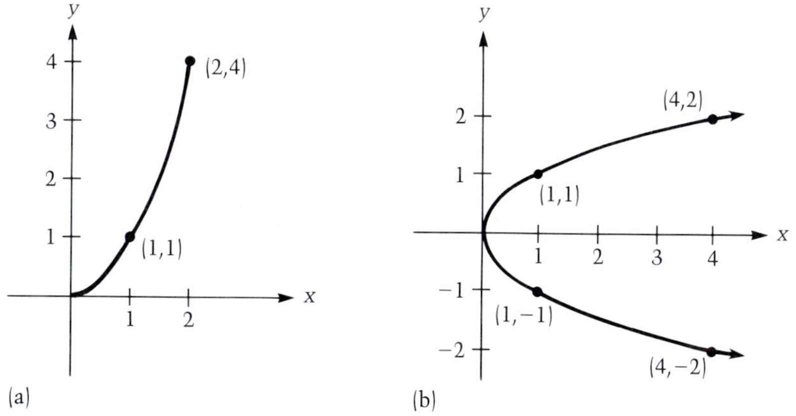

FIG. 2.18. (a) $y = x^2$, $x \geq 0$. (b) $y^2 = x$, $x \geq 0$.

The relation defined by the equation $Ax + By = C$ where x is any real number and $B \neq 0$ is a function. This function has the desirable property that the equation defining it can be solved for y in terms of x. Solving $Ax + By = C$ for y, we obtain

$$y = -\frac{A}{B}x + \frac{C}{B}.$$

linear function

Such an equation is said to give y *explicitly* in terms of x. Furthermore, this function, called a **linear function,** is often defined by the equation

$$y = mx + b.$$

As we know, the graph of this linear function is a nonvertical line in the coordinate plane having m as its slope and b as its y-intercept.

For any function f, the number y in the range paired with a given x in the domain is often denoted by $f(x)$ (read, "f of x," or "the value of f at x"). For example,

$$y = 3x + 5 \qquad \text{and} \qquad f(x) = 3x + 5$$

define the same function. Since $f(1)$ represents the number in the range paired with 1 in the domain, for the linear function $f(x) = 3x + 5$ we have $f(1) = 8$.

Linear function values

EXAMPLE 1 Consider the function defined by $f(x) = 5x - 6$, where x is any real number.

(a) What is $f(2)$?

(b) What is $f(-3)$?

(c) Is $f(2)f(-3) = f(-6)$?

Solution

(a) $f(2) = 5(2) - 6 = 4$.

(b) $f(-3) = 5(-3) - 6 = -21$.

(c) $f(-6) = 5(-6) - 6 = -36$. Since $f(2)f(-3) = -84$, the answer is "no."

---

independent variable
dependent variable

A letter, such as x, used to denote numbers in the domain of a function is called the **independent variable,** and a letter, such as y, used to denote numbers in the range is called the **dependent variable.** Generally, a function will be defined by an equation in the independent and dependent variables with the domain either stated or implied by the nature of the problem.

When functions are constructed to solve practical problems, it is often desirable to use letters other than x and y for the independent and dependent variables (domain and range variables). This practice helps us remember what each variable represents. Consider the following examples.

Area of a Circle: If A represents the area of a circle with radius r, then

$$A(r) = \pi r^2, \quad r > 0,$$

defines the corresponding area function; r is the independent variable and A is the dependent variable. (See Figure 2.19.)

FIG. 2.19. $A(r) = \pi r^2$.

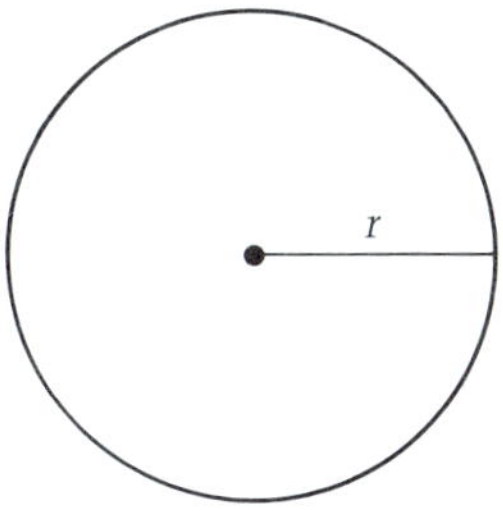

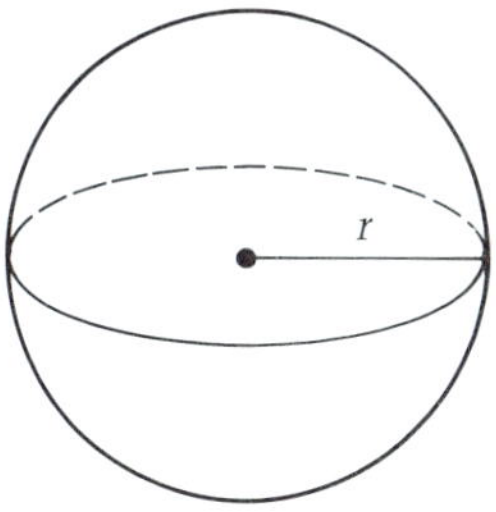

FIG. 2.20. $V(r) = \frac{4}{3}\pi r^3$.

Volume of a Sphere: If V represents the volume of a sphere with radius r, then

$$V(r) = \frac{4}{3}\pi r^3, \quad r > 0,$$

defines the corresponding volume function; r is the independent variable and V is the dependent variable. (See Figure 2.20.)

Perimeter of a Square: If P represents the perimeter of a square with s as the length of one side, then

$$P(s) = 4s, \quad s > 0,$$

defines the corresponding perimeter function; s is the independent variable and P is the dependent variable. (See Figure 2.21.)

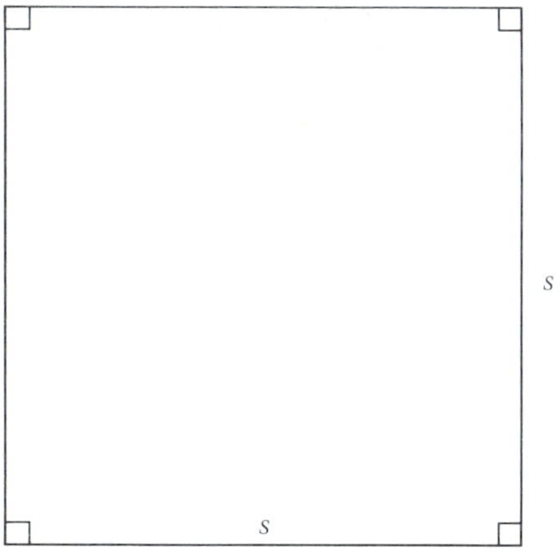

FIG. 2.21. $P(s) = 4s$.

The volume V of a right circular cylinder (tin can) having h as altitude and r as radius of the top and bottom is given by

$$V(r, h) = \pi r^2 h.$$

In this case, the dependent variable V is determined by the *two* independent variables r and h, and V is said to be a *function of two variables.* In Chapters 4 and 14, we shall investigate functions of more than one variable, but, unless stated otherwise, "function" will refer only to a function of one variable.

Consider the function defined by $f(x) = 2x - 4$ where $3 \le x \le 8$. Since the graph of $f(x) = 2x - 4$ is a line, it is not hard to show that the graph of this function is the line segment given in Figure 2.22. Note that

$$f(3) \le f(x) \quad \text{for each} \quad x \text{ in } [3, 8]$$

and

$$f(8) \ge f(x) \quad \text{for each} \quad x \text{ in } [3, 8].$$

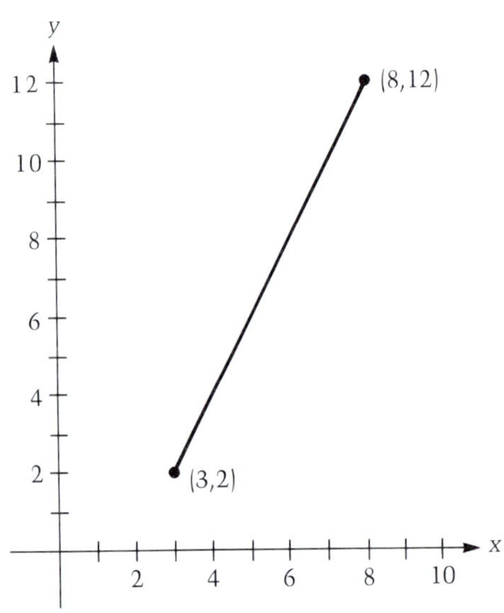

FIG. 2.22. $f(x) = 2x - 5$, $3 \le x \le 8$.

We call $f(3) = 2$ the *(absolute) minimum value* of the function and $f(8) = 12$ the *(absolute) maximum value* of the function. In general, for any function f, if there exists a number v in the domain such that $f(v) \ge f(x)$ for all x in the domain, then $f(v)$ is the **maximum value** of the function. Similarly, if there exists a number u in the domain such that $f(u) \le f(x)$ for all x in the domain, then $f(u)$ is the **minimum value** of the function.

maximum value

minimum value

Maximum and minimum values

EXAMPLE 2 Consider the function defined by $f(x) = x^2$ where $-2 \le x \le 2$. (See Figure 2.23.)

(a) What is its maximum value?

(b) What is its minimum value?

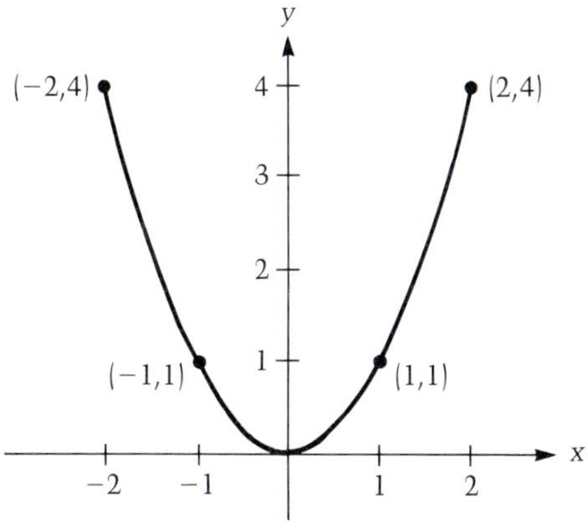

FIG. 2.23. $f(x) = x^2$, $-2 \le x \le 2$.

Solution

(a) It is clear from Figure 2.23 that 4 is the maximum value of the function.

(b) Since $x^2 \ge 0$ for any real number x and since $f(0) = 0$, the minimum value of the function is 0. (This can also be seen in Figure 2.23.)

Although a function need not have an absolute maximum value (or an absolute minimum value), if it has one, it cannot have more than one. However, as in Example 2, a function can have a maximum value at more than one point in its domain. In fact, for the constant function defined by $f(x) = 6$, which has a horizontal line as its graph, the maximum and minimum value 6 is attained at each number in its domain.

Maximum and minimum values

EXAMPLE 3

(a) Graph the function defined by $f(x) = 2x + 4$ where $-3 < x \le 2$.

(b) Give the maximum value of the function if it exists.

(c) Give the minimum value of the function if it exists.

Solution

(a) See Figure 2.24.

(b) The maximum value of the function is $f(2) = 8$.

(c) The function has no minimum value. (The range consists of all real numbers y such that $-2 < y \leq 8$; this set contains no least number.)

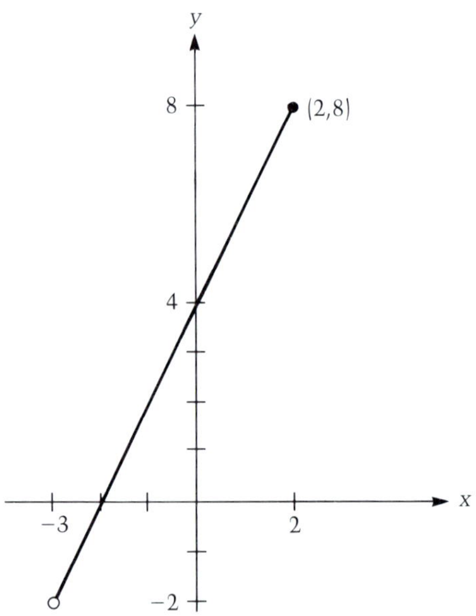

FIG. 2.24. $f(x) = 2x + 4$, $-3 < x \leq 2$.

Average cost function

EXAMPLE 4 The cost in dollars of producing q units of a certain product is given by

$$C(q) = 2q^2 + 4q + 50 \qquad \text{for} \quad q \geq 1.$$

(a) What is the cost of producing one unit?

(b) What is the cost of producing 10 units?

(c) What is the average cost per unit for producing 10 units?

(d) What is the average cost function?

(e) Graph the average cost function.

(f) How many units will minimize the average cost per unit? (What is the minimum value of the average cost function?)

Solution

(a) $C(1) = 2(1)^2 + 4(1) + 50 = \$56$ is the cost of producing the first unit.

(b) $C(10) = 2(10)^2 + 4(10) + 50 = \$290$ is the cost of producing the first 10 units.

(c) Since $C(10) = 290$ is the cost of the first 10 units,

$$\frac{C(10)}{10} = \$29$$

is their average cost.

average cost function

(d) Since $C(q) = 2q^2 + 4q + 50$ is the cost function, the *average cost function* $\overline{C}(q)$ is defined by

$$\overline{C}(q) = \frac{C(q)}{q} = \frac{2q^2 + 4q + 50}{q} = 2q + 4 + \frac{50}{q} \qquad \text{for} \quad q \geq 1.$$

(e) Note that for "large" q

$$2q + 4 + \frac{50}{q} \doteq 2q + 4 \qquad \text{(i.e., the difference is "small").}$$

Thus, we see that for "large" q the function values for the average cost function are approximated by the function values of the linear function defined by

$$L(q) = 2q + 4.$$

This straight line with slope 2 and y-intercept 4 is called an *asymptote* for the graph of $\overline{C}(q)$. (See Figure 2.25.) A table of values for the average cost function is given below.

Average Cost Function

q	1	2	3	4	5	6	7	8	9
$\overline{C}(q)$	56	33	$26\frac{2}{3}$	$24\frac{1}{2}$	24	$24\frac{1}{3}$	$25\frac{1}{7}$	$26\frac{1}{4}$	$27\frac{5}{9}$

(f) It appears from the table of values and the graph of the average cost function that $\overline{C}(5) = \$24$ is the minimum average cost. If this is the case, producing 5 units would minimize aver-

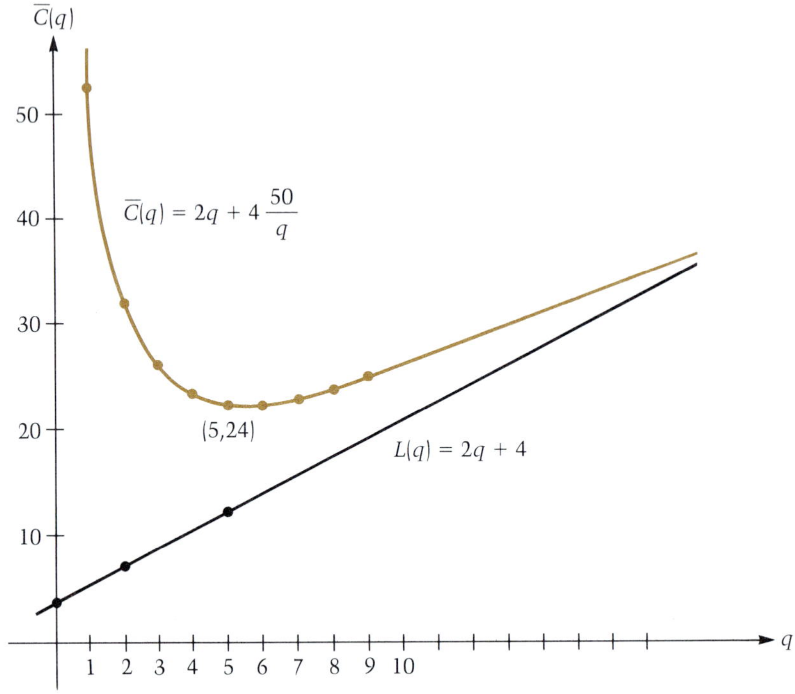

FIG. 2.25.

age cost. (In Chapter 10 we shall find a way to prove that $\overline{C}(5) = 24$ is the minimum without resorting to a table of values or a graph of the function.)

Before turning to the exercises, let us point out that we will continue to learn more about functions, graphs of functions, and applications of functions throughout the remainder of the text.

EXERCISES

Set A–Function Values

In Exercises 1 through 8, answer the following questions. (a) What is $f(2)$? (b) What is $f(3)$? (c) Is $f(2) + f(3) = f(5)$? (d) Is $f(2)f(3) = f(6)$?

1. $f(x) = 8x + 3$, x a real number
2. $f(x) = 5x - 7$, x a real number
3. $f(x) = 3x^2$, x a real number
4. $f(x) = -x^2$, x a real number

5. $f(x) = x^2 + 4$, x a real number
6. $f(x) = x^2 - x$, x a real number
7. $f(x) = x^3$, x a real number
8. $f(x) = x^3 - 1$, x a real number

Set B–Maximum and Minimum Values of Functions and Their Graphs

In Exercises 9 through 22, do each of the following. (a) Graph the function. (b) Give the maximum value of the function if one exists. (c) Give the minimum value of the function if one exists.

9. $f(x) = 7x - 5,\ -1 \le x \le 8$
10. $f(x) = 3x + 4,\ -3 \le x \le 9$
11. $f(x) = -6x + 9,\ -2 < x < 7$
12. $f(x) = -5x + 8,\ -4 < x < 5$
13. $f(x) = -4x + 7,\ -3 \le x < 5$
14. $f(x) = -3x + 2,\ -2 < x \le 6$
15. $f(x) = x^2 + 4,\ -1 \le x \le 6$
16. $f(x) = x^2 + 1,\ -3 \le x \le 8$
17. $f(x) = -x^2 + 9,\ -3 < x \le 3$
18. $f(x) = -x^2 - 3,\ -2 \le x < 4$
19. $f(x) = x + \dfrac{1}{x},\ 1 \le x \le 8$
20. $f(x) = 4 + \dfrac{2}{x},\ 1 \le x \le 10$
21. $f(x) = 4x + 2 + \dfrac{36}{x}$, x a positive real number
22. $f(x) = 2x + 3 + \dfrac{72}{x}$, x a positive real number

2.7 APPLICATIONS OF LINEAR FUNCTIONS

Suppose a new automobile costing \$7200 depreciates at a constant amount of \$800 per year. Its *initial value* $V(0)$ is \$7200, and at the end of 2 years its value is

$$\begin{aligned} V(2) &= 7200 - 2(800) \\ &= 7200 - 1600 \\ &= \$5600. \end{aligned}$$

At the end of t years, its value is

$$V(t) = 7200 - 800t \text{ dollars.}$$

Since $V(t)$ must be greater than or equal to 0,

$$7200 - 800t \geq 0,$$
$$-800t \geq -7200,$$
$$t \leq 9.$$

Therefore, the domain of the function V is $0 \leq t \leq 9$. (See Figure 2.26.) If we assume that this formula holds for any number in the interval [0, 9], we find, for example, that

$$V\left(\frac{7}{2}\right) = 7200 - 800\left(\frac{7}{2}\right)$$
$$= 7200 - 2800$$
$$= \$4400$$

linear depreciation

is the value of the car after it is $3\frac{1}{2}$ years old. (*Linear depreciation* is a standard type of depreciation often used by businesses for machinery and other equipment.)

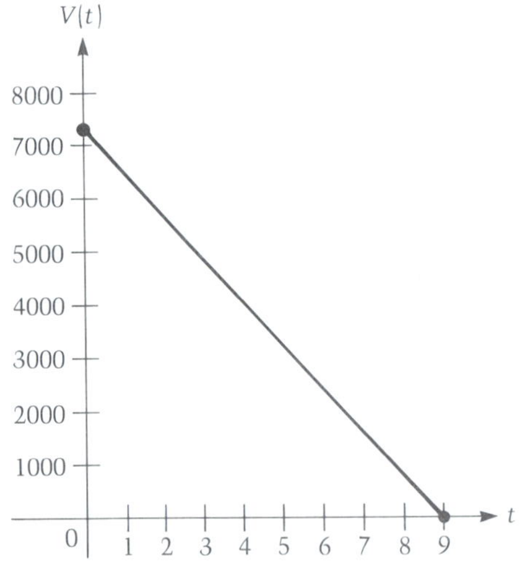

FIG. 2.26. $V(t) = 7200 - 800t$, $0 \leq t \leq 9$.
(*Note:* Different scales are used on the axes.)

The cost function C for buying x light bulbs selling for \$3.00 apiece is given by

$$C(x) = 3x$$

where x is a positive integer. The cost $C(x)$ is said to be *directly proportional to* x, the number of bulbs bought. If x is doubled, for example, then $C(x)$ is doubled. In general, if

$$y = kx,$$

directly proportional

then y is said to be **directly proportional** to x and k is called the *constant of proportionality.* The graph of such a direct proportionality function is a straight line through the origin with slope k.

The **average cost** $\overline{C}(x)$ of x items, such as the \$3.00 light bulbs, is given by

average cost

$$\overline{C}(x) = \frac{C(x)}{x}$$

where $C(x)$ is the cost of x items. If $C(x) = 3x$, then $\overline{C}(x) = 3$ which, of course, is the expected result.

Now let us suppose that the company from which \$3.00 light bulbs are to be purchased will ship any number of the bulbs for an \$8.00 packaging and mailing fee. The cost function for the person who plans to purchase the bulbs and have them delivered by mail is

$$C(x) = 3x + 8 \text{ dollars.}$$

We should observe the following points:

(1) Although the cost function is a linear function, the cost is not directly proportional to the number of bulbs bought. For example, $C(1) = 11$ and $C(2) = 14$, and we see that although the number of bulbs is doubled, the cost is not doubled.

(2) The average cost function is defined by

$$\overline{C}(x) = \frac{C(x)}{x} = 3 + \frac{8}{x}$$

which is neither a constant function nor a linear function.

(3) We can see that the average cost decreases, since as x increases, $8/x$ decreases. However, since $8/x$ is positive for positive x,

$$\overline{C}(x) = 3 + \frac{8}{x} > 3,$$

and the average cost is always greater than \$3. In fact, $\overline{C}(x)$ "approaches" 3 as x gets large, and $y = 3$ is a horizontal asymptote for the graph of the average cost function.

(4) Since the domain is the set of positive integers, we conclude that $C(1) = 11$ is the minimum value of $C(x)$ and that the cost function has no maximum value. (See Figure 2.27.) But for the average cost function, $\overline{C}(x) = 11$ is its maximum value and it has no minimum value.

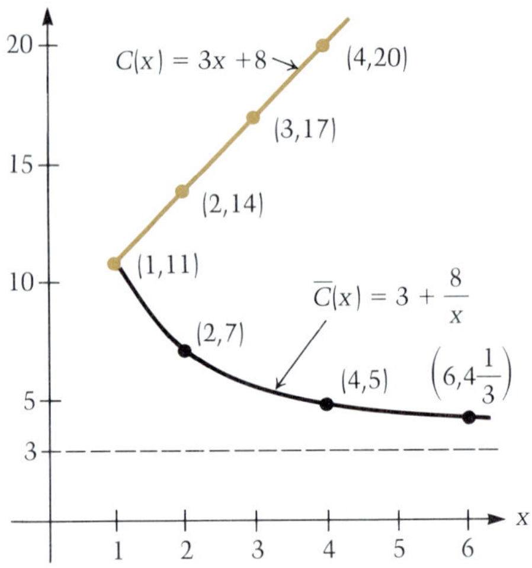

FIG. 2.27. $C(x) = 3x + 8$ and $\overline{C}(x) = 3 + \frac{8}{x}$, *where* $x \geq 1$.

(5) Although the domain of $C(x)$ and $\overline{C}(x)$ is the set of positive integers, it is common practice to graph the functions as though each domain was the set of real numbers x where $x \geq 1$.

Suppose a college organization decides to sell popcorn at a local fair in order to raise money for charity. If space at the fair rents for \$60 and if the cost to produce each sack of popcorn is 20 cents, then the cost C of producing x sacks of popcorn is

$$C(x) = 0.20x + 60 \text{ dollars}$$

where x is an integer greater than or equal to zero. Since

$$C(100) = 0.2(100) + 60 = 80,$$

the cost of producing 100 sacks of popcorn is \$80. Although the graph of this function contains only those points on the line with

60 as y-intercept and $0.20 = \frac{1}{5}$ as slope for which x is an integer, this fact is ignored when we graph the function. (See Figure 2.28.) Of course, when solving any practical problem associated with such a graph, we should not forget that the domain contains only integers.

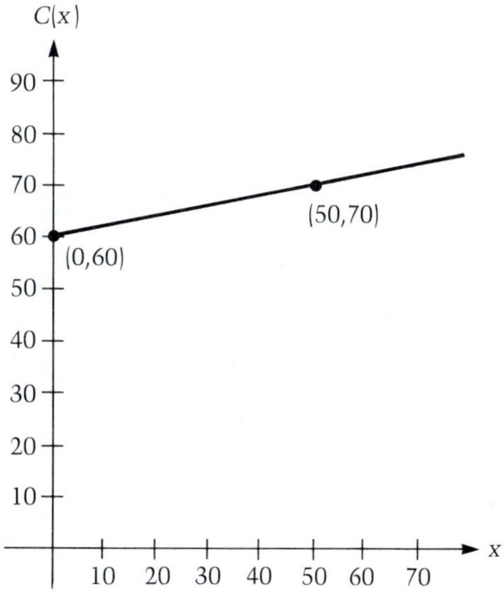

FIG. 2.28.
$C(x) = 0.20x + 60$, $x \geq 0$.

Now suppose each sack of popcorn is sold for 50 cents. The total revenue R from selling x sacks is

$$R(x) = \$0.50x.$$

Therefore, for example, $R(100) = 0.5(100) = \$50$ would be received from selling 100 sacks of popcorn. The profit P from selling x sacks of popcorn is the difference between the revenue and cost for x sacks; it is given by

$$P(x) = R(x) - C(x).$$

If $R(x) > C(x)$, then $P(x) > 0$ and a profit is made. If $R(x) < C(x)$ then $P(x) < 0$ and a loss is incurred. For example, since

$$P(100) = R(100) - C(100) = 50 - 80 = -30,$$

the organization loses \$30 if only 100 sacks of popcorn are sold.

The point at which $P(x) = 0$ is called the **break-even point**; that is, the break-even point is the value of x for which

$$R(x) = C(x).$$

If the revenue and cost functions are graphed in the same coordinate system, the break-even point is where their graphs intersect. (See Figure 2.29.) For the organization selling popcorn at the fair, the break-even point can be found as follows. Since $R(x) = 0.50x$ and $C(x) = 0.20x + 60$, $R(x) = C(x)$ if

$$0.50x = 0.20x + 60,$$
$$0.30x = 60,$$
$$x = 200.$$

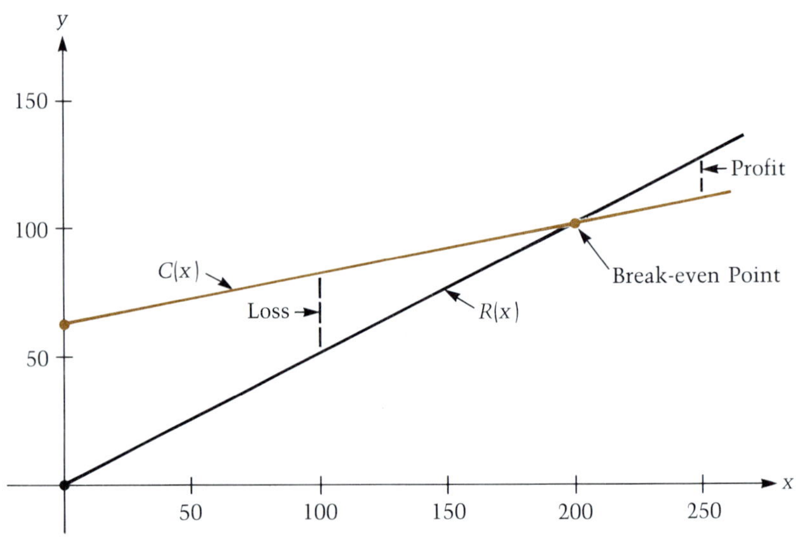

FIG. 2.29. $P(x) = R(x) - C(x)$.

Therefore, to make a profit, the organization would have to sell more than 200 sacks of popcorn. If the organization wants to make \$300, it would be necessary that $P(x) = 300$. Solving $R(x) - C(x) = 300$, we obtain

$$0.50x - (0.20x + 60) = 300,$$
$$0.30x = 360,$$
$$x = 1200.$$

Therefore a profit of \$300 is realized by selling 1200 sacks of popcorn.

PRACTICE PROBLEM Suppose the organization selling popcorn at the fair could rent space for \$50. If popcorn costing 15 cents per sack to produce could be sold for 35 cents, what would be the break-even point?

Answer: 250 sacks. ■

Sales function

EXAMPLE 1

(a) A manufacturer of television sets concludes from previous introductions of new models that 15,000 sets can be sold with no advertising. Also from past sales records, it is assumed that for each \$1000 in advertising an additional 50 sets will be sold. How many sets will be sold with an advertising budget of x thousand dollars?

Solution (a): If N is the number of sets sold, then

$$N(x) = 15{,}000 + 50x.$$

Since, for example, $N(0) = 15{,}000$ and $N(3) = 15{,}150$, 15,000 sets will be sold with no advertising budget and 15,150 sets will be sold with a \$3000 advertising budget.

Revenue function

(b) Suppose each television set sells for \$650. What is the revenue function R in terms of x?

Solution (b): Since $N(x)$ is the number sold with an advertising budget of x thousand dollars and since each set sells for \$650,

$$\begin{aligned} R(x) &= 650N(x) \\ &= 650(50x + 15{,}000) \\ &= 32{,}500x + 9{,}750{,}000. \end{aligned}$$

Note that the total revenue with no advertising budget is

$$R(0) = \$9{,}750{,}000.$$

Cost function

(c) Suppose the initial fixed costs to produce the television sets are \$4,500,000 and suppose each set costs \$500 to manufacture. What is the total cost function C in terms of x?

Solution (c): The total cost would include

the fixed costs: \$4,500,000,

the costs to manufacture $N(x)$ sets: \$500$N(x)$,

the costs for x thousand dollars of advertising: \$1000x.

Consequently,

$$\begin{aligned} C(x) &= 1000x + 500(15{,}000 + 50x) + 4{,}500{,}000 \\ &= 1000x + 7{,}500{,}000 + 25{,}000x + 4{,}500{,}000 \\ &= 26{,}000x + 12{,}000{,}000. \end{aligned}$$

Note that if $x = 0$, then no advertising is done, and the cost to produce the first 15,000 sets would be

$$C(0) = \$12{,}000{,}000.$$

Profit function

(d) Find the profit function P in terms of x.

Solution (d): Since the profit function P is given by $P(x) = R(x) - C(x)$,

$$\begin{aligned} P(x) &= 32{,}500x + 9{,}750{,}000 - 26{,}000x - 12{,}000{,}000 \\ &= 6{,}500x - 2{,}250{,}000. \end{aligned}$$

Since $P(0) = -2{,}250{,}000$, a loss of \$2,250,000 would result if the company did not advertise.

Break-even point

(e) What is the break-even point?

Solution (e): The break-even point is where $P(x) = 0$; it is found as follows.

$$\begin{aligned} 6500x - 2{,}250{,}000 &= 0, \\ 6500x &= 2{,}250{,}000, \\ x &\doteq 346.15385. \end{aligned}$$

Since x is in thousands of dollars, an advertising budget of approximately \$346,154 is required to break even; that is, the break-even point is reached when $N(346.2) = 32{,}310$ sets are sold.

Profit from advertising

(f) What would be the profit with a million dollar advertising budget?

Solution (f): Since 1,000,000 = 1000 thousands, the profit from a million dollar advertising budget is

$$\begin{aligned} P(1000) &= 6500(1000) - 2{,}250{,}000 \\ &= \$4{,}250{,}000. \end{aligned}$$

---

demand function

supply function

equilibrium price
equilibrium quantity

Assume $D(p)$ is the number of units of a certain product that consumers will demand at a certain price p. If we assume D is determined solely by p, then a **demand function** with p as independent variable can be constructed as a model for market demand. Since consumers are reluctant to pay high prices, it is reasonable to expect that the demand decreases as price p increases. On the other side of this market are the producers who supply the product. Since their goal is to make a profit, it is reasonable to expect that as price increases, profit increases and the number of units S the producers are willing to supply increases. If supply S is considered to be determined solely by price p, then a **supply function** $S(p)$ with p as independent variable can be constructed as a model for market supply. For a given price p, if supply $S(p)$ is less than demand $D(p)$, then a shortage exists. If supply $S(p)$ is greater than demand $D(p)$, then a surplus results. The value $\bar{p}$ such that $D(\bar{p}) = S(\bar{p})$ is called the **equilibrium quantity,** and $(\bar{p}, \bar{q})$ is called the **market equilibrium point.** (See Figure 2.30.)

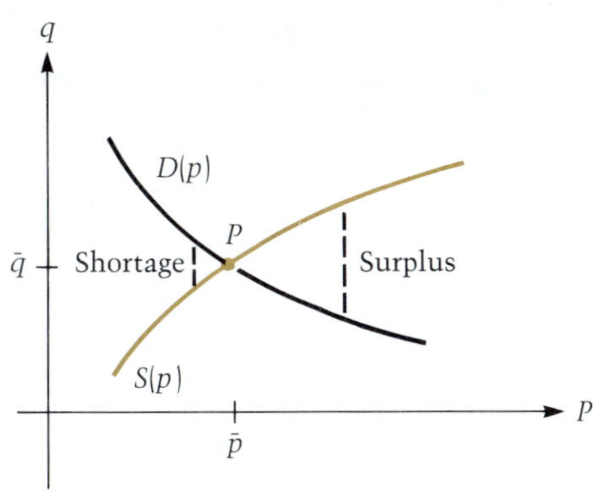

FIG. 2.30. P, equilibrium point; $\bar{p}$, equilibrium price.

Demand function

EXAMPLE 2 Suppose the demand function for a certain product is given by

$$D(p) = 1000(82 - 8p) \text{ units}$$

where $p \geq \$4.50$.

(a) How many units will be sold if the price is \$4.50?

(b) How many units will be sold if the price is \$6.00?

(c) At what price would the consumer not be willing to buy the product?

Solution

(a) The number of units that will be sold at \$4.50 is

$$\begin{aligned} D(4.50) &= 1000[82 - 8(4.5)] \\ &= 1000(82 - 36) \\ &= 46{,}000 \text{ units.} \end{aligned}$$

(b) The number of units that will be sold at \$6.00 is

$$D(6) = 1000(82 - 48) = 34{,}000 \text{ units.}$$

(c) The price at which the consumer will not be willing to buy the product is the p for which $D(p) = 0$. Since $D(p) = 0$ when

$$\begin{aligned} 1000(82 - 8p) &= 0 \\ 82 - 8p &= 0 \\ -8p &= -82 \\ p &= 10.25, \end{aligned}$$

demand will be zero if the price is \$10.25.

Supply function

EXAMPLE 3 Suppose the supply function for the product discussed in Example 2 is given by

$$S(p) = 2000p \text{ units}$$

where $p \geq \$4.50$.

(a) How many units will be supplied if the price is \$4.50?

(b) How many units will be supplied if the price is \$6.00?

(c) What is the equilibrium price for the product?

(d) What is the equilibrium quantity?

Solution

(a) $S(4.50) = 2000(4.50) = 9000$ units will be supplied if $p =$ \$4.50.

(b) $S(6) = 2000(6) = 12{,}000$ units will be supplied if $p = \$6.00$.

(c) Since the equilibrium price is the p for which $S(p) = D(p)$,

$$\begin{aligned} 2000p &= 1000(82 - 8p), \\ 2p &= 82 - 8p, \\ 10p &= 82, \\ p &= 8.2. \end{aligned}$$

The equilibrium price is \$8.20.

(d) The equilibrium quantity is $S(8.2) = 2000(8.2) = 16{,}400$ units.

EXERCISES

Set A–Applications of Linear Functions

Linear depreciation

1. Assume that an automobile costs \$8000 and that it depreciates linearly at \$1200 per year.
   (a) Express the value V of the car in terms of t years.
   (b) What is $V(2)$?
   (c) For what value of t is $V(t) = 0$?
   (d) What is the value of the car after 30 months?

Linear depreciation

2. Assume that an automobile costs \$6000 and that it depreciates linearly at \$800 per year.
   (a) Express the value V of the car in terms of t years.
   (b) What is $V(2)$?
   (c) For what value of t is $V(t) = 0$?
   (d) What is the value of the car after 33 months?

Demand function

3. Suppose the demand function for a certain product is given by

$$D(p) = 1000(76 - 8p) \text{ units}$$

   if $p \geq \$3.50$.
   (a) How many units will be sold if the price is \$3.50?
   (b) How many units will be sold if the price is \$6.00?
   (c) At what price would the consumer not be willing to buy the product?
   (d) What is $D(p + 1) - D(p)$? What does this number represent?

Demand function

4. Suppose the demand function for a certain product is given by

$$D(p) = 200(84 - 6p) \text{ units}$$

if $p \geq \$2.50$.
(a) How many units will be sold if the price is \$2.50?
(b) How many units will be sold if the price is \$5.00?
(c) At what price would the consumer not be willing to buy the product?
(d) What is $D(p + 1) - D(p)$? What does this number represent?

Supply function

5. Suppose the supply function for the product discussed in Exercise 3 is given by

$$S(p) = 8000p \text{ units}$$

if $p \geq \$3.50$.
(a) How many units will be supplied if the price is \$3.50?
(b) How many units will be supplied if the price is \$6.00?
(c) What is the equilibrium price for the product?
(d) What is the equilibrium quantity?

Supply function

6. Suppose the supply function for the product discussed in Exercise 4 is given by

$$S(p) = 1200p \text{ units}$$

if $p \geq \$2.50$.
(a) How many units will be supplied if the price is \$2.50?
(b) How many units will be supplied if the price is \$5.00?
(c) What is the equilibrium price for the product?
(d) What is the equilibrium quantity?

Repair charges

7. An automobile repair agency states that its repair charges C are directly proportional to the time t in hours spent working on a car.
(a) If the repair shop charges \$43.50 for $1\frac{1}{2}$ hours of work, what is the constant of proportionality?
(b) What does the constant of proportionality represent in this problem?
(c) What is the function associated with the repair charges?
(d) What would a repair job taking $4\frac{3}{4}$ hours cost?

Production function

8. Assume that the number of units N of a product that can be produced by a given machine is directly proportional to the number of hours t the machine is allowed to run.

(a) If the machine produces 4880 units in 8 hours, what is the constant of proportionality?
(b) What does the constant of proportionality represent in this problem?
(c) What is the production function in terms of the number of hours t that the machine runs?
(d) How many units would be produced by the machine in 100 hours?
(e) How many hours would it take for the machine to produce 128,100 units?

Charges for checking account

9. For a checking account, a bank charges \$3.50 per month and 15 cents for each check.
(a) What are the charges for writing 20 checks in one month?
(b) If x checks are written in one month, what are the charges C in terms of x?
(c) Use part (b) to find the charges for writing 35 checks.
(d) What is the average cost function per month for writing checks?
(e) What is the average cost for writing 35 checks in one month?
(f) Graph the cost and average cost functions in the same coordinate system.

Charges for checking account

10. For a checking account, a bank charges \$5.00 per month and 10 cents for each check.
(a) What are the charges for writing 20 checks in one month?
(b) If x checks are written in one month, what are the charges C in terms of x?
(c) Use part (b) to find the charges for writing 35 checks.
(d) What is the average cost function per month for writing checks?
(e) What is the average cost for writing 35 checks in one month?
(f) Graph the cost and average cost functions in the same coordinate system.

11. For the checking account charges discussed in Exercises 9 and 10, how many checks would provide equal charges under the two charge schedules?

Use the following information in Exercises 12 through 17. A manufacturer of television sets decides from previous experiences in introducing new television models that 12,000 sets of a given model can be sold with no advertising. Also on the basis of past sales records, it is anticipated that for each \$1000 in advertising an additional 40 sets will be sold.

Sales function

12. (a) How many sets N will be sold with an advertising budget of x thousand dollars?
    (b) Use part (a) to determine how many sets will be sold with an advertising budget of $90,000.

Revenue function

13. (a) Suppose each television set sells for $600. What is the total revenue function R in terms of x?
    (b) Use part (a) to determine the revenue received with an advertising budget of $90,000.

Cost function

14. (a) Suppose the fixed costs to produce the television sets are $3,500,000 and suppose each set costs $450 to manufacture. What is the total cost function in terms of x?
    (b) Use part (a) to determine the total cost with an advertising budget of $90,000.

Profit function

15. (a) Find the profit function in terms of x.
    (b) Use part (a) to determine the profit (or loss) with an advertising budget of $90,000.

Break-even point

16. (a) What is the advertising budget for the break-even point?
    (b) How many sets have to be sold to break even?

Profit from advertising

17. What is the profit (loss) with a one-million dollar advertising budget?

Use the following information in Exercises 18 through 23. A manufacturer of television sets decides from previous experiences in introducing new television models that 8000 sets can be sold with no advertising. Also from past sales records, it is anticipated that for each $1000 in advertising an additional 30 sets will be sold.

Sales function

18. (a) How many sets will be sold with an advertising budget of x thousand dollars?
    (b) Use part (a) to determine how many sets will be sold with an advertising budget of $100,000.

Revenue function

19. (a) Suppose each television set sells for $500. What is the total revenue function R in terms of x?
    (b) Use part (a) to determine the revenue received with an advertising budget of $100,000.

Cost function

20. (a) Suppose the fixed costs to produce the television sets are $2,500,000 and suppose each set costs $400 to manufacture. What is the total cost function in terms of x?
    (b) Use part (a) to determine the total cost with an advertising budget of $100,000.

Profit function

21. (a) Find the profit function in terms of x.
    (b) Use part (a) to determine the profit with an advertising budget of $100,000.

Break-even point

22. (a) What is the advertising budget for the break-even point?
(b) How many sets have to be sold to break even?

Profit from advertising

23. What is the profit (loss) with a ten-million dollar advertising budget?

2.8 SUMMARY AND REVIEW EXERCISES

1. BASIC INEQUALITY PROPERTIES

(a) If $a < b$, then

$$a + c < b + c \quad \text{and} \quad a - c < b - c.$$

(b) If $a < b$ and $c > 0$, then

$$ac < bc \quad \text{and} \quad \frac{a}{c} < \frac{b}{c}.$$

(c) If $a < b$ and $c < 0$, then

$$ac > bc \quad \text{and} \quad \frac{a}{c} > \frac{b}{c}.$$

(d) If $a < b$ and $c < d$, then

$$a + c < b + d.$$

2. SLOPES OF LINES

(a) If (x_1, y_1) and (x_2, y_2) are two points in the coordinate plane such that $x_1 \neq x_2$, then

$$m = \frac{y_2 - y_1}{x_2 - x_1}$$

is the **slope** of the line containing the two points.

(b) Two nonvertical lines with slopes m_1 and m_2 are **parallel** if and only if $m_1 = m_2$.

(c) Two nonvertical lines with slopes m_1 and m_2 are **perpendicular** if and only if $m_1 = -1/m_2$.

3. EQUATIONS OF LINES

(a) An equation of a line containing the point (x_1, y_1) having slope m is

$$y - y_1 = m(x - x_1). \quad \textbf{(Point-slope form)}$$

(b) An equation of the line with m as slope and b as y-intercept is

$$y = mx + b. \quad \textbf{(Slope-intercept form)}$$

(c) Any linear equation in two variables can be written in the form $Ax + By = C$. If $B \neq 0$, its slope is $-A/B$.

4. SOLUTION OF LINEAR SYSTEMS

The solution set of the linear system in two variables

$$Ax + By = E,$$
$$Cx + Dy = F,$$

provided $AD - CB \neq 0$ is given by

$$x = \frac{\begin{vmatrix} E & B \\ F & D \end{vmatrix}}{\begin{vmatrix} A & B \\ C & D \end{vmatrix}} = \frac{ED - FB}{AD - CB}, \qquad y = \frac{\begin{vmatrix} A & E \\ C & F \end{vmatrix}}{\begin{vmatrix} A & B \\ C & D \end{vmatrix}} = \frac{AF - CE}{AD - CB}.$$

5. FUNCTIONS

(a) A **function** is a set of ordered pairs of real numbers such that no two ordered pairs have the same first element. The **domain** is the set of all first elements and the **range** is the set of all second elements.

(b) A letter, such as x, used to denote numbers in the domain of a function is called the **independent variable,** and a letter, such as y, used to denote numbers in the range is called the **dependent variable.**

(c) For any function f, if there exists a number v in the domain such that $f(v) \geq f(x)$ for all x in the domain, then $f(v)$ is the **maximum value** of f. If there exists a number u in the domain such that $f(u) \leq f(x)$ for all x in the domain, then $f(u)$ is the **minimum value** of f.

6. LINEAR FUNCTIONS

(a) A **linear function** is defined by $y(x) = mx + b$ where x is any real number. Its graph is the line having m as slope and b as y-intercept.

(b) For the linear function defined by $y(x) = kx$ where k is a constant, $y(x)$ is said to be **directly proportional** to x and k is called the **constant of proportionality.**

7. SOME GENERAL FUNCTIONS

(a) If $C(x)$ is the total cost function for x units of a given product, then

$$\overline{C}(x) = \frac{C(x)}{x}$$

is the **average cost function.**

(b) If $R(x)$ is the revenue function for x units of a given product and $C(x)$ is the total cost function, then $P(x) = R(x) - C(x)$ is the **profit function.** The **break-even point** is the x for which $P(x) = 0$.

(c) If $D(p)$ is a demand function and $S(p)$ is a supply function for a certain commodity, then the value $\overline{p}$ such that $D(\overline{p}) = S(\overline{p})$ is called the **equilibrium price;** $\overline{q} = D(\overline{p}) = S(\overline{p})$ is called the **equilibrium quantity** and $(\overline{p}, \overline{q})$ is the **market equilibrium point.**

8. REVIEW EXERCISES

1. Find the solution set:

$$\frac{4x-3}{5} - \frac{2x+1}{2} \geq \frac{5x+6}{4}$$

2. Find the solution set:

$$1.35x + 4.0 < 5.45x - 8.3$$

3. Find an equation of the line containing $(-2, 3)$ and $(4, 7)$.

4. Find an equation of the line containing $(-5, 2)$ and $(1, -3)$.

5. Find an equation of the perpendicular bisector of the line segment with $(-4, 7)$ and $(8, 5)$ as endpoints.

6. Find an equation of the perpendicular bisector of the line segment with $(-3, -4)$ and $(5, 8)$ as endpoints.

7. Solve the linear system:

$$2x - 3y = 9,$$
$$4x + 7y = -8$$

8. Solve the linear system:

$$5x + 3y = 6,$$
$$9x - 4y = -55$$

9. Find the maximum and minimum values of the function defined by $f(x) = -6x + 2$, $1 \leq x \leq 5$.

10. Find the maximum and minimum values of the function defined by $f(x) = x^2 + 2x + 5$, $-2 \leq x \leq 6$. [*Hint:* $x^2 + 2x + 5 = (x^2 + 2x + 1) + 4$.]

11. A large department store receives a shipment of 225 radios containing two makes. One costs \$35 apiece and the other costs \$62 apiece. If the total bill for the shipment is \$11,088, how many \$35 radios are in the shipment?

12. The total amount collected from a parking meter containing 419 coins made up of nickels and dimes is \$28.25. How many dimes were in the meter?

13. A person bought a total of 76 gallons of gasoline for two different cars in one month and received a bill for \$91.44. If one car uses regular gas costing \$1.17 a gallon and the other car uses unleaded costing \$1.23 a gallon, how many gallons of the regular were bought?

14. Two trucks must make deliveries over a route totaling 600 miles. Truck *A* gets 12 miles per gallon and Truck *B* gets 9 miles per gallon. If the total fuel consumption for the deliveries cannot exceed 56 gallons, what is the minimum number of miles that Truck *A* must travel?

15. A company has three outlet stores which are located from company headquarters as follows: Outlet *A*, 300 miles east and 400 miles south; Outlet *B*, 500 miles west and 200 miles north; Outlet *C*, 100 miles east and 400 miles north. At what location from company headquarters should a warehouse be located to be equidistant from each of the outlet stores?

16. A retailer pays \$67.50 for a shipment containing nut bars and chocolate bars. He sells the nut bars for 65 cents each and the chocolate bars for 45 cents each and receives a total of \$99.50 for the sale of the shipment. If the nut bars cost 45 cents apiece and the chocolate bars cost 30 cents apiece, how many of each were in the shipment?

17. A salesman sells two types of motors priced at \$328 and \$450 apiece. In addition to a salary, he gets a 5% commission on each \$328 motor he sells and a 7% commission on each \$450 motor. In one month his total sales for the two motors amount to \$20,190 and his total commission is \$1216.50. How many of each type of motor did he sell?

18. The cost function for x units of a particular product is $C(x) = 3x^2 + 2x + 15$ dollars and the revenue function is $R(x) = \$48x$. Find the break-even point for this product.

19. An automobile repair shop states that its repair charges C are directly proportional to the time t in hours working on a car. If the charge for a 1 hour and 40 minute job is \$62.50, what is the repair function C in terms of t?

20. For the cost function $C(x) = 3x^2 + 2x + 27$, what is the average cost function $\overline{C}(x)$? Determine what integer x minimizes $\overline{C}(x)$. What is the minimum average cost?

CHAPTER

3

MATRICES AND LINEAR SYSTEMS

3

3.1 INTRODUCTION

Problems which arise in the "real world" can frequently be expressed by several linear equations. These equations in several unknowns form a model of the problem. It is often possible to solve the practical problem by finding solutions for the equations. The study of linear systems (equations) leads naturally to another concept known as *matrices*. This chapter deals with these ideas and their applications. When the material has been mastered, you will be able to deal with problems of the following kinds.

1. How should production be scheduled in a plant so that the time resources of various departments are fully utilized?
2. How should a given amount of money be invested so that certain predetermined objectives are realized? ■

3.2 SOLVING LINEAR SYSTEMS

Gaussian elimination

The problem of solving two linear equations in two unknowns was considered in Chapter 2. In this chapter we investigate the problem of solving systems with more than two equations, or more than two variables, or both. One process for finding solutions of such systems involves changing (or reducing) a linear system to an equivalent system from which the solution is easily obtained. This technique is called *Gaussian elimination.* To illustrate this method consider the following example:

$$\begin{aligned} &\text{(a)} \qquad x - y = -2, \\ &\text{(b)} \qquad x + y = 2. \end{aligned} \qquad \text{(System 1)}$$

To find the solution of System 1 by Gaussian elimination our objective is to replace Eqs. (a) and (b) by two other equations having the same simultaneous solutions as (a) and (b) but whose solutions are immediately evident. As a first step we replace Eq. (b) by the equation obtained from adding corresponding members of Eqs. (a) and (b) to obtain a new second equation (b_1):

$$\begin{aligned} &\text{(a)} \qquad x - y = -2, \\ &(\text{b}_1) \qquad 2x \phantom{{}-y} = 0. \end{aligned}$$

If (b_1) is multiplied by $\frac{1}{2}$, we get

$$\begin{array}{ll} (a) & x - y = -2, \\ (b_2) & x \qquad = 0. \end{array}$$

Now we subtract (b_2) from (a) to obtain

$$\begin{array}{ll} (a_1) & \quad - y = -2, \\ (b_2) & x \qquad = 0. \end{array}$$

Finally, we multiply (a_1) by -1:

$$\begin{array}{ll} (a_2) & \qquad y = 2, \\ (b_2) & x \qquad = 0. \end{array}$$

It is obvious that the simultaneous solution of (a_2), (b_2) is $x = 0$, $y = 2$. This solution is also denoted by $(0, 2)$. The problem is now solved because the solution of the system (a_2), (b_2) is the same as the solution of the system (a), (b) with which we started.

The process of solving System 1 which began with Eqs. (a) and (b) and ended with Eqs. (a_2) and (b_2) made use of two of the following standard operations on equations:

1. Add the same number to both sides of an equation.
2. Add two equations or subtract one equation from the other.
3. Multiply each side of an equation by the same nonzero number.

It will not be proved here, but it is not difficult to show that operations 1, 2, and 3 do not change the solutions of a system of equations.

As another example of this reduction process we find the simultaneous solutions of the following pair of equations:

$$\begin{aligned} 2x - y &= 1, \\ -4x + 3y &= 2. \end{aligned} \qquad \text{(System 2)}$$

The sequence of steps is shown below with explanations at each step.

System	Operation
$2x - y = 1$ $-4x + 3y = 2$	Multiply the first equation by 3.
$6x - 3y = 3$ $-4x + 3y = 2$	Replace the second equation by the sum of the first and second equations.
$6x - 3y = 3$ $2x + 0 = 5$	Divide the first equation by 3.
$2x - y = 1$ $2x = 5$	Replace the first equation by the difference of the first and second equations.
$0 - y = -4$ $2x = 5$	Multiply the first equation by -1 and multiply the second equation by $\frac{1}{2}$.
$y = 4$ $x = \frac{5}{2}$	

Therefore, the simultaneous solution is $(\frac{5}{2}, 4)$. It is easy to check this solution in the original equations by replacing x by $\frac{5}{2}$ and y by 4:

$$2x - y = 2(\tfrac{5}{2}) - 4 = 1,$$

$$-4x + 3y = -4(\tfrac{5}{2}) + 3(4) = 2.$$

In the example just completed observe that the first step was to multiply the first equation by 3 and the third step was to divide the same equation by 3. Thus at the end of Step 3 the first equation is unchanged. With this observation Steps 1, 2, and 3 may be shortened to a single step:

> Replace the second equation by the result of multiplying the first equation by 3 and adding the result to the second.

Then the first step in the above example is as follows:

	System	Operation
(a) (b)	$2x - y = 1$ $-4x + 3y = 2$	Replace the second equation by $3(a) + (b)$.
(a) (b_1)	$2x - y = 1$ $2x = 5$	

In the preceding explanation the notation "3(a) + (b)" means that Eq. (a) is multiplied by 3 and the result is added to Eq. (b).

tranformation process

The technique we have used for determining simultaneous solutions of linear equations *transforms* one pair of equations into another pair having the same solutions. This *transformation process* aims at replacing one of the two equations by one which involves x only and replacing the other by an equation which involves y only. If this result is achieved, the original pair of equations has a unique simultaneous solution.

Solving two linear equations in two unknowns

EXAMPLE 1 Use the transformation process to determine the simultaneous solution of the system

$$3y - 4x + 2 = 0,$$
$$2x + 5y - 14 = 0.$$

Solution: The following is only one of several ways this problem may be solved. First, write the equations as

$$-4x + 3y = -2,$$
$$2x + 5y = 14.$$

Then the equations may be transformed as follows:

	Equations	Operation
(a)	$-4x + 3y = -2$	
(b)	$2x + 5y = 14$	
		Replace (a) by 2(b) + (a).
(a_1)	$13y = 26$	
(b)	$2x + 5y = 14$	
		Replace (a_1) by $(\frac{1}{13})(a_1)$.
(a_2)	$y = 2$	
(b)	$2x + 5y = 14$	
		Replace (b) by $-5(a_2) + (b)$.
(a_2)	$y = 2$	
(b_1)	$2x = 4$	
		Replace (b_1) by $(\frac{1}{2})(b_1)$.
(a_2)	$y = 2$	
(b_2)	$x = 2$	

Thus the solution is $x = 2$, $y = 2$.

PRACTICE PROBLEM Use the transformation process to solve the system

$$x - 2y = 5,$$
$$2x + 5y = 1.$$

Answer

$$\begin{array}{|rcr|}\hline x - 2y &=& 5 \\ 2x + 5y &=& 1 \\ \hline \end{array}$$
$$\begin{array}{|rcr|}\hline x - 2y &=& 5 \\ 9y &=& -9 \\ \hline \end{array}$$
$$\begin{array}{|rcr|}\hline x - 2y &=& 5 \\ y &=& -1 \\ \hline \end{array}$$
$$\begin{array}{|rcr|}\hline x &=& 3 \\ y &=& -1 \\ \hline \end{array}$$

(Other sequences are possible.) ■

All the examples considered so far have had a unique solution. This is not the case for all systems. For example, suppose we have the system

$$\text{(a)} \qquad 2x - y = 1,$$
$$\text{(b)} \qquad -4x + 2y = 1.$$

In an attempt to obtain a solution we might start our transformation process by trying to eliminate y from the second equation. This can be done by multiplying (a) by 2 and adding the result to (b).

$$\begin{array}{l|rcr|}\cline{2-4} \text{(a)} & 2x - y &=& 1 \\ \text{(b)} & -4x + 2y &=& 1 \\ \cline{2-4} \text{(a)} & 2x - y &=& 1 \\ \text{(b}_1\text{)} & 0 + 0 &=& 3 \\ \cline{2-4} \end{array} \qquad \text{Replace Eq. (b) by 2(a) + (b).}$$

Equation (b_1) is the contradiction "0 = 3," *indicating that this pair of equations has no simultaneous solution.*

Now consider the system

$$\begin{aligned} &\text{(a)} & 2x - y &= 1, \\ &\text{(b)} & -4x + 2y &= -2. \end{aligned}$$

If we apply the same transformation as before, the result is:

$$\begin{array}{ll} \text{(a)} & 2x - y = 1 \\ \text{(b)} & -4x + 2y = -2 \\ \\ \text{(a)} & 2x - y = 1 \\ \text{(b}_1\text{)} & 0 + 0 = 0 \end{array}$$

Replace Eq. (b) by 2(a) + (b).

This time the transformation has led to the true statement "$0 = 0$." The implication here is that *the system has infinitely many solutions.* To see why this is so, remember that Eqs. (a) and (b) have the same solutions as Eqs. (a) and (b_1). But the equation "$0 = 0$" is true independently of x and y and in equation $2x - y = 1$ *any* value for x may be chosen and a corresponding value for y determined.

If a system of equations has no simultaneous solutions, then the system is said to be **inconsistent**; if there are infinitely many solutions, the system is called **dependent.** Inconsistent and de-

inconsistent

dependent

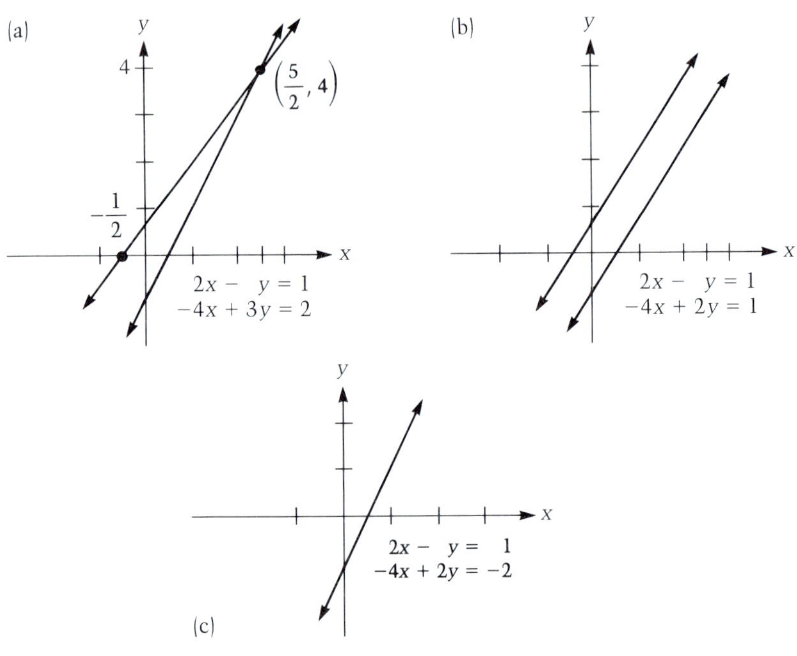

FIG. 3.1. (a) Unique solution. (b) Inconsistent system. (c) Dependent system.

pendent systems in two unknowns have convenient geometric interpretations. The graphs of an inconsistent system in x and y are two parallel lines while the graph of a dependent system is a single line. Given a system of two equations in x and y, the system has a unique solution, is independent or inconsistent. These three cases are illustrated in the graphs of Fig. 3.1.

Solving two equations in two unknowns

EXAMPLE 2 Determine whether each of the following systems has a unique solution, is inconsistent, or dependent. Use the transformation process in each case and if a system is dependent, give two solutions.

(i) $-4x + 3y = 14,\quad 2x + 5y = 6.$ (ii) $4x + 10y = 12,\quad 2x + 5y = 6.$ (iii) $4x + 10y = 11,\quad 2x + 5y = 6.$

Solution

(i)

$$\begin{array}{ll} \text{(a)} & -4x + 3y = 14 \\ \text{(b)} & 2x + 5y = 6 \end{array}$$

$(b) \to 2(b) + (a).$

$$\begin{array}{ll} \text{(a)} & -4x + 3y = 14 \\ (b_1) & 13y = 26 \end{array}$$

$(b_1) \to (\frac{1}{13})(b_1).$

$$\begin{array}{ll} \text{(a)} & -4x + 3y = 14 \\ (b_2) & y = 2 \end{array}$$

$(a) \to -3(b_2) + (a).$

$$\begin{array}{ll} (a_1) & -4x = 8 \\ (b_2) & y = 2 \end{array}$$

$(a_1) \to (-\frac{1}{4})(a_1).$

$$\begin{array}{ll} (a_2) & x = -2 \\ (b_2) & y = 2 \end{array}$$

This system has the unique solution $(-2, 2)$. To check, substitute $x = -2$ and $y = 2$ in the original equations:

$$-4x + 3y = -4(-2) + 3(2) = 8 + 6 = 14, \surd$$
$$2x + 5y = 2(-2) + 5(2) = -4 + 10 = 6. \surd$$

(ii)

$$\begin{array}{ll} \text{(a)} & 4x + 10y = 12 \\ \text{(b)} & 2x + 5y = 6 \end{array}$$

$(a) \to -2(b) + (a).$

$$\begin{array}{ll} (a_1) & 0 = 0 \\ \text{(b)} & 2x + 5y = 6 \end{array}$$

Equation (a_1) indicates that this system is dependent and has infinitely many solutions. To find two of these solutions, we let x be 0 and 1 in Eq. (b) and solve the resulting equations for y:

For $x = 0$:

$$\begin{aligned} 2(0) + 5y &= 6, \\ 5y &= 6, \\ y &= \frac{6}{5}. \end{aligned}$$

For $x = 1$:

$$\begin{aligned} 2(1) + 5y &= 6, \\ 2 + 5y &= 6, \\ 5y &= 4, \\ y &= \frac{4}{5}. \end{aligned}$$

Therefore, two solutions are $(0, \frac{6}{5})$ and $(1, \frac{4}{5})$.

(iii)

$$\begin{array}{ll} (a) & 4x + 10y = 11 \\ (b) & 2x + 5y = 6 \end{array} \qquad (a) \to -2(b) + (a).$$

$$\begin{array}{ll} (a_1) & 0 = -1 \\ (b) & 2x + 5y = 6 \end{array}$$

In this case, Eq. (a_1) indicates the system is inconsistent and hence has no solution.

EXERCISES

Set A–Solving Two Equations in Two Unknowns

In Exercises 1 through 20, use the transformation process to solve the systems given.

1. $x - y = 0$
   $x + y = 6$

2. $x - 2y = 2$
   $x + 2y = 6$

3. $x + 2y = 1$
   $-2x + y = 3$

4. $2x - y = 3$
   $x - 3y = -1$

5. $3x - 2y = 1$
   $5x + y = 6$

6. $4x + 2y = 18$
   $x - 5y = -12$

7. $\begin{aligned} 2x - y &= -11 \\ 4x - 7y &= -27 \end{aligned}$

8. $\begin{aligned} -5x + 6y &= 3 \\ 3x - y &= -7 \end{aligned}$

9. $\begin{aligned} 2x + 3y &= 5 \\ 4x - 5y &= -1 \end{aligned}$

10. $\begin{aligned} -3x + 2y &= -14 \\ -9x + 8y &= -26 \end{aligned}$

11. $\begin{aligned} -7x + 3y &= 29 \\ 2y - 14x &= 38 \end{aligned}$

12. $\begin{aligned} 12x - 7y &= -61 \\ 5y + 6x &= -73 \end{aligned}$

13. $\begin{aligned} 5x + 8y &= 36 \\ 3y - 25x &= 121 \end{aligned}$

14. $\begin{aligned} -18y + 7x &= 125 \\ 9x - 36y &= 225 \end{aligned}$

15. $\begin{aligned} 8x + 7y &= 1 \\ -3x + 2y &= -5 \end{aligned}$

16. $\begin{aligned} 5x - 3y &= -30 \\ 2x + 5y &= 19 \end{aligned}$

17. $\begin{aligned} 15x + 2y &= -17 \\ -18x - 5y &= 23 \end{aligned}$

18. $\begin{aligned} 4x + 5y &= 7 \\ 3x + 7y &= 15 \end{aligned}$

19. $\begin{aligned} 5x + 5y &= -10 \\ 7x + 2y &= -19 \end{aligned}$

20. $\begin{aligned} -8x + 9y &= -112 \\ -5x + 6y &= -73 \end{aligned}$

Set B–Solving Two Equations in Two Unknowns, Some of Which Do Not Have Unique Solutions

In Exercises 21 through 32, use the transformation process to solve the equations if possible. If a system is dependent, give two solutions.

21. $\begin{aligned} 7x - 3y &= 4 \\ 14x - 6y &= 8 \end{aligned}$

22. $\begin{aligned} 5x - 2y &= 3 \\ 10x - 4y &= 5 \end{aligned}$

23. $\begin{aligned} 5x - 2y &= 3 \\ 6x + 3y &= 9 \end{aligned}$

24. $\begin{aligned} 7x - 3y &= 10 \\ 8x + 5y &= 3 \end{aligned}$

25. $\begin{aligned} 12x + 5y &= 7 \\ 6x + (\tfrac{5}{2})y &= 1 \end{aligned}$

26. $\begin{aligned} 12x + 5y &= 6 \\ 6x + (\tfrac{5}{2})y &= 3 \end{aligned}$

27. $\begin{aligned} -8x + 3y &= 5 \\ 8x + 3y &= 3 \end{aligned}$

28. $\begin{aligned} -13x + 4y &= -1 \\ -4y + 13x &= 1 \end{aligned}$

29. $\begin{aligned} 6x - 6y &= 0 \\ 3x - 3y &= 1 \end{aligned}$

30. $\begin{aligned} (\tfrac{5}{3})x + (\tfrac{8}{3})y &= 1 \\ 5x + 8y &= 3 \end{aligned}$

31. $\begin{aligned} 20x + 8y &= -28 \\ 5x + 2y &= -7 \end{aligned}$

32. $\begin{aligned} 0.8x + 0.9y &= 1.7 \\ 8x + 9y &= 17 \end{aligned}$

3.3 LINEAR SYSTEMS IN MORE THAN TWO UNKNOWNS

So far we have only considered equations in two unknowns. Let us now consider equations with three or more unknowns.

The equation

$$x - 2y + 3z = 4 \tag{1}$$

is a linear equation in three unknowns (variables) x, y, z and

$$7x_1 + 4x_2 - 8x_3 + x_4 + 11x_5 = 0 \tag{2}$$

is a linear equation in five unknowns: x_1, x_2, x_3, x_4, x_5. By a *solution* of Eq. (1) we mean three numbers which, if substituted for x, y, z, will result in a true statement. A similar definition holds for the *solution* of Eq. (2). For example, (0, 1, 2) is a solution of (1) because

$$0 - 2(1) + 3(2) = 4$$

is a true statement.

It is always a simple matter to find solutions for linear equations such as (1) and (2): we may assign any values we please to all the unknowns *except one* and then solve for the remaining one. To illustrate, in (2), let $x_1 = 1, x_2 = 2, x_3 = 1, x_5 = 0$. This assigns values to all variables except x_4. Now substitute these values in (2):

$$\begin{aligned} 7(1) + 4(2) - 8(1) + x_4 + 11(0) &= 0, \\ 7 + 8 - 8 + x_4 + 0 &= 0, \\ 7 + x_4 + 0 &= 0, \\ x_4 &= -7. \end{aligned}$$

Hence, (1, 2, 1, −7, 0) is a solution of (2).

Finding solutions of an equation in several unknowns

EXAMPLE 1 Given the equation

$$2x + 5y - 6z + 3w = 0,$$

find a solution for which (a) $x = 0$, (b) $x = 1$ and $y = -1$, (c) $x = 1$, $y = -1$, and $z = -2$.

Solution

(a) If we take $x = 0$ in the given equation we have

$$2(0) + 5y - 6z + 3w = 0,$$

or

$$5y - 6z + 3w = 0.$$

From this equation it is easy to see that each of the variables y, z, w may be taken to be 0 to obtain a solution. Thus one solution with $x = 0$ is $(0, 0, 0, 0)$.

(b) Taking $x = 1$ and $y = -1$ in our equation gives

$$\begin{aligned} 2(1) + 5(-1) - 6z + 3w &= 0, \\ 2 - \quad 5 - 6z + 3w &= 0, \\ -3 - 6z + 3w &= 0, \\ -6z + 3w &= 3, \\ -2z + w &= 1. \end{aligned}$$

In this equation, we may take $z = 0$ and $w = 1$ and obtain the solution $(1, -1, 0, 1)$.

(c) As in part (b) the values $x = 1$, $y = -1$ may be substituted in the equation

$$2x + 5y - 6z + 3w = 0$$

to obtain

$$-2z + w = 1.$$

Now the value $z = -2$ may be substituted into this equation:

$$\begin{aligned} -2(-2) + w &= 1, \\ 4 + w &= 1, \\ w &= -3. \end{aligned}$$

Therefore $(1, -1, -2, -3)$ is a solution of the original equation for which $x = 1$, $y = -2$, and $z = -2$.

PRACTICE PROBLEM Find a solution for the equation

$$2x + 5y - 6z + 3w = 0$$

in which $x = 2$, $y = 1$, $z = 1$.

Answer: $(2, 1, 1, -1)$. ■

System (3) given below is an example of a system of two linear equations in three unknowns:

$$\begin{array}{ll} \text{(a)} & x - 2y + 3z = 4 \\ \text{(b)} & 2x + y + z = -2 \end{array} \qquad \text{(System 3)}$$

simultaneous solutions

The problem which usually concerns us with such systems is to find all numbers that are solutions to both equations. Such numbers are said to be **simultaneous solutions of the system.** If simultaneous solutions exist, they may be found by using the transformation process. The technique will be illustrated on System 3:

$$\begin{array}{lrcl} \text{(a)} & x - 2y + 3z &=& 4 \\ \text{(b)} & 2x + y + z &=& -2 \end{array} \qquad \text{(b)} \to -2\text{(a)} + \text{(b)}.$$

$$\begin{array}{lrcl} \text{(a)} & x - 2y + 3z &=& 4 \\ (\text{b}_1) & 5y - 5z &=& -10 \end{array} \qquad (\text{b}_1) \to (\tfrac{1}{5})(\text{b}_1).$$

$$\begin{array}{lrcl} \text{(a)} & x - 2y + 3z &=& 4 \\ (\text{b}_2) & y - z &=& -2 \end{array} \qquad \text{(a)} \to 2(\text{b}_2) + \text{(a)}.$$

$$\begin{array}{lrcl} (\text{a}_1) & x \qquad + z &=& 0 \\ (\text{b}_2) & y - z &=& -2 \end{array}$$

At this point it might seem natural to try to eliminate z from one of the two equations (a_1) or (b_2). We could, for example, replace (a_1) by $(\text{a}_1) + (\text{b}_2)$:

$$\begin{array}{ll} (\text{a}_2) & x + y \qquad = -2, \\ (\text{b}_2) & \qquad y - z = -2. \end{array}$$

The result has been to remove z from the first equation, but in so doing we have brought y back into the equation. We cannot achieve any further elimination without such a difficulty. If we stop with (a_1) and (b_2), we have

$$\begin{array}{r} x \qquad + z = 0, \\ y - z = -2. \end{array}$$

and these two equations may be written as

$$\begin{array}{ll} (\text{a}_3) & x = -z, \\ (\text{b}_3) & y = z - 2. \end{array}$$

From (a_3) and (b_3) we see that z may be assigned *any* value we like and then x and y may be determined to give a solution. For example, if $z = 0$, then from (a_3) and (b_3)

$$x = -z = 0,$$
$$y = z - 2 = 0 - 2 = -2.$$

parameter

Therefore, one of the solutions to the system is the triple $(0, -2, 0)$. The number z in (a_3), (b_3) is called a *parameter,* and every solution of System 3 may be expressed in terms of this parameter. If (x, y, z) is any solution of System 3, it may be expressed as $(-z, z - 2, z)$. The replacements shown below lead to the given expression:

$$\begin{array}{ccc} (x, & y, & z) \\ \downarrow & \downarrow & \\ \text{from } (a_3) & \text{from } (b_3) & \\ \downarrow & \downarrow & \\ (-z, & z - 2, & z) \end{array}$$

To verify that $(-z, z - 2, z)$ is a solution for any z, we substitute into System 3:

$$\text{(a)} \qquad \begin{aligned} x - 2y + 3z &= (-z) - 2(z - 2) + 3z \\ &= -z - 2z + 4 + 3z \\ &= 4. \checkmark \end{aligned}$$

$$\text{(b)} \qquad \begin{aligned} 2x + y + z &= 2(-z) + (z - 2) + z \\ &= -2z + z - 2 + z \\ &= -2. \checkmark \end{aligned}$$

general solution

We say that $(-z, z - 2, z)$ is a *general solution* of System 3. It is worth pointing out that System 3 has more unknowns than equations and this accounts for the fact there are infinitely many solutions.

Suppose we are given several linear equations in a number of unknowns and are asked to find all the simultaneous solutions—if there are any. One of the standard methods for attacking this problem is the one which has just been used to solve System 3.

pivot element equation

Briefly stated, the method consists of trying to write the equations in a reduced form by eliminating as many unknowns from the system as possible. To accomplish this, we systematically choose one unknown at a time to eliminate from all equations except one. In solving System 3, we started by choosing x in the first equation—x is called the *pivot element* and Eq. (a) is called the *pivot equation.* Then the pivot equation was used to eliminate x from the other equation. It is helpful to circle pivot elements as we have done below in solving System 3:

(a)	$\textcircled{x} - 2y + 3z = 4$	
(b)	$2x + y + z = -2$	$(b) \to -2(a) + (b).$
(a)	$x - 2y + 3z = 4$	
(b_1)	$5y - 5z = -10$	$(b_1) \to (\frac{1}{5})(b_1).$
(a)	$x - 2y + 3z = 4$	
(b_2)	$\textcircled{y} - z = -2$	$(a) \to 2(b_2) + (a).$
(a_1)	$x + z = 0$	
(b_2)	$y - z = -2$	

In this solution of System 3, x was the pivot element chosen first (in the first equation). The Gaussian elimination process was applied by using Eq. (a) to remove the x-term from Eq. (b). When this had been accomplished, a new pivot equation and a new pivot element were chosen, and the elimination process applied again. The second pivot element chosen was y, but it could have been z just as well.

Solving two equations in three unknowns

EXAMPLE 2 Solve the following system by using the pivot process. Obtain a general solution in two different ways.

$$x + y + z = 4,$$
$$2x + 3y + z = 9.$$

Solution: For the first solution we will start by choosing x as a pivot in the first equation (y or z could have been chosen):

$$\begin{array}{ll} \text{(a)} & \textcircled{x} + y + z = 4 \\ \text{(b)} & 2x + 3y + z = 9 \end{array}$$

(b) → −2(a) + (b).

$$\begin{array}{ll} \text{(a)} & x + y + z = 4 \\ (\text{b}_1) & y - \textcircled{z} = 1 \end{array}$$

(a) → (b_1) + (a).

$$\begin{array}{l} x + 2y = 5 \\ y - z = 1 \end{array}$$

Solve for x and z.

$$\begin{array}{l} x = 5 - 2y \\ z = -1 + y \end{array}$$

For the second solution we start with y as pivot in the first equation:

$$\begin{array}{ll} \text{(a)} & x + \textcircled{y} + z = 4 \\ \text{(b)} & 2x + 3y + z = 9 \end{array}$$

(b) → −3(a) + (b).

$$\begin{array}{ll} \text{(a)} & x + y + z = 4 \\ (\text{b}_1) & \textcircled{-x} - 2z = -3 \end{array}$$

(a) → (b_1) + (a) and
(b_1) → −(b_1).

$$\begin{array}{l} y - z = 1 \\ x + 2z = 3 \end{array}$$

Solve for y and x.

$$\begin{array}{l} y = 1 + z \\ x = 3 - 2z \end{array}$$

Therefore, one general solution is $(5 - 2y, y, -1 + y)$ and another is $(3 - 2z, 1 + z, z)$.

PRACTICE PROBLEM Find a general solution for the system of the above example which expresses y and z in terms of x.

Answer: $y = 0.5(5 - x)$, $z = 0.5(3 - x)$. ■

Solving three equations in three unknowns

EXAMPLE 3 Solve the following system by using the pivot process.

$$\begin{aligned} x_1 + x_2 - x_3 &= 3, \\ 2x_1 + x_2 + x_3 &= 5, \\ 3x_1 - 2x_2 + x_3 &= 4. \end{aligned}$$

Solution

$$
\begin{array}{ll}
\text{(a)} & x_1 + x_2 - \boxed{x_3} = 3 \\
\text{(b)} & 2x_1 + x_2 + x_3 = 5 \\
\text{(c)} & 3x_1 - 2x_2 + x_3 = 4
\end{array}
\qquad \text{(b)} \to \text{(a)} + \text{(b) and (c)} \to \text{(a)} + \text{(c)}.
$$

$$
\begin{array}{ll}
\text{(a)} & x_1 + x_2 - x_3 = 3 \\
(\text{b}_1) & 3x_1 + 2x_2 = 8 \\
(\text{c}_1) & 4x_1 - \boxed{x_2} = 7
\end{array}
\qquad \text{(a)} \to (\text{c}_1) + \text{(a) and } (\text{b}_1) \to 2(\text{c}_1) + (\text{b}_1).
$$

$$
\begin{array}{ll}
(\text{a}_1) & 5x_1 - x_3 = 10 \\
(\text{b}_2) & \boxed{11x_1} = 22 \\
(\text{c}_1) & 4x_1 - x_2 = 7
\end{array}
\qquad (\text{b}_2) \to (\tfrac{1}{11})(\text{b}_2).
$$

$$
\begin{array}{ll}
(\text{a}_1) & 5x_1 - x_3 = 10 \\
(\text{b}_3) & \boxed{x_1} = 2 \\
(\text{c}_1) & 4x_1 - x_2 = 7
\end{array}
\qquad (\text{a}_1) \to -5(\text{b}_3) + (\text{a}_1) \text{ and } (\text{c}_1) \to -4(\text{b}_3) + (\text{c}_1).
$$

$$
\begin{array}{ll}
(\text{a}_2) & -x_3 = 0 \\
(\text{b}_3) & x_1 = 2 \\
(\text{c}_2) & -x_2 = -1
\end{array}
$$

Therefore,

$$x_3 = 0, \qquad x_1 = 2, \qquad x_2 = 1,$$

and the unique solution is (2, 1, 0).

PRACTICE PROBLEM Solve the following system by using the pivot process.

$$
\begin{aligned}
x + y - z &= -1, \\
x \qquad + z &= 2, \\
-y + z &= 4.
\end{aligned}
$$

Answer: (3, −5, −1). ■

SUMMARY OF THE PIVOT PROCESS FOR SOLVING EQUATIONS

1. Choose a pivot equation and a pivot element in the pivot equation.

2. Use the pivot equation and the Gaussian elimination process to eliminate the chosen element (unknown) in each equation other than the pivot equation.

3. Repeat Steps 1 and 2 for a new pivot equation and a new pivot element (if possible). Continue until all equations or all unknowns have been used as pivots.

4. The process terminates when all equations or all unknowns have been used as pivots (no unknown or equation should be used more than once as a pivot).

EXERCISES

Set A–Finding Solutions for Equations in Several Unknowns

1. Find a solution of the equation $3x - 5y + 2z = 12$ in which
   (a) $x = 0,\ y = 0$ (b) $y = 1,\ z = 1$
2. Find a solution of the equation $-4x + 3y + 7z = 10$ in which
   (a) $x = 0,\ y = 1$ (b) $x = 2,\ z = 0$
3. Find a solution of the equation $x - 2y - z + w = 1$ in which
   (a) $x = 2,\ y = 2,\ z = 3$ (b) $y = \frac{1}{2},\ z = 7,\ w = -4$
4. Find a solution of the equation $2x + 3y + z - w = 10$ in which
   (a) $x = 4,\ y = -1,\ z = -2$ (b) $x = -1,\ y = 2,\ w = 4$
5. Verify that (1, 1, 1) is not a simultaneous solution of the system

$$\begin{aligned} 2x - 3y + z &= 0, \\ 3x + y - z &= 3, \\ x - y + 2z &= 1, \end{aligned}$$

but is a simultaneous solution of the system

$$\begin{aligned} 2x - 3y + z &= 0, \\ 3x + y - z &= 3. \end{aligned}$$

6. Verify that $(1 - 2y, y, 2 + y)$ is a general solution of the system

$$\begin{aligned} x + y + z &= 3, \\ 2x + y + 3z &= 8. \end{aligned}$$

7. Verify that $(x_4, 1 - x_4, 2 + 2x_4, x_4)$, where x_4 is any number, is a general solution of the system

$$6x_1 + 2x_2 + x_3 - 6x_4 = 4,$$
$$5x_1 + 2x_2 + x_3 - 5x_4 = 4,$$
$$3x_1 + x_2 + x_3 - 4x_4 = 3.$$

8. Verify that $(\frac{6}{7} - 3x_2, x_2, \frac{9}{7} - 4x_2, -\frac{5}{7} + 3x_2)$ is a general solution of the system

$$x_1 - 2x_2 + x_3 + 3x_4 = 0,$$
$$2x_1 + 3x_2 + 3x_3 + 5x_4 = 2,$$
$$3x_1 + x_2 + x_3 + 4x_4 = 1.$$

Set B–Solving Two Equations in Three Unknowns

In Exercises 9 through 20, solve the systems given by using the pivot process.

9. Solve the system so that y is a parameter, i.e. the solutions are expressed in terms of y:

$$2x - y + z = -2,$$
$$x - 2y + z = -3.$$

10. Solve the system of Exercise 9 so that
(a) z is a parameter (b) x is a parameter.

11. Solve this system so that y is a parameter:

$$3x - y + 2z = 1,$$
$$4x + 2y + 3z = 1.$$

12. Solve the system of Exercise 11 so that
(a) z is a parameter. (b) x is a parameter.

13. Solve this system so that y is a parameter:

$$y - 4x + 5z = 7,$$
$$3x + 2z + 3y = 10.$$

14. Solve the system of Exercise 13 so that
(a) z is a parameter. (b) x is a parameter.

15. Solve this system so that y is a parameter:

$$x - 4y + 2z = 5,$$
$$2x - 7y - 2z = -2.$$

16. Solve the system of Exercise 15 so that
(a) z is a parameter. (b) x is a parameter.

17. Solve this system so that y is a parameter:

$$10x + 3y + 6z = 18,$$
$$6x + 2y + 3z = 13.$$

18. Solve the system of Exercise 17 so that
(a) z is a parameter. (b) x is a parameter.

19. Solve this system so that y is a parameter:

$$-8y + 3z - 5x = -8,$$
$$2z - 4y + 3x = -9.$$

20. Solve the system of Exercise 19 so that
(a) z is a parameter. (b) x is a parameter.

Set C–Solving Three Equations in Three Unknowns

In Exercises 21 through 32, use the pivot process to solve the equations.

21. $\begin{aligned} x - 2y + 9z &= -16 \\ -4x + 2y + 3z &= 16 \\ y &= 8 \end{aligned}$

22. $\begin{aligned} -4x + y - 5z &= 58 \\ 8x + y + 3z &= -74 \\ 4x - y - 3z &= -10 \end{aligned}$

23. $\begin{aligned} x - 7y + 6z &= 23 \\ x + 5y + 4z &= -31 \\ -2x + 5y + z &= -40 \end{aligned}$

24. $\begin{aligned} 4x + 5y + 3z &= -4 \\ 5x + 6y + 4z &= -4 \\ -x - y + 3z &= -4 \end{aligned}$

25. $\begin{aligned} -x + 8y - 2z &= -15 \\ -9y + 6z &= 42 \\ 3x - 8y + 7z &= 17 \end{aligned}$

26. $\begin{aligned} 3x + 5y + z &= -10 \\ 9x - 9y - z &= -42 \\ y - 3z &= 10 \end{aligned}$

27. $\begin{aligned} -8x + 9y - 7z &= 157 \\ 8x + 8y + 5z &= 4 \\ -6x + 4y - z &= 76 \end{aligned}$

28. $\begin{aligned} 8x + 3y - 9z &= 10 \\ -9x + 3y + 4z &= -24 \\ -4x - 3y + z &= -2 \end{aligned}$

29. $\begin{aligned} 2x + y + 5z &= -12 \\ 2x + 3y - 4z &= -8 \\ x - 9z &= -7 \end{aligned}$

30. $\begin{aligned} -9x - 3y + 7z &= -11 \\ 3x + 4y + 3z &= 9 \\ x - z &= 1 \end{aligned}$

31. $\begin{aligned} 5x + 4y &= 12 \\ 5x + 2y + 2z &= -4 \\ 9x + 5y + 3z &= 4 \end{aligned}$

32. $\begin{aligned} 7x + 9y - 4z &= 81 \\ 4y - 9z &= 7 \\ 5x + 5y + 6z &= 61 \end{aligned}$

Set D–Solving Systems with More Than Three Unknowns

In Exercises 33 through 38, solve the systems using the pivot process and the indicated parameter(s).

33. Solve, using x as a parameter.
$$\begin{aligned} 10x + y + 2z + w &= -4, \\ 5x - y + z + 2w &= -8, \\ 20x + y + 4z + 3w &= -12. \end{aligned}$$

34. Solve, using y as a parameter.
$$\begin{aligned} -x - 2z &= 5, \\ 3x - y - z + 2w &= 10, \\ x - z + w &= 6. \end{aligned}$$

35. Solve, using z as a parameter.
$$\begin{aligned} 2x + y - 20z + w &= 0, \\ x + 3y - 21z + 2w &= 1, \\ x + y - 12z + w &= 1 \end{aligned}$$

36. Solve, using w as a parameter.
$$\begin{aligned} 2x + 3y - 2z - 17w &= 11, \\ x + 2y - z - 8w &= 1, \\ 2x + y + z - 13w &= 44. \end{aligned}$$

37. Solve, using w and v as parameters.
$$\begin{aligned} x - y + z - 9w + 11v &= 0 \\ -x - 3y + 2z - 36w + 7v &= -16 \\ -x - 2y + z - 24w + 2v &= -12 \end{aligned}$$

38. Solve, using z and v as parameters.
$$\begin{aligned} x - 2y + 12z + w - 25v &= -23 \\ -x + 2y + 18z + w - 7v &= -9 \\ x - y - 12z - 2w - v &= -2 \end{aligned}$$

3.4 MATRICES AND LINEAR SYSTEMS

After working the exercises of Sections 3.2 and 3.3, the student may feel that the amount of writing required was excessive. This is indeed the case. In fact, the essential features of a linear system are the coefficients of the unknowns and the constants—other symbols may be suppressed. For this reason it is possible to solve a linear system by writing only the coefficients and constants. To illustrate how to proceed consider the following system:

$$\begin{aligned} x + y - z &= -1, \\ x + z &= 2, \\ -y + z &= 4. \end{aligned}$$

The system may be represented without using the symbols x, y, z,

+, =. These will be deleted from the system and 0s inserted where an unknown is missing. For example, the equation

$$x \quad + z = 2$$

is the same as

$$1 \cdot x + 0 \cdot y + 1 \cdot z = 2,$$

which suggests that we simply write

$$1 \quad 0 \quad 1 \quad 2$$

for this equation. When we treat the entire system in this manner we have the following array of numbers:

$$\begin{bmatrix} 1 & 1 & -1 & -1 \\ 1 & 0 & 1 & 2 \\ 0 & -1 & 1 & 4 \end{bmatrix}.$$

matrix
augmented matrix

An array of numbers such as this is called a **matrix.** In this case, the matrix above is called the **augmented matrix** of the system

$$\begin{aligned} x + y - z &= -1, \\ x \qquad + z &= 2, \\ -y + z &= 4, \end{aligned}$$

and

$$\begin{bmatrix} 1 & 1 & -1 \\ 1 & 0 & 1 \\ 0 & -1 & 1 \end{bmatrix}$$

matrix of coefficients

is the **matrix of coefficients** of the same system. Note that the augmented matrix is obtained from the matrix of coefficients by adjoining the column of constants to the matrix of coefficients. It is also important to remember that each column of the matrix of coefficients corresponds to a single unknown. This is sometimes emphasized by writing the unknowns above the appropriate columns:

$$\begin{matrix} x & y & z \end{matrix}$$
$$\begin{bmatrix} 1 & 1 & -1 \\ 1 & 0 & 1 \\ 0 & -1 & 1 \end{bmatrix}.$$

Writing the matrix of a linear system

EXAMPLE 1 Write the matrix of coefficients and augmented matrix for the system

$$\begin{aligned} 2y - x + z + w &= 0, \\ z + y + w - x &= 5, \\ 3y + z - x + 2w &= 1. \end{aligned}$$

Solution: The system should be written so that each unknown occupies the same column in each equation:

$$\begin{aligned} -x + 2y + z + w &= 0, \\ -x + y + z + w &= 5, \\ -x + 3y + z + 2w &= 1. \end{aligned}$$

Then the matrix of coefficients is

$$\begin{bmatrix} -1 & 2 & 1 & 1 \\ -1 & 1 & 1 & 1 \\ -1 & 3 & 1 & 2 \end{bmatrix},$$

and

$$\left[\begin{array}{cccc:c} -1 & 2 & 1 & 1 & 0 \\ -1 & 1 & 1 & 1 & 5 \\ -1 & 3 & 1 & 2 & 1 \end{array}\right]$$

is the augmented matrix of this system.

PRACTICE PROBLEM Write the matrix of coefficients of the following system and also the augmented matrix:

$$\begin{aligned} x - 2y + 3z &= 4, \\ 2x + y + z &= -2. \end{aligned}$$

Answer

$$\begin{bmatrix} 1 & -2 & 3 \\ 2 & 1 & 1 \end{bmatrix}, \quad \left[\begin{array}{ccc:c} 1 & -2 & 3 & 4 \\ 2 & 1 & 1 & -2 \end{array}\right]. \blacksquare$$

Each matrix has a number of rows and a number of columns. For example, the matrix

$$A = \begin{bmatrix} -1 & 2 & 1 & 1 \\ -1 & 1 & 1 & 1 \\ -1 & 3 & 1 & 2 \end{bmatrix}$$

has three rows and four columns and these are denoted by R_1, R_2, R_3, C_1, C_2, C_3, C_4:

$$A = \begin{bmatrix} -1 & 2 & 1 & 1 \\ -1 & 1 & 1 & 1 \\ -1 & 3 & 1 & 2 \\ \uparrow & \uparrow & \uparrow & \uparrow \end{bmatrix} \begin{matrix} \leftarrow R_1 \\ \leftarrow R_2. \\ \leftarrow R_3 \\ \end{matrix}$$
$$\quad C_1 \quad C_2 \quad C_3 \quad C_4$$

We say that A is a *3-by-4 matrix,* meaning that it has three rows and four columns. The numbers in a matrix are called the *elements* or *entries* of the matrix, and each entry is identified by its row position and column position. In the matrix

$$B = \begin{bmatrix} 0 & 1 \\ -1 & 2 \end{bmatrix}$$

0 is in the *first* row and *first* column,
1 is in the *first* row and *second* column,
−1 is in the *second* row and *first* column,
2 is in the *second* row and *second* column.

There is a special notation which indicates the position of each entry in a matrix. It is the *i, j*-notation. We write, for example, a_{ij} to represent a matrix entry in the ith row and jth column. For the 2-by-2 matrix B above we have

$$a_{11} = 0, \qquad a_{12} = 1, \qquad a_{21} = -1, \qquad a_{22} = 2.$$

These four equations say the same thing as the four lines given just below the matrix B.

In subsequent sections several applications of matrices will be made (*matrices* is the plural of matrix). At this point we will be concerned only with the problem of solving linear systems. To solve a system of linear equations by using matrices, we simply use the elements and rows of the matrices rather than the equations themselves. To do this it is necessary to translate the operations involving pivot equations and elements into the appropriate operations for matrices.

PIVOT OPERATIONS FOR MATRICES

1. Choose a pivot row R_i of the augmented matrix and a pivot element a_{ij} of the row R_i (a_{ij} cannot be 0). Do not choose a_{ij} in the last column.
2. Use the row R_i and Gaussian elimination to obtain 0s in the jth column above and below the pivot element a_{ij} (after this has been completed, the jth column will contain only 0s except for the ith row).
3. Repeat Steps 1 and 2 for a new pivot row and a new pivot element (if possible). The process should continue until all rows have been used as pivots or until a pivot element has been chosen from each column (except the last).
4. The process terminates as soon as each row has been used as a pivot or when a pivot element has been chosen from each column.

To illustrate the technique described in the pivot operations for matrices the following system will be solved.

$$\begin{aligned} x + y - z &= -1, \\ x \phantom{{}+y} + z &= 2, \\ - y + z &= 4. \end{aligned}$$

The augmented matrix of this system is

$$\left[\begin{array}{ccc:c} 1 & 1 & -1 & -1 \\ 1 & 0 & 1 & 2 \\ 0 & -1 & 1 & 4 \end{array}\right].$$

In the following steps, the notation $R_1 \to R_3 + R_1$, for example, means that row 1 is replaced by the result of adding row 3 to row 1.

$$\left[\begin{array}{ccc:c} 1 & 1 & -1 & -1 \\ 1 & 0 & 1 & 2 \\ 0 & \textcircled{-1} & 1 & 4 \end{array}\right] \quad R_1 \to R_3 + R_1.$$

$$\left[\begin{array}{ccc:c} \textcircled{1} & 0 & 0 & 3 \\ 1 & 0 & 1 & 2 \\ 0 & -1 & 1 & 4 \end{array}\right] \quad R_2 \to -R_1 + R_2.$$

$$\left[\begin{array}{ccc:c} 1 & 0 & 0 & 3 \\ 0 & 0 & \textcircled{1} & -1 \\ 0 & -1 & 1 & 4 \end{array}\right] \quad R_3 \to -R_2 + R_3.$$

$$\left[\begin{array}{ccc:c} 1 & 0 & 0 & 3 \\ 0 & 0 & 1 & -1 \\ 0 & -1 & 0 & 5 \end{array}\right] \quad R_3 \to -R_3.$$

$$\left[\begin{array}{ccc:c} 1 & 0 & 0 & 3 \\ 0 & 0 & 1 & -1 \\ 0 & 1 & 0 & -5 \end{array}\right]$$

The last matrix in this sequence is

$$\left[\begin{array}{ccc:c} 1 & 0 & 0 & 3 \\ 0 & 0 & 1 & -1 \\ 0 & 1 & 0 & -5 \end{array}\right]$$

and this is the augmented matrix of the linear system

$$\begin{aligned} 1\cdot x + 0\cdot y + 0\cdot z &= 3, \\ 0\cdot x + 0\cdot y + 1\cdot z &= -1, \\ 0\cdot x + 1\cdot y + 0\cdot z &= -5, \end{aligned}$$

so that the solution is $x = 3$, $y = -5$, $z = -1$.

Solving linear systems by matrices

EXAMPLE 2 Solve the following linear system by using matrices:

$$\begin{aligned} x - 2y + 3z &= 4, \\ 2x + y + z &= -2. \end{aligned}$$

Solution

$$\left[\begin{array}{ccc|c} \textcircled{1} & -2 & 3 & 4 \\ 2 & 1 & 1 & -2 \end{array}\right] \quad R_2 \to R_2 + (-2)R_1.$$

$$\left[\begin{array}{ccc|c} 1 & -2 & 3 & 4 \\ 0 & 5 & -5 & -10 \end{array}\right] \quad R_2 \to (\tfrac{1}{5})R_2.$$

$$\left[\begin{array}{ccc|c} 1 & -2 & 3 & 4 \\ 0 & \textcircled{1} & -1 & -2 \end{array}\right] \quad R_1 \to R_1 + 2R_2.$$

$$\left[\begin{array}{ccc|c} 1 & 0 & 1 & 0 \\ 0 & 1 & -1 & -2 \end{array}\right]$$

This process must terminate at this point because there are no more pivot rows to choose which have not been already used. The final matrix corresponds to the system

$$\begin{aligned} x \quad + z &= 0, \\ y - z &= -2, \end{aligned}$$

and therefore,

$$\begin{aligned} x &= -z, \\ y &= z - 2. \end{aligned}$$

This gives the general solution $(-z, z - 2, z)$.

PRACTICE PROBLEM Use matrices to solve the system

$$\begin{aligned} 2x + 3y - z &= -3, \\ 3x - 4y - 2z &= 3. \end{aligned}$$

Answer

$$x = -9 - 10y, \qquad z = -15 - 17y$$

or

$$y = -\frac{1}{10}(9 + x), \qquad z = \frac{1}{10}(3 + x),$$

or

$$y = -\frac{(15 + z)}{17}, \qquad x = \frac{(3 + 10z)}{17}. \blacksquare$$

Solving systems by using matrices

EXAMPLE 3 Solve the following system:

$$\begin{aligned} x_1 - 2x_2 + x_3 + 3x_4 &= 0, \\ 2x_1 + 3x_2 + 3x_3 + 5x_4 &= 2, \\ 3x_1 + x_2 + x_3 + 4x_4 &= 1. \end{aligned}$$

Solution: Using matrices we have

$$\left[\begin{array}{cccc:c} \textcircled{1} & -2 & 1 & 3 & 0 \\ 2 & 3 & 3 & 5 & 2 \\ 3 & 1 & 1 & 4 & 1 \end{array}\right]$$

$R_2 \to R_2 + (-2)R_1$ and $R_3 \to R_3 + (-3)R_1$.

$$\left[\begin{array}{cccc:c} 1 & -2 & 1 & 3 & 0 \\ 0 & 7 & \textcircled{1} & -1 & 2 \\ 0 & 7 & -2 & -5 & 1 \end{array}\right]$$

$R_1 \to R_1 + (-1)R_2$ and $R_3 \to R_3 + 2R_2$.

$$\left[\begin{array}{cccc:c} 1 & -9 & 0 & 4 & -2 \\ 0 & 7 & 1 & -1 & 2 \\ 0 & 21 & 0 & -7 & 5 \end{array}\right]$$

$R_3 \to (-\tfrac{1}{7})R_3$.

$$\left[\begin{array}{cccc:c} 1 & -9 & 0 & 4 & -2 \\ 0 & 7 & 1 & -1 & 2 \\ 0 & -3 & 0 & \textcircled{1} & -\tfrac{5}{7} \end{array}\right]$$

$R_1 \to R_1 + (-4)R_3$ and $R_2 \to R_2 + R_3$.

$$\left[\begin{array}{cccc:c} 1 & 3 & 0 & 0 & \tfrac{6}{7} \\ 0 & 4 & 1 & 0 & \tfrac{9}{7} \\ 0 & -3 & 0 & 1 & -\tfrac{5}{7} \end{array}\right]$$

This cannot be carried any further since there are no more rows to choose as pivots. The last matrix is the augmented matrix of

$$\begin{aligned} x_1 + 3x_2 &= \frac{6}{7}, \\ 4x_2 + x_3 &= \frac{9}{7}, \\ -\,3x_2 + x_4 &= -\frac{5}{7}. \end{aligned}$$

This may be expressed in simplified form as

$$x_1 = \frac{6}{7} - 3x_2,$$

$$x_3 = \frac{9}{7} - 4x_2,$$

$$x_4 = -\frac{5}{7} + 3x_2,$$

and a general solution is $(\frac{6}{7} - 3x_2, x_2, \frac{9}{7} - 4x_2, -\frac{5}{7} + 3x_2)$.

PRACTICE PROBLEM Solve the following system by using matrices:

$$\begin{aligned} x + 2y + z - 3w &= 0, \\ 2x - 3y + z + 5w &= 5, \\ 3x + y + 3z - 4w &= 3. \end{aligned}$$

Answers

$$x = 2 - 3w, \qquad y = -\frac{3}{5} + w, \qquad z = -\frac{4}{5} + 4w,$$

or

$$y = \frac{1}{15} - \frac{x}{3}, \qquad z = \frac{28}{15} - \frac{4x}{3}, \qquad w = \frac{2 - x}{3},$$

or

$$x = \frac{1}{5} - 3y, \qquad z = \frac{8}{5} + 4y, \qquad w = \frac{1}{5} - 3y,$$

or

$$x = \frac{1}{5} - \frac{3z}{4}, \qquad y = -\frac{2}{5} + \frac{z}{4}, \qquad w = \frac{1}{5} + \frac{z}{4}.$$ ■

Solving systems by using matrices

EXAMPLE 4 Solve the following system:

$$\begin{aligned} 2x_1 + 4x_2 + x_3 &= -3, \\ -3x_1 - x_2 + x_3 &= 5, \\ -4x_1 + 2x_2 + 3x_3 &= 2. \end{aligned}$$

Solution

$$\begin{bmatrix} 2 & 4 & 1 & \vdots & -3 \\ -3 & \textcircled{-1} & 1 & \vdots & 5 \\ -4 & 2 & 3 & \vdots & 2 \end{bmatrix} \quad R_1 \to R_1 + 4R_2 \text{ and } R_3 \to R_3 + 2R_2.$$

$$\begin{bmatrix} -10 & 0 & 5 & \vdots & 17 \\ -3 & -1 & 1 & \vdots & 5 \\ -10 & 0 & 5 & \vdots & 12 \end{bmatrix} \quad R_1 \to \tfrac{1}{5}R_1.$$

$$\begin{bmatrix} -2 & 0 & \textcircled{1} & \vdots & \frac{17}{5} \\ -3 & -1 & 1 & \vdots & 5 \\ -10 & 0 & 5 & \vdots & 12 \end{bmatrix} \quad R_2 \to R_2 + (-1)R_1 \text{ and } R_3 \to R_3 + (-5)R_1.$$

$$\begin{bmatrix} -2 & 0 & 1 & \vdots & \frac{17}{5} \\ -1 & -1 & 0 & \vdots & \frac{8}{5} \\ 0 & 0 & 0 & \vdots & -5 \end{bmatrix}$$

At this point R_1 and R_2 have already been used as pivot rows, which leaves only R_3 as a final choice. But since a pivot element may not be 0 and must not be chosen in the last column, there is no pivot element to choose in R_3. Thus the process must terminate. Since R_3 of the last matrix represents the equation "$0 = -5$," the system is inconsistent and has no solution.

PRACTICE PROBLEM Solve the following system:

$$\begin{aligned} 4x - 3y + 7z &= -1, \\ -x + 2y - 3z &= 3, \\ x + y \phantom{{}-3z} &= 0. \end{aligned}$$

Answer: No solution. ■

SOME IMPORTANT POINTS IN USING MATRICES AND THE PIVOT PROCESS

1. *Any* row of a matrix may be chosen as a pivot row if it has not been used as a pivot row in a previous step (for example, row 2 can be used only once).
2. *Any* element a_{ij} of the pivot row R_i may be chosen as the pivot element of that row if $a_{ij} \neq 0$ and if a_{ij} is not in the last column.
3. If a_{ij} has been chosen as a pivot element in R_i and a_{ij} is not 1

or -1, then it is usually convenient to replace R_i by $(1/a_{ij})R_i$; this operation will replace a_{ij} by a 1 in the ith row and jth column.

EXERCISES

Set A–Matrices of Linear Systems

In Exercises 1 through 6, write the matrix of coefficients and the augmented matrix of the given systems.

1. $\begin{aligned} x + y &= 1 \\ 2x - 3y &= 2 \end{aligned}$

2. $\begin{aligned} -7x &= 7 \\ -x - 7y &= -6 \end{aligned}$

3. $\begin{aligned} 3x - 5y - 6z &= 4 \\ -7x + 9y - 6z &= -8 \end{aligned}$

4. $\begin{aligned} -2x - 2y - 4z &= -7 \\ 9x + 3y + 3z &= -3 \end{aligned}$

5. $\begin{aligned} -z + 9x + 5z &= 0 \\ x - 3w + 6y + 6z &= -6 \\ -9w + 4x - 7z + 3y &= -8 \\ 2x + 9w + 4y - z &= 8 \end{aligned}$

6. $\begin{aligned} 3x - 9w - 3z + 7y &= 6 \\ 2w - 7x - y - 7z &= 0 \\ -y + 6z - 9w + 6x &= 4 \\ -z + 5y - 7x + 8w &= -9 \end{aligned}$

Set B–Solving Systems by Matrices

In Exercises 7 through 32, use matrices to solve the systems given.

7. $\begin{aligned} 2x - 4y &= -4 \\ 7x - y &= -1 \end{aligned}$

8. $\begin{aligned} 5x - 6y &= -7 \\ 3x + 3y &= 9 \end{aligned}$

9. $\begin{aligned} 2x + 5y &= 2 \\ 8x + 6y &= 8 \end{aligned}$

10. $\begin{aligned} 8x + 3y &= 5 \\ 4x - 6y &= 10 \end{aligned}$

11. $\begin{aligned} 2x - y + z &= -2 \\ x - 2y + z &= -3 \end{aligned}$

12. $\begin{aligned} 3x - y + 2z &= 1 \\ 4x + 2y + 3z &= 1 \end{aligned}$

13. $\begin{aligned} y - 4x + 5z &= 7 \\ 3x + 2z + 3y &= 10 \end{aligned}$

14. $\begin{aligned} x - 4y + 2z &= 5 \\ 2x - 7y + 2z &= -2 \end{aligned}$

15. $\begin{aligned} 10x + 3y + 6z &= 18 \\ 6x + 2y + 3z &= 13 \end{aligned}$

16. $\begin{aligned} -8y + 3z - 5x &= -8 \\ 2z - 4y + 3x &= -9 \end{aligned}$

17. $\begin{aligned} x - 2y + 9z &= -16 \\ -4x + 2y + 3z &= 16 \\ y &= 8 \end{aligned}$

18. $\begin{aligned} -4x + y - 5z &= 58 \\ 8x + y + 3z &= -74 \\ 4x - y - 3z &= -10 \end{aligned}$

19. $\begin{aligned} x - 7y + 6z &= 23 \\ x + 5y + 4z &= -31 \\ -2x + 5y + z &= -40 \end{aligned}$

20. $\begin{aligned} 4x + 5y + 3z &= -4 \\ 5x + 6y + 4z &= -4 \\ -x - y + 3z &= -4 \end{aligned}$

21. $\begin{aligned} -x + 8y - 2z &= -15 \\ - 9y + 6z &= 42 \\ 3x - 8y + 7z &= 17 \end{aligned}$

22. $\begin{aligned} 3x + 5y + z &= -10 \\ 9x - 9y - z &= -42 \\ y - 3z &= 10 \end{aligned}$

23. $$\begin{aligned} -8x + 9y - 7z &= 157 \\ 8x + 8y + 5z &= 4 \\ -6x + 4y - z &= 76 \end{aligned}$$

24. $$\begin{aligned} 8x + 3y - 9z &= 10 \\ -9x + 3y + 4z &= -24 \\ -4x - 3y + z &= -2 \end{aligned}$$

25. $$\begin{aligned} 2x + y + 5z &= -12 \\ 2x + 3y - 4z &= -8 \\ x - 9z &= -7 \end{aligned}$$

26. $$\begin{aligned} -9x - 3y + 7z &= -11 \\ 3x + 4y + 3z &= 9 \\ x - z &= 1 \end{aligned}$$

27. $$\begin{aligned} 5x + 4y &= 12 \\ 5x + 2y + 2z &= -4 \\ 9x + 5y + 3z &= 4 \end{aligned}$$

28. $$\begin{aligned} 7x + 9y - 4z &= 81 \\ 4y - 9z &= 7 \\ 5x + 5y + 6z &= 61 \end{aligned}$$

29. Solve, using x as a parameter.
$$\begin{aligned} 10x + y + 2z + w &= -4 \\ 5x - y + z + 2w &= -8 \\ 20x + y + 4z + 3w &= -12 \end{aligned}$$

30. Solve, using y as a parameter.
$$\begin{aligned} -x - 2z &= 5 \\ 3x - y - z + 2w &= 10 \\ x - z + w &= 6 \end{aligned}$$

31. Solve, using z as a parameter.
$$\begin{aligned} 2x + y - 20z + w &= 0 \\ x + 3y - 21z + 2w &= 1 \\ x + y - 12z + w &= 1 \end{aligned}$$

32. Solve, using w as a parameter.
$$\begin{aligned} 2x + 3y - 2z - 17w &= 11 \\ x + 2y - z - 8w &= 1 \\ 2x + y + z - 13w &= 44 \end{aligned}$$

3.5 SOME MATRIX COMPUTATIONS

The usefulness of matrices for solving systems of linear equations has been demonstrated in Section 3.4. Some of the additional applications to be made here require new concepts which will be considered in this section.

For mathematics problems it is helpful to choose a notation as convenient and streamlined as possible. Matrix notation is frequently a convenient and economical way of expressing certain kinds of problems which are otherwise somewhat cumbersome. Many of these uses require that appropriate definitions be made.

Recall that the notation a_{ij} has been used to represent the element in the ith row and jth column of a matrix. This notation is helpful in expressing an entire matrix. We write

$$A = [a_{ij}]_{3\times 4}$$

for example, to mean a 3-by-4 matrix having elements a_{ij}. Thus

$$A = [a_{ij}]_{3\times 4} = \begin{bmatrix} a_{11} & a_{12} & a_{13} & a_{14} \\ a_{21} & a_{22} & a_{23} & a_{24} \\ a_{31} & a_{32} & a_{33} & a_{34} \end{bmatrix}.$$

dimensions

If $A = [a_{ij}]_{m\times n}$, then A is a matrix of m rows and n columns, and we say that A is a matrix having *dimensions* m and n or A is an m-by-n matrix.

Writing the entries of a matrix from a formula

EXAMPLE 1 Write the 3-by-2 matrix whose elements are given by the formula

$$a_{ij} = 2i + j.$$

Solution: The meaning of the problem is that the matrix $A = [a_{ij}]_{3\times 2}$ should be constructed so that for each appropriate i, j, $a_{ij} = 2i + j$. Since A is 3 by 2, we must compute a_{ij} for $i = 1, 2, 3$ and $j = 1, 2$:

$$\begin{aligned} a_{11} &= 2\cdot 1 + 1 = 3, & a_{12} &= 2\cdot 1 + 2 = 4,\\ a_{21} &= 2\cdot 2 + 1 = 5, & a_{22} &= 2\cdot 2 + 2 = 6,\\ a_{31} &= 2\cdot 3 + 1 = 7, & a_{32} &= 2\cdot 3 + 2 = 8. \end{aligned}$$

Therefore,

$$[a_{ij}]_{3\times 2} = \begin{bmatrix} 3 & 4\\ 5 & 6\\ 7 & 8 \end{bmatrix}.$$

PRACTICE PROBLEM Construct the 2-by-3 matrix whose entries are given by the formula $a_{ij} = 3i + j - 5$.

Answer

$$\begin{bmatrix} -1 & 0 & 1\\ 2 & 3 & 4 \end{bmatrix}. \blacksquare$$

DEFINITION 1

(a) Let $A = [a_{ij}]_{m\times n}$ and $B = [b_{ij}]_{m\times n}$. Then the *sum of A and B*, denoted by $A + B$, is the matrix

matrix sum

$$C = [c_{ij}]_{m\times n}$$

where c_{ij} is obtained by the formula

$$c_{ij} = a_{ij} + b_{ij}, \qquad \text{for all} \quad i, j.$$

scalar product

(b) Let $A = [a_{ij}]_{m \times n}$ and let d be any number. Then the *product of d and A*, called the *scalar product* and denoted by dA, is the matrix $C = [c_{ij}]_{m \times n}$, where c_{ij} is given by the formula

$$c_{ij} = d \cdot a_{ij}, \qquad \text{for all} \quad i, j.$$

Several points concerning Definition 1 should be emphasized. First, part (a) requires that two matrices A, B have the same dimensions before we can form $A + B$. For example, we cannot add a 3-by-2 matrix to a 3-by-3. Also, part (a) states that the sum of A and B is obtained by adding corresponding elements of A and B. Consider

$$A = \begin{bmatrix} 0 & 1 \\ -1 & 2 \end{bmatrix} \quad \text{and} \quad B = \begin{bmatrix} 2 & 1 \\ 3 & -2 \end{bmatrix}.$$

In this case, $A + B$ can be computed from Definition 1(a) because A and B are both 2-by-2 matrices:

$$\begin{aligned} A + B &= \begin{bmatrix} 0 & 1 \\ -1 & 2 \end{bmatrix} + \begin{bmatrix} 2 & 1 \\ 3 & -2 \end{bmatrix} \\ &= \begin{bmatrix} 0+2 & 1+1 \\ -1+3 & 2+(-2) \end{bmatrix} \\ &= \begin{bmatrix} 2 & 2 \\ 2 & 0 \end{bmatrix}. \end{aligned}$$

To apply part (b) of Definition 1 we multiply each element of a matrix A by the given number d in order to form the product dA. If $d = 3$ and A is the 2-by-2 matrix in the above sum, we have

$$dA = 3 \cdot \begin{bmatrix} 0 & 1 \\ -1 & 2 \end{bmatrix} = \begin{bmatrix} 3 \cdot 0 & 3 \cdot 1 \\ 3 \cdot (-1) & 3 \cdot 2 \end{bmatrix} = \begin{bmatrix} 0 & 3 \\ -3 & 6 \end{bmatrix}.$$

Adding matrices

EXAMPLE 2 Compute $2A + 3B$ for the matrices A, B given below:

$$A = \begin{bmatrix} 1 & -1 & 2 \\ 0 & 4 & -1 \end{bmatrix}, \qquad B = \begin{bmatrix} 0 & \frac{1}{3} & -1 \\ \frac{2}{3} & 0 & 2 \end{bmatrix}.$$

Solution

$$\begin{aligned}2A + 3B &= 2\begin{bmatrix}1 & -1 & 2\\ 0 & 4 & -1\end{bmatrix} + 3\begin{bmatrix}0 & \frac{1}{3} & -1\\ \frac{2}{3} & 0 & 2\end{bmatrix}\\ &= \begin{bmatrix}2 & -2 & 4\\ 0 & 8 & -2\end{bmatrix} + \begin{bmatrix}0 & 1 & -3\\ 2 & 0 & 6\end{bmatrix}\\ &= \begin{bmatrix}2 & -1 & 1\\ 2 & 8 & 4\end{bmatrix}.\end{aligned}$$

Therefore, $2A + 3B$ is the 2-by-3 matrix

$$\begin{bmatrix}2 & -1 & 1\\ 2 & 8 & 4\end{bmatrix}.$$

PRACTICE PROBLEM If A and B are the matrices in Example 2, compute $5A + 6B$.

Answer

$$\begin{bmatrix}5 & -3 & 4\\ 4 & 20 & 7\end{bmatrix}. \quad ■$$

equality

DEFINITION 2 If $A = [a_{ij}]_{m\times n}$ and $B = [b_{ij}]_{m\times n}$, then $A = B$ if and only if

$$a_{ij} = b_{ij}, \qquad \text{for all} \quad i, j.$$

Observe that Definition 2 implies that two matrices A, B *cannot* be equal if they have different dimensions. If two matrices A, B do have the same dimensions, they are equal only if *all* corresponding elements are equal. By Definition 2 we may write

$$\begin{bmatrix}1 & 0\\ 2 & -1\end{bmatrix} = \begin{bmatrix}\frac{4}{4} & 0\cdot 4\\ 5-3 & (-1)^3\end{bmatrix}$$

since each of these matrices is 2 by 2 and since

$$1 = \tfrac{4}{4}, \qquad 0 = 0\cdot 4, \qquad 2 = 5 - 3, \qquad -1 = (-1)^3.$$

On the other hand,

$$\begin{bmatrix}1 & 0\\ 2 & -1\end{bmatrix} \neq \begin{bmatrix}1 & 0 & 0\\ 2 & -1 & 0\end{bmatrix}$$

because these have different dimensions, and

$$\begin{bmatrix} 1 & 0 \\ 2 & -1 \end{bmatrix} \neq \begin{bmatrix} 1 & 0 \\ 2 & 1 \end{bmatrix}$$

because $-1 \neq 1$.

Solving a matrix equation

EXAMPLE 3 Solve the following matrix equation for x, y, z, w:

$$\begin{bmatrix} x-1 & x+y \\ x-y+z & z+5w \end{bmatrix} = \begin{bmatrix} 1 & 0 \\ 0 & 1 \end{bmatrix}.$$

Solution: From Definition 2 we must have

(a) $x - 1 = 1,$

(b) $x + y = 0,$

(c) $x - y + z = 0,$

(d) $z + 5w = 1.$

Solving Equation (a) gives $x = 2$. Substituting $x = 2$ into (b) yields $y = -2$. Then from (c), by substituting $x = 2$ and $y = -2$, we have

$$\begin{aligned} x - y + z &= 0, \\ 2 - (-2) + z &= 0, \\ 4 + z &= 0, \\ z &= -4. \end{aligned}$$

Finally, we use the value obtained for z in Eq. (d):

$$\begin{aligned} -4 + 5w &= 1, \\ 5w &= 5, \\ w &= 1. \end{aligned}$$

Therefore, the solution is $x = 2$, $y = -2$, $z = -4$, $w = 1$.

PRACTICE PROBLEM Solve the following matrix equation for x, y, z, w:

$$\begin{bmatrix} x+1 & x+2y \\ 2x-y+z & 5z+w \end{bmatrix} = \begin{bmatrix} 2 & -1 \\ 5 & 8 \end{bmatrix}.$$

Answer: $x = 1$, $y = -1$, $z = 2$, $w = -2$. ■

Solving a matrix equation

EXAMPLE 4 Solve the following matrix equation:

$$2 \cdot \begin{bmatrix} x & y \\ z & w \end{bmatrix} + \begin{bmatrix} 0 & 1 \\ 1 & 0 \end{bmatrix} = \begin{bmatrix} 0 & 0 \\ 0 & 0 \end{bmatrix}.$$

Solution: Using Definition 1, we have

$$2 \cdot \begin{bmatrix} x & y \\ z & w \end{bmatrix} + \begin{bmatrix} 0 & 1 \\ 1 & 0 \end{bmatrix} = \begin{bmatrix} 2x & 2y \\ 2z & 2w \end{bmatrix} + \begin{bmatrix} 0 & 1 \\ 1 & 0 \end{bmatrix} = \begin{bmatrix} 2x & 2y+1 \\ 2z+1 & 2w \end{bmatrix}.$$

Therefore, the equation may be expressed as

$$\begin{bmatrix} 2x & 2y+1 \\ 2z+1 & 2w \end{bmatrix} = \begin{bmatrix} 0 & 0 \\ 0 & 0 \end{bmatrix},$$

and by Definition 2,

$$\begin{aligned} 2x &= 0, \\ 2y + 1 &= 0, \\ 2z + 1 &= 0, \\ 2w &= 0. \end{aligned}$$

These four equations give $x = 0$, $y = -\frac{1}{2}$, $z = -\frac{1}{2}$, and $w = 0$.

PRACTICE PROBLEM Solve the following matrix equation for x, y, z, w:

$$\begin{bmatrix} x-1 & y+1 \\ z & 1-w \end{bmatrix} + \begin{bmatrix} 1 & 0 \\ 0 & 1 \end{bmatrix} = \begin{bmatrix} -2 & 0 \\ 0 & -2 \end{bmatrix}.$$

Answer: $x = -2$, $y = -1$, $z = 0$, $w = 4$. ■

If each element of a matrix is the number 0, the matrix is called a *zero matrix*. The 2-by-2 zero matrix is, for example,

$$\begin{bmatrix} 0 & 0 \\ 0 & 0 \end{bmatrix}.$$

This matrix may also be expressed as $0_{2\times 2}$ and generally $0_{m\times n}$ means the zero matrix of dimensions m and n.

matrix difference

If A and B are matrices having the same dimensions, the *difference $A - B$* is defined as follows:

$$A - B = A + (-1) \cdot B.$$

To illustrate this definition, let

$$A = \begin{bmatrix} 3 & 2 & -1 \\ 0 & 1 & 8 \end{bmatrix}, \qquad B = \begin{bmatrix} 1 & 1 & 0 \\ 2 & -1 & 4 \end{bmatrix}.$$

Then

$$\begin{aligned} A - B &= A + (-1) \cdot B \\ &= \begin{bmatrix} 3 & 2 & -1 \\ 0 & 1 & 8 \end{bmatrix} + (-1) \cdot \begin{bmatrix} 1 & 1 & 0 \\ 2 & -1 & 4 \end{bmatrix} \\ &= \begin{bmatrix} 3 & 2 & -1 \\ 0 & 1 & 8 \end{bmatrix} + \begin{bmatrix} -1 & -1 & 0 \\ -2 & 1 & -4 \end{bmatrix} \\ &= \begin{bmatrix} 3 + (-1) & 2 + (-1) & -1 + 0 \\ 0 + (-2) & 1 + 1 & 8 + (-4) \end{bmatrix} \\ &= \begin{bmatrix} 2 & 1 & -1 \\ -2 & 2 & 4 \end{bmatrix}. \end{aligned}$$

From the definition of $A - B$ one may say that to *subtract B from A we subtract each element of B from the corresponding element of A.*

If $A = [a_{ij}]_{m \times n}$, it is not difficult to see that

$$A - A = 0_{m \times n}.$$

For example, if $m = n = 2$,

$$\begin{aligned} A - A = A + (-1)A &= \begin{bmatrix} a_{11} & a_{12} \\ a_{21} & a_{22} \end{bmatrix} + (-1) \begin{bmatrix} a_{11} & a_{12} \\ a_{21} & a_{22} \end{bmatrix} \\ &= \begin{bmatrix} a_{11} & a_{12} \\ a_{21} & a_{22} \end{bmatrix} + \begin{bmatrix} -a_{11} & -a_{12} \\ -a_{21} & -a_{22} \end{bmatrix} \\ &= \begin{bmatrix} a_{11} + (-a_{11}) & a_{12} + (-a_{12}) \\ a_{21} + (-a_{21}) & a_{22} + (-a_{22}) \end{bmatrix} \\ &= \begin{bmatrix} 0 & 0 \\ 0 & 0 \end{bmatrix} \\ &= 0_{2 \times 2}. \end{aligned}$$

Solving a matrix equation

EXAMPLE 5 Let

$$A = \begin{bmatrix} 3 & 2 & -1 \\ 0 & 1 & 8 \end{bmatrix}, \qquad B = \begin{bmatrix} 1 & 1 & 0 \\ 2 & -1 & 4 \end{bmatrix}.$$

Solve the matrix equation $2A + X = B$.

Solution: A and $2A$ are 2-by-3 matrices and hence X must also have these dimensions. Therefore, we let

$$X = \begin{bmatrix} x_{11} & x_{12} & x_{13} \\ x_{21} & x_{22} & x_{23} \end{bmatrix}.$$

Then

$$\begin{aligned} 2A + X &= 2\begin{bmatrix} 3 & 2 & -1 \\ 0 & 1 & 8 \end{bmatrix} + \begin{bmatrix} x_{11} & x_{12} & x_{13} \\ x_{21} & x_{22} & x_{23} \end{bmatrix} \\ &= \begin{bmatrix} 6 & 4 & -2 \\ 0 & 2 & 16 \end{bmatrix} + \begin{bmatrix} x_{11} & x_{12} & x_{13} \\ x_{21} & x_{22} & x_{23} \end{bmatrix} \\ &= \begin{bmatrix} 6 + x_{11} & 4 + x_{12} & -2 + x_{13} \\ x_{21} & 2 + x_{22} & 16 + x_{23} \end{bmatrix}. \end{aligned}$$

To solve the given equation $2A + X = B$, we must have

$$\begin{bmatrix} 6 + x_{11} & 4 + x_{12} & -2 + x_{13} \\ x_{21} & 2 + x_{22} & 16 + x_{23} \end{bmatrix} = \begin{bmatrix} 1 & 1 & 0 \\ 2 & -1 & 4 \end{bmatrix}.$$

Then the matrix equation here leads to the following six equations:

$$\begin{aligned} 6 + x_{11} &= 1, & 4 + x_{12} &= 1, & -2 + x_{13} &= 0, \\ x_{21} &= 2; & 2 + x_{22} &= -1; & 16 + x_{23} &= 4. \end{aligned}$$

Solving these equations, we get

$$\begin{aligned} x_{11} &= -5, & x_{12} &= -3, & x_{13} &= 2, \\ x_{21} &= 2, & x_{22} &= -3, & x_{23} &= -12. \end{aligned}$$

Thus,

$$X = \begin{bmatrix} -5 & -3 & 2 \\ 2 & -3 & -12 \end{bmatrix}.$$

PRACTICE PROBLEM Using the matrices A, B of this example, solve the equation $X + 2B = A$.

Answer

$$X = \begin{bmatrix} 1 & 0 & -1 \\ -4 & 3 & 0 \end{bmatrix}. \blacksquare$$

EXERCISES

Set A–Constructing Matrices from Formulas

1. Write down the 2-by-2 matrix whose entries a_{ij} are given by the formula

$$a_{ij} = i \cdot j.$$

2. Write down the 3-by-2 matrix whose entries a_{ij} are given by the formula

$$a_{ij} = 2i - 3j + 4.$$

3. Write down the 4-by-2 matrix whose entries a_{ij} are given by the formula

$$a_{ij} = i^2 + j + 1.$$

4. Write down the 2-by-3 matrix whose entries a_{ij} are given by the formula

$$a_{ij} = \frac{i^2 + j^2}{i + j}.$$

5. Write down the 1-by-5 matrix whose entries a_{ij} are given by the formula

$$a_{ij} = i^2 - j^2.$$

6. Write down the 2-by-4 matrix whose entries a_{ij} are given by the formula

$$a_{ij} = (i + j)^2 - 1.$$

7. Write down the 3-by-3 matrix whose entries a_{ij} are given by the formula

$$a_{ij} = \begin{cases} 1 & \text{if} \quad i = j, \\ 0 & \text{if} \quad i \neq j. \end{cases}$$

8. Write down the 4-by-4 matrix whose entries a_{ij} are given by the formula

$$a_{ij} = \begin{cases} i + j & \text{if } i < j, \\ 0 & \text{if } i \geq j. \end{cases}$$

9. Write down the 4-by-4 matrix whose entries a_{ij} are given by the formula

$$a_{ij} = \begin{cases} i + 1 & \text{if } i \geq j, \\ i - 1 & \text{if } i < j. \end{cases}$$

10. Write down the 5-by-5 matrix whose entries a_{ij} are given by the formula

$$a_{ij} = \begin{cases} i^2 & \text{if } i > j, \\ 1 & \text{if } i = j, \\ i^2 - j & \text{if } i < j. \end{cases}$$

Set B–Matrix Computations: Addition, Subtraction and Multiplication

In Exercises 11 through 20, use the matrices

$$A = \begin{bmatrix} 1 & 0 & -1 \\ 3 & 1 & -2 \\ 3 & 1 & 0 \end{bmatrix}, \qquad B = \begin{bmatrix} 0 & -1 & 2 \\ -1 & 5 & 1 \\ -2 & 7 & 3 \end{bmatrix}$$

to make the computations.

11. $A + B$
12. $B - A$
13. $2A - B$
14. $A + 2(A - B)$
15. $B - (B + 2A)$
16. $2(A + B) + 3(B - A)$
17. $[2(A - B) + B] - A$
18. $A - (3B - 2A)$
19. $2[2(B - A) - B] - 2(A - B)$
20. $A - 2(A - B) - 3(A - 2B) - 4(A - 3B)$

Set C–Solving Matrix Equations

In Exercises 21 through 26 solve the equations for the unknowns.

21. $\begin{bmatrix} x + 4 & y - 5 \\ z - 1 & w \end{bmatrix} = \begin{bmatrix} 3 & 2 \\ 1 & 1 \end{bmatrix}$

22. $\begin{bmatrix} x - 2 & 3 - y \\ 1 - 3z & 2w + 1 \end{bmatrix} = \begin{bmatrix} 4 & -6 \\ 8 & -10 \end{bmatrix}$

23. $\begin{bmatrix} 1-x & 2y+3 \\ 5-2z & 3w \end{bmatrix} = \begin{bmatrix} 0.5 & 0 \\ 0 & 1 \end{bmatrix}$

24. $\begin{bmatrix} x-1 & 2 \\ 1-z & 0 \end{bmatrix} + \begin{bmatrix} x & y+4 \\ z & -w \end{bmatrix} = \begin{bmatrix} 1 & 0 \\ 0 & 1 \end{bmatrix}$

25. $2\begin{bmatrix} x & y-1 \\ z-1 & w \end{bmatrix} - \begin{bmatrix} x & 1-y \\ z & 1-w \end{bmatrix} = \begin{bmatrix} -1 & 1 \\ -1 & 2 \end{bmatrix}$

26. $3\begin{bmatrix} 1-2x & y+z \\ 2-z & 2w \end{bmatrix} - 2\begin{bmatrix} 1-x & 2y+1 \\ 5z+1 & 3w+2 \end{bmatrix} = 4\begin{bmatrix} x & y \\ z & w \end{bmatrix}$

Set D–Solving Matrix Equations
Use the following matrices to solve the equations of Exercises 27 through 32.

$$A = \begin{bmatrix} 0 & -1 & -2 \\ 2 & 0 & -3 \\ 2 & 0 & -1 \end{bmatrix}, \qquad B = \begin{bmatrix} 1 & 0 & 3 \\ 0 & 6 & 2 \\ -1 & 8 & 4 \end{bmatrix}$$

27. $X + A = B$

28. $A - X = B$

29. $2X - 3A = B$

30. $X + 2A + 3B = 0_{3\times 3}$

31. $2(A - X) = 3(B - X)$

32. $2[X - (A + B)] + 3[(B - A) - X] = 0_{3\times 3}$

3.6 MATRIX MULTIPLICATION

row vector
column vector

A matrix having only a single row is called a *row vector* and a matrix having only one column is called a *column vector*. For example,

$$[1 \quad 1 \quad 1], \qquad [0 \quad -1 \quad 2 \quad \tfrac{1}{2} \quad -5]$$

are row vectors and

$$\begin{bmatrix} 0 \\ 1 \end{bmatrix}, \qquad \begin{bmatrix} -1 \\ 1 \\ 2 \end{bmatrix}$$

are column vectors.

Suppose we have two linear equations such as

$$3x - 2y = 1,$$
$$5x + 4y = 3.$$

These equations may be expressed by using column vectors as follows. Let

$$A = \begin{bmatrix} 3 \\ 5 \end{bmatrix}, \quad B = \begin{bmatrix} -2 \\ 4 \end{bmatrix}, \quad C = \begin{bmatrix} 1 \\ 3 \end{bmatrix}.$$

Then the single equation

$$xA + yB = C$$

expresses the two given equations as we can see by computing

$$xA + yB = C,$$
$$x\begin{bmatrix} 3 \\ 5 \end{bmatrix} + y\begin{bmatrix} -2 \\ 4 \end{bmatrix} = \begin{bmatrix} 1 \\ 3 \end{bmatrix},$$
$$\begin{bmatrix} 3x \\ 5x \end{bmatrix} + \begin{bmatrix} -2y \\ 4y \end{bmatrix} = \begin{bmatrix} 1 \\ 3 \end{bmatrix},$$
$$\begin{bmatrix} 3x - 2y \\ 5x + 4y \end{bmatrix} = \begin{bmatrix} 1 \\ 3 \end{bmatrix}.$$

By definition of matrix equality, the last equation is equivalent to the two given equations.

Here is another way in which vectors are useful. Suppose a record club has recordings available as LPs, cassettes, cartridges, and reel-to-reel tapes. The club's inventory of these different kinds of recordings may be conveniently expressed as a row vector. For example,

$$\begin{matrix} & \text{LP} & \text{CAS} & \text{CAR} & \text{R-R} \\ & \downarrow & \downarrow & \downarrow & \downarrow \\ A = & [101 & 54 & 38 & 73\,] \end{matrix}$$

would mean that the club has on hand 101 LPs, 54 cassettes, 38 cartridges, and 73 tapes. (These figures are not intended to be realistic.) The club may keep a running inventory by the appropriate additions and subtractions. At the beginning of a working day, suppose the club's inventory is given by vector A above. During the day the club makes sales which are given by the sales vector

$$B = [10 \quad 3 \quad 0 \quad 12],$$

which means that during the day 10 LPs were sold, 3 cassettes, 0 cartridge, and 12 tapes. At the end of the day the vector $A - B$ may be computed to determine the new inventory vector:

$$\begin{aligned} A - B &= [101 \quad 54 \quad 38 \quad 73] - [10 \quad 3 \quad 0 \quad 12] \\ &= [101 - 10 \quad 54 - 3 \quad 38 - 0 \quad 73 - 12] \\ &= [91 \quad 51 \quad 38 \quad 61]. \end{aligned}$$

The record club could compute the total amount of money realized from sales by using sales vectors. If the vector B above represents sales on a given day, and the prices of individual items are

LP	CAS	CAR	R-R
5.95	7.50	8.85	10.30

then the amount of money, C, received on this day may be calculated as follows:

$$\begin{aligned} B = \;& [10 \qquad 3 \qquad 0 \qquad 12] \\ & \;\downarrow \qquad\; \downarrow \qquad \downarrow \qquad \downarrow \\ C = \;& 10(5.95) + 3(7.50) + 0(8.85) + 12(10.30) \\ = \;& 59.50 + 22.50 + 0.00 + 123.60 \\ = \;& 205.60. \end{aligned}$$

In the above computation, the cost of each item has been multiplied by the number of items sold. In this way we obtain the amount from the sale of each kind of recording: LP, Cas, Car, R-R. Adding these amounts results in the total sales.

vector multiplication

It is advantageous to formalize the kind of computation which has just been carried out by defining *vector multiplication.*

DEFINITION 1 Let $A = [a_1 \quad a_2 \quad a_3 \ldots a_n]$ and let

$$B = \begin{bmatrix} b_1 \\ b_2 \\ b_3 \\ \vdots \\ b_n \end{bmatrix}.$$

dot product
vector product

Then the *dot product* or *vector product* $A \cdot B$ is defined as follows:

$$A \cdot B = a_1 \cdot b_1 + a_2 \cdot b_2 + \cdots + a_n \cdot b_n.$$

Definition 1 requires the first factor A of $A \cdot B$ to be a row vector and the second to be a column vector. Also note that the number of elements in A and B must be the same. To illustrate this definition consider the following:

$$[1 \quad 0 \quad -1] \cdot \begin{bmatrix} 2 \\ -3 \\ 1 \end{bmatrix} = 1 \cdot 2 + 0 \cdot (-3) + (-1) \cdot 1 = 1.$$

In our example about the record club, the prices of individual items are conveniently expressed by the price vector

$$A = [\underset{\substack{\uparrow \\ \text{LP}}}{5.95} \quad \underset{\substack{\uparrow \\ \text{CAS}}}{7.50} \quad \underset{\substack{\uparrow \\ \text{CAR}}}{8.85} \quad \underset{\substack{\uparrow \\ \text{R-R}}}{10.30}].$$

If

$$B = \begin{bmatrix} 13 \\ 5 \\ 3 \\ 8 \end{bmatrix} \begin{matrix} \leftarrow \text{LP} \\ \leftarrow \text{CAS} \\ \leftarrow \text{CAR} \\ \leftarrow \text{R-R} \end{matrix}$$

is the vector which represents the sales of these items on a given day, then $A \cdot B$ is the amount of money which results from these sales:

$$A \cdot B = [5.95 \quad 7.50 \quad 8.85 \quad 10.30] \cdot \begin{bmatrix} 13 \\ 5 \\ 3 \\ 8 \end{bmatrix}$$

$$= (5.95) \cdot 13 + (7.50) \cdot 5 + (8.85) \cdot 3 + (10.30) \cdot 8$$
$$= 77.35 + 37.50 + 26.55 + 82.40$$
$$= 223.80.$$

We now consider matrix multiplication. Although the following definition may seem strange at first, it will become less so as we discuss its many applications in subsequent sections. While vector multiplication results in a number, the result of matrix multiplication is a matrix. We now explain how to obtain the matrix which is the product of two given matrices. Consider each matrix as being composed of row vectors and column vectors; each row of a matrix is a row vector and each column is a column vector. For example, the matrix

$$\begin{bmatrix} -1 & 2 & 1 & 1 \\ -1 & 1 & 1 & 1 \\ -1 & 3 & 1 & 2 \end{bmatrix}$$

is composed of the row vectors

$$R_1 = [-1 \quad 2 \quad 1 \quad 1],$$
$$R_2 = [-1 \quad 1 \quad 1 \quad 1],$$
$$R_3 = [-1 \quad 3 \quad 1 \quad 2],$$

and the column vectors

$$C_1 = \begin{bmatrix} -1 \\ -1 \\ -1 \end{bmatrix}, \quad C_2 = \begin{bmatrix} 2 \\ 1 \\ 3 \end{bmatrix}, \quad C_3 = \begin{bmatrix} 1 \\ 1 \\ 1 \end{bmatrix}, \quad C_4 = \begin{bmatrix} 1 \\ 1 \\ 2 \end{bmatrix}.$$

DEFINITION 2 Let $A = [a_{ij}]_{m \times n}$ and let $B = [b_{ij}]_{n \times k}$. Then the *product AB* is defined to be the matrix $D = [d_{ij}]_{m \times k}$ where each d_{ij} is obtained by the formula

matrix multiplication

$$d_{ij} = R_i \cdot C_j,$$

with R_i the ith row vector of A, and C_j the jth column vector of B.

Before we look at some examples of Definition 2 it is appropriate to point out the restrictions on the dimensions of A and B. A is an m-by-n matrix and B is an n-by-k matrix, that is, *the*

Mnemonic

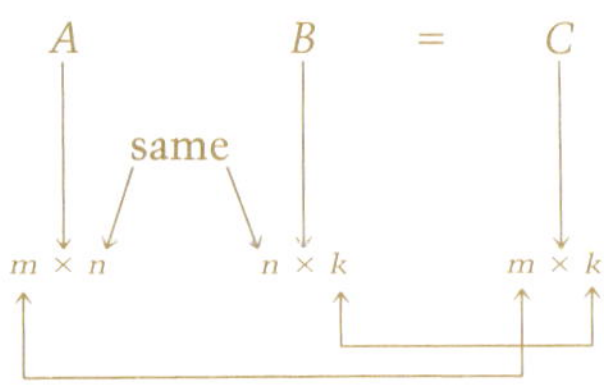

number of columns of A must be the same as the number of rows of B. If this condition is not met, the product AB is *not* defined. If this condition is satisfied, AB is defined to be an m-by-k matrix. Also, Definition 2 requires that the element in the ith row and jth column of AB be the number obtained by forming the vector product of the ith row vector of A and the jth column vector of B.

Consider the following two matrices:

$$A = \begin{bmatrix} 2 & 0 \\ -1 & 1 \end{bmatrix} \quad \text{and} \quad B = \begin{bmatrix} 1 & 1 & 1 \\ -1 & 0 & 4 \end{bmatrix}.$$

A is a 2-by-2 and B is a 2-by-3 matrix; hence, we may form AB. The result will be a 2-by-3 matrix which may be obtained by using the row vectors of A and the column vectors of B. If we let

$$AB = [d_{ij}]_{2\times 3} = \begin{bmatrix} d_{11} & d_{12} & d_{13} \\ d_{21} & d_{22} & d_{23} \end{bmatrix},$$

then by Definition 2, the d_{ij} are obtained by computing the following:

$$\begin{aligned} d_{11} &= R_1 \cdot C_1, \qquad & d_{12} &= R_1 \cdot C_2, \qquad & d_{13} &= R_1 \cdot C_3, \\ d_{21} &= R_2 \cdot C_1, \qquad & d_{22} &= R_2 \cdot C_2, \qquad & d_{23} &= R_2 \cdot C_3. \end{aligned}$$

The row vectors R_i of A are

$$R_1 = [2 \quad 0], \qquad R_2 = [-1 \quad 1],$$

and the column vectors of B are

$$C_1 = \begin{bmatrix} 1 \\ -1 \end{bmatrix}, \qquad C_2 = \begin{bmatrix} 1 \\ 0 \end{bmatrix}, \qquad C_3 = \begin{bmatrix} 1 \\ 4 \end{bmatrix}.$$

Hence, we have

$$d_{11} = R_1 \cdot C_1 = [2 \quad 0] \cdot \begin{bmatrix} 1 \\ -1 \end{bmatrix} = 2 \cdot 1 + 0 \cdot (-1) = 2,$$

$$d_{12} = R_1 \cdot C_2 = [2 \quad 0] \cdot \begin{bmatrix} 1 \\ 0 \end{bmatrix} = 2 \cdot 1 + 0 \cdot 0 = 2,$$

$$d_{13} = R_1 \cdot C_3 = [2 \quad 0] \cdot \begin{bmatrix} 1 \\ 4 \end{bmatrix} = 2 \cdot 1 + 0 \cdot 4 = 2,$$

$$d_{21} = R_2 \cdot C_1 = [-1 \quad 1] \cdot \begin{bmatrix} 1 \\ -1 \end{bmatrix} = (-1) \cdot (1) + 1 \cdot (-1) = -2,$$

$$d_{22} = R_2 \cdot C_2 = [-1 \quad 1] \cdot \begin{bmatrix} 1 \\ 0 \end{bmatrix} = (-1) \cdot (1) + 1 \cdot 0 = -1,$$

$$d_{23} = R_2 \cdot C_3 = [-1 \quad 1] \cdot \begin{bmatrix} 1 \\ 4 \end{bmatrix} = (-1) \cdot 1 + 1 \cdot 4 = 3.$$

Therefore, our product AB is

$$\begin{aligned} AB &= \begin{bmatrix} 2 & 0 \\ -1 & 1 \end{bmatrix} \begin{bmatrix} 1 & 1 & 1 \\ -1 & 0 & 4 \end{bmatrix} \\ &= \begin{bmatrix} d_{11} & d_{12} & d_{13} \\ d_{21} & d_{22} & d_{23} \end{bmatrix} \\ &= \begin{bmatrix} 2 & 2 & 2 \\ -2 & -1 & 3 \end{bmatrix}. \end{aligned}$$

Multiplying matrices

EXAMPLE 1 Multiply the matrices A, B given below:

$$A = \begin{bmatrix} 3 & 0 \\ -1 & 1 \\ 0 & 2 \end{bmatrix}, \qquad B = \begin{bmatrix} -1 & 0 & 0 \\ 0 & 2 & 1 \end{bmatrix}.$$

Solution: The product AB is defined since A has 2 columns and B has 2 rows and the result will be a 3-by-3 matrix. If

$$AB = [d_{ij}]_{3\times 3},$$

then

$$d_{11} = R_1 \cdot C_1 = [\ 3 \quad 0] \cdot \begin{bmatrix} -1 \\ 0 \end{bmatrix} = 3 \cdot (-1) + 0 \cdot 0 = -3,$$

$$d_{12} = R_1 \cdot C_2 = [\ 3 \quad 0] \cdot \begin{bmatrix} 0 \\ 2 \end{bmatrix} = 3 \cdot 0 + 0 \cdot 2 = 0,$$

$$d_{13} = R_1 \cdot C_3 = [\ 3 \quad 0] \cdot \begin{bmatrix} 0 \\ 1 \end{bmatrix} = 3 \cdot 0 + 0 \cdot 1 = 0,$$

$$d_{21} = R_2 \cdot C_1 = [-1 \quad 1] \cdot \begin{bmatrix} -1 \\ 0 \end{bmatrix} = (-1) \cdot (-1) + 1 \cdot 0 = 1,$$

$$d_{22} = R_2 \cdot C_2 = [-1 \quad 1] \cdot \begin{bmatrix} 0 \\ 2 \end{bmatrix} = (-1) \cdot 0 + 1 \cdot 2 = 2,$$

$$d_{23} = R_2 \cdot C_3 = [-1 \quad 1] \cdot \begin{bmatrix} 0 \\ 1 \end{bmatrix} = (-1) \cdot 0 + 1 \cdot 1 = 1,$$

$$d_{31} = R_3 \cdot C_1 = [\ 0 \quad 2] \cdot \begin{bmatrix} -1 \\ 0 \end{bmatrix} = 0 \cdot (-1) + 2 \cdot 0 = 0,$$

$$d_{32} = R_3 \cdot C_2 = [\ 0 \quad 2] \cdot \begin{bmatrix} 0 \\ 2 \end{bmatrix} = 0 \cdot 0 + 2 \cdot 2 = 4,$$

$$d_{33} = R_3 \cdot C_3 = [\ 0 \quad 2] \cdot \begin{bmatrix} 0 \\ 1 \end{bmatrix} = 0 \cdot 0 + 2 \cdot 1 = 2.$$

Then

$$AB = \begin{bmatrix} -3 & 0 & 0 \\ 1 & 2 & 1 \\ 0 & 4 & 2 \end{bmatrix}.$$

PRACTICE PROBLEM Find the product AB, where A, B are given as follows:

$$A = \begin{bmatrix} 1 & 2 \\ -1 & 3 \end{bmatrix} \quad \text{and} \quad B = \begin{bmatrix} 0 & -1 \\ 2 & 5 \end{bmatrix}.$$

Answer

$$AB = \begin{bmatrix} 4 & 9 \\ 6 & 16 \end{bmatrix}. \blacksquare$$

Multiplying matrices

EXAMPLE 2 If A and B are the matrices given below, form the matrix products AB and BA if possible:

$$A = \begin{bmatrix} 0 \\ 1 \\ -1 \end{bmatrix}, \quad B = \begin{bmatrix} 5 & -3 & 1 \\ -1 & 0 & 2 \end{bmatrix}.$$

Solution: A is a 3-by-1 and B is a 2-by-3 matrix. Thus AB is not defined, but BA is defined and is a 2-by-1 matrix. If $BA = [d_{ij}]_{2\times 1}$,

then

$$d_{11} = R_1 \cdot C_1 = [5 \quad -3 \quad 1] \cdot \begin{bmatrix} 0 \\ 1 \\ -1 \end{bmatrix}$$

$$= 5 \cdot 0 + (-3)1 + 1 \cdot (-1) = -4,$$

$$d_{21} = R_2 \cdot C_1 = [-1 \quad 0 \quad 2] \cdot \begin{bmatrix} 0 \\ 1 \\ -1 \end{bmatrix}$$

$$= (-1) \cdot 0 + 0 \cdot 1 + 2 \cdot (-1) = -2,$$

and

$$BA = \begin{bmatrix} 5 & -3 & 1 \\ -1 & 0 & 2 \end{bmatrix} \begin{bmatrix} 0 \\ 1 \\ -1 \end{bmatrix} = \begin{bmatrix} -4 \\ -2 \end{bmatrix}.$$

PRACTICE PROBLEM If X and Y are the matrices given below, form the products XY and YX if possible:

$$X = \begin{bmatrix} 2 & 1 \\ -1 & 3 \end{bmatrix} \quad \text{and} \quad Y = [1 \quad -1].$$

Answer: XY is not defined; $YX = [3 \quad -2]$. ■

Suppose we are given the following linear system:

$$\begin{aligned} 3x_1 + 5x_2 &= 8, \\ 2x_1 + x_2 &= -1. \end{aligned}$$

If we let

$$A = \begin{bmatrix} 3 & 5 \\ 2 & 1 \end{bmatrix}, \quad X = \begin{bmatrix} x_1 \\ x_2 \end{bmatrix}, \quad C = \begin{bmatrix} 8 \\ -1 \end{bmatrix},$$

then the matrix equation

$$AX = C$$

is an equivalent way to express the system. To see why, multiply

$$AX = \begin{bmatrix} 3 & 5 \\ 2 & 1 \end{bmatrix} \begin{bmatrix} x_1 \\ x_2 \end{bmatrix} = \begin{bmatrix} 3x_1 + 5x_2 \\ 2x_1 + x_2 \end{bmatrix}.$$

Then by the definition of matrix equality,

$$\begin{bmatrix} 3x_1 + 5x_2 \\ 2x_1 + x_2 \end{bmatrix} = \begin{bmatrix} 8 \\ -1 \end{bmatrix}$$

is equivalent to the original two equations. Matrix notation used in this fashion is not restricted to two equations in two unknowns; it may be used equally well for expressing any system of linear equations.

EXERCISES

Set A–Computing Vector Products

In Exercises 1 through 8, compute the vector products using the given vectors.

$$A = [0 \quad 1 \quad -1], \qquad B = [1 \quad -2 \quad 3]$$

$$C = \begin{bmatrix} 5 \\ -2 \\ 1 \end{bmatrix}, \qquad D = \begin{bmatrix} -4 \\ 0 \\ 2 \end{bmatrix}, \qquad E = [7 \quad -4 \quad -9 \quad -2],$$

$$F = [-4 \quad 7 \quad 1 \quad -7], \qquad G = \begin{bmatrix} -7 \\ -4 \\ -5 \\ -3 \end{bmatrix}, \qquad H = \begin{bmatrix} 2 \\ 1 \\ 9 \\ -4 \end{bmatrix}$$

1. $A \cdot C$
2. $A \cdot D$
3. $B \cdot C$
4. $B \cdot D$
5. $E \cdot G$
6. $E \cdot H$
7. $F \cdot G$
8. $F \cdot H$

Set B–Solving Vector Equations

In Exercises 9 through 12 you are given two vectors A and X. In each exercise, you are to solve the equations

$$\text{(a) } A \cdot X = 0 \qquad \text{and} \qquad \text{(b) } A \cdot X = 5.$$

9. $A = [0 \quad 1 \quad -1], \quad X = \begin{bmatrix} 5 \\ y \\ 2 \end{bmatrix}$

10. $A = [-7 \quad 6 \quad 7]$, $X = \begin{bmatrix} y \\ -5 \\ -1 \end{bmatrix}$

11. $A = [-1 \quad 1 \quad 3 \quad -5]$, $X = \begin{bmatrix} 2 \\ 2y \\ 8 \\ 0 \end{bmatrix}$

12. $A = [3 \quad y \quad 3 \quad 6]$, $X = \begin{bmatrix} 9 \\ y \\ -4 \\ 0 \end{bmatrix}$

Set C–Questions About Dimensions

In Exercises 13 through 17, A and B are matrices.

13. If $A \cdot B$ is defined, what can you conclude about A and B?

14. If AB is defined, what can you conclude about A and B?

15. If AB is a 5-by-7 matrix, what can you conclude about A and B?

16. If AB and BA are *both* defined, what can you conclude about A and B?

17. If AA is defined, what can you conclude about A?

Set D–Computing Individual Entries of a Matrix Product

In Exercises 18 through 26, use the given matrix A and compute the entry requested:

$$A = \begin{bmatrix} 1 & -1 & 2 \\ 0 & 1 & 3 \\ -1 & 5 & 1 \end{bmatrix}.$$

18. The entry in the first row, first column of AA

19. The entry in the first row, second column of AA

20. The entry in the first row, third column of AA

21. The entry in the second row, first column of AA

22. The entry in the second row, second column of AA

23. The entry in the second row, third column of AA

24. The entry in the third row, first column of AA

25. The entry in the third row, second column of AA

26. The entry in the third row, third column of AA

Set E–Multiplying Matrices

Find the products in Exercises 27 through 34.

27. $\begin{bmatrix} 1 & 2 \\ -3 & 5 \end{bmatrix}\begin{bmatrix} 0 & 7 \\ 5 & 2 \end{bmatrix}$

28. $\begin{bmatrix} 1 & 4 \end{bmatrix}\begin{bmatrix} 2 & -1 \\ -2 & 4 \end{bmatrix}$

29. $\begin{bmatrix} -2 & 4 \\ -1 & 2 \end{bmatrix}\begin{bmatrix} -1 \\ -3 \end{bmatrix}$

30. $\begin{bmatrix} 1 & -1 \\ 0 & 1 \\ -1 & 2 \end{bmatrix}\begin{bmatrix} 4 & 6 & 0 \\ 5 & -2 & 3 \end{bmatrix}$

31. $\begin{bmatrix} 0 & -5 \\ -9 & 4 \end{bmatrix}\begin{bmatrix} 0 & -1 & -3 \\ 6 & 7 & -8 \end{bmatrix}$

32. $\begin{bmatrix} -9 & -1 & -1 \\ 4 & 1 & -9 \\ 1 & -7 & -1 \end{bmatrix}\begin{bmatrix} 1 \\ -4 \\ -5 \end{bmatrix}$

33. $\begin{bmatrix} -3 & 3 & 4 \\ -3 & -6 & -1 \\ 7 & -4 & 3 \end{bmatrix}\begin{bmatrix} 1 & -1 \\ 0 & -9 \\ -3 & 2 \end{bmatrix}$

34. $\begin{bmatrix} -8 & -4 & -3 \\ 8 & -7 & -1 \\ 0 & -9 & -8 \end{bmatrix}\begin{bmatrix} 1 & -8 & -6 \\ 7 & 4 & -5 \\ -9 & -9 & -5 \end{bmatrix}$

Set F–Multiplying Matrices

In Exercises 35 through 41 compute the matrix products where

$$A = \begin{bmatrix} 1 & 0 & 1 \\ 0 & 1 & 1 \\ 1 & 1 & 0 \end{bmatrix} \quad \text{and} \quad B = \begin{bmatrix} 2 & -1 & 3 \\ 0 & 5 & 3 \\ -2 & 0 & 1 \end{bmatrix}.$$

35. $A^2(AA)$ 36. B^2 37. AB 38. BA

39. A^3 40. B^2A 41. $(AB)A$

Set G–Solving Matrix Equations

In Exercises 42 through 45 solve the equations given.

42. $\begin{bmatrix} x & y \end{bmatrix}\begin{bmatrix} -1 & 2 \\ 1 & -2 \end{bmatrix} = \begin{bmatrix} 0 & 0 \end{bmatrix}$ for x, y.

43. $\begin{bmatrix} 2 & -1 \\ 4 & 3 \end{bmatrix}\begin{bmatrix} x \\ y \end{bmatrix} = \begin{bmatrix} 0 \\ 10 \end{bmatrix}$ for x, y.

44. $\begin{bmatrix} 1 & 2 \end{bmatrix}\begin{bmatrix} a & b \\ b & a \end{bmatrix} = \begin{bmatrix} -2 & 2 \end{bmatrix}$ for a, b.

45. $\begin{bmatrix} 2 & -3 \\ 7 & 2 \end{bmatrix}\begin{bmatrix} x \\ y \end{bmatrix} = \begin{bmatrix} -1 \\ 9 \end{bmatrix}$ for x, y.

Set H—Representing Linear Systems as Matrix Equations

In Exercises 46 through 51 matrices A, X and C are given. In each case you are to determine the linear system of equations expressed by the matrix equation $AX = C$.

46. $A = \begin{bmatrix} 1 & 1 \\ 1 & -1 \end{bmatrix}, \quad X = \begin{bmatrix} x \\ y \end{bmatrix}, \quad C = \begin{bmatrix} 0 \\ 0 \end{bmatrix}$

47. $A = \begin{bmatrix} 2 & -3 \\ 1 & 4 \end{bmatrix}, \quad X = \begin{bmatrix} x \\ y \end{bmatrix}, \quad C = \begin{bmatrix} 10 \\ 8 \end{bmatrix}$

48. $A = \begin{bmatrix} 2 & -3 & 6 \\ -3 & 5 & 6 \end{bmatrix}, \quad X = \begin{bmatrix} x \\ y \\ z \end{bmatrix}, \quad C = \begin{bmatrix} -2 \\ 1 \end{bmatrix}$

49. $A = \begin{bmatrix} -8 & 6 & -6 \\ 3 & 4 & 6 \\ 0 & -7 & -2 \end{bmatrix}, \quad X = \begin{bmatrix} x \\ y \\ z \end{bmatrix}, \quad C = \begin{bmatrix} -4 \\ 3 \\ 1 \end{bmatrix}$

50. $A = \begin{bmatrix} 9 & -7 & 6 \\ -7 & 4 & -8 \\ 4 & 0 & 1 \end{bmatrix}, \quad X = \begin{bmatrix} x \\ y \\ z \end{bmatrix}, \quad C = \begin{bmatrix} 5 \\ -9 \\ 0 \end{bmatrix}$

51. $A = \begin{bmatrix} 3 & 0 & 1 & -2 \\ 0 & -1 & 0 & 5 \\ 5 & 3 & 2 & -4 \end{bmatrix}, \quad X = \begin{bmatrix} x \\ y \\ z \\ w \end{bmatrix}, \quad C = \begin{bmatrix} 1 \\ -3 \\ 4 \end{bmatrix}$

3.7 INVERSES OF SQUARE MATRICES

The 2-by-2 matrix

$$I_2 = \begin{bmatrix} 1 & 0 \\ 0 & 1 \end{bmatrix}$$

identity matrix

is called the *second-order multiplicative identity matrix.* The word "identity" is used in this instance because I_2 has the following property:

$$AI_2 = I_2A = A$$

where A is *any* 2-by-2 matrix. To demonstrate this fact, let

$$A = [a_{ij}]_{2\times 2}.$$

Then

$$\begin{aligned} AI_2 &= \begin{bmatrix} a_{11} & a_{12} \\ a_{21} & a_{22} \end{bmatrix} \begin{bmatrix} 1 & 0 \\ 0 & 1 \end{bmatrix} \\ &= \begin{bmatrix} a_{11} \cdot 1 + a_{12} \cdot 0 & a_{11} \cdot 0 + a_{12} \cdot 1 \\ a_{21} \cdot 1 + a_{22} \cdot 0 & a_{21} \cdot 0 + a_{22} \cdot 1 \end{bmatrix} \\ &= \begin{bmatrix} a_{11} & a_{12} \\ a_{21} & a_{22} \end{bmatrix} \\ &= A. \end{aligned}$$

In the same way we could show that $I_2A = A$.

More generally, the *nth-order multiplicative identity matrix* $I_n = [a_{ij}]_{n \times n}$ is the n-by-n matrix defined as follows:

$$\begin{aligned} a_{ij} &= 1, \quad \text{if } i = j, \\ a_{ij} &= 0, \quad \text{if } i \neq j. \end{aligned}$$

For example, I_3 is the matrix whose entries are given by

$$\begin{aligned} a_{11} &= 1, & a_{12} &= 0, & a_{13} &= 0, \\ a_{21} &= 0, & a_{22} &= 1, & a_{23} &= 0, \\ a_{31} &= 0, & a_{32} &= 0, & a_{33} &= 1. \end{aligned}$$

Thus

$$I_3 = \begin{bmatrix} 1 & 0 & 0 \\ 0 & 1 & 0 \\ 0 & 0 & 1 \end{bmatrix}.$$

Matrix identity products

EXAMPLE 1 Show that if A and B are the matrices given below, then the products AB and BA are each the 2-by-2 multiplicative identity matrix; that is, show that $AB = I_2$ and $BA = I_2$.

$$A = \begin{bmatrix} 2 & 1 \\ -1 & 3 \end{bmatrix}, \quad B = \begin{bmatrix} \frac{3}{7} & -\frac{1}{7} \\ \frac{1}{7} & \frac{2}{7} \end{bmatrix}.$$

Solution

$$AB = \begin{bmatrix} 2 & 1 \\ -1 & 3 \end{bmatrix}\begin{bmatrix} \frac{3}{7} & -\frac{1}{7} \\ \frac{1}{7} & \frac{2}{7} \end{bmatrix} = \begin{bmatrix} \frac{6}{7} + \frac{1}{7} & -\frac{2}{7} + \frac{2}{7} \\ -\frac{3}{7} + \frac{3}{7} & \frac{1}{7} + \frac{6}{7} \end{bmatrix}$$

$$= \begin{bmatrix} 1 & 0 \\ 0 & 1 \end{bmatrix}$$

$$= I_2,$$

$$BA = \begin{bmatrix} \frac{3}{7} & -\frac{1}{7} \\ \frac{1}{7} & \frac{2}{7} \end{bmatrix}\begin{bmatrix} 2 & 1 \\ -1 & 3 \end{bmatrix} = \begin{bmatrix} \frac{6}{7} + \frac{1}{7} & \frac{3}{7} - \frac{3}{7} \\ \frac{2}{7} - \frac{2}{7} & \frac{1}{7} + \frac{6}{7} \end{bmatrix} = I_2.$$

PRACTICE PROBLEM Let

$$A = \begin{bmatrix} 1 & 0 \\ 0 & 2 \end{bmatrix} \quad \text{and} \quad B = \begin{bmatrix} 1 & 0 \\ 0 & \frac{1}{2} \end{bmatrix}.$$

Show that AB and BA are both I_2. ■

multiplicative inverse

In the preceding example we showed that $AB = I_2$ and $BA = I_2$. In this case, we say that A and B are *multiplicative inverses* of each other. If A is any n-by-n matrix, the matrix B is said to be the **multiplicative inverse** of A if $AB = I_n$ and $BA = I_n$. Note that this implies that the inverse of B must also be an n-by-n matrix. The problem we now consider is that of determining the inverse of a matrix if there is one.

If A is a square matrix and A has an inverse, it is denoted by A^{-1}. Thus by definition,

$$AA^{-1} = A^{-1}A = I_n.$$

As was implied, not all square matrices have inverses. For example, the matrix

$$A = \begin{bmatrix} 0 & 1 \\ 0 & 0 \end{bmatrix}$$

has no inverse. To see why this is so, let $B = [b_{ij}]_{2\times 2}$ be any 2-by-2 matrix. Then

$$AB = \begin{bmatrix} 0 & 1 \\ 0 & 0 \end{bmatrix} \begin{bmatrix} b_{11} & b_{12} \\ b_{21} & b_{22} \end{bmatrix} = \begin{bmatrix} b_{21} & b_{22} \\ 0 & 0 \end{bmatrix}.$$

Now AB cannot be the matrix I_2 since by definition of equality of matrices we cannot have

$$\begin{bmatrix} b_{21} & b_{22} \\ 0 & 0 \end{bmatrix} = \begin{bmatrix} 1 & 0 \\ 0 & 1 \end{bmatrix}.$$

singular
nonsingular

If a square matrix A has no inverse, it is said to be **singular** and if A^{-1} exists, A is called **nonsingular.**

The inverse of a nonsingular matrix can be determined in several ways but one of the most efficient ways is to use elementary row operations. This means that the same kinds of operations which we use to solve systems of equations may be used to determine inverses of square matrices. The procedure may be described as follows. Let A be an n-by-n matrix whose inverse we wish to find (if it has one).

Step 1. Write

$$A \mid I$$

where I is the n-by-n identity matrix.

Step 2. Use row transformations on A to convert A into I and for each transformation used on A, carry out the same operation on I as indicated below:

Transform A into I (if possible).

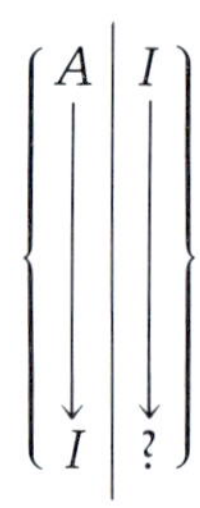

Transformations applied to A are also applied to I.

Step 3: If the transformation process applied to A breaks down and it is not possible to transform A into I, then A has no inverse and is singular. Otherwise, if the transformation process is successful in converting A into I, then the matrix that results from I at the last step is the inverse of A:

$$\begin{array}{c|c} A & I \\ \downarrow & \downarrow \\ I & A^{-1} \end{array}$$

As an example we compute the inverse of the matrix

$$A = \begin{bmatrix} 2 & 1 \\ -1 & 3 \end{bmatrix}$$

$$\begin{array}{c|c|l}
A & I_2 & \\
\downarrow & \downarrow & \\
\begin{bmatrix} 2 & 1 \\ -1 & 3 \end{bmatrix} & \begin{bmatrix} 1 & 0 \\ 0 & 1 \end{bmatrix} & \\
\downarrow & \downarrow & T_1\colon R_1 \to \tfrac{1}{2}R_1 \\
\begin{bmatrix} 1 & \tfrac{1}{2} \\ -1 & 3 \end{bmatrix} & \begin{bmatrix} \tfrac{1}{2} & 0 \\ 0 & 1 \end{bmatrix} & \\
\downarrow & \downarrow & T_2\colon R_2 \to R_1 + R_2 \\
\begin{bmatrix} 1 & \tfrac{1}{2} \\ 0 & \tfrac{7}{2} \end{bmatrix} & \begin{bmatrix} \tfrac{1}{2} & 0 \\ \tfrac{1}{2} & 1 \end{bmatrix} & \\
\downarrow & \downarrow & T_3\colon R_2 \to \tfrac{2}{7}R_2 \\
\begin{bmatrix} 1 & \tfrac{1}{2} \\ 0 & 1 \end{bmatrix} & \begin{bmatrix} \tfrac{1}{2} & 0 \\ \tfrac{1}{7} & \tfrac{2}{7} \end{bmatrix} & \\
\downarrow & \downarrow & T_4\colon R_1 \to -\tfrac{1}{2}R_2 + R_1 \\
\begin{bmatrix} 1 & 0 \\ 0 & 1 \end{bmatrix} & \begin{bmatrix} \tfrac{3}{7} & -\tfrac{1}{7} \\ \tfrac{1}{7} & \tfrac{2}{7} \end{bmatrix} & \\
\downarrow & \downarrow & \\
I_2 & A^{-1} &
\end{array}$$

Therefore, the inverse of the matrix

$$\begin{bmatrix} 2 & 1 \\ -1 & 3 \end{bmatrix}$$

is

$$\begin{bmatrix} \frac{3}{7} & -\frac{1}{7} \\ \frac{1}{7} & \frac{2}{7} \end{bmatrix}$$

The inverse of a matrix

EXAMPLE 2 Find the inverse of the matrix

$$A = \begin{bmatrix} 3 & -5 \\ -1 & 2 \end{bmatrix}.$$

Solution

$$\begin{array}{c|c|l}
A & I_2 & \\
\downarrow & \downarrow & \\
\begin{bmatrix} 3 & -5 \\ -1 & 2 \end{bmatrix} & \begin{bmatrix} 1 & 0 \\ 0 & 1 \end{bmatrix} & \\
\downarrow & \downarrow & R_1 \to \tfrac{1}{3}R_1 \\
\begin{bmatrix} 1 & -\frac{5}{3} \\ -1 & 2 \end{bmatrix} & \begin{bmatrix} \frac{1}{3} & 0 \\ 0 & 1 \end{bmatrix} & \\
\downarrow & \downarrow & R_2 \to R_1 + R_2 \\
\begin{bmatrix} 1 & -\frac{5}{3} \\ 0 & \frac{1}{3} \end{bmatrix} & \begin{bmatrix} \frac{1}{3} & 0 \\ \frac{1}{3} & 1 \end{bmatrix} & \\
\downarrow & \downarrow & R_2 \to 3R_2 \\
\begin{bmatrix} 1 & -\frac{5}{3} \\ 0 & 1 \end{bmatrix} & \begin{bmatrix} \frac{1}{3} & 0 \\ 1 & 3 \end{bmatrix} & \\
\downarrow & \downarrow & R_1 \to \tfrac{5}{3}R_2 + R_1 \\
\begin{bmatrix} 1 & 0 \\ 0 & 1 \end{bmatrix} & \begin{bmatrix} 2 & 5 \\ 1 & 3 \end{bmatrix} & \\
\downarrow & \downarrow & \\
I_2 & A^{-1} &
\end{array}$$

Check

$$\begin{bmatrix} 3 & -5 \\ -1 & 2 \end{bmatrix}\begin{bmatrix} 2 & 5 \\ 1 & 3 \end{bmatrix} = \begin{bmatrix} 6-5 & 15-15 \\ -2+2 & -5+6 \end{bmatrix} = \begin{bmatrix} 1 & 0 \\ 0 & 1 \end{bmatrix}$$

PRACTICE PROBLEM Find the inverse of the matrix

$$A = \begin{bmatrix} 1 & 1 \\ 1 & 2 \end{bmatrix}.$$

Answer: $A^{-1} = \begin{bmatrix} 2 & -1 \\ -1 & 1 \end{bmatrix}$. ■

The inverse of a matrix

EXAMPLE 3 Find the inverse of the matrix

$$A = \begin{bmatrix} 2 & 4 & 1 \\ -3 & -1 & 1 \\ -4 & 2 & 3 \end{bmatrix}.$$

Solution

$$\left[\begin{array}{rrr} 2 & 4 & 1 \\ -3 & -1 & 1 \\ -4 & 2 & 3 \end{array}\right] \left\| \left[\begin{array}{rrr} 1 & 0 & 0 \\ 0 & 1 & 0 \\ 0 & 0 & 1 \end{array}\right]\right. \qquad R_1 \to \tfrac{1}{2}R_1$$

$$\left[\begin{array}{rrr} 1 & 2 & \frac{1}{2} \\ -3 & -1 & 1 \\ -4 & 2 & 3 \end{array}\right] \left\| \left[\begin{array}{rrr} \frac{1}{2} & 0 & 0 \\ 0 & 1 & 0 \\ 0 & 0 & 1 \end{array}\right]\right. \qquad \begin{array}{l} R_2 \to 3R_1 + R_2 \\ R_3 \to 4R_1 + R_3 \end{array}$$

$$\left[\begin{array}{rrr} 1 & 2 & \frac{1}{2} \\ 0 & 5 & \frac{5}{2} \\ 0 & 10 & 5 \end{array}\right] \left\| \left[\begin{array}{rrr} \frac{1}{2} & 0 & 0 \\ \frac{3}{2} & 1 & 0 \\ 2 & 0 & 1 \end{array}\right]\right. \qquad R_2 \to \tfrac{1}{5}R_2$$

$$\left[\begin{array}{rrr} 1 & 2 & \frac{1}{2} \\ 0 & 1 & \frac{1}{2} \\ 0 & 10 & 5 \end{array}\right] \left\| \left[\begin{array}{rrr} \frac{1}{2} & 0 & 0 \\ \frac{3}{10} & \frac{1}{5} & 0 \\ 2 & 0 & 1 \end{array}\right]\right. \qquad \begin{array}{l} R_1 \to -2R_2 + R_1 \\ R_3 \to -10R_2 + R_3 \end{array}$$

$$\left[\begin{array}{rrr} 1 & 0 & -\frac{1}{2} \\ 0 & 1 & \frac{1}{2} \\ 0 & 0 & 0 \end{array}\right] \left\| \left[\begin{array}{rrr} -\frac{1}{10} & -\frac{2}{5} & 0 \\ \frac{3}{10} & \frac{1}{5} & 0 \\ -1 & -2 & 1 \end{array}\right]\right.$$

The last step in the process brings us to the matrix

$$\begin{bmatrix} 1 & 0 & -\frac{1}{2} \\ 0 & 1 & \frac{1}{2} \\ 0 & 0 & 0 \end{bmatrix}$$

whose last row is

$$[0 \quad 0 \quad 0]$$

This indicates that A has no inverse since it will be impossible to obtain a 1 in row 3 and column 3 while keeping the other two zeros in row 3.

PRACTICE PROBLEM Find A^{-1} if it exists for

$$A = \begin{bmatrix} 2 & 3 & 3 \\ 1 & -2 & 1 \\ 0 & 7 & 1 \end{bmatrix}.$$

Answer: A^{-1} does not exist. ■

We have already seen that systems of linear equations may be expressed by matrix equations. The system

$$2x - 3y = 1$$
$$x + 2y = 5$$

can be expressed as the single matrix equation

$$AX = C$$

where

$$A = \begin{bmatrix} 2 & -3 \\ 1 & 2 \end{bmatrix}, \quad X = \begin{bmatrix} x \\ y \end{bmatrix}, \quad C = \begin{bmatrix} 1 \\ 5 \end{bmatrix}$$

If A is nonsingular, A^{-1} exists and we may solve for X as follows:

$$A^{-1}AX = A^{-1}C$$
$$I_2X = A^{-1}C$$
$$X = A^{-1}C$$

Therefore, if the equation $AX = C$ has a solution, it is given as $X = A^{-1}C$. We can see that the solution is obtained by computing A^{-1} and then multiplying by C on the right. To illustrate how this works the computations will be performed for the given system. First compute A^{-1}:

$$\left[\begin{array}{rr|rr} 2 & -3 & 1 & 0 \\ 1 & 2 & 0 & 1 \end{array}\right] \xrightarrow{R_1 \to \frac{1}{2}R_1} \left[\begin{array}{rr|rr} 1 & -\frac{3}{2} & \frac{1}{2} & 0 \\ 1 & 2 & 0 & 1 \end{array}\right] \xrightarrow{R_2 \to -R_1 + R_2} \left[\begin{array}{rr|rr} 1 & -\frac{3}{2} & \frac{1}{2} & 0 \\ 0 & \frac{7}{2} & -\frac{1}{2} & 1 \end{array}\right]$$

$$\xrightarrow{R_2 \to \frac{2}{7}R_2} \left[\begin{array}{rr|rr} 1 & -\frac{3}{2} & \frac{1}{2} & 0 \\ 0 & 1 & -\frac{1}{7} & \frac{2}{7} \end{array}\right] \xrightarrow{R_1 \to \frac{3}{2}R_2 + R_1} \left[\begin{array}{rr|rr} 1 & 0 & \frac{2}{7} & \frac{3}{7} \\ 0 & 1 & -\frac{1}{7} & \frac{2}{7} \end{array}\right]$$

$$\qquad\qquad I_2 \qquad\qquad\quad A^{-1}$$

The solution for this system is

$$X = A^{-1}C = \begin{bmatrix} \frac{2}{7} & \frac{3}{7} \\ -\frac{1}{7} & \frac{2}{7} \end{bmatrix} \begin{bmatrix} 1 \\ 5 \end{bmatrix} = \begin{bmatrix} \frac{2}{7} + \frac{15}{7} \\ -\frac{1}{7} + \frac{10}{7} \end{bmatrix} = \begin{bmatrix} \frac{17}{7} \\ \frac{9}{7} \end{bmatrix}$$

that is, $x = \frac{17}{7}$, $y = \frac{9}{7}$. To check we have

$$2x - 3y = 2 \cdot \tfrac{17}{7} - 3 \cdot \tfrac{9}{7} = \tfrac{34}{7} - \tfrac{27}{7} = \tfrac{7}{7} = 1$$

$$x + 2y = \tfrac{17}{7} + 2 \cdot \tfrac{9}{7} = \tfrac{17}{7} + \tfrac{18}{7} = \tfrac{35}{7} = 5$$

Solving a matrix equation using inverses

EXAMPLE 4 Solve the matrix equation $AX = C$ where

$$A = \begin{bmatrix} 1 & 1 & 0 \\ 1 & 0 & 1 \\ 0 & 1 & 1 \end{bmatrix}, \qquad X = \begin{bmatrix} x_1 \\ x_2 \\ x_3 \end{bmatrix}, \qquad C = \begin{bmatrix} 1 \\ -1 \\ 3 \end{bmatrix}.$$

Solution

$$
\begin{array}{c|c}
A & I_3 \\
\begin{bmatrix} 1 & 1 & 0 \\ 1 & 0 & 1 \\ 0 & 1 & 1 \end{bmatrix} & \begin{bmatrix} 1 & 0 & 0 \\ 0 & 1 & 0 \\ 0 & 0 & 1 \end{bmatrix}
\end{array}
$$

$R_2 \rightarrow -R_1 + R_2$

$$
\begin{array}{c|c}
\begin{bmatrix} 1 & 1 & 0 \\ 0 & -1 & 1 \\ 0 & 1 & 1 \end{bmatrix} & \begin{bmatrix} 1 & 0 & 0 \\ -1 & 1 & 0 \\ 0 & 0 & 1 \end{bmatrix}
\end{array}
$$

$R_1 \rightarrow R_2 + R_1$
$R_3 \rightarrow R_2 + R_3$

$$
\begin{array}{c|c}
\begin{bmatrix} 1 & 0 & 1 \\ 0 & -1 & 1 \\ 0 & 0 & 2 \end{bmatrix} & \begin{bmatrix} 0 & 1 & 0 \\ -1 & 1 & 0 \\ -1 & 1 & 1 \end{bmatrix}
\end{array}
$$

$R_3 \rightarrow (\frac{1}{2})R_3$
$R_2 \rightarrow -R_2$

$$
\begin{array}{c|c}
\begin{bmatrix} 1 & 0 & 1 \\ 0 & 1 & -1 \\ 0 & 0 & 1 \end{bmatrix} & \begin{bmatrix} 0 & 1 & 0 \\ 1 & -1 & 0 \\ -\frac{1}{2} & \frac{1}{2} & \frac{1}{2} \end{bmatrix}
\end{array}
$$

$R_1 \rightarrow -R_3 + R_1$
$R_2 \rightarrow R_3 + R_2$

$$
\begin{array}{c|c}
\begin{bmatrix} 1 & 0 & 0 \\ 0 & 1 & 0 \\ 0 & 0 & 1 \end{bmatrix} & \begin{bmatrix} \frac{1}{2} & \frac{1}{2} & -\frac{1}{2} \\ \frac{1}{2} & -\frac{1}{2} & \frac{1}{2} \\ -\frac{1}{2} & \frac{1}{2} & \frac{1}{2} \end{bmatrix} \\
I_3 & A^{-1}
\end{array}
$$

Therefore, the solution is

$$
X = A^{-1}C = \frac{1}{2}\begin{bmatrix} 1 & 1 & -1 \\ 1 & -1 & 1 \\ -1 & 1 & 1 \end{bmatrix}\begin{bmatrix} 1 \\ -1 \\ 3 \end{bmatrix}
= \frac{1}{2}\begin{bmatrix} (1 - 1 - 3) \\ (1 + 1 + 3) \\ (-1 - 1 + 3) \end{bmatrix} = \begin{bmatrix} -\frac{3}{2} \\ \frac{5}{2} \\ \frac{1}{2} \end{bmatrix}
$$

or $x_1 = -\frac{3}{2}$, $x_2 = \frac{5}{2}$, $x_3 = \frac{1}{2}$. To check this in the original equations gives

$$
AX = \begin{bmatrix} 1 & 1 & 0 \\ 1 & 0 & 1 \\ 0 & 1 & 1 \end{bmatrix}\begin{bmatrix} -\frac{3}{2} \\ \frac{5}{2} \\ \frac{1}{2} \end{bmatrix} = \begin{bmatrix} -\frac{3}{2} + \frac{5}{2} + 0 \\ -\frac{3}{2} + 0 + \frac{1}{2} \\ 0 + \frac{5}{2} + \frac{1}{2} \end{bmatrix} = \begin{bmatrix} 1 \\ -1 \\ 3 \end{bmatrix} = C
$$

PRACTICE PROBLEM Solve the matrix equation $AX = C$, where

$$A = \begin{bmatrix} 1 & -1 & 1 \\ 2 & -3 & 3 \\ 1 & -2 & 1 \end{bmatrix}, \quad X = \begin{bmatrix} x \\ y \\ z \end{bmatrix}, \quad C = \begin{bmatrix} 0 \\ -2 \\ -1 \end{bmatrix}.$$

Answer

$$A^{-1} = \begin{bmatrix} 3 & -1 & 0 \\ 1 & 0 & -1 \\ -1 & 1 & -1 \end{bmatrix}, \quad X = \begin{bmatrix} 2 \\ 1 \\ -1 \end{bmatrix}. \blacksquare$$

EXERCISES

Set A–Computing Inverses of Matrices

In Exercises 1 through 34, find the inverse of each of the given matrices.

1. $\begin{bmatrix} 1 & 0 \\ 1 & 1 \end{bmatrix}$ **2.** $\begin{bmatrix} 0 & 1 \\ 1 & 0 \end{bmatrix}$ **3.** $\begin{bmatrix} 1 & 5 \\ 0 & 1 \end{bmatrix}$

4. $\begin{bmatrix} 1 & -2 \\ 0 & 1 \end{bmatrix}$ **5.** $\begin{bmatrix} 1 & 4 \\ 0 & -1 \end{bmatrix}$ **6.** $\begin{bmatrix} -1 & 0 \\ -2 & 1 \end{bmatrix}$

7. $\begin{bmatrix} 1 & 5 \\ 7 & 36 \end{bmatrix}$ **8.** $\begin{bmatrix} 1 & -2 \\ -7 & 15 \end{bmatrix}$ **9.** $\begin{bmatrix} 1 & 4 \\ 3 & 11 \end{bmatrix}$

10. $\begin{bmatrix} 1 & 5 \\ -5 & -24 \end{bmatrix}$ **11.** $\begin{bmatrix} 1 & -2 \\ -8 & 17 \end{bmatrix}$ **12.** $\begin{bmatrix} 1 & 4 \\ 8 & 31 \end{bmatrix}$

13. $\begin{bmatrix} -1 & 1 \\ -1 & 1 \end{bmatrix}$ **14.** $\begin{bmatrix} 1 & 2 \\ 3 & 4 \end{bmatrix}$ **15.** $\begin{bmatrix} 1 & 2 \\ -2 & -1 \end{bmatrix}$

16. $\begin{bmatrix} 5 & 7 \\ 3 & -2 \end{bmatrix}$ **17.** $\begin{bmatrix} 1 & 0 & 0 \\ 0 & 1 & 0 \\ 0 & 0 & 1 \end{bmatrix}$ **18.** $\begin{bmatrix} 1 & 0 & 0 \\ 0 & 1 & 0 \\ 1 & 0 & 1 \end{bmatrix}$

19. $\begin{bmatrix} 1 & 0 & 1 \\ 0 & 1 & 0 \\ 0 & 1 & 1 \end{bmatrix}$ **20.** $\begin{bmatrix} -1 & -5 & 6 \\ 0 & -1 & -2 \\ 0 & 0 & -1 \end{bmatrix}$ **21.** $\begin{bmatrix} -1 & 1 & 0 \\ 0 & -1 & -3 \\ 0 & 0 & 1 \end{bmatrix}$

22. $\begin{bmatrix} -1 & -5 & 6 \\ 8 & 39 & -50 \\ 9 & 45 & -55 \end{bmatrix}$ **23.** $\begin{bmatrix} -1 & 1 & 0 \\ -7 & 6 & -3 \\ -8 & 8 & 1 \end{bmatrix}$ **24.** $\begin{bmatrix} -1 & -5 & 6 \\ 0 & -1 & -2 \\ 0 & 3 & 5 \end{bmatrix}$

25. $\begin{bmatrix} -1 & 1 & 0 \\ 5 & -6 & -3 \\ -6 & 6 & 1 \end{bmatrix}$ 26. $\begin{bmatrix} -1 & -5 & 6 \\ 0 & -1 & -2 \\ -4 & -20 & 23 \end{bmatrix}$ 27. $\begin{bmatrix} -7 & 7 & 1 \\ -1 & 0 & -2 \\ -6 & 6 & 1 \end{bmatrix}$

28. $\begin{bmatrix} 2 & 2 & 1 \\ 1 & 1 & 1 \\ 1 & 2 & 3 \end{bmatrix}$ 29. $\begin{bmatrix} 3 & 2 & 1 \\ 4 & 3 & 2 \\ 5 & 4 & 3 \end{bmatrix}$ 30. $\begin{bmatrix} 1 & -1 & 3 \\ 2 & 0 & -2 \\ -1 & -2 & -3 \end{bmatrix}$

31. $\begin{bmatrix} 0 & 1 & 0 & 1 \\ 1 & 0 & 1 & 0 \\ 0 & 1 & 0 & 1 \\ 1 & 0 & 1 & 0 \end{bmatrix}$ 32. $\begin{bmatrix} 1 & 1 & 1 & 0 \\ 1 & 1 & 0 & 1 \\ 1 & 0 & 1 & 1 \\ 0 & 1 & 1 & 1 \end{bmatrix}$

33. $\begin{bmatrix} 1 & 1 & 1 & 1 \\ -1 & 1 & -1 & 0 \\ -1 & 0 & 1 & 1 \\ -1 & -1 & -1 & 1 \end{bmatrix}$ 34. $\begin{bmatrix} 1 & 1 & 1 & 1 \\ -1 & 1 & -1 & 0 \\ -1 & 0 & 1 & -1 \\ -1 & 1 & -1 & 1 \end{bmatrix}$

Set B–Solving Systems of Linear Equations by Using Matrices

In Exercises 35 through 50 you are to express each system as $AX = B$, where A, X, B are matrices and solve by computing A^{-1}.

35. $\begin{aligned} x - y &= -9 \\ -7x + 8y &= 68 \end{aligned}$

36. $\begin{aligned} -13x + 7y &= 29 \\ -2x + y &= 4 \end{aligned}$

37. $\begin{aligned} x + 2y &= -1 \\ 8x + 15y &= -12 \end{aligned}$

38. $\begin{aligned} 14x + 3y &= -43 \\ 5x + y &= -15 \end{aligned}$

39. $\begin{aligned} x - y &= 7 \\ 7x - 6y &= 44 \end{aligned}$

40. $\begin{aligned} x - 5y &= 13 \\ 6x - 31y &= 80 \end{aligned}$

41. $\begin{aligned} -x + 4y &= 23 \\ 8x + 33y &= 206 \end{aligned}$

42. $\begin{aligned} 32x + 3y &= -79 \\ 11x + y &= -16 \end{aligned}$

43. $\begin{aligned} x + 9z &= -9 \\ 5x - y + 53z &= -12 \\ 4x + 36z &= -36 \end{aligned}$

44. $\begin{aligned} 6x + 21y + 7z &= 25 \\ 14x - 17y - 6z &= 36 \\ x - 3y - z &= 2 \end{aligned}$

45. $\begin{aligned} x + 2y + 3z &= -12 \\ 2x + 5y + 6z &= -23 \\ 5x + 10y + 16z &= -63 \end{aligned}$

46. $\begin{aligned} x + 8y - 5z &= 12 \\ 5x + 41y - 17z &= 130 \\ 9x + 72y - 46z &= 100 \end{aligned}$

47. $\begin{aligned} -x - 9y + 7z &= -22 \\ -4x - 35y + 32z &= -52 \\ 8x + 72y - 57z &= 169 \end{aligned}$

48. $\begin{aligned} 13x + 24y - 6z &= -250 \\ 10x + 31y - 8z &= -343 \\ -2x - 4y + z &= 47 \end{aligned}$

49. $\begin{aligned} 5x + 3y + 3z + 3w &= 2 \\ 5x + 3y + 2z + 2w &= 2 \\ 3x + y + w &= 1 \\ 2x + y + z + w &= 1 \end{aligned}$

50. $\begin{aligned} x - y + 2z - w &= 1 \\ 2x - y + 7z - w &= 5 \\ 3x - 3y + 5z + w &= -6 \\ 4x - 4y + 8z - 5w &= 6 \end{aligned}$

3.8 SOME APPLICATIONS OF MATRICES

Matrices are useful in the solutions of many practical problems. This section and the next are devoted to some of their applications.

As indicated earlier, matrices can be helpful in representing, in a compact and systematic way, the data which are related to a given problem. This is especially true for data stored in computers. As an example of this notational convenience we consider the inventory of mathematics texts in a college bookstore. To keep the problems simple suppose the bookstore stocks only the following texts in mathematics:

Mathematics 100

Mathematics 200

Mathematics 210

Mathematics 220

Mathematics 250

The inventory for mathematics texts may be represented by an "inventory vector":

$$M = [x_1 \quad x_2 \quad x_3 \quad x_4 \quad x_5].$$

The components of M give the number of texts which the bookstore has in stock of each kind:

$$M = \begin{array}{c} \\ [\end{array}\begin{array}{ccccc} M100 & M200 & M210 & M220 & M250 \\ x_1 & x_2 & x_3 & x_4 & x_5 \end{array}\begin{array}{c} \\ ]. \end{array}$$

At the close of business on a certain day, for example, M may be given as follows:

$$M = [49 \quad 18 \quad 32 \quad 97 \quad 30].$$

From this, one may quickly see, for instance, that the number of Math 210 texts on hand is 32.

In a similar way the number of each text sold on a certain day can be represented by a "sales vector," S. An example of S is

$$S = [2 \quad 1 \quad -1 \quad 3 \quad 0].$$

The vector S contains the following information:

2 Math 100 texts were sold

1 Math 200 text was sold

1 Math 210 text was *returned*

3 Math 220 texts were sold

0 Math 250 text was sold

Note that the third component of S is negative, and that this is interpreted to mean that a text was returned.

During the week of June 8 the bookstore recorded the following sales of mathematics texts:

$$\begin{aligned}
\text{Monday: } S_1 &= [2 \quad 1 \quad -1 \quad 3 \quad 0] \\
\text{Tuesday: } S_2 &= [0 \quad 2 \quad 0 \quad 1 \quad 4] \\
\text{Wednesday: } S_3 &= [2 \quad 0 \quad 1 \quad 5 \quad 2] \\
\text{Thursday: } S_4 &= [-1 \quad 3 \quad 2 \quad 1 \quad 5] \\
\text{Friday: } S_5 &= [0 \quad 0 \quad 1 \quad 2 \quad -2]
\end{aligned}$$

The total sales of mathematics texts for this week is the vector

$$\begin{aligned}
W &= S_1 + S_2 + S_3 + S_4 + S_5 \\
&= [3 \quad 6 \quad 3 \quad 12 \quad 9].
\end{aligned}$$

If the bookstore began the week with the inventory vector

$$M = [49 \quad 18 \quad 32 \quad 97 \quad 30],$$

then the vector

$$\begin{aligned}
R &= M - W \\
&= [49 \quad 18 \quad 32 \quad 97 \quad 30] - [3 \quad 6 \quad 3 \quad 12 \quad 9] \\
&= [46 \quad 12 \quad 29 \quad 85 \quad 21]
\end{aligned}$$

is the inventory vector at the end of the week.

Suppose the two vectors C and A given below contain the cost per text and the sales price per text of the given books:

$$C = \begin{bmatrix} 12.50 \\ 14.95 \\ 15.85 \\ 10.50 \\ 23.50 \end{bmatrix} \begin{matrix} \text{Math 100} \\ \text{Math 200} \\ \text{Math 210} \\ \text{Math 220} \\ \text{Math 250} \end{matrix} \qquad A = \begin{bmatrix} 18.75 \\ 22.43 \\ 23.78 \\ 15.75 \\ 35.25 \end{bmatrix} \begin{matrix} \text{Math 100} \\ \text{Math 200} \\ \text{Math 210} \\ \text{Math 220} \\ \text{Math 250} \end{matrix}$$

If $S = [151 \quad 126 \quad 83 \quad 91 \quad 118]$ is the sales vector for the month of September, then the number SC is the cost to the bookstore of the texts sold in that month and SA is the gross sales amount for these texts:

$$SC = [151 \quad 126 \quad 83 \quad 91 \quad 118]\begin{bmatrix} 12.50 \\ 14.95 \\ 15.85 \\ 10.50 \\ 23.50 \end{bmatrix}$$

$$= \$8815.25,$$

$$SA = [151 \quad 126 \quad 83 \quad 91 \quad 118]\begin{bmatrix} 18.75 \\ 22.43 \\ 23.78 \\ 15.75 \\ 35.25 \end{bmatrix}$$

$$= \$13{,}223.92.$$

From these figures, the profit P from mathematics texts can be determined for September:

$$\begin{aligned} P &= SA - SC \\ &= \$13{,}223.92 - \$8815.25 \\ &= \$4408.67. \end{aligned}$$

Matrices for storing data

EXAMPLE The Strong Desk Company manufactures three models of desks: the Executive, the Secretary, and the Student. The materials required for the construction of these desks are wood, metal, glue and finish. The following matrix R shows the amount (given in some convenient units) of each material needed for each desk.

$$R = \begin{array}{c} \begin{array}{cccc} \text{Wood} & \text{Metal} & \text{Glue} & \text{Finish} \end{array} \\ \begin{bmatrix} 20 & 2 & 32 & 3 \\ 15 & 1 & 20 & 2 \\ 10 & 3 & 0 & 1 \end{bmatrix} \end{array} \begin{array}{l} \\ \text{Executive} \\ \text{Secretary} \\ \text{Student} \end{array}$$

For example, it takes 20 units (row 2, column 3) of glue to construct one Secretary model. An order for these desks may be expressed as an "order vector"

$$X = [x_1 \quad x_2 \quad x_3]$$

where

x_1 is the number of Executive models ordered,

x_2 is the number of Secretary models ordered,

x_3 is the number of Student models ordered.

Let us now consider the product XR and its interpretation: XR is a 1-by-4 matrix, and the entry in the first row, first column is

$$y_1 = 20x_1 + 15x_2 + 10x_3,$$

which gives the total amount of wood needed to fill the order X. The other entries of XR are obtained in a similar way:

$$XR = [y_1 \quad y_2 \quad y_3 \quad y_4]$$

where

$y_1 =$ the amount of wood needed to fill order X,

$y_2 =$ the amount of metal needed to fill order X,

$y_3 =$ the amount of glue needed to fill order X,

$y_4 =$ the amount of finish needed to fill order X.

If an order vector

$$X = [20 \quad 35 \quad 50],$$

then

$$\begin{aligned} XR &= [20 \quad 35 \quad 50] \begin{bmatrix} 20 & 2 & 32 & 3 \\ 15 & 1 & 20 & 2 \\ 10 & 3 & 0 & 1 \end{bmatrix} \\ &= [1425 \quad 225 \quad 1340 \quad 180]. \end{aligned}$$

Thus, filling an order for 20 Executive, 35 Secretary and 50 Student models will require 1425 units of wood, 225 units of metal, 1340 units of glue, and 180 units of finish.

It is helpful to arrange the costs of the raw materials for these desks in a cost vector

$$Y = \begin{bmatrix} z_1 \\ z_2 \\ z_3 \\ z_4 \end{bmatrix}$$

where

$$z_1 = \text{cost/unit of wood},$$
$$z_2 = \text{cost/unit of metal},$$
$$z_3 = \text{cost/unit of glue},$$
$$z_4 = \text{cost/unit of finish}.$$

Then the product RY expresses the cost of raw materials for each kind of desk. For example, suppose

$$Y = \begin{bmatrix} 15.00 \\ 3.50 \\ 5.25 \\ 6.75 \end{bmatrix}.$$

Now the costs of raw materials for the desks are

$$RY = \begin{bmatrix} 20 & 2 & 32 & 3 \\ 15 & 1 & 20 & 2 \\ 10 & 3 & 0 & 1 \end{bmatrix} \begin{bmatrix} 15.00 \\ 3.50 \\ 5.25 \\ 6.75 \end{bmatrix} = \begin{bmatrix} 495.25 \\ 347.00 \\ 167.25 \end{bmatrix}.$$

These are the costs of raw materials for each kind of desk and do not include other possible costs such as labor, etc. We may now compute the total cost of raw materials to fill the order $X = [20 \quad 35 \quad 50]$. This may be accomplished by computing XRY:

$$X(RY) = [20 \quad 35 \quad 50] \begin{bmatrix} 495.25 \\ 347.00 \\ 167.25 \end{bmatrix} = \$30{,}412.50.$$

PRACTICE PROBLEM

(a) If $X = [15 \quad 25 \quad 40]$ is an order vector for desks, use the matrix R of this example to determine the amount of raw materials required for this order. Express this as an appropriate matrix product.

Answer: $XR = [1075 \quad 175 \quad 980 \quad 135]$.

(b) If Y is the cost vector and

$$Y = \begin{bmatrix} 13.50 \\ 3.25 \\ 4.95 \\ 6.20 \end{bmatrix}$$

use the matrix R of this example to determine the cost of raw materials for each kind of desk. Express your answer as an appropriate matrix product.

Answer:

$$RY = \begin{bmatrix} 453.50 \\ 317.15 \\ 150.95 \end{bmatrix}.$$

(c) Use X as in (a) and Y as in (b) and determine the total costs of raw materials to fill the order X. Use an appropriate matrix product.

Answer: $XRY = \$20{,}769.25$. ■

EXERCISES

Matrices for storing data

1. The Strong Desk Company produces its three models of desks–Executive, Secretary, Student–in four finishes: maple, oak, walnut and teak. The matrix D given below shows the numbers of desks produced in the month of December, where the rows indicate models (Executive, Secretary, Student in that order) and columns indicate finishes (maple, oak, walnut, teak in that order).

$$D = \begin{bmatrix} 12 & 8 & 9 & 4 \\ 15 & 5 & 10 & 2 \\ 34 & 29 & 3 & 0 \end{bmatrix}.$$

(a) How many Executive models were made of walnut?
(b) How many Student models were made of teak?
(c) How many Executive models were made in December?
(d) How many Secretary models were made of oak?
(e) How many Secretary models were made in December?
(f) How many Student models were made in December?

(g) How many desks were made of teak?
(h) How many desks were made of oak?

Matrices for storing data

2. In the month of December the Strong Desk Company received the following order for its three models in the four finish types available:

Number Ordered

Finish	**Exec.**	**Sec.**	**Stu.**
Maple	30	36	50
Oak	20	24	40
Walnut	20	20	42
Teak	10	8	6

The Company estimates that it will take 2 months to fill such a large order. If half the order is filled in January, use a matrix of the type given in Exercise 1 to represent the numbers of desks to be produced in January.

Inventory control

3. The College Bookstore represents its inventory of foreign language texts as a 1-by-5 row vector

$$T = [t_1 \quad t_2 \quad t_3 \quad t_4 \quad t_5]$$

where

t_1 = the number of French texts,

t_2 = the number of German texts,

t_3 = the number of Spanish texts,

t_4 = the number of Russian texts,

t_5 = the number of Italian texts.

(a) If the inventory vector at the close of business on June 4 is

$$T = [151 \quad 212 \quad 318 \quad 108 \quad 132]$$

and if

$$S = [35 \quad 20 \quad 42 \quad 15 \quad 12]$$

represents sales of the texts on June 5, what is the inventory vector at the close of business on June 5 if no new texts were added to the bookstore's stock?

(b) On June 6 the bookstore's sales vector of foreign language texts was

$$S = [12 \quad 15 \quad 18 \quad 10 \quad 3],$$

and new stock was received as indicated by the following vector:

$$N = [300 \quad 325 \quad 450 \quad 200 \quad 150].$$

What is the inventory vector at the close of business on June 6?

Matrices for storing data

4. The Strong Desk Company produces desks at two plants, one in Filmore and the other in Warrenton. The two matrices A and B given below show the Company's production of desks in the first half of 1984 (A) and the second half of that year (B). The rows of these matrices denote the Company's desk models: Executive, Secretary, Student respectively. The first column of the matrices gives information about the Filmore plant and the second about the Warrenton plant.

$$A = \begin{bmatrix} 157 & 218 \\ 309 & 227 \\ 684 & 459 \end{bmatrix}, \qquad B = \begin{bmatrix} 207 & 312 \\ 283 & 232 \\ 511 & 423 \end{bmatrix}$$

(a) Use an appropriate matrix to show the Company's desk production for the year 1984.
(b) How many Student desks did the Company produce at the Filmore plant in 1984?
(c) How many Executive models did the Company produce in 1984?
(d) How does the production of desks at the Filmore plant in the first half compare with that plant's production in the second half?
(e) How many desks were produced at the Warrenton Plant in 1984?
(f) How many desks did the Company produce in 1984?
(g) The Company wishes to increase its production in 1985 by 10% over that of 1984. Represent the Company's production schedule of desks by an appropriate matrix.

Matrices for storing data

5. Three departments are involved in making desks for the Strong Desk Company: the cabinet department, the finishing department, and the crating department. The matrix W shows the man-hours of work/desk in each of these departments, and the matrix C shows the hourly wages in these

departments at the Company's two plants in Filmore and Warrenton.

$$W = \begin{array}{c} \\ \\ \\ \end{array}\begin{bmatrix} 15 & 3 & 0.5 \\ 10 & 2.5 & 0.5 \\ 2 & 0.4 & 0.2 \end{bmatrix}\begin{array}{l} \text{Executive} \\ \text{Secretary} \\ \text{Student} \end{array}$$

(columns: Cabinet, Finishing, Crating)

$$C = \begin{bmatrix} 12.50 & 11.75 \\ 7.50 & 6.25 \\ 4.25 & 3.75 \end{bmatrix}\begin{array}{l} \text{Cabinet} \\ \text{Finishing} \\ \text{Crating} \end{array}$$

(columns: Filmore, Warrenton)

(a) Compute the labor costs of producing an Executive desk at the Filmore plant.
(b) Compute the labor costs of producing a Student desk at the Warrenton plant.
(c) Let A be the matrix product WC. What are the dimensions of A?
(d) If $A = WC = [a_{ij}]$, interpret each of the entries a_{ij}. For example, $a_{11}(= \$212.13)$ is the labor cost of producing an Executive desk at the Filmore plant.
(e) Suppose that labor costs increase at both plants by 8%. Represent the new labor costs by an appropriate matrix.
(f) If C' is the new labor cost matrix after the increase of 8%, compute WC'.

Matrices for storing data

Use the following information in Exercises 6 through 8. John Troll plans some improvements to his home which will require lumber, cement, and nails. John estimates the quantities needed to be as follows:

200 board-feet of lumber,
100 lb of cement,
10 lb of nails.

6. Represent John's construction needs as a row vector.
7. Two home builders' supply stores are conveniently located relative to John's house: Ace Home and Supply and Buildarama. The costs of the materials needed by John at the two stores are shown below:

	Ace	**Buildarama**
Lumber (per board-foot)	\$3.95	\$4.05
Cement (per pound)	0.23	0.18
Nails (per pound)	0.33	0.31

Represent the costs at the two stores by a matrix.

8. Show how to represent John's costs for this construction project by an appropriate matrix product. Determine the total cost at each store and decide which store will supply the material at the lower cost.

Data storage: nutrition

9. The matrix shows the vitamin content of the indicated foods (in some convenient units).

$$X = \begin{array}{c} \begin{array}{cccc} \text{A} & \text{B} & \text{C} & \text{D} \end{array} \\ \begin{bmatrix} 0.01 & 0 & 0.1 & 0.1 \\ 0.2 & 0 & 0.2 & 0 \\ 0.1 & 0 & 0.05 & 0 \\ 0.3 & 0.05 & 0.1 & 0.2 \end{bmatrix} \end{array} \begin{array}{l} \text{Tomatoes} \\ \text{Squash} \\ \text{Lettuce} \\ \text{Beets} \end{array}$$

(a) Suppose the following amounts of these foods are eaten:

1.5 units of tomatoes,

1 unit of squash,

0.5 units of lettuce,

2 units of beets.

Represent the above data as a matrix and determine the amount of each vitamin ingested by an appropriate matrix product.

(b) Vitamin costs in dollars are:

0.20/unit of A

0.50/unit of B

0.30/unit of C

0.40/unit of D

If the amounts of foods given in (a) are eaten, what is the cost for each vitamin? What is the total cost for vitamins?

Data storage: nutrition

10. The following matrix shows the number of units of the nutrients protein (P), carbohydrates (C), fat (F) contained in the soups X, Y, Z.

$$A = \begin{array}{c} \begin{array}{ccc} \text{P} & \text{C} & \text{F} \end{array} \\ \begin{bmatrix} 2 & 20 & 2 \\ 2 & 11 & 8 \\ 12 & 32 & 5 \end{bmatrix} \end{array} \begin{array}{l} X \\ Y \\ Z \end{array}$$

(a) If a meal consists of 30 units of X, 40 units of Y, and 20 units of Z, how much of each of the nutrients P, C, F will be consumed?

(b) What is the cost of each nutrient in the meal in part (a) if

P cost 0.05/unit,

C cost 0.01/unit,

F cost 0.005/unit?

(c) Determine the total cost of the nutrients in the meal described in part (a).

3.9 SOME FURTHER APPLICATIONS OF LINEAR SYSTEMS—MATRICES

Several applications of systems of linear equations were discussed in Chapter 2. In this section we consider some additional examples. These examples will give us the opportunity to examine some problems which are slightly different from those encountered before and at the same time will permit us to use matrix methods in their solutions.

Optimum scheduling

EXAMPLE 1 The Strong Desk Company—as we have seen—manufactures three models of desks: Executive, Secretary, Student. Each desk must be scheduled in each of the following three departments for processing: cabinet, finishing, and crating. The matrix P below shows the number of man-hours per desk required in each department:

$$\begin{array}{cccc} \text{Exec.} & \text{Sec.} & \text{Stu.} & \\ \end{array}$$
$$P = \begin{bmatrix} 15 & 3 & 0.5 \\ 10 & 2.5 & 0.5 \\ 2 & 0.4 & 0.2 \end{bmatrix} \begin{array}{l} \text{Cabinet} \\ \text{Finishing} \\ \text{Crating} \end{array}$$

Suppose that 70 hours, 55 hours, and 12 hours are the maximum times available in the cabinet, finishing, and crating departments, respectively. This information may be conveniently displayed in a *data box* as shown below:

	Exec.	Sec.	Stu.	Restrictions
Cabinet	15	3	0.5	70
Finishing	10	2.5	0.5	55
Crating	2	0.4	0.2	12

If the plant is to run at full capacity, how many desks of each kind should be scheduled each day? To answer this question let x_1, x_2, x_3 be the number of Executive, Secretary, and Student desks, respectively, to be produced each day. The variables x_1, x_2, x_3 may be placed in the data box like this:

x_1	x_2	x_3	
15	3	0.5	70
10	2.5	0.5	55
2	0.4	0.2	12

From the data box, it is easy to see that

$$15x_1 + 3x_2 + 0.5x_3$$

is the number of hours required in the cabinet department to make x_1 Executive, x_2 Secretary, and x_3 Student desks. The cabinet department has only 70 hours available and the plant must operate at full capacity. Therefore,

$$15x_1 + 3x_2 + 0.5x_3 = 70.$$

In a similar way we obtain two other equations. This leads to the following linear system

$$\begin{aligned} 15x_1 + 3x_2 \quad + 0.5x_3 &= 70, \\ 10x_1 + 2.5x_2 + 0.5x_3 &= 55, \\ 2x_1 + 0.4x_2 + 0.2x_3 &= 12. \end{aligned}$$

This system will be solved by using the augmented matrix.

$$\left[\begin{array}{ccc|c} 15 & 3 & 0.5 & 70 \\ 10 & 2.5 & 0.5 & 55 \\ 2 & 0.4 & 0.2 & 12 \end{array}\right]$$

$R_1 \to 10R_1$
$R_2 \to 10R_2$
$R_3 \to 10R_3$

$$\left[\begin{array}{ccc|c} 150 & 30 & 5 & 700 \\ 100 & 25 & 5 & 550 \\ 20 & 4 & 2 & 120 \end{array}\right]$$

$R_1 \to (\frac{1}{5})R_1$

$$\left[\begin{array}{ccc|c} 30 & 6 & 1 & 140 \\ 100 & 25 & 5 & 550 \\ 20 & 4 & 2 & 120 \end{array}\right]$$

$R_2 \to -5R_1 + R_2$
$R_3 \to -2R_1 + R_3$

$$\left[\begin{array}{rrr|r} 30 & 6 & 1 & 140 \\ -50 & -5 & 0 & -150 \\ -40 & -8 & 0 & -160 \end{array}\right] \quad R_2 \to \left(-\tfrac{1}{5}\right)R_2$$

$$\left[\begin{array}{rrr|r} 30 & 6 & 1 & 140 \\ 10 & 1 & 0 & 30 \\ -40 & -8 & 0 & -160 \end{array}\right] \quad \begin{array}{l} R_1 \to -6R_2 + R_1 \\ R_3 \to 8R_2 + R_3 \end{array}$$

$$\left[\begin{array}{rrr|r} -30 & 0 & 1 & -40 \\ 10 & 1 & 0 & 30 \\ 40 & 0 & 0 & 80 \end{array}\right] \quad R_3 \to \left(\tfrac{1}{40}\right)R_3$$

$$\left[\begin{array}{rrr|r} -30 & 0 & 1 & -40 \\ 10 & 1 & 0 & 30 \\ 1 & 0 & 0 & 2 \end{array}\right] \quad \begin{array}{l} R_2 \to -10R_3 + R_2 \\ R_1 \to 30R_3 + R_1 \end{array}$$

$$\left[\begin{array}{rrr|r} 0 & 0 & 1 & 20 \\ 0 & 1 & 0 & 10 \\ 1 & 0 & 0 & 2 \end{array}\right]$$

Hence, $x_1 = 2$, $x_2 = 10$, $x_3 = 20$. Each day the plant should schedule the production of 2 Executive, 10 Secretary, and 20 Student desks.

Optimum scheduling

EXAMPLE 2 The plant manager for the Strong Desk Company wishes to determine the proper schedule for desks so that the cabinet and finishing departments will operate at full capacity (because of changes in the departments, the crating department can now process essentially all the desks produced in the other departments). Assuming the hours available in the cabinet and finishing departments to be the same as in Example 1 and using x_1, x_2, x_3 as before leads to the following data box:

x_1	x_2	x_3	
15	3	0.5	70
10	2.5	0.5	55

The corresponding equations are

$$15x_1 + 3x_2 + 0.5x_3 = 70,$$
$$10x_1 + 2.5x_2 + 0.5x_3 = 55,$$

and the augmented matrix is

$$\left[\begin{array}{ccc|c} 15 & 3 & 0.5 & 70 \\ 10 & 2.5 & 0.5 & 55 \end{array}\right].$$

To solve the system it is convenient to start by multiplying both rows by 10:

$$\left[\begin{array}{ccc|c} 150 & 30 & 5 & 700 \\ 100 & 25 & 5 & 550 \end{array}\right] \xrightarrow{R_1 \to (\frac{1}{5})R_1} \left[\begin{array}{ccc|c} 30 & 6 & 1 & 140 \\ 100 & 25 & 5 & 550 \end{array}\right] \xrightarrow{R_2 \to -5R_1 + R_2} \left[\begin{array}{ccc|c} 30 & 6 & 1 & 140 \\ -50 & -5 & 0 & -150 \end{array}\right]$$

$$\xrightarrow{R_2 \to (-\frac{1}{5})R_2} \left[\begin{array}{ccc|c} 30 & 6 & 1 & 140 \\ 10 & 1 & 0 & 30 \end{array}\right] \xrightarrow{R_1 \to -6R_2 + R_1} \left[\begin{array}{ccc|c} -30 & 0 & 1 & -40 \\ 10 & 1 & 0 & 30 \end{array}\right]$$

The pivot process terminates here since both rows have been used as pivots. The resulting equations are

$$\begin{aligned} -30x_1 + x_3 &= -40, \\ 10x_1 + x_2 &= 30, \end{aligned}$$

or

$$x_2 = 30 - 10x_1, \tag{1}$$

$$x_3 = -40 + 30x_1. \tag{2}$$

constraints

The two equations show that x_1 can be *any* real number and if x_2 and x_3 are calculated as indicated, a solution to the system will result. If we are only interested in solutions without regard to their practical implications, then nothing more need be done. But we are interested in more. The variables represent numbers of physical objects and as such are subject to appropriate *constraints.* We can produce neither negative nor fractional parts of desks. Therefore, we impose two kinds of constraints on x_1, x_2, x_3:

Constraint 1: $x_1 \geq 0$, $x_2 \geq 0$, $x_3 \geq 0$.

Constraint 2: x_1, x_2, x_3 must be whole numbers.

Using constraint 1 and Eqs. (1) and (2) gives

$$x_1 \geq 0,$$
$$30 - 10x_1 \geq 0,$$
$$-40 + 30x_1 \geq 0.$$

When these inequalities are solved, we have

$$x_1 \geq 0,$$
$$3 \geq x_1,$$
$$x_1 \geq \frac{4}{3},$$

or

$$\frac{4}{3} \leq x_1 \leq 3. \tag{3}$$

By constraint 2, x_1 must be a whole number, and when this is combined with (3) we see that

$$x_1 = 2 \qquad \text{or} \qquad x_1 = 3.$$

There are only two possible solutions to the system which are of interest to the plant manager:

Solution 1	*Solution 2*
$x_1 = 2$	$x_1 = 3$
$x_2 = 30 - 10x_1 = 10$	$x_2 = 30 - 10x_1 = 0$
$x_3 = -40 + 30x_1 = 20$	$x_3 = -40 + 30x_1 = 50$

Thus the manager has two possible schedule vectors:

$$[2 \quad 10 \quad 20] \quad \text{and} \quad [3 \quad 0 \quad 50].$$

Optimum scheduling

EXAMPLE 3 For the Strong Desk Company the matrix P shows the man-hours required in the three departments to make each of its three desks:

$$P = \begin{array}{c} \text{Exec.} \quad \text{Sec.} \quad \text{Stu.} \\ \begin{bmatrix} 15 & 3 & 0.5 \\ 10 & 2.5 & 0.5 \\ 2 & 0.4 & 0.2 \end{bmatrix} \end{array} \begin{array}{l} \text{Cabinet} \\ \text{Finishing} \\ \text{Crating} \end{array}$$

The Company operates three plants A, B, C, and the matrix P reflects time requirements in all three plants. The times available in the various departments at each plant are given below.

	Plant		
	A	B	C
Cabinet	50	105	95
Finishing	40	85	75
Crating	8	18	18

For example, in the cabinet department of plant B there are 105 hours available each day. All this information is conveniently summarized in a data box.

	Exec.	Sec.	Stu.	A	B	C
Cabinet	15	3	0.5	50	105	95
Finishing	10	2.5	0.5	40	85	75
Crating	2	0.4	0.2	8	18	18

We now wish to know the manufacturing schedule of desks at each plant for full production. In other words, we need to know how many desks of each kind should be scheduled for production each day at each of the three plants. First, consider plant A and let x_1, x_2, x_3 be the number of Executive, Secretary, and Student desks to be scheduled at plant A each day for full production. Just as in Example 1, we are led to the following system:

Plant A
$$\begin{aligned} 15x_1 + 3x_2 + 0.5x_3 &= 50, \\ 10x_1 + 2.5x_2 + 0.5x_3 &= 40, \\ 2x_1 + 0.4x_2 + 0.2x_3 &= 8. \end{aligned}$$

For plants B and C, the systems to solve are

Plant B
$$\begin{aligned} 15x_1 + 3x_2 + 0.5x_3 &= 105, \\ 10x_1 + 2.5x_2 + 0.5x_3 &= 85, \\ 2x_1 + 0.4x_2 + 0.2x_3 &= 18, \end{aligned}$$

Plant C
$$\begin{aligned} 15x_1 + 3x_2 + 0.5x_3 &= 95, \\ 10x_1 + 2.5x_2 + 0.5x_3 &= 75, \\ 2x_1 + 0.4x_2 + 0.2x_3 &= 18. \end{aligned}$$

Observe that the left-hand sides of all three systems are the same and that P is the coefficient matrix for each. Now let

$$X = \begin{bmatrix} x_1 \\ x_2 \\ x_3 \end{bmatrix}, \qquad A = \begin{bmatrix} 50 \\ 40 \\ 8 \end{bmatrix}, \qquad B = \begin{bmatrix} 105 \\ 85 \\ 18 \end{bmatrix}, \qquad C = \begin{bmatrix} 95 \\ 75 \\ 18 \end{bmatrix}.$$

Then the three systems (A), (B), (C) may be written as

$$PX = A, \qquad PX = B, \qquad PX = C.$$

If P^{-1} exists, the solutions to the three equations are

$$X = P^{-1}A, \qquad X = P^{-1}B, \qquad X = P^{-1}C. \tag{4}$$

Thus the problem can be solved by computing P^{-1} and carrying out the multiplications in (4). Calculating P^{-1} gives

$$P^{-1} = \begin{bmatrix} 0.3 & -0.4 & 0.25 \\ -1 & 2 & -2.5 \\ -1 & 0 & 7.5 \end{bmatrix}.$$

Then

$$P^{-1}A = \begin{bmatrix} 0.3 & -0.4 & 0.25 \\ -1 & 2 & -2.5 \\ -1 & 0 & 7.5 \end{bmatrix}\begin{bmatrix} 50 \\ 40 \\ 8 \end{bmatrix} = \begin{bmatrix} 1 \\ 10 \\ 10 \end{bmatrix},$$

$$P^{-1}B = \begin{bmatrix} 0.3 & -0.4 & 0.25 \\ -1 & 2 & -2.5 \\ -1 & 0 & 7.5 \end{bmatrix}\begin{bmatrix} 105 \\ 85 \\ 18 \end{bmatrix} = \begin{bmatrix} 2 \\ 20 \\ 30 \end{bmatrix},$$

$$P^{-1}C = \begin{bmatrix} 0.3 & -0.4 & 0.25 \\ -1 & 2 & -2.5 \\ -1 & 0 & 7.5 \end{bmatrix}\begin{bmatrix} 95 \\ 75 \\ 18 \end{bmatrix} = \begin{bmatrix} 3 \\ 10 \\ 40 \end{bmatrix}.$$

Therefore, for plant A we read the matrix

$$P^{-1}A = \begin{bmatrix} 1 \\ 10 \\ 10 \end{bmatrix} \begin{matrix} \text{Exec.} \\ \text{Sec.} \\ \text{Stu.} \end{matrix}$$

To operate at full capacity, plant A should schedule 1 Executive, 10 Secretary, and 10 Student desks for production each day. In the same way we can read the production schedule for the other two plants.

EXERCISES

For these exercises matrix methods should be used whenever appropriate.

The following information should be used in Exercises 1 through 6.

Optimum scheduling

The Ace Sporting Company makes three models of tennis racquets: the Junior, the Competition, the Pro. Three departments are involved in processing these racquets: cutting, laminating, and finishing. The man-hours required to make the Junior model are 1, 1, 1.5 in the cutting, laminating, and finishing departments, respectively. For the Competition model the man-hours required are 1, 1.5, 2 and for the Pro model the time requirements are 1.5, 2, 2.5.

1. Construct a data box for this information with the rows assigned to departments and the columns to models.
2. Suppose that each day 150 man-hours are available in the cutting department, 200 man-hours in the laminating department, and 260 man-hours in the finishing department. Add this information to your data box of Exercise 1 and determine how many racquets of each kind should be scheduled each day so that the plant operates at full capacity.
3. Assume that the plant manager is interested in scheduling for the cutting and finishing departments only and that the man-hours available are those of Exercise 2. What are the possible numbers of each kind of racquet that can be scheduled so that the departments produce at full capacity? What is the maximum number of Pro models that can be made each day?
4. Suppose that the cutting, laminating, and finishing departments have 76, 103, and 140 man-hours available, respectively. Work Exercise 2 with this new information.
5. Assume that the times available in cutting and laminating are those given in Exercise 4. If the plant manager is interested in scheduling for these two departments only, what are the possible schedules that will ensure operation at full capacity?
6. Suppose that the Ace Sporting Company has three plants D, E, F and the man-hours available in the three departments of each plant are as follows:

	D	E	F
Cutting	37	45	48
Laminating	48	57	60
Finishing	65	77	81

By computing the inverse of an appropriate matrix determine schedules for each plant so that full capacity is achieved.

Investment

7. An investment firm offers three kinds of investment certificates: I, II, III. Each certificate is composed of three grades of stocks: HR (high risk), MR (moderate risk), and LR (little risk). The data box below shows the percent composition for each kind of certificate.

	I	II	III
HR	10	70	20
MR	30	20	60
LR	60	10	20

A customer wishes to invest \$100,000 in these certificates so that 36% is in HR, 32% is in MR, and 32% is in LR. How much of the \$100,000 should be invested in each of I, II, III to achieve this?

Optimum scheduling

8. The Executive Training Personnel Agency (ETPA) trains four types of business executives:

L/R: Low profile —reserved

L/A: Low profile —aggressive

H/R: High profile—reserved

H/A: High profile—aggressive

Each student at ETPA must receive training in each of three departments: Ethics, Social Poise, Business Science. The following data box indicates the units of training in the departments for each kind of student.

	L/R	L/A	H/R	H/A
E	1	1	0.5	0
SP	2	2	3	4
BS	2	2.5	1	6

Each month department E can deliver 23 units, department SP can deliver 110 units of instruction, and department BS can deliver 99 units. ETPA receives more applications than it can accept and must therefore control the admissions process so that departments are fully utilized without being overcrowded. To achieve this goal, how many students of each kind should be accepted per month?

3.10 FURTHER APPLICATIONS—INPUT-OUTPUT ANALYSIS (OPTIONAL)

input-output

In 1973 the Nobel Prize for Economics was awarded to Wassily Leontief of Harvard University for his studies of the American economy. Leontief's analysis focuses on the interrelation of the various sectors of the economy and makes use of so-called *input-output* matrices. No attempt will be made here to supply all the details associated with this subject, but a simple and brief version is presented in an attempt to explain some of the important concepts.

Since the economy is very complex, a proper study would involve a large number of sectors (Leontief's original work used 500 sectors). For simplicity's sake our examples will be confined to small units of the economy and involve no more than four sectors.

Each sector of an economy may be viewed as both a producer and a consumer. For example, the energy sector produces energy (oil, electricity, etc.) and consumes various products as a result of its production. In order to produce electricity from a hydroelectric dam, the energy sector consumes quantities of steel and concrete to build the dam. But energy must be consumed in order to produce energy. Thus the energy sector consumes some of its own production.

We will speak of a sector's production as *output* and it's consumption as *input*.

$$\text{Production} \leftrightarrow \text{Output}$$
$$\text{Consumption} \leftrightarrow \text{Input}$$

As an example consider the following hypothetical (and oversimplified) case in which the sectors chosen are energy and steel. The table below shows the inputs and outputs of these two relative to each other; it does not show how they relate to the other sectors of the economy.

TABLE 1

Outputs \ Inputs	Energy	Steel
Energy	25	40
Steel	30	10

The numbers in the table are assumed to represent worth in some convenient units such as megabucks. The entries are interpreted as follows. From the output of steel, for example, 30 units were consumed by the energy sector and 10 units were consumed by steel itself.

Now consider the parts of the economy other than energy and steel. Demands from this part of the economy are made on energy and steel and should be reflected in our table. It must also be realized that energy and steel are consumers outside of the energy-steel sector. Thus our table should take into account some additional production and consumption from energy and steel. Additional production will be called *other production factors* and additional consumption will be called *final demand.* With these elements considered, Table 1 now appears as Table 2.

TABLE 2

Outputs \ Inputs	Energy	Steel	Final demand	Total
Energy	25	40	325	390
Steel	30	10	120	160
Other production factors	335	110		
Total	390	160		

Note in Table 2 that the totals in each sector's row and column are the same. This reflects the fact that we are constructing a model which assumes that *all production* is *consumed.*

Assumption: All production is consumed.

The entry in row 1, column 2 of Table 2 is 40, which means that energy produced 40 units of output that was consumed as input by steel. In other words, for each dollar that steel spends on input, \$0.25 $(\frac{40}{160})$ goes for energy. Other calculations of this sort from Table 2 lead to the following matrix which is called the *technological matrix* for these two sectors:

technological matrix

$$A = \begin{bmatrix} 0.064 & 0.250 \\ 0.077 & 0.063 \end{bmatrix}.$$

Now let X be the total output matrix:

$$X = \begin{bmatrix} 390 \\ 160 \end{bmatrix}.$$

The matrix product AX gives the total output of energy and steel which goes to energy and steel as input:

$$\begin{aligned} AX &= \begin{bmatrix} 0.064 & 0.250 \\ 0.077 & 0.063 \end{bmatrix} \begin{bmatrix} 390 \\ 160 \end{bmatrix} \\ &= \begin{bmatrix} (0.064)390 + (0.250)160 \\ (0.077)390 + (0.063)160 \end{bmatrix} \\ &= \begin{bmatrix} 64.96 \\ 40.11 \end{bmatrix}. \end{aligned}$$

The entries in this final matrix round to 65 and 40—the correct numbers from Table 2. The matrix AX is called *internal production.* The word "internal" here means internal to the energy-steel sector.

We can see from Table 2 that the quantities total production, internal production, and final demand are related by the equation

Total production = Internal production + Final demand.

If we let D be the final demand matrix, then the above equation may be expressed in terms of A, X, and D:

$$X = AX + D. \tag{1}$$

Equation (1) is a goal which we have been seeking. Its importance will be explained by solving for X:

$$\begin{aligned} X - AX &= D, \\ IX - AX &= D \quad (I \text{ is the identity matrix}), \\ (I - A)X &= D, \\ X &= (I - A)^{-1}D. \end{aligned} \tag{2}$$

In the above solution we assume that $I - A$ has an inverse.

Leontief matrix

Equation (2) expresses X (total production) in terms of $I - A$ and D (final demand). The matrix $I - A$ is called the *Leontief matrix.* The importance of Eq. (2) stems from its use as a method of forecasting total production. If the technological matrix A remains the same but final demand D changes, then (2) can be used to compute the total production X needed for this new demand.

The energy-steel example is an instance of an *open Leontief model* (closed models are not considered here) and is typical of the open model which may be summarized as follows:

THE OPEN LEONTIEF MODEL

1. Sectors of an economy are described by a technological matrix

$$A = [a_{ij}]$$

where the entry in the ith row, jth column gives the amount/dollar which sector j consumes as input from the production (output) of sector i.

2. X is the total production vector of the given sectors.
3. D is the final demand of the sectors.
4. The matrices A, X, D are related by the equation

$$X = (I - A)^{-1}D.$$

Technological matrix

EXAMPLE 1 Assume we have three sectors U, V, W of an economy which are related as shown in the table:

Outputs \ Inputs	U	V	W	Final demand	Total
U	60	80	75	125	340
V	20	15	25	80	140
W	30	35	20	105	190
Other production factors	230	10	70		
Total	340	140	190		

Compute (a) the technological matrix A and (b) the Leontief matrix $I - A$.

Solution

(a) The matrix A comes from the U, V, W part of the given table only, that is from

$$\begin{bmatrix} 60 & 80 & 75 \\ 20 & 15 & 25 \\ 30 & 35 & 20 \end{bmatrix}.$$

The dollar amounts of this table must be converted to amounts/dollar of input to the three sectors. Since sector U, for example, has input of 340, each entry of column 1 must be divided by 340. Similarly column 2 is divided by 140 and column 3 by 190. All entries are rounded to three places:

$$A = \begin{bmatrix} 0.176 & 0.571 & 0.395 \\ 0.059 & 0.107 & 0.132 \\ 0.088 & 0.250 & 0.105 \end{bmatrix}.$$

(b) The Leontief matrix $I - A$ is

$$I - A = \begin{bmatrix} 1 & 0 & 0 \\ 0 & 1 & 0 \\ 0 & 0 & 1 \end{bmatrix} - \begin{bmatrix} 0.176 & 0.571 & 0.395 \\ 0.059 & 0.107 & 0.132 \\ 0.088 & 0.250 & 0.105 \end{bmatrix}$$
$$= \begin{bmatrix} 0.824 & -0.571 & -0.395 \\ -0.059 & 0.893 & -0.132 \\ -0.088 & -0.250 & 0.895 \end{bmatrix}.$$

PRACTICE PROBLEM Given the following table compute (a) the technological matrix and (b) the Leontief matrix.

Outputs \ Inputs	U	V	W	Final demand	Total
U	50	80	90	390	610
V	60	90	10	380	540
W	50	50	20	450	570
Other production factors	450	320	450		
Total	610	540	570		

Answer

(a)
$$A = \begin{bmatrix} 0.082 & 0.148 & 0.158 \\ 0.098 & 0.167 & 0.018 \\ 0.082 & 0.093 & 0.035 \end{bmatrix}.$$

(b)
$$I - A = \begin{bmatrix} 0.918 & -0.148 & -0.158 \\ -0.098 & 0.833 & -0.018 \\ -0.082 & -0.093 & 0.965 \end{bmatrix}. \blacksquare$$

Forecasting output

EXAMPLE 2 Assume we have two sectors U, V, of an economy. Their interrelation is shown in the table below.

Outputs \ Inputs	U	V	Final demand	Total
U	85	95	390	570
V	50	80	550	680
Other production factors	435	505		
Total	570	680		

(a) Compute the technological matrix A.

(b) Compute the Leontief matrix.

(c) Find the final demand vector.

(d) Find the total production vector.

(e) If final demand changes from 390 to 300 for U and from 550 to 600 for V (and A remains the same), what is the total output required?

Solution

(a) The technological matrix A is computed from the matrix

$$\begin{bmatrix} 85 & 95 \\ 50 & 80 \end{bmatrix}.$$

Dividing column 1 by 570 and column 2 by 680, we have (rounded to three places)

$$A = \begin{bmatrix} 0.149 & 0.140 \\ 0.088 & 0.118 \end{bmatrix}.$$

(b) The Leontief matrix is $I - A$:

$$\begin{aligned} I - A &= \begin{bmatrix} 1 & 0 \\ 0 & 1 \end{bmatrix} - \begin{bmatrix} 0.149 & 0.140 \\ 0.088 & 0.118 \end{bmatrix} \\ &= \begin{bmatrix} 0.851 & -0.140 \\ -0.088 & 0.882 \end{bmatrix}. \end{aligned}$$

(c) From the table the final demand vector is

$$D = \begin{bmatrix} 390 \\ 550 \end{bmatrix}.$$

(d) From the table the total production vector is

$$X = \begin{bmatrix} 570 \\ 680 \end{bmatrix}.$$

(e) In order to determine the new output needed for the demand

$$D' = \begin{bmatrix} 300 \\ 600 \end{bmatrix}.$$

we use Eq. (2):

$$X = (I - A)^{-1}D'.$$

In this equation we know D' and A, and we need to compute

$$(I - A)^{-1}.$$

Use the matrix $I - A$ obtained in part (b):

$$\left[\begin{array}{rr|rr} 0.851 & -0.140 & 1 & 0 \\ -0.088 & 0.882 & 0 & 1 \end{array}\right] \quad R_1 \to R_1/0.851$$

$$\left[\begin{array}{rr|rr} 1 & -0.165 & 1.175 & 0 \\ -0.088 & 0.882 & 0 & 1 \end{array}\right] \quad R_2 \to 0.088R_1 + R_2$$

$$\left[\begin{array}{rr|rr} 1 & -0.165 & 1.175 & 0 \\ 0 & 0.867 & 0.103 & 1 \end{array}\right]$$
$$R_2 \to R_2/0.867$$
$$\left[\begin{array}{rr|rr} 1 & -0.165 & 1.175 & 0 \\ 0 & 1 & 0.119 & 1.153 \end{array}\right]$$
$$R_1 \to 0.165R_2 + R_1$$
$$\left[\begin{array}{rr|rr} 1 & 0 & 1.195 & 0.190 \\ 0 & 1 & 0.119 & 1.153 \end{array}\right]$$

Therefore,

$$(I - A)^{-1} = \begin{bmatrix} 1.195 & 0.190 \\ 0.119 & 1.153 \end{bmatrix},$$

and

$$\begin{aligned} X &= (I - A)^{-1}D' \\ &= \begin{bmatrix} 1.195 & 0.190 \\ 0.119 & 1.153 \end{bmatrix}\begin{bmatrix} 300 \\ 600 \end{bmatrix} \\ &= \begin{bmatrix} 472.5 \\ 727.5 \end{bmatrix}. \end{aligned}$$

EXERCISES

Set A–Finding Technological Matrices, Final Demand Vectors and Production Vectors

In each of Exercises 1 through 8, determine
(a) the technological matrix,
(b) the final demand vector, and
(c) the total production vector.

1.

Outputs \ Inputs	U	V	Final demand	Total
U	20	40	700	760
V	40	90	910	1040
Other production factors	700	910		
Total	760	1040		

2.

Inputs / Outputs	U	V	Final demand	Total
U	60	90	580	730
V	30	50	500	580
Other production factors	640	440		
Total	730	580		

3.

Inputs / Outputs	U	V	Final demand	Total
U	20	60	240	320
V	30	10	870	910
Other production factors	270	840		
Total	320	910		

4.

Inputs / Outputs	U	V	Final demand	Total
U	10	70	820	900
V	80	70	680	830
Other production factors	810	690		
Total	900	830		

5.

Inputs / Outputs	U	V	Final demand	Total
U	90	10	430	530
V	20	80	670	770
Other production factors	420	680		
Total	530	770		

6.

Outputs \ Inputs	*U*	*V*	*W*	Final demand	Total
U	36	56	23	218	333
V	19	14	20	256	309
W	34	30	43	494	601
Other production factors	244	209	515		
Total	333	309	601		

7.

Outputs \ Inputs	*U*	*V*	*W*	Final demand	Total
U	87	18	67	498	670
V	26	45	67	845	983
W	33	40	31	470	574
Other production factors	524	880	409		
Total	670	983	574		

8.

Outputs \ Inputs	*U*	*V*	*W*	Final demand	Total
U	96	17	45	566	724
V	60	92	49	316	517
W	93	23	43	459	618
Other production factors	475	385	481		
Total	724	517	618		

Set B—Calculating the Inverse of the Leontief Matrix and Forecasting Production

In each of Exercises 9 through 16, you are given a technological matrix and a final demand vector. You are to calculate the inverse of the Leontief matrix and determine the total production vector X.

9. $A = \begin{bmatrix} 0.035 & 0.120 \\ 0.088 & 0.053 \end{bmatrix}, \quad D = \begin{bmatrix} 460 \\ 660 \end{bmatrix}$

10. $A = \begin{bmatrix} 0.010 & 0.030 \\ 0.090 & 0.080 \end{bmatrix}, \quad D = \begin{bmatrix} 720 \\ 530 \end{bmatrix}$

11. $A = \begin{bmatrix} 0.110 & 0.110 \\ 0.020 & 0.110 \end{bmatrix}, \quad D = \begin{bmatrix} 360 \\ 230 \end{bmatrix}$

12. $A = \begin{bmatrix} 0.110 & 0.050 \\ 0.130 & 0.080 \end{bmatrix}, \quad D = \begin{bmatrix} 460 \\ 640 \end{bmatrix}$

13. $A = \begin{bmatrix} 0.200 & 0.160 \\ 0.120 & 0.070 \end{bmatrix}, \quad D = \begin{bmatrix} 110 \\ 490 \end{bmatrix}$

14. $A = \begin{bmatrix} 0.045 & 0.023 & 0.045 \\ 0.030 & 0.023 & 0.080 \\ 0.104 & 0.091 & 0.034 \end{bmatrix}, \quad D = \begin{bmatrix} 580 \\ 770 \\ 400 \end{bmatrix}$

15. $A = \begin{bmatrix} 0.067 & 0.048 & 0.144 \\ 0.047 & 0.048 & 0.158 \\ 0.069 & 0.043 & 0.116 \end{bmatrix}, \quad D = \begin{bmatrix} 384 \\ 472 \\ 469 \end{bmatrix}$

16. $A = \begin{bmatrix} 0.026 & 0.230 & 0.032 \\ 0.020 & 0.303 & 0.059 \\ 0.051 & 0.221 & 0.038 \end{bmatrix}, \quad D = \begin{bmatrix} 496 \\ 167 \\ 584 \end{bmatrix}$

3.11 SUMMARY AND REVIEW EXERCISES

REVIEW OF CONCEPTS

A. Linear Systems Systems of linear equations may be solved by the pivot process:

1. Choose a pivot equation and a pivot element in the pivot equation.
2. Use the pivot equation and the Gaussian elimination process to eliminate the chosen element (unknown).
3. Repeat Steps 1 and 2 for a new pivot equation and a new pivot element (if possible). Continue until all equations or all unknowns have been used as pivots.
4. The process terminates when all equations or all unknowns have been used as pivots (no unknown or equation should be used more than once as a pivot).

B. Solving Linear Systems by Matrices

1. *Using the Pivot Process Applied to the Augmented Matrix*
   (a) Write the system as an augmented matrix.
   (b) Apply the pivot process to the augmented matrix:
   - Choose a pivot row R_i of the augmented matrix and a nonzero pivot element a_{ij} of row i. Do not choose a_{ij} in the last column.
   - Use row i and Gaussian elimination to obtain 0s in the jth column above and below the pivot element (after this has been completed, the jth column will contain only 0s except in the ith row).
   - Repeat the above two steps for a new pivot row and a new pivot element (if possible). The process should continue until all rows have been used as pivots or until a pivot element has been chosen from each column (except the last).
   - The process terminates as soon as each row has been used as a pivot or when a pivot element has been chosen from each column.

2. *Solving by Using Matrix Inverses*
   (a) Write the linear system as a matrix equation:

$$AX = B,$$

   where A is the square matrix of coefficients, X is the column of unknowns, and B is the column of constants.
   (b) Find the inverse of A and solve as

$$X = A^{-1}B.$$

C. Matrix Computations

1. *Equality.* Two matrices are equal if they have the same dimensions and corresponding entries are equal.

2. *Addition.* Matrices having the same dimensions are added by adding corresponding entries.

3. *Subtraction.* Matrices are subtracted by the rule $A - B = A + (-B)$.

4. *Multiplication by Numbers.* A matrix is multiplied by a number by multiplying each entry by that number.

5. *Multiplication of Matrices*
   (a) Vector Multiplication. Two vectors A and B may be multiplied to obtain a number if A is a row vector, B is a column vector, and A and B have the same number of entries. The product $A \cdot B$ is obtained by multiplying corresponding entries of A and B and adding the results.
   (b) Multiplication of Two Matrices. Two matrices A and B may be multiplied to obtain a matrix if their dimensions satisfy the following requirements:

$$\underset{n \times k}{A} \qquad \underset{k \times m}{B}$$

$$\text{(the inner dimensions } k \text{ — Same)}$$

   ↑---Same---↑

   The product AB is the matrix $[d_{ij}]_{n\times m}$, where d_{ij} is given by the following rule:

$$d_{ij} = R_i \cdot C_j,$$

   where R_i is the ith row of A, and C_j is the jth column of B.

6. *Inverses of Square Matrices.* If A is a square matrix, its inverse A^{-1} (if it exists) is a matrix having the same dimensions as A such that

$$AA^{-1} = A^{-1}A = I,$$

   where I is the identity matrix. If A^{-1} exists, it may be found as follows:

$$(A \mid I) \longrightarrow (I \mid A^{-1})$$

   This scheme indicates that A and I are written side by side. Then A is transformed into I by the pivot process. At the same time, the transformations applied to A are also applied to I. The matrix that results from I at the last step is A^{-1}.

REVIEW EXERCISES

1. Solve these systems.
   (a) $\begin{aligned} 2x + y &= 4 \\ 3x + 5y &= -15 \end{aligned}$
   (b) $\begin{aligned} 4x + 3y &= 23 \\ -5x + 7y &= -61 \end{aligned}$

2. Solve these systems by the pivot process.

(a) $\begin{aligned} 2x + y + 3z &= 12 \\ x + y + z &= 3 \\ 3x + 5y + 2z &= 0 \end{aligned}$ (b) $\begin{aligned} 8x + 7y + z &= 9 \\ 5x - 8y + 3z &= 17 \end{aligned}$

3. Solve these systems by using augmented matrices.

(a) $\begin{aligned} 2x + 5y &= 10 \\ 8x - y &= -44 \end{aligned}$ (b) $\begin{aligned} x \quad\quad + 6z &= 8 \\ -6x + y + 6z &= -9 \\ -5x + 5y + 8z &= -17 \end{aligned}$

(c) $\begin{aligned} -7x + y + 6z &= 11 \\ 3x - 5y + 4z &= 91 \end{aligned}$

4. Make the following matrix computations where the matrices A and B are given below.

$$A = \begin{bmatrix} -1 & 0 & 2 \\ -5 & 9 & 1 \\ -5 & 2 & -8 \end{bmatrix} \qquad B = \begin{bmatrix} 1 & -5 & -7 \\ 6 & 0 & 2 \\ -5 & 9 & -6 \end{bmatrix}$$

(a) $A + B$ (b) $B - A$ (c) $-3A + 2B$ (d) AB (e) BA

5. Find the inverse of the following matrices.

(a) $\begin{bmatrix} -6 & 6 \\ -5 & -1 \end{bmatrix}$

(b) $\begin{bmatrix} 1 & -2 & 3 \\ 2 & -5 & 10 \\ -1 & 2 & -3 \end{bmatrix}$

6. Solve the equation $2A - 3X = B$, where A, B are the matrices of Exercise 4.

7. Write each of the following systems in the form $AX = B$, where A is the matrix of coefficients and B is the column of constants. Then solve the system by calculating A^{-1}.

(a) $\begin{aligned} 2x + 9y &= -38 \\ x + 5y &= -21 \end{aligned}$ (b) $\begin{aligned} 3x - 7y + 13z &= 90 \\ 2x - 5y + 10z &= 66 \\ 2x - 4y + 7z &= 52 \end{aligned}$

CHAPTER

4

LINEAR PROGRAMMING IN THE PLANE

4

4.1 INTRODUCTION
4.2 GRAPHS OF LINEAR INEQUALITIES
4.3 LINEAR PROGRAMMING IN THE PLANE
4.4 LP MODELS—APPLICATIONS
4.5 SUMMARY AND REVIEW EXERCISES

4.1 INTRODUCTION

The linear programming model may be used to solve many problems which arise in diverse situations, and some of these applications are indicated throughout this chapter. Examples 1 and 2 listed below give an indication of the kinds of problems the student will be able to solve when the material of this chapter has been mastered.

1. A certain oil company produces gasoline and diesel fuel and operates under the following restrictions.
   (a) The company must produce at least 4 million barrels of diesel fuel each day.
   (b) The amount of diesel must not be more than 80% of total production.
   (c) Total production cannot exceed 10 million barrels per day.

   If the profit on a barrel of gasoline is \$2 and that on a barrel of diesel is \$1, how many barrels of each kind of fuel should the company produce each day for maximum profit?

2. A banker is arranging an investment program for a customer who has a total of \$15,000 to invest in stocks and bonds. The customer insists that the banker adhere to the following rules:
   (a) The total investment in stocks and bonds must be at least \$2000.
   (b) The amount invested in stocks is not to exceed \$10,000.
   (c) The amount invested in bonds must be at least 75% of \$6000 less than the amount invested in stocks.

   If stocks yield 18% and bonds yield 15%, how much should be invested in each kind of security for maximum return? ■

4.2 GRAPHS OF LINEAR INEQUALITIES

We begin this chapter with an investigation of linear inequalities in the plane. This is our starting point because the easiest examples of linear programming problems occur in two dimensions where graphs greatly aid in understanding.

Recall that the equation

$$x = 3 \tag{1}$$

is an example of a linear equation and is represented in the xy-plane as a line parallel to the y-axis intersecting the x-axis at the point (3, 0). (See Figure 4.1.) Equation (1) is a short way of describing all those points (x, y) in the plane for which $x = 3$. With Eq. (1) we can associate several inequalities:

$$x \leq 3, \tag{2a}$$

$$x \geq 3, \tag{2b}$$

$$x < 3, \tag{2c}$$

$$x > 3. \tag{2d}$$

Inequality (2a) is a short way of describing all the points (x, y) in the plane for which $x \leq 3$, that is, all the points whose first coordinate is 3 or less. For example, all the points (3, 0), (2, 1), (2.9, 5),

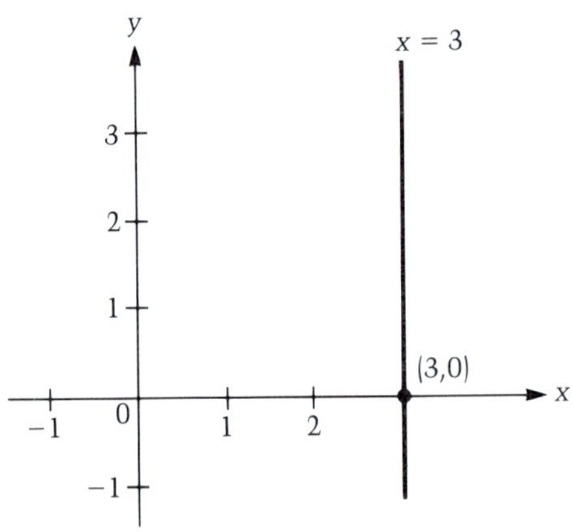

FIG. 4.1.

FIG. 4.2.

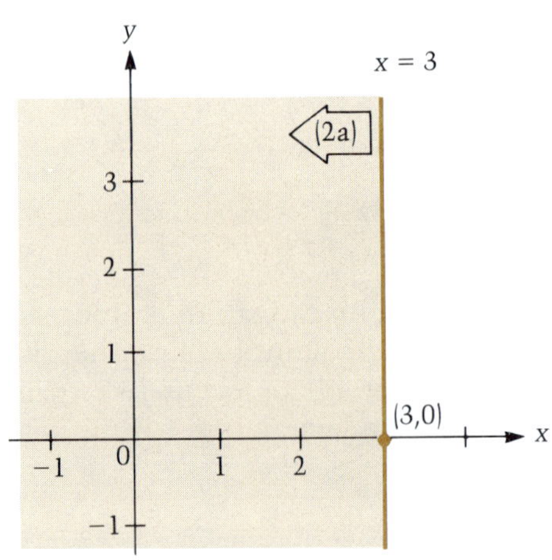

FIG. 4.3.

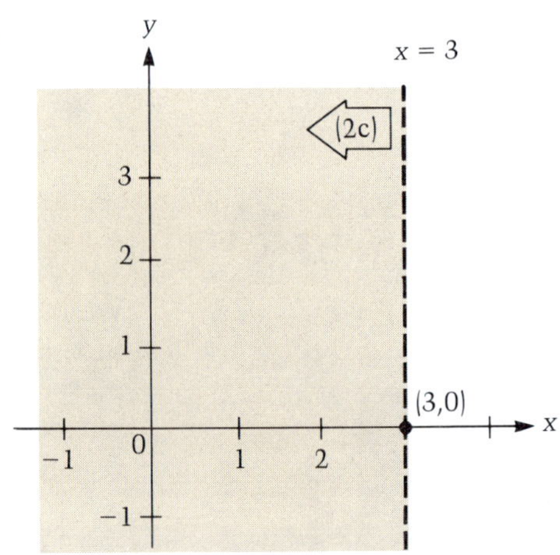

$(-1, 1)$ satisfy this condition. In fact, every point which is on or to the left of the line $x = 3$ satisfies this condition. Therefore, the graph of (2a), that is the graph of $x \leq 3$, is the shaded area of Figure 4.2 including the line $x = 3$. The graph of $x < 3$ is the shaded area of Figure 4.3, which does *not* include the line $x = 3$. (This line is dashed to indicate that it is not included in the graph.) The graphs of $x \geq 3$ are shown in Figures 4.4 and 4.5, respectively.

FIG. 4.4.

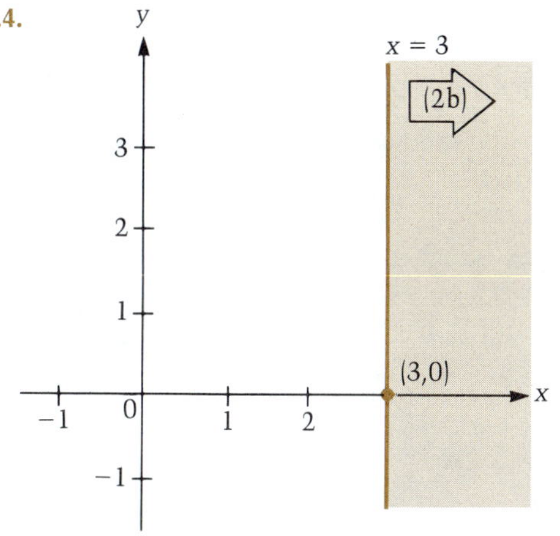

FIG. 4.5.

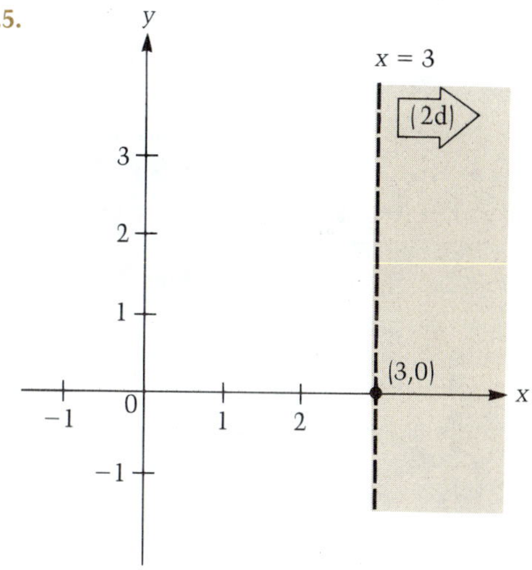

Graphing a linear inequality

EXAMPLE 1 Graph in two dimensions the linear inequality $y > -1$.

Solution: The inequality is a short way of expressing all those points (x, y) in the plane whose second coordinate y is greater than -1. Hence, its graph is the shaded area of Figure 4.6. (Note

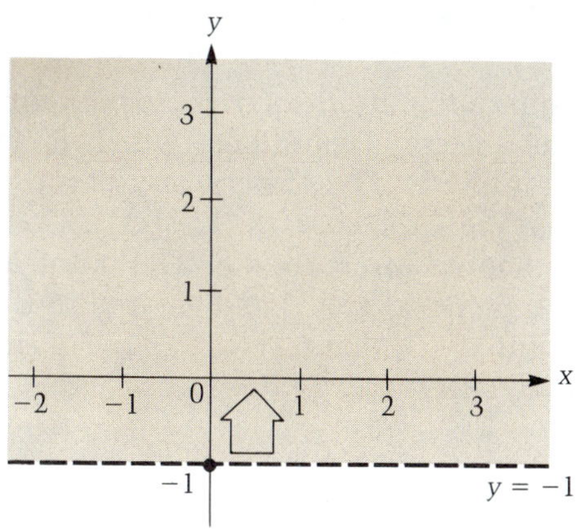

FIG. 4.6.

that the line $y = -1$ is dashed in the figure to indicate that it is not part of the graph.)

PRACTICE PROBLEM Graph in two dimensions the inequality $x \le 1$.

Answer

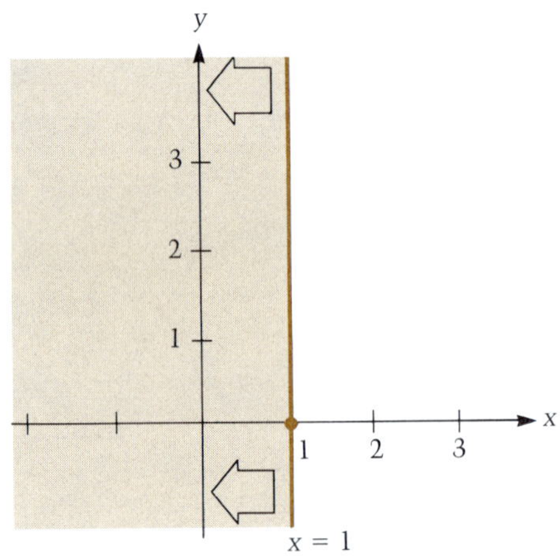

We will now look at inequalities which are slightly more complicated. Consider the linear equation

$$x - y = 1. \tag{3}$$

The graph of Eq. (3) appears in Figure 4.7. If we solve for y in terms of x, we get

$$y = x - 1. \tag{4}$$

Equation (4) tells us that every point (x, y) on the graph of Eq. (3) has a second coordinate which is equal to 1 less than the first coordinate. Therefore, every point (x, y) on the graph of Eq. (3) may be expressed as $(x, x - 1)$. Consider a typical point $P(x, y)$ as shown in Figure 4.8. The point $P(x, y)$ lies on a line through the point $(x, x - 1)$ and parallel to the y-axis, which is above the point $x - 1$, that is, $y > x - 1$. This indicates that the graph of the inequality

$$y > x - 1$$

consists of those points (x, y) *above* the line which is the graph of

FIG. 4.7.

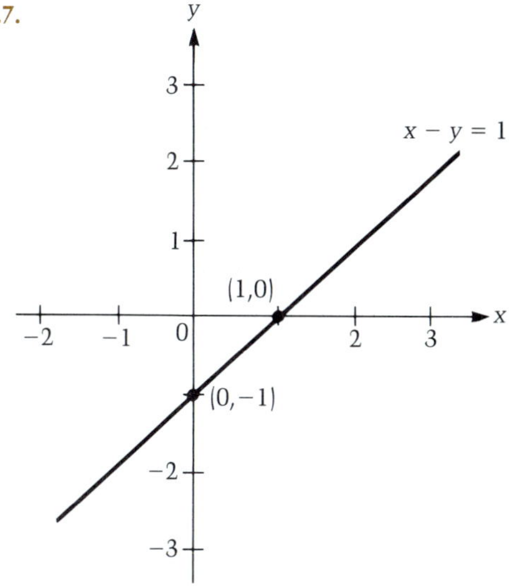

FIG. 4.8.

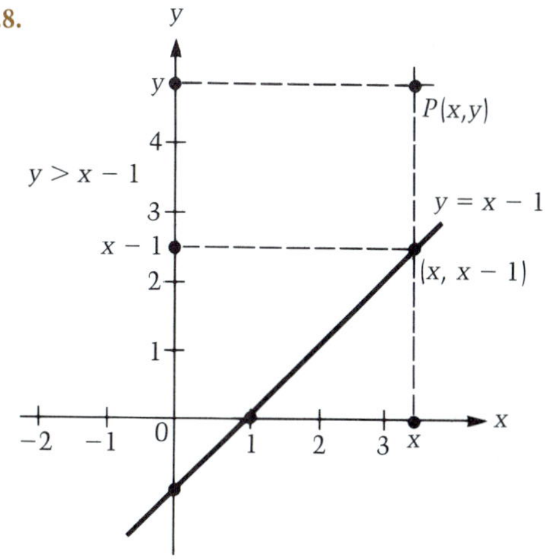

$y = x - 1$. The graph of the inequality

$$y > x - 1 \tag{4a}$$

is the shaded area shown in Figure 4.9. Figure 4.10 shows another point $Q(x, y)$ on the line which is parallel to the y-axis and passes through $(x, x - 1)$. In this case, the point (x, y) is *below* the point $(x, x - 1)$, and , as the figure indicates, y is less than $x - 1$. Therefore, points (x, y) which are below the line $y = x - 1$ satisfy the inequality

$$y < x - 1. \tag{4b}$$

FIG. 4.9.

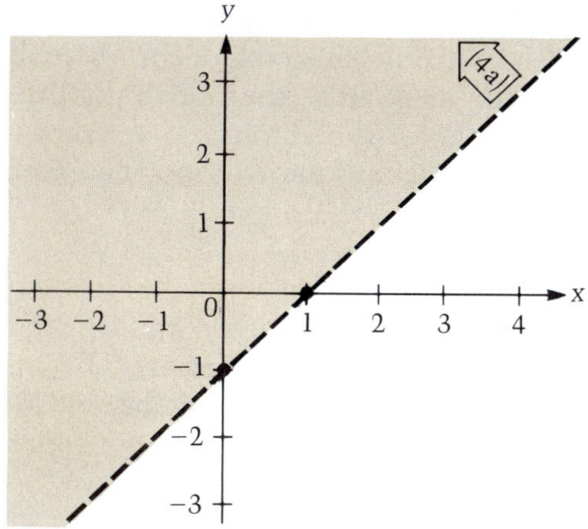

FIG. 4.10.

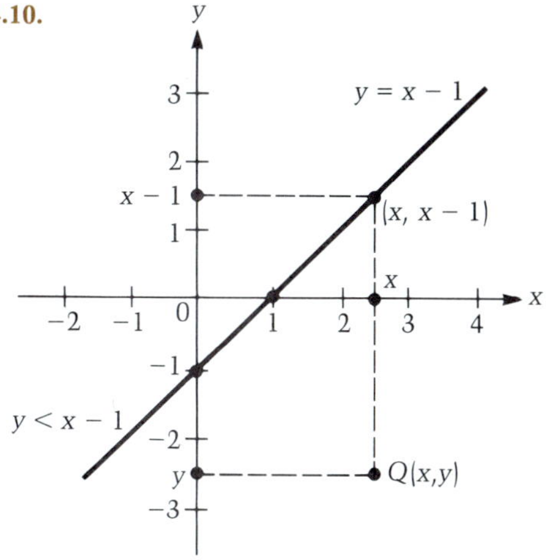

The graph of this inequality is the shaded area shown in Figure 4.11. The graph of the inequality $y \leq x - 1$ is the same as the graph of $y < x - 1$ except that the line $y = x - 1$ is included.

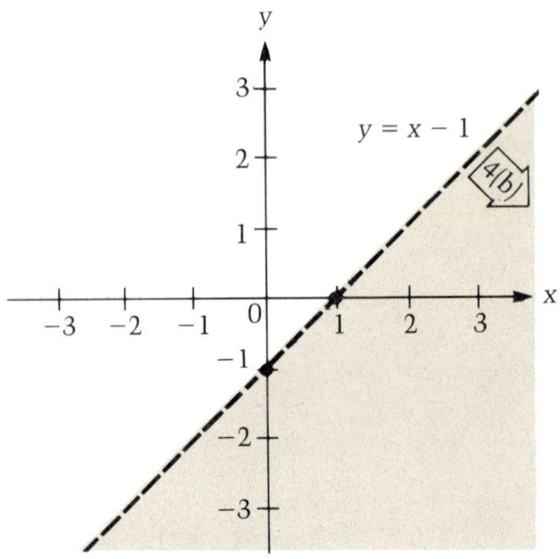

FIG. 4.11.

The kind of analysis just carried out shows in general that the graph of a linear inequality consists of all points in the plane which are on one side or the other of the line which is the graph of the linear equation associated with the inequality. This information greatly simplifies the problem of graphing linear inequalities. If we are given a linear inequality, the first step should be to graph the linear equation associated with the inequality. Then the graph of the inequality is contained on one side of this line only. To determine which side of the line is correct for the graph, we simply choose any point (u, v) *not* on the line and use this as a test point: if (u, v) satisfies the inequality, then the side of the line containing (u, v) is contained in the graph of the inequality, and if (u, v) does not satisfy the inequality, the side of the line opposite to (u, v) is the correct side for the graph.

As an example of this technique we find the graph of

$$x + y \geq 0. \tag{5}$$

The graph of $x + y = 0$ is first constructed and this is shown in Figure 4.12. As a test point (u, v), we choose $(1, 1)$. This point is *not* on the line $x + y = 0$ (since $1 + 1 \neq 0$), but it does satisfy inequality (5) since

$$1 + 1 \geq 0$$

is a true statement. Hence the graph of (5) is the line shown in Figure 4.12 and all the points above it. The graph is shown in Figure 4.13.

FIG. 4.12.

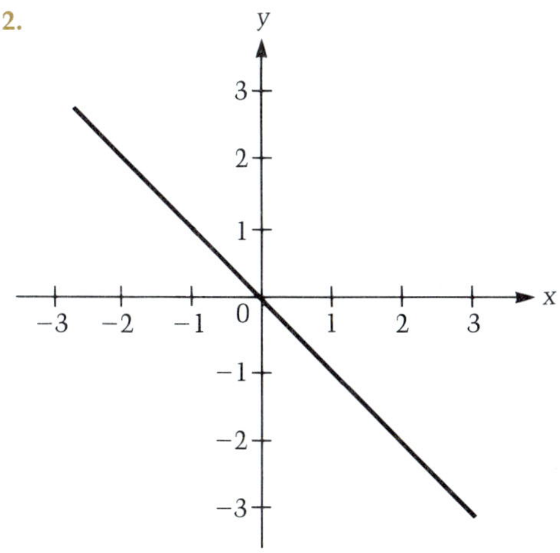

FIG. 4.13.

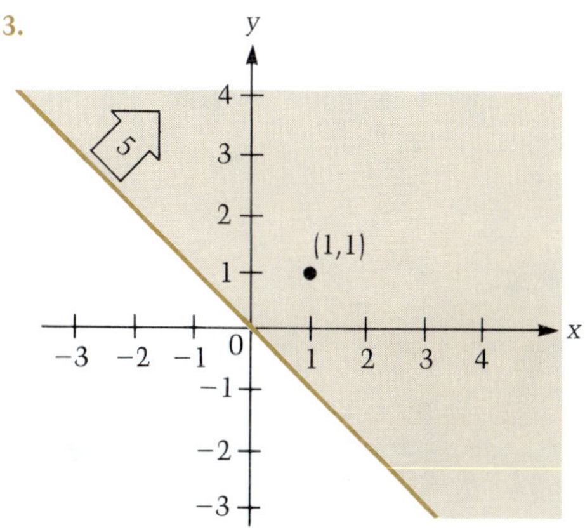

Graph of a linear inequality

EXAMPLE 2 Graph the linear inequality

$$2x - 3y < 6.$$

Solution: We graph the equation $2x - 3y = 6$ by finding the intercepts. These are $(3, 0)$ and $(0, -2)$, and the graph is shown in Figure 4.14. Now the graph of the inequality is contained on one side of this line. In this case it is convenient to use the origin as

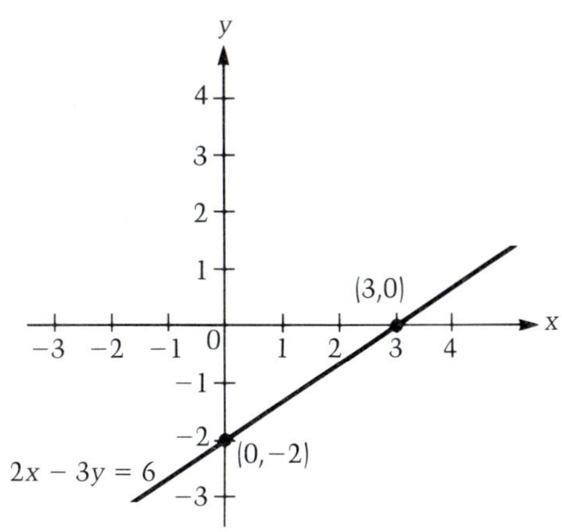

FIG. 4.14.

the test point. The point (0, 0) satisfies the inequality since

$$2 \cdot 0 - 3 \cdot 0 < 6$$

is a true statement. Thus, the graph is that shown in Figure 4.15.

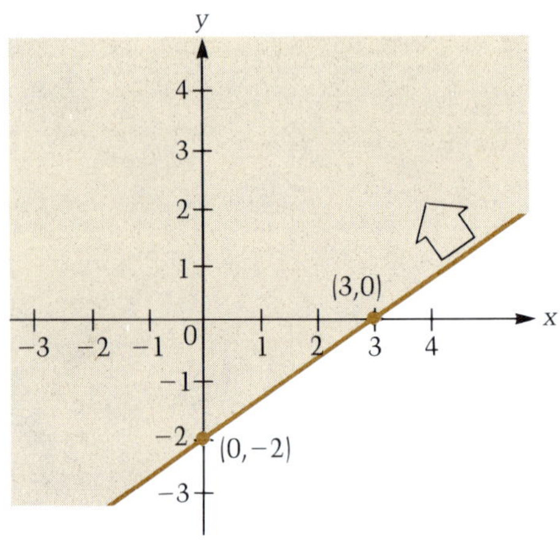

FIG. 4.15.

From this point we will be concerned with **systems** of linear inequalities. Given several inequalities, we will be interested in those points which satisfy the inequalities simultaneously. To illustrate this consider the inequalities

$$x \geq 0,$$
$$y \geq 0,$$
$$x - y \leq 0,$$
$$y \leq 5.$$

The set of points which satisfy the first two of these inequalities includes the region to the right of the y-axis ($x \geq 0$) and above the x-axis ($y \geq 0$) as shown in Figure 4.16. To satisfy the inequality $y \leq 5$, we must take the points on and below the line $y = 5$. This will further restrict the graph as shown in Figure 4.17. Now the graph of

$$x - y \leq 0$$

may be constructed in the region already obtained. The graph of $x - y = 0$ is given in Figure 4.18, and the graph of the inequality

FIG. 4.16.

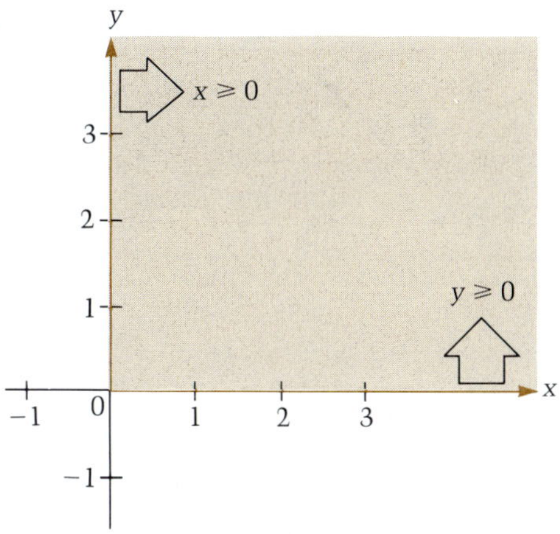

FIG. 4.17.

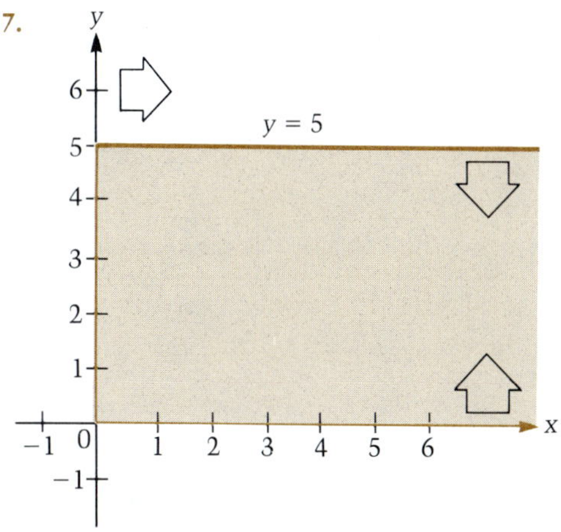

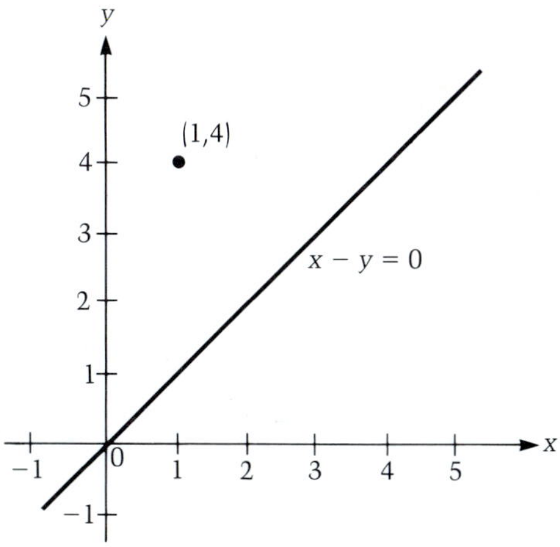

FIG. 4.18.

is on one side of this line. The point (1, 4) satisfies the inequality

$$x - y \leq 0$$

since $1 - 4 \leq 0$ is a true statement, and this shows that the graph of the set of inequalities is the region indicated in Figure 4.19. Therefore, the triangular region of Figure 4.19 contains all the points which satisfy the inequalities

$$x \geq 0,$$
$$y \geq 0,$$
$$x - y \leq 0,$$
$$y \leq 5.$$

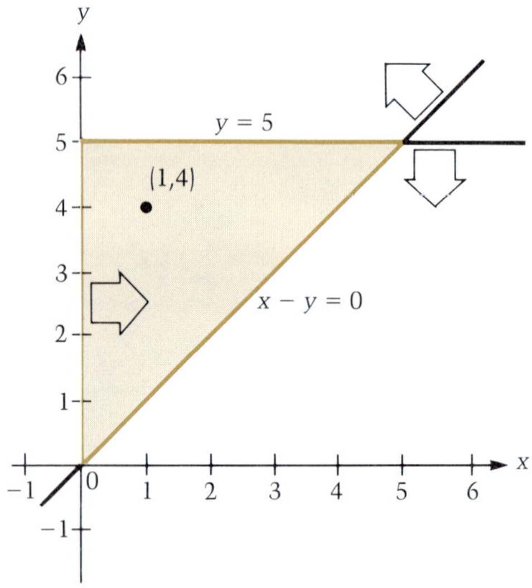

FIG. 4.19.

Note that in this case the inequality $y \geq 0$ is redundant since the points (x, y) satisfying the other three inequalities automatically satisfy $y \geq 0$.

Graphing a set of linear inequalities

EXAMPLE 3 Construct the graph of the following set of linear equalities:

$$x \geq -2,$$
$$-x + y \leq 2,$$
$$x - y \leq 0,$$
$$x + y \leq 1.$$

Solution: Construct first the graph of each of the equations

$$x = -2,$$
$$-x + y = 2,$$
$$x - y = 0,$$
$$x + y = 1.$$

The graphs of these are shown in Figure 4.20. The point $(0, -1)$ is not on any of these lines and may be used as a test point for each of the inequalities. Using the following computations the proper sides of the lines are determined. In Figure 4.21 the information contained in the last column of Table 4.1 is indicated by arrows.

Therefore, the graph of the system of inequalities is the shaded region shown in Figure 4.22.

FIG. 4.20.

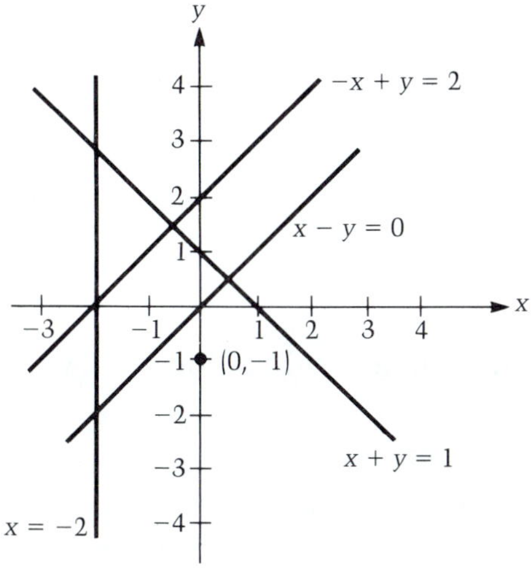

FIG. 4.21.

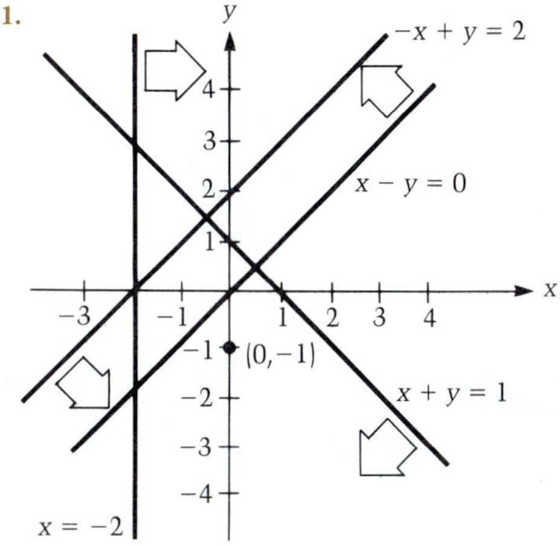

TABLE 4.1

Inequality	Test point	Test computation	True or false	Same or opposite side of test point
$x \geq -2$	$(0, -1)$	$0 \geq -2$	True	Same
$-x + y \leq 2$	$(0, -1)$	$-0 + (-1) \leq 2$	True	Same
$x - y \leq 0$	$(0, -1)$	$0 - (-1) \leq 0$	False	Opposite
$x + y \leq 1$	$(0, -1)$	$0 + (-1) \leq 1$	True	Same

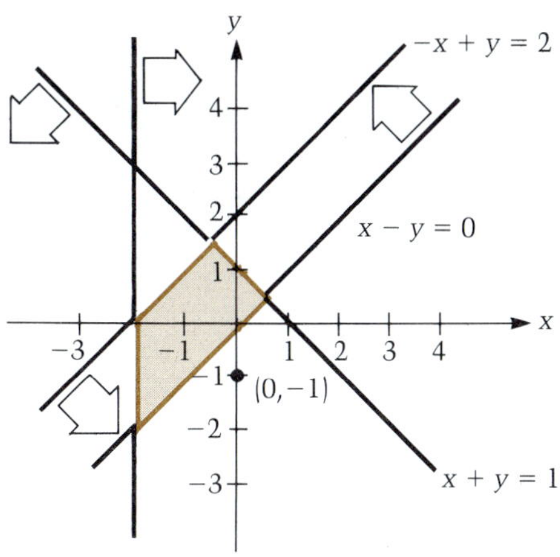

FIG. 4.22.

EXERCISES

Set A–Graphs of Single Linear Inequalities in One Unknown

In Exercises 1 through 10, construct in the xy-plane the graph of each of the linear inequalities.

1. $x \geq 0$
2. $y < 0$
3. $y \leq -3$
4. $x \geq 3.5$
5. $y \leq -2.5$
6. $x < 1.75$
7. $2x \geq -5$
8. $2(x + 1) \geq 6$
9. $0.5(y - 1) < -2$
10. $0.25(y + 2) \geq -1$

Set B–Graphs of Single Linear Inequalities in Two Unknowns

In Exercises 11 through 24, construct the graph of each linear inequality in the xy-plane.

11. $x + y < 1$
12. $x - y \geq 2$
13. $x - 1 \geq y$
14. $y + 2 < x$
15. $2x - y \geq 2$
16. $3y - x < -3$
17. $x + 2y \leq -2$
18. $4y - x > 4$
19. $2x - 5y \leq 10$
20. $5x + 3y > -15$
21. $7x + 2y < -14$
22. $3x - 7y \geq 21$
23. $2x - 3y > -3$
24. $5x + 4y \geq -25$

Set C–Graphs of Systems of Linear Inequalities

In Exercises 25 through 34, construct the graph of each system of inequalities in the xy-plane.

25. $\begin{aligned} x - y &\leq -1 \\ x &\geq -2 \end{aligned}$
26. $\begin{aligned} x - y &> -1 \\ x &\geq -2 \end{aligned}$
27. $\begin{aligned} x - y &\leq -1 \\ x &< -2 \end{aligned}$
28. $\begin{aligned} x - y &\geq -1 \\ x &\leq -2 \end{aligned}$
29. $\begin{aligned} 2x - y &\geq -2 \\ 2x + y &\leq 2 \end{aligned}$
30. $\begin{aligned} 2x - y &\geq -2 \\ 2x + y &\leq 2 \\ y &\geq -3 \end{aligned}$
31. $\begin{aligned} 2x - y &\geq -2 \\ 2x + y &\leq 2 \\ y &\geq 3 \end{aligned}$
32. $\begin{aligned} x &\geq -2 \\ x &\leq 2 \\ y &\geq -3 \\ y &\leq 1 \end{aligned}$
33. $\begin{aligned} x - y &\leq 1 \\ x + y &\leq 1 \\ -x + y &\leq 1 \\ -x - y &\leq 1 \end{aligned}$
34. $\begin{aligned} y &\leq 7 \\ 5x + 3y &\geq 15 \\ -2x + 3y &\leq 15 \\ 3x + y &\leq 31 \\ x - 7y &\leq 3 \end{aligned}$

4.3 LINEAR PROGRAMMING PROBLEMS IN THE PLANE

LP

This section explains the anatomy of **Linear Programming** (LP) problems and at the same time shows how these problems arise naturally in practical situations.

It is difficult to overestimate the importance of LP in solving practical problems. The following quotation from a recent text* on LP will indicate to some extent just how powerful the method is.

> I first heard about linear programming around 1958. I had just come to Caltech as an associate professor. I was making a trip to New York with my boss, Professor Gilbert McCann, who was the director of Caltech's computing center. We were making a survey of large industrial computer installations to find out what were the principal industrial uses of computers. One of the companies we planned to visit was the Mobil Oil Corporation.
>
> When we arrived at Mobil, a secretary told us that we would be meeting with Dr. Albert Sokolov. . .
>
> We had a nice chat about computers at Mobil. The company had a huge new computing center. I knew it must have cost millions of dollars. Was it worth it? I asked him:
>
> 'Al, you've got a huge installation here, millions of dollars worth. I know oil companies have a lot of money, but they don't like to waste it, either. How long will it take the company to pay off your investment in computers?'
>
> He thought for a moment, apparently making a rough mental calculation. Then he answered:
>
> 'We paid it off in about two weeks.'
>
> 'That's amazing,' I said. 'What kind of problems do you do with this computer?'
>
> 'Mainly linear programming.'
>
> Al explained at length. Using linear programming, they were able to make optimal production decisions that had formerly been made—not so well—by vice presidents. The result was a great gain for the company. Other big oil companies were doing the same; it was good for business. It was also good for the consumer, who was getting more of what he wanted at less cost."

Now that you are convinced that LP is an important mathematical tool, we will look at a simple example which is easy to analyze.

Several problems were discussed in the previous chapter which concerned production schedules. The question considered there was how to schedule production so that various depart-

---

* Franklin, Joel; *Methods of Mathematical Economics*, Springer-Verlag, 1980.

ments would operate at full capacity. For our first LP problem we will return to the Strong Desk Company and again consider production schedules. This time it will be assumed that the company makes only two desk models: Executive and Secretary. These desks are processed in the cabinet, finishing, and crating departments. The Executive model requires the following number of hours in the cabinet, finishing and crating departments, respectively: 2, 1, 1. The time requirements for the Secretary model are 1, 2, 1. The numbers of hours available in these departments each day are 16, 16, 10. As usual it is helpful to place this information in a data box:

	Exec.	Sec.	Hours available/day
Cabinet	2	1	16
Finishing	1	2	16
Crating	1	1	10

We would like to know how many desks of each kind should be scheduled so that the three departments will operate at full capacity. Let

x = the number of Executive models to schedule per day

and

y = the number of Secretary models to schedule per day.

For full capacity the following system must have a solution:

$$\begin{aligned} 2x + y &= 16, \\ x + 2y &= 16, \\ x + y &= 10. \end{aligned}$$

Using the augmented matrix we try for a solution:

$$\begin{bmatrix} 2 & 1 & 16 \\ 1 & 2 & 16 \\ 1 & 1 & 10 \end{bmatrix} \quad \begin{matrix} R_1 \to -2R_3 + R_1 \\ R_2 \to -R_3 + R_2 \end{matrix}$$

$$\begin{bmatrix} 0 & -1 & -4 \\ 0 & 1 & 6 \\ 1 & 1 & 10 \end{bmatrix} \quad R_1 \to R_2 + R_1$$

$$\begin{bmatrix} \boxed{0 \quad 0 \quad 2} \\ 0 \quad 1 \quad 6 \\ 1 \quad 1 \quad 10 \end{bmatrix}$$

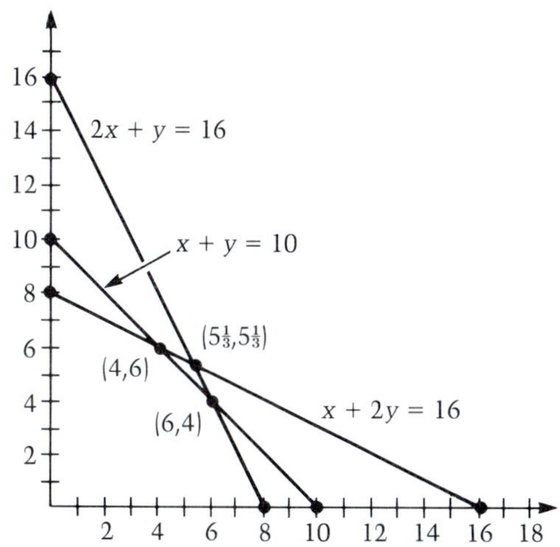

FIG. 4.23.

The first row of the last matrix shows that the system has no solution. We can conclude that *it is not possible to schedule production for full capacity.* The graphs of the three equations are given in Figure 4.23 and show that the three lines intersect in three points rather than in a single point (which would indicate a solution). Therefore, no production schedule will utilize all the capacity of every department. For example, suppose we decide to take $x = 4$ and $y = 6$ (schedule 4 Executive and 6 Secretary models). This schedule would require the following hours in the three departments:

$$\begin{aligned} \text{Cabinet:}\quad & 2x + y = 2(4) + 6 = 8 + 6 = 14, \\ \text{Finishing:}\quad & x + 2y = 4 + 2(6) = 4 + 12 = 16, \\ \text{Crating:}\quad & x + y = 4 + 6 = 10. \end{aligned}$$

Here we see that Finishing and Crating operate at full capacity but Cabinet is two hours short (16 hours are available but only 14 are used).

What should be done in the face of such a problem? Every possible schedule is going to waste some time in at least one department. Is there any schedule that will be better than the others in some respect? There are many different ways to answer this question but possibly the most natural one involves calculating the profit for the production schedule. Assume that the profit from each Executive model is \$150 and from each Secretary model is \$125. Then

$$P = 150x + 125y \tag{1}$$

is the profit from x Executive models and y Secretary models.

Our problem may now be stated in terms of P:

> How should the production schedule be set up so that P will be as large as possible?

From Eq. (1) it is evident that P increases as x and y increase, which shows that if we try to maximize P, we are also likely to push the departments toward full capacity. At any rate, if we succeed in choosing x and y so that P is as large as possible, any wasted time will be less harmful to the Company than it otherwise would.

The original problem was to solve the system

$$\begin{aligned} 2x + y &= 16, \\ x + 2y &= 16, \\ x + y &= 10 \end{aligned}$$

which does not have a solution. Thus, it is not possible to choose x and y so that *equality* occurs in all three equations. However, x and y must be chosen so that the left-hand sides of these equations do not exceed the right-hand sides. Therefore, the system of equations is replaced by a system of inequalities:

$$\begin{aligned} 2x + y &\leq 16, \\ x + 2y &\leq 16, \\ x + y &\leq 10. \end{aligned}$$

constraints

These inequalities along with $x \geq 0$, $y \geq 0$ (which indicate that we cannot produce negative numbers of desks) are the *constraints* of the problem. We have now arrived at one kind of LP problem which may be stated as follows:

> Maximize $P = 150x + 125y$
> subject to the constraints
>
> $$\begin{aligned} 2x + y &\leq 16, \\ x + 2y &\leq 16, \\ x + y &\leq 10, \\ x \geq 0, \quad y &\geq 0. \end{aligned}$$

objective function

The function $P = 150x + 125y$ is called the *objective function* and $x \geq 0$, $y \geq 0$ are called *nonnegativity constraints.* All the LP problems which we consider here will have nonnegativity constraints.

CHARACTERISTICS OF LP PROBLEMS

Each LP problem consists of two parts:

1. *An objective function.* This function is always linear and is to be either maximized or minimized.
2. *A set of constraints.* These are always linear inequalities.

feasible solution

optimal solution

A set of values of the variables of the LP problem which satisfies the constraints is said to be a *feasible solution.* A feasible solution which maximizes or minimizes the objective function is called an *optimal solution.*

set of feasible solutions

corner points

Before we discuss further examples of LP problems, we will obtain the solution to the problem of the Strong Desk Company. The constraints have been graphed in Figure 4.24. The set of points which satisfy the constraints (the shaded region of Figure 4.24) is the graph of the set of feasible solutions. Conceptually, we wish to evaluate the objective function P at *every* point (x, y) in the set of feasible solutions and from the resulting numbers choose the largest one. This would give us the optimal solution. As a practical matter this is of course not possible. Fortunately, the function P needs to be evaluated only at the corner points of the set of feasible solutions to provide the optimal solution. Let us explain why this is the case. Figure 4.25 shows the *corner points* with their coordinates. In Table 4.2, P has been evaluated at each of these points. Figure 4.26 shows the set of feasible solutions with lines passing through the corner points. Each line is the graph of $P = 150x + 125y$ for some particular value of P. Consider, for example, the line $150x + 125y = 1000$. If the objective function is evaluated at any point on this line, the result will be 1000. A similar remark applies to the other lines through the set

FIG. 4.24.

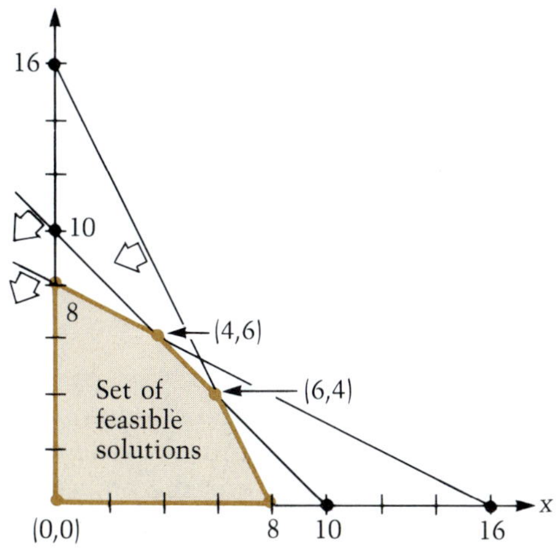

FIG. 4.25.

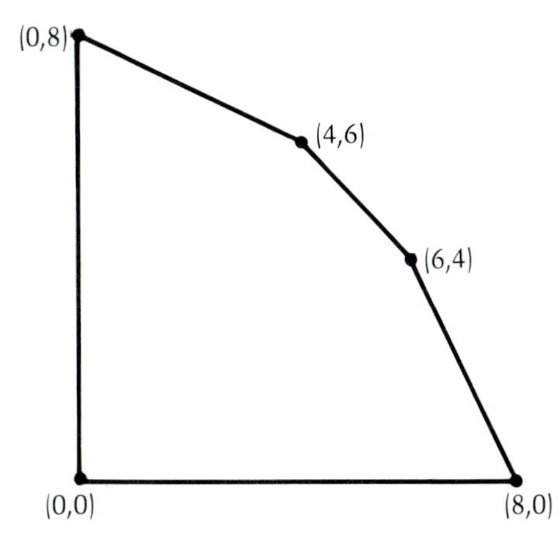

TABLE 4.2

Corner Points	Value of P
(0, 0)	$P = 150(0) + 125(0) = 0$
(8, 0)	$P = 150(8) + 125(0) = 1200$
(6, 4)	$P = 150(6) + 125(4) = 1400$
(4, 6)	$P = 150(4) + 125(6) = 1350$
(0, 8)	$P = 150(0) + 125(8) = 1000$

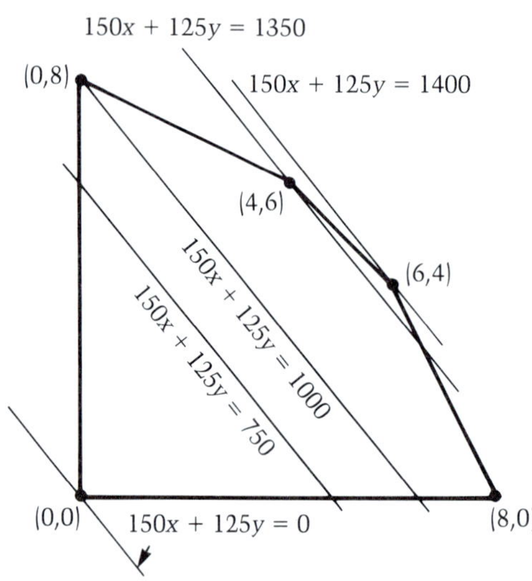

FIG. 4.26.

of feasible solutions. These lines are all parallel to one another (all have the same slope), and if we imagine moving them through the set of feasible solutions—starting with (0, 0)—then the value of P *increases* from 0 [at (0, 0)] to 1400 [at (6, 4)]. Since (6, 4) is the last point of the set of feasible solutions through which the line moves, P reaches its maximum value there. This discussion illustrates the following important result.

THEOREM If the objective function of an LP problem has a maximum (or minimum) value, then this value occurs at a corner point (or on a boundary line) of the set of feasible solutions.

As the theorem indicates, it is possible for an objective function not to have a maximum (or minimum) value on the set of feasible solutions. For example, if $P = x$ is the objective function and $x \geq 0$, $y \geq 0$ are the only constraints, then P has no maximum value since x may be taken arbitrarily large. We will not be interested in problems of this type and will deal only with LP problems which *always* have optimal solutions, i.e. maximum or minimum values.

STEPS IN SOLVING LP PROBLEMS IN THE PLANE

1. Graph the set of constraints to obtain the set of feasible solutions.
2. Determine the coordinates of the corner points of the set of feasible solutions. (This step may be completed by solving pairs of equations simultaneously.)
3. Evaluate the objective function at the corner points.
4. (a) If the objective function is to be maximized, the largest value from Step 3 gives the optimal solution of the problem.

   (b) If the objective function is to be minimized, the smallest value from Step 3 gives the optimal solution of the problem.

We can see from Table 1 that the Strong Desk Company problem is solved by scheduling 6 Executive and 4 Secretary models each day. The profit which results is \$1400, and this is the largest possible under the given conditions. What are the time requirements in the three departments for this production schedule? For an answer, we should use the times given in the original data box:

	Exec.	Sec.
Cabinet	2	1
Finishing	1	2
Crating	1	1

Time requirements for production schedule $x = 6$, $y = 4$

Cabinet	$2x + y = 2(6) + 4 = 16$
Finishing	$x + 2y = 6 + 2(4) = 14$
Crating	$x + y = 6 + 4 = 10$

Observe that our solution requires the cabinet and crating departments to operate at full capacity but wastes two hours in the finishing department.

In the next section we will return to practical problems which can be solved as LP problems. In the remainder of this section we deal with the details of solving an LP problem.

Finding the max and min of a linear function on a set of feasible solutions

EXAMPLE 1 Evaluate the function

$$P = 2x + y$$

on the set of feasible solutions shown below and find its maximum and minimum values.

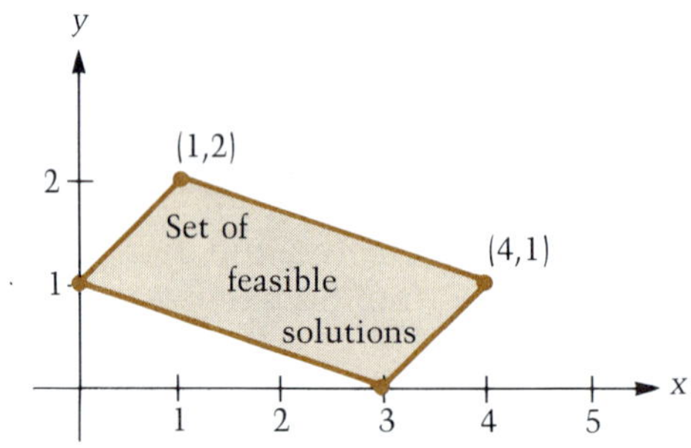

Solution: The function $P = 2x + y$ must be evaluated at each of the corner points (0, 1), (1, 2), (4, 1), (3, 0).

$$(0, 1)\text{: } P = 2x + y = 2(0) + 1 = 1 \quad \text{smallest}$$

$$(1, 2)\text{: } P = 2x + y = 2(1) + 2 = 4$$

$$(4, 1)\text{: } P = 2x + y = 2(4) + 1 = 9 \quad \text{largest}$$

$$(3, 0)\text{: } P = 2x + y = 2(3) + 0 = 6$$

We conclude that 1 is the minimum value and 9 is the maximum value of P on this set of feasible solutions.

PRACTICE PROBLEM Find the maximum and minimum values of $P = 3x + 5y$ on the set of feasible solutions in this example.

Answer: P has the minimum value of 5 at $x = 0$, $y = 1$. P has the maximum value of 17 at $x = 4$, $y = 1$. ■

Solving an LP problem graphically

EXAMPLE 2 Solve the following LP problem.

Maximize $P = 25x + 12y$
subject to the constraints

$$x \le 3,$$

$$y \le 4,$$

$$2x + y \le 8,$$

$$x \ge 0, \quad y \ge 0.$$

Solution

Step 1. Graph the set of feasible solutions. This means that we must graph the given inequalities as is done for this case in the accompanying figure.

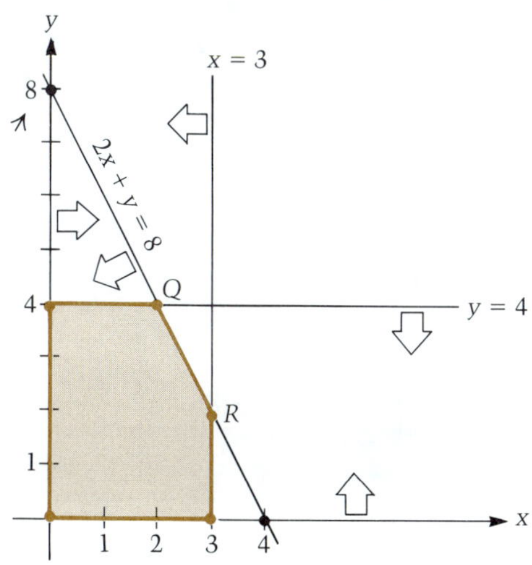

Step 2. Determine the coordinates of the corner points. The set of feasible solutions is shown in the figure. Except for Q and R the

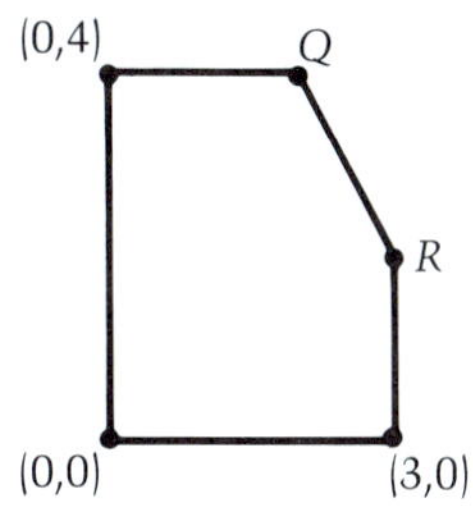

coordinates are obvious. But Q is the intersection point of the lines $y = 4$ and $2x + y = 8$. Solving for x gives

$$\begin{aligned} 2x + y &= 8, \\ 2x + 4 &= 8, \\ 2x &= 4, \\ x &= 2. \end{aligned}$$

Therefore, $Q = (2, 4)$.

The corner point R is the intersection of $x = 3$ and $2x + y = 8$:

$$\begin{aligned} 2x + y &= 8, \\ 2(3) + y &= 8, \\ 6 + y &= 8, \\ y &= 2. \end{aligned}$$

Therefore, $R = (3, 2)$.

Step 3. Evaluate the objective function P at the corner points:

Corner points	**Value of $P = 25x + 12y$**
(0, 0)	$P = 25(0) + 12(0) = 0$
(3, 0)	$P = 25(3) + 12(0) = 75$
(3, 2)	$P = 25(3) + 12(2) = 99$ ←
(2, 4)	$P = 25(2) + 12(4) = 98$
(0, 4)	$P = 25(0) + 12(4) = 48$

Step 4. Choose the largest value in Step 3 since P was to be maximized. From Step 3, we see that the largest value of P is 99 which occurs at $x = 3$, $y = 2$.

PRACTICE PROBLEM Solve the following:

Maximize $P = 2x - y$
subject to the constraints

$$\begin{aligned} 2x + y &\le 6, \\ -x + 2y &\le 2, \\ x \ge 0, \quad y &\ge 0. \end{aligned}$$

Answer: Maximum of 6 at (3, 0).

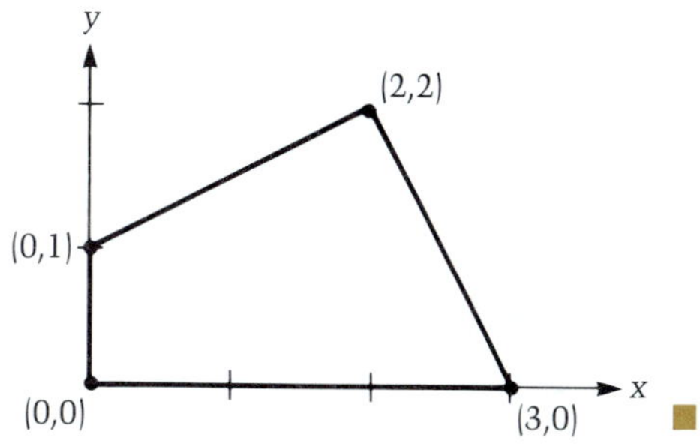

Solving an LP problem graphically

EXAMPLE 3 Solve the following:

Minimize $C = 8x - y$
subject to the constraints

$$2x + y \geq 8,$$
$$4x + y \geq 12,$$
$$2x + 3y \geq 12,$$
$$x \geq 0, \quad y \geq 0.$$

Solution

Step 1. Graph the inequalities:

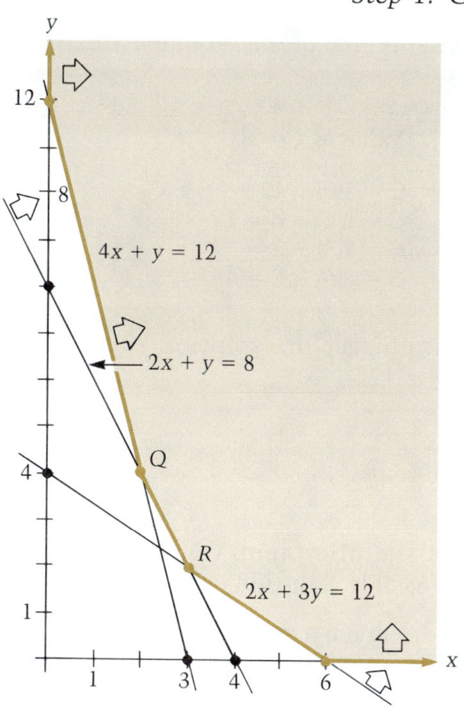

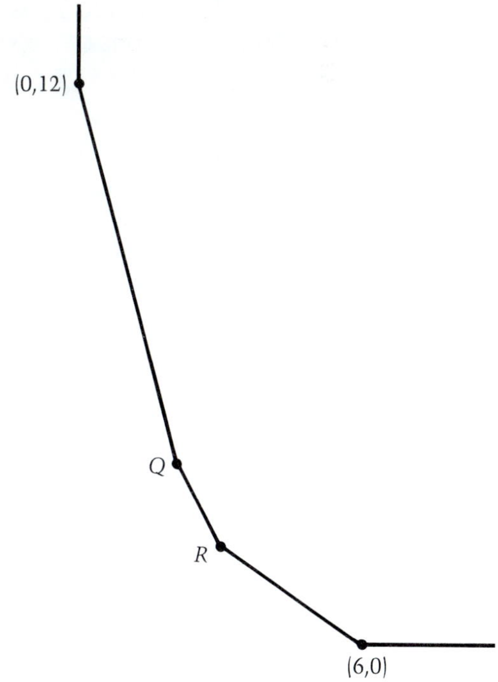

Step 2. Determine the coordinates of the corner points. Q and R are the only corner points whose coordinates are not obvious. Q is the intersection of the lines $4x + y = 12$ and $2x + y = 8$:

$$\begin{bmatrix} 4x + y = 12 \\ 2x + y = 8 \end{bmatrix} \quad R_1 \to -R_2 + R_1$$

$$\begin{bmatrix} 2x \quad = 4 \\ 2x + y = 8 \end{bmatrix} \quad \begin{matrix} R_1 \to 0.5R_1 \\ R_2 \to -R_1 + R_2 \end{matrix}$$

$$\begin{bmatrix} x \quad = 2 \\ y = 4 \end{bmatrix}$$

Therefore, $Q = (2, 4)$. R is the intersection of the lines $2x + y = 8$ and $2x + 3y = 12$:

$$\begin{bmatrix} 2x + 3y = 12 \\ 2x + y = 8 \end{bmatrix} \quad R_1 \to -R_2 + R_1$$

$$\begin{bmatrix} 2y = 4 \\ 2x + y = 8 \end{bmatrix} \quad R_1 \to 0.5R_1$$

$$\begin{bmatrix} y = 2 \\ x \quad = 3 \end{bmatrix}$$

Therefore, $R = (3, 2)$.

Step 3. Evaluate C at the corner points of the set of feasible solutions.

Corner points	Value of $C = 8x - y$
(0, 12)	$C = 8(0) - 12 = -12$
(2, 4)	$C = 8(2) - 4 = 12$
(3, 2)	$C = 8(3) - 2 = 22$
(6, 0)	$C = 8(6) - 0 = 48$

Step 4. Since C must be minimized, the solution is $C = -12$ which occurs for $x = 0$ and $y = 12$.

EXERCISES

Set A–Finding the Maximum and Minimum Values of a Linear Function on a Given Set of Feasible Solutions

In Exercises 1 through 4, you are given a set of feasible solutions. For each function determine the maximum and minimum value on this set.

1. (a) $F = x + y$ (b) $F = -x + y$ (c) $F = 5x - y$
(d) $F = -3x + 4y$

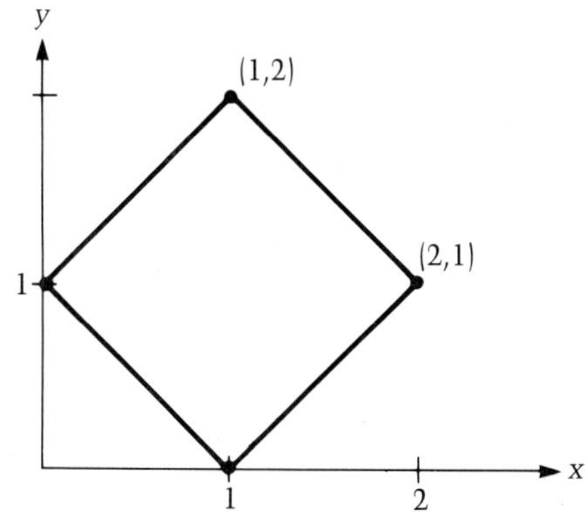

2. (a) $F = 10x + 3y$ (b) $F = x - 4y$ (c) $F = -5x + 2y$
(d) $F = -x + 4y$

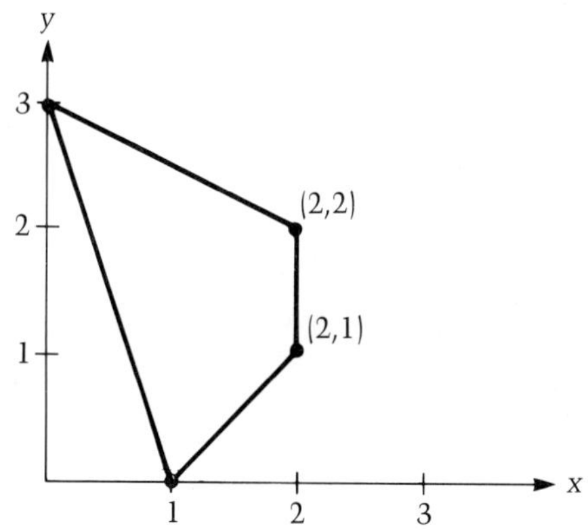

3. (a) $F = x + y$ (b) $F = -x - y$ (c) $F = x - 2y$
(d) $F = -x + 2y$

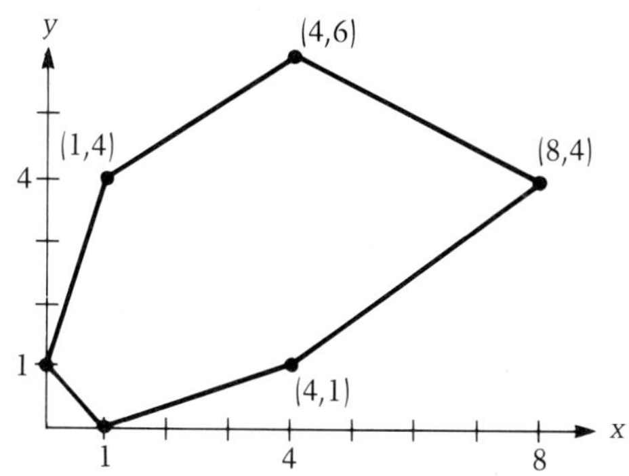

4. (a) $F = 2x + 3y$ (b) $F = x + 3y$ (c) $F = 2x + y$
(d) $F = 8x + 5y$

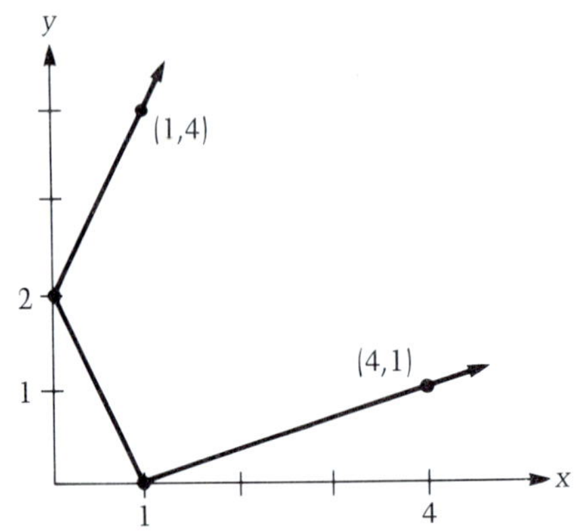

Set B–Solving LP Problems Graphically

In Exercises 5 through 14, graphically solve the LP problems given.

5. Maximize $P = 10x + 5y$ subject to the constraints

$$2x + y \geq 6,$$
$$4x + y \leq 12,$$
$$x \geq 0, \quad y \geq 0.$$

6. Maximize $P = 15x + 32y$ subject to the constraints

$$x + 2y \leq 10,$$
$$2x + y \leq 8,$$
$$x \geq 0, \quad y \geq 0.$$

7. Maximize $P = 25x - 13y$ subject to the constraints

$$x + y \leq 9,$$
$$2x + y \leq 12.$$
$$x \geq 0, \quad y \geq 0.$$

8. Minimize $C = 32x + 12y$ subject to the constraints

$$x + y \geq 7,$$
$$2x + 5y \geq 20,$$
$$x \geq 0, \quad y \geq 0.$$

9. Minimize $C = -5x + 18y$ subject to the constraints

$$x + y \geq 5,$$
$$x + 2y \geq 7,$$
$$x \geq 0, \quad y \geq 0.$$

10. Minimize $C = 3x - 5y$ subject to the constraints

$$x + y \geq 5,$$
$$2x + 5y \geq 16,$$
$$5x + 2y \geq 16,$$
$$x \geq 0, \quad y \geq 0.$$

11. Maximize $P = -4x + 7y$ subject to the constraints

$$\begin{aligned} 3x + 2y &\le 24, \\ x + y &\le 9, \\ x + 4y &\le 24, \\ x \ge 0, \quad y &\ge 0. \end{aligned}$$

12. Minimize $C = 4x + 5y$ subject to the constraints

$$\begin{aligned} -x + 2y &\le 6, \\ x + y &\le 9, \\ 3x - 2y &\le 12, \\ x \ge 0, \quad y &\ge 0. \end{aligned}$$

13. Minimize $C = -12x - 3y$ subject to the constraints

$$\begin{aligned} y &\le 8, \\ x + y &\ge 3, \\ 3x + y &\le 33, \\ x + 2y &\le 21, \\ x \ge 0, \quad y &\ge 0. \end{aligned}$$

14. Maximize $P = 3x - 2y$ subject to the constraints

$$\begin{aligned} x + 2y &\le 21, \\ -2x + 5y &\le 6, \\ 4x + y &\ge 6, \\ 2x + 3y &\ge 8, \\ 6x - y &\le 48, \\ x \ge 0, \quad y &\ge 0. \end{aligned}$$

4.4 LP MODELS—APPLICATIONS

In Section 4.2 a hypothetical problem concerning the Strong Desk Company was solved by modeling the situation as an LP problem. Now that the details of solving an LP problem have been mastered, several additional practical situations will be considered which also lead to the LP model.

---

*Investment management**

EXAMPLE 1 Suppose a bank has 100 million dollars to invest and that some of this money must be invested in loans and some in securities. A decision must be made: How much should be invested in loans and how much in securities? The unknowns have now been identified. Let

$$x = \text{the amount invested in loans,}$$

$$y = \text{the amount invested in securities.}$$

---

* Broaddus, Alfred, "Linear Programming: A New Approach to Bank Portfolio Management," *Monthly Review of the Federal Reserve Bank of Richmond,* 1972.

It is clear that the bank cannot invest negative amounts of money and so the usual nonnegativity constraints must be present: $x \geq 0$, $y \geq 0$.

The rate of return on loans is 15% and that on securities is 10%, and if rate of return on investment is the only consideration, then a decision is easy to make: put all the money into loans. Unfortunately, life is not this simple.

While money invested in loans pays a good return, it has the disadvantage of not being liquid. This means that money invested in a loan cannot be readily recovered. For example, if a bank invests money in a 30-year loan, it will be a long time before that money is again available for other purposes. On the other hand, securities do not pay such a high return, but they are liquid; securities can be sold at market value any time. In this problem, we will assume that *the bank has decided that at least* 25% *of its total investment must be liquid.* Let us translate this into a mathematical statement. The statement is about the *total investment* and about the *liquid part* of the investment. Since securities are liquid, y represents the part which is liquid and $x + y$ represents total investment:

Liquid part	Total investment
y	$x + y$

To complete the translation, $0.25(x + y)$ is 25% of the total investment and the liquid part, y, must be at least $0.25(x + y)$; i.e.

$$y \geq 0.25(x + y).$$

This last inequality may be simplified:

$$y \geq \left(\frac{1}{4}\right)(x + y),$$

$$4y \geq x + y,$$

$$3y \geq x,$$

$$x - 3y \leq 0.$$

One final element enters into this problem: the bank wishes to invest at least 30 million in loans. Thus

$$x \geq 30.$$

Now all constraints have been identified:

1. Nonnegativity: $x \geq 0, \quad y \geq 0$
2. Liquidity: $x - 3y \leq 0$
3. Loan: $x \geq 30$
4. Total investment: $x + y \leq 100$

What is the objective function? It is clear that the bank wishes to obtain the largest possible return on investment. Since loans pay 15% and securities 10%, the bank will receive a return of

$$0.15x + 0.10y$$

millions of dollars on the investment. Therefore, the bank wishes to maximize

$$P = 0.15x + 0.10y.$$

This is the objective function and the LP problem is

$$\begin{aligned} &\text{Maximize } P = 0.15x + 0.10y \\ &\text{subject to the constraints} \\ &\quad x + \ \ y \leq 100, \\ &\quad x - 3y \leq \ \ 0, \\ &\qquad\quad x \geq \ 30, \\ &\quad x \geq 0, \quad y \geq 0. \end{aligned}$$

The graph of the set of feasible solutions is shown in Figure 4.27. The corner points are $Q = (30, 70)$, $R = (30, 10)$, $S = (75, 25)$.

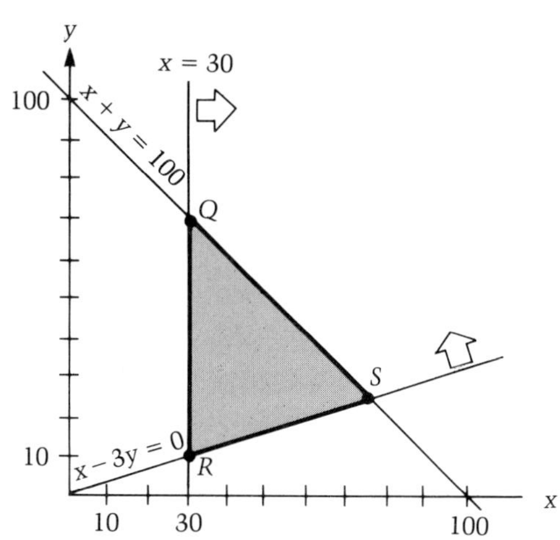

FIG. 4.27.

Corner points	Value of $P = 0.15x + 0.10y$
(30, 70)	$P = 0.15(30) + 0.10(70) = 11.5$
(30, 10)	$P = 0.15(30) + 0.10(10) = 5.5$
(75, 25)	$P = 0.15(75) + 0.10(25) = 13.75$

From the table we conclude that the bank should invest 75 million in loans and 25 million in securities. This will produce a maximum return of 13.75 million dollars.

Now that we have solved some problems by using the LP model it may be helpful to list some of the steps which are usually necessary in this process.

SUGGESTIONS FOR ORGANIZING LP PROBLEMS

1. Read the problem to determine its general nature and to *identify the unknowns.*
2. Assign letters to the unknowns.
3. *Identify the constraints.* This usually means translating the English statements about the unknowns into mathematical statements (inequalities).
4. *Identify the objective function.*
5. *Identify the LP problem.*
6. *Solve the LP problem.* Obtain the optimal solution.
7. Interpret the solution of the LP problem in the original setting.

To carry out Step 3 of these suggestions it is often helpful to construct a data box with the unknowns appearing across the top. This will facilitate the identification of the constraints and the objective function. Example 2 illustrates the use of a data box.

Cost control

EXAMPLE 2 The Buildarama Lumber Company owns forests of pine and fir trees. From these trees the company produces pulpwood, resin, and lumber. An acre of fir trees will yield 1 unit of pulpwood, 1 unit of resin, and 2 units of lumber. An acre of pine yields 3, 2, 1 units, respectively, of pulpwood, resin, and lumber. Buildarama estimates that it costs \$1000 to harvest one acre of pine and \$1500 to harvest one acre of fir. The company has contracts requiring the delivery of 6 units of pulpwood, 5 units of resin, and 4 units of lumber. Therefore, at least the quantities ordered must be produced by Buildarama to fulfill the contracts. In order to do this and *minimize costs* how many acres of pine and fir should be harvested?

Solution

Step 1. Identify the unknowns. The unknowns are the number of acres of pine and fir to be harvested.

Step 2. Assign letters to the unknowns. Let

x = the number of acres of pine to be harvested,

y = the number of acres of fir to be harvested.

Step 3. Identify the constraints. Construct a data box with x, y across the top:

	x Pine	y Fir	Units needed
Pulpwood/acre	3	1	6
Resin/acre	2	1	5
Lumber/acre	1	2	4
Harvesting costs/acre	$1000	$1500	

Using the data box, we see that

$$3x + y$$

represents the units of pulpwood produced from x acres of pine and y acres of fir. Therefore, one of our constraints is

$$3x + y \geq 6.$$

The other constraints are determined in the same way:

$$2x + y \geq 5,$$
$$x + 2y \geq 4.$$

Step 4. Identify the objective function. The data box shows that

$$C = 1000x + 1500y$$

is the cost of harvesting x acres of pine and y acres of fir.

Step 5. Identify the LP problem. The LP model is

$$\begin{aligned}&\text{Minimize } C = 1000x + 1500y\\&\text{subject to the constraints}\\&3x + y \geq 6,\\&2x + y \geq 5,\\&x + 2y \geq 4,\\&x \geq 0, \quad y \geq 0.\end{aligned}$$

Step 6. Solve the LP problem. The graph of the set of feasible solutions is shown below. Now evaluate the objective function at the corner points (0, 6), (1, 3), (2, 1), and (4, 0):

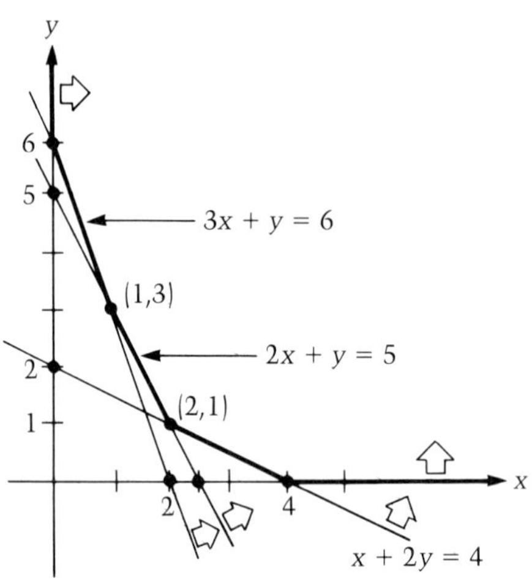

Corner points	Value of $C = 1000x + 1500y$
(0, 6)	$C = 1000(0) + 1500(6) = 9000$
(1, 3)	$C = 1000(1) + 1500(3) = 5500$
(2, 1)	$C = 1000(2) + 1500(1) = 3500$
(4, 0)	$C = 1000(4) + 1500(0) = 4000$

From the table, the solution is $x = 2$, $y = 1$, which gives a minimum value of $C = 3500$.

Step 7. Interpret the solution. From Step 6 we see that Buildarama should harvest 2 acres of pine and 1 acre of fir to meet the contract requirements at the lowest possible cost of \$3500.

EXERCISES

Set A–Variations on Example 1, Investment Management

1. In Example 1 of this section, the bank desires to invest at least 25% of its funds in liquid assets. This requirement leads to the constraint $x - 3y \le 0$. Write (in simple form) the constraint which would result if the bank desired its liquid assets to be at least
(a) 50% (b) 30%
2. Rework Example 1 with the following change: the bank will invest at least \$30 million but no more than \$60 million in loans.
3. Rework Example 1 with the following changes:
(a) The bank has \$150 million to invest.
(b) Loans pay 20% interest and securities pay 15%.
(c) The bank will invest at least \$35 million in loans.

Set B–Variations in Example 2, Cost Control

4. Rework Example 2, assuming that it costs Buildarama \$1000/acre to harvest pine and \$4000/acre to harvest fir.
5. Rework Example 2, with the costs of harvesting trees changed to \$6000/acre for pine and \$1000/acre for fir.

Set C–Setting Up and Solving LP Problems of Various Types

Maximizing profits

6. The Stillon Oil Company produces only gasoline and diesel fuel and naturally desires to make the largest profit possible. The Company must operate under the following conditions:
(a) The refinery can produce no more than 4 million barrels of gasoline per day.
(b) The refinery can produce no more than 5 million barrels of diesel fuel per day.
(c) Government regulations allow Stillon to produce no more than 6 million barrels per day of gasoline and diesel fuel together.
(d) The profit per barrel of gasoline is \$1.10 and that per barrel of diesel fuel is \$1.00.
How many barrels of gasoline and diesel fuel should Stillon produce each day so as to realize its largest profit?

Minimizing costs

7. The Stillon Oil Company operates two refineries, A and B, and each produces regular, high-test, and unleaded gasoline. It costs \$10,000/day to operate refinery A and \$12,000/day to operate B. Every day A produces 1000 gallons of regular, 3000 gallons of high-test and 2000 gallons of unleaded. The daily production at refinery B is 4000 gallons of regular, 5000 gal-

lons of high-test, and 1000 gallons of unleaded. Stillon has orders from its own stations for 32,000 gallons of regular, 47,000 gallons of high-test, and 22,000 gallons of unleaded (any gasoline produced in excess of these amounts can be sold to independent dealers). How many days should each of the refineries operate to fill these orders and minimize the costs of operation?

Maximizing return on investment

8. Mr. Jones has \$100,000 to invest. He plans to invest at least \$20,000 in stocks and bonds. However, the amount invested in stocks must not exceed the amount invested in bonds. For maximum return, how much should Mr. Jones invest in each kind of security if
   (a) stocks pay 20% and bonds 15%?
   (b) stocks pay 15% and bonds 18%?

Minimizing costs of operation

9. The Thor Coal Company owns two mines: Abe I and Abe II. In one day's operation Abe I produces 3 tons of hard, 1 ton of ordinary, and 1 ton of soft coal. Abe II produces 2, 2, and 4 tons of hard, ordinary, and soft coal per day. The costs of operating these mines one day are \$250 for Abe I and \$310 for Abe II. In order to fulfill contracts, the Thor Co. must produce at least 10 tons of hard, 6 tons of ordinary, and 8 tons of soft coal. To meet these requirements at minimum cost, how many days should the two mines be operated?

Maximizing return on investment

10. Janet Chance has \$12,000 to invest. She plans to invest some (possibly all) of her money in stocks and bonds according to the following rules:
   (a) No more than \$6000 is to be put into bonds.
   (b) At least \$3000 is to be invested in stocks and bonds.
   (c) The amount invested in bonds must not exceed the amount invested in stocks by more than \$3000.
   (d) The amount invested in stocks cannot exceed half the amount invested in bonds by more than \$6000.

   If stocks pay 18% and bonds 15%, how much should be invested in each kind of security for maximum return?

4.5 SUMMARY AND REVIEW EXERCISES

REVIEW OF CONCEPTS

I. Graphs of Linear Inequalities. Steps in graphing linear inequalities are:

1. Convert the linear inequality to the corresponding equality.
2. Graph the resulting linear equation.

3. Choose a test point which is not on the line just obtained.
4. Use the test in the original linear inequality.
5. If the inequality is true, the graph is contained in the half-plane containing the test point.
6. If the inequality is not true, the graph is contained in the half-plane not containing the test point.
7. The graph of the inequality may or may not contain the graph of the equation (this depends on the inequality).

II. Characteristics of Linear Programming Problems. Each LP problem consists of two parts:

1. *An Objective Function.* This function is always linear and is to be either maximized or minimized.
2. *A Set of Constraints.* These are always linear inequalities.

III. Steps in Solving LP Problems in the Plane

1. Graph the set of constraints to obtain the set of feasible solutions.
2. Determine the coordinates of the corner points of the set of feasible solutions. (This step may be completed by solving pairs of equations simultaneously.)
3. Evaluate the objective function at the corner points.
4. (a) If the objective function is to be maximized, the largest value from Step 3 is the optimal solution of the problem.
   (b) If the objective function is to be minimized, the smallest value from Step 3 is the optimal solution of the problem.

IV. Suggestions for Applying the LP Model to Real-world Problems

1. Read the problem to determine its general nature and to identify the unknowns.
2. Assign letters to the unknowns.
3. Identify the constraints. This usually means translating the English statements about the unknowns into mathematical statements (inequalities).
4. Identify the objective function.

5. Identify the LP problem.
6. Solve the LP problem. Obtain the optimal solution.
7. Interpret the solution of the LP problem in the original setting.

REVIEW EXERCISES

1. Graph these linear inequalities.
   (a) $-x + y \le 9$
   (b) $4x + 3y > 7$
   (c) $-8x + 5y \le -3,\quad 5x + y \ge 6$
   (d) $2x - 3y < 1,\quad x - 6y \le -4,\quad x \ge 0,\quad y \le 3$

2. Let the set of feasible solutions be the points on and interior to the triangle with vertices (1, 1), (2, 1), and (4, 5). Find the maximum and minimum value of each function on this set.
   (a) $F = 7x + 2y$
   (b) $F = x + 5y$
   (c) $F = -2x - 4y$
   (d) $F = -5x + 9y$

3. Solve the following LP problems graphically.
   (a) Maximize $P = 6x - 3y$ subject to the constraints

$$x - 3y + 2 \le 0,$$
$$2x + 3y \le 14,$$
$$x \ge 1, \quad x \ge 0, \quad y \ge 0.$$

   (b) Minimize $P = 4x + 8y$ subject to the constraints

$$2x + 3y \le 14,$$
$$x + y \ge 3,$$
$$x \le 2y,$$
$$x \ge 1, \quad x \ge 0, \quad y \ge 0.$$

4. John Blue has \$10,000 to invest. He will invest some (possibly all) the money in stock and bonds subject to the following restrictions.
   (a) No more than \$5000 is to be invested in stocks.
   (b) The amount invested in stocks cannot exceed by more than \$3000 the amount invested in bonds.
   (c) The amount invested in bonds and four-thirds the amount in stocks must be at least \$4000.
   (d) Twice the amount in bonds cannot exceed the amount in stocks by more than \$8000.

   If stocks yield 19% and bonds 18%, how much should be invested in each for maximum return?

CHAPTER

5

THE SIMPLEX ALGORITHM

5

5.1 INTRODUCTION

Chapter 4 was devoted to a study of LP problems and their solutions from a geometric point of view. The present chapter expands the scope of these problems by introducing an algebraic technique which is quite general and does not rely on a geometric interpretation. At the conclusion of this chapter the reader will be able to work the following kinds of problems.

Transportation problem

1. The Jones Furniture Company has three retail outlets: A, B, C. Each of these outlets has ordered 20 student desks. The desks must be supplied by the Company's two warehouses I and II. Each warehouse has 30 of these desks in stock. To ship a desk from warehouse I to outlet A or to B costs \$5/desk; to ship a desk from II to B or to A costs \$4. To ship a desk from I or II to C costs \$6. How should the Jones Company schedule shipments so as to keep costs as low as possible?

Diet problem

2. A dietitian wishes to use chicken and eggs to meet certain minimum weekly dietary requirements. Specifically, these two foods are to produce at least 1500 kcal of energy and 50 gm of protein per week. The table below gives information about energy, protein, and costs of these foods per serving.

Food	Energy	Protein	Costs (cents)
Chicken	200	100	25
Eggs	30	10	10

The dietitian feels that no more than 10 servings per week of these two foods should be consumed. To meet these requirements at lowest cost, how many weekly servings of each food should the dietitian plan? ■

5.2 THE SIMPLEX ALGORITHM, PART I

In the previous chapter LP problems have been studied in the plane. Consideration was confined to the plane for good reason: the problems are relatively simple there and it is easier to understand them in this setting. Unfortunately, LP problems which arise in the real world hardly ever occur in the plane. The geometry of LP problems having three unknowns is usually very tedious, and for problems with more than three it is impossible. What is needed, then, is a general method which can be used to solve an LP problem no matter how many unknowns are involved. Fortunately, there is such a method. It is called the *simplex algorithm* and is due to the mathematician George Dantzig.

simplex algorithm

This section is devoted to an explanation of the simplex method for simple problems. Although the method is quite general, it is easier to understand if we start with just two unknowns. This section also treats only problems in which the objective function is to be maximized. Other types will be dealt with later. We also require that constraints (other than nonnegativity constraints) have the form

$$L \leq c$$

where L is a linear function and c is a nonnegative constant. For example,

$$2x - 3y \leq 20$$

is acceptable, but

$$2x - 3y \leq -20$$

is not.

ASSUMPTIONS FOR THIS SECTION

1. Objective functions are to be maximized.
2. All unknowns are nonnegative.
3. All constraints other than those covered in (2) have the standard form

$$L \leq c$$

where L is linear and $c \geq 0$.

In Chapter 3 systems of linear equations were studied in some detail, and we learned there how to use the Gaussian elimination method to solve such systems. It is natural, therefore, to try to use this method on LP problems by turning them into systems of equations. For illustration consider the problem:

Maximize $P = 2x + 3y$
subject to the constraints

$$-x + y \leq 1,$$

$$2x + y \leq 4,$$

$$x \geq 0, \quad y \geq 0.$$

The constraint $-x + y \leq 1$ may be replaced by an equation by reasoning that $-x + y$ may be *increased* (if necessary) to bring its value up to 1, i.e. we may choose a nonnegative number t such that

$$-x + y + t = 1.$$

slack variable

The letter t is called a *slack variable.* The second constraint is also replaced by an equation by choosing another slack variable s:

$$2x + y + s = 4.$$

The original problem may now be written like this

$$-x + y + t = 1, \tag{1a}$$

$$2x + y + s = 4, \tag{1b}$$

$$-2x - 3y + P = 0. \tag{1c}$$

This system has five unknowns and since there are more unknowns than equations, there are infinitely many solutions. From among all these solutions, we wish to choose one in which all unknowns are nonnegative and *P is as large as possible.*

Observe that it is very easy to choose a solution to system (1). If we choose $x = 0$, $y = 0$, then $t = 1$, $s = 4$, $P = 0$. Therefore,

$$x = 0, \quad y = 0, \quad t = 1, \quad s = 4, \quad P = 0 \tag{2}$$

is a solution. Note that it was natural to choose $x = 0$, $y = 0$ because x, y occur in *all* equations, whereas t, s, P occur in only *one equation each.* There is no reason to believe that solution (2) has given us a maximum value for P. In fact, since

$$P = 2x + 3y, \tag{3}$$

we see that we have probably made the worst choice possible. If we choose either x or y to be positive, P will be larger than it is in solution (2). From (3) we see that increases in y will increase P faster than increases in x will. Therefore, let us try to improve solution (2) by keeping $x = 0$ but increasing y.

How much can y be increased? We can answer this by solving (1a) and (1b) for t, s (remember we are still taking $x = 0$):

$$t = 1 - y \qquad (1a)'$$

$$s = 4 - y. \qquad (1b)'$$

Equation (1a)′ shows that we must not choose y larger than 1 because if we do, then t becomes negative (a no-no). From (1b)′, y cannot be larger than 4. Thus the limit for y is set by Eq. (1a). Hence, we eliminate y from all equations (1) except (1a):

$$\begin{array}{rrrrrl} -x + & y + & t & & & = 1 \\ 2x + & y & & + s & & = 4 \\ -2x - & 3y & & & + P & = 0 \end{array} \quad \begin{array}{l} R_2 \rightarrow -R_1 + R_2 \\ R_3 \rightarrow 3R_1 + R_3 \end{array}$$

$$\begin{array}{rrrrrl} -x + & y + & t & & & = 1 \qquad (4a) \\ 3x & - & t & + s & & = 3 \qquad (4b) \\ -5x & & + 3t & & + P & = 3 \qquad (4c) \end{array}$$

Now we can read a new solution by choosing $x = 0$, $t = 0$:

$$x = 0, \quad y = 1, \quad t = 0, \quad s = 3, \quad P = 3. \qquad (5)$$

We still do not know whether (5) is our best solution, so we solve for P in (4c) (remember that $t = 0$):

$$P = 3 + 5x. \qquad (6)$$

Equation (6) shows that the value of P can still be increased if x is increased. To determine how much it is possible to increase x, rewrite (4a) and (4b):

$$y = 1 + x, \qquad (4a)''$$

$$s = 3 - 3x. \qquad (4b)''$$

Equation (4a)″ shows no restriction on x but (4b)″ indicates that x must not be increased by more than 1. So Eq. (4b) is critical, and this indicates that we should eliminate x from all equations except (4b):

$$
\begin{aligned}
-x + y + t \qquad\qquad &= 1 \\
3x \quad - t + s \quad &= 3 \\
-5x \quad + 3t \qquad + P &= 3
\end{aligned}
\qquad R_2 \to \left(\frac{1}{3}\right)R_2
$$

$$
\begin{aligned}
-x + y + t \qquad\qquad &= 1 \\
x \quad - \left(\frac{t}{3}\right) + \left(\frac{s}{3}\right) \quad &= 1 \\
-5x \quad + 3t \qquad + P &= 3
\end{aligned}
\qquad
\begin{aligned}
R_1 &\to R_2 + R_1 \\
R_3 &\to 5R_2 + R_3
\end{aligned}
$$

$$y + \left(\frac{2t}{3}\right) + \left(\frac{s}{3}\right) = 2 \tag{7a}$$

$$x - \left(\frac{t}{3}\right) + \left(\frac{s}{3}\right) = 1 \tag{7b}$$

$$\left(\frac{4t}{3}\right) + \left(\frac{5s}{3}\right) + P = 8 \tag{7c}$$

The easy solution is read by taking $t = 0$, $s = 0$:

$$x = 1, \quad y = 2, \quad t = 0, \quad s = 0, \quad P = 8. \tag{8}$$

To see how good solution (8) is we solve (7c) for P:

$$P = 8 - \left(\frac{4t}{3}\right) - \left(\frac{5s}{3}\right). \tag{9}$$

From Eq. (9) it is clear that if t or s has any positive value, then P is *smaller* than 8. Therefore, solution (8) is the best we can do: $x = 1$, $y = 2$, and P has a maximum value of 8.

Converting an LP problem to a system of equations

EXAMPLE 1 Express the following LP problem as a system of equations by using slack variables:

Maximize $P = 50x + 60y$
subject to the constraints

$$
\begin{aligned}
5x + 8y &\le 20, \\
x + 15y &\le 15, \\
2x + y &\le 30, \\
x \ge 0, \quad y &\ge 0.
\end{aligned}
$$

Solution: A slack variable is needed for each constraint except the two nonnegativity constraints. Since there are three such constraints we need three slack variables. We may choose any suitable letters for these variables. The letters t, s, u will be used here. Then the system is

$$\begin{aligned} 5x + 8y + t \qquad\qquad &= 20, \\ x + 15y \quad + s \qquad &= 15, \\ 2x + y \qquad\quad + u \quad &= 30, \\ -50x - 60y \qquad\qquad + P &= 0. \end{aligned}$$

When an LP problem is given—as in Example 1—it can always be expressed as a system of equations by using slack variables. When this is done, the objective function should be written as the last equation with the unknowns all on the left-hand side of the equation.

initial simplex tableau

The first step in the Simplex Algorithm is to convert the LP problem into a system of equations and then write the augmented matrix for this system. The result of doing this is called the *initial simplex tableau.* In our first example we started with this LP problem:

Maximize $P = 2x + 3y$
subject to the constraints

$$-x + y \leq 1,$$
$$2x + y \leq 4,$$
$$x \geq 0, \quad y \geq 0.$$

This was next expressed as a system of equations:

$$\begin{aligned} -x + y + t \qquad\quad &= 1, \\ 2x + y \quad + s \quad &= 4, \\ -2x - 3y \qquad\quad + P &= 0. \end{aligned} \tag{1}$$

The augmented matrix is the initial simplex tableau and begins the simplex algorithm. The system of equations (1) will be used to illustrate the steps in the simplex algorithm.

STEP 1 of the Simplex Algorithm Construct the initial simplex tableau.

For system (1) the initial tableau is

$$\begin{array}{cc|ccc|c} x & y & t & s & P & \\ \hline -1 & 1 & 1 & 0 & 0 & 1 \\ 2 & 1 & 0 & 1 & 0 & 4 \\ \hline -2 & -3 & 0 & 0 & 1 & 0 \end{array}.$$

Before continuing with the next step in the simplex method it is necessary to explain a technique which allows one to read obvious solutions from a simplex tableau. Going back to system (1), we recall that we began by choosing an obvious solution. The obvious solution was obtained by taking $x = 0$, $y = 0$ and then solving for the remaining variables. This gave us solution (2):

$$x = 0, \quad y = 0, \quad t = 1, \quad s = 4, \quad P = 0.$$

This solution is reflected in the initial simplex tableau in the following way: *the columns for* x *and* y *have more than one nonzero entry, but the columns for t, s, P all have exactly one nonzero entry.* Solutions may always be read from a simplex tableau by using the principle involved here.

read-out principle

READ-OUT PRINCIPLE (ROP) FOR A SIMPLEX TABLEAU

1. If a column for a variable has more than one nonzero entry, then that variable is set equal to zero.
2. If a column for a variable has exactly one nonzero entry (not in last row) and this is 1, then the variable is set equal to the constant in the row containing the 1.

Now return to the solution to system (1). After solution (2) was chosen, it was observed that this solution did not give a value of P that was optimal. Such a determination was made by looking at the objective function $P = 2x + 3y$. We decided to eliminate y from all equations except one because we could see that increasing y caused P to increase more rapidly than increasing x did. This fact is reflected in the simplex tableau: the -3 in the y-column is *more negative* than the -2 in the x-column. In the tableau, the y-column is singled out because of all the columns corresponding to variables, the y-column contains the most negative entry in the last row. It is called the *pivot column.*

pivot column

STEP 2 of the Simplex Algorithm Choose the pivot column. In the last row of the tableau look for the most negative number which is in a column corresponding to a variable. If there are no such numbers, the algorithm terminates and the solution has been obtained. This optimal solution is read by using ROP. If there are negative numbers in the last row and also in a column corresponding to a variable, choose the most negative one. The column containing this negative number is the *pivot column.*

pivot element

After the pivot column has been determined, the *pivot row* and *pivot element* must be chosen. This step corresponds to deciding which equation is critical for the increase of the variable chosen in the pivot column. The pivot row is obtained by the following procedure: for each nonzero element a of the pivot column, not in the last row, form the quotient c/a, where c is the constant in the same row as a. Determine which of these quotients is the smallest positive number. When this smallest positive quotient has been found, the pivot row is the row corresponding to this quotient. The element in the pivot row and pivot column is called the *pivot element.* If a is the pivot element, the pivot row should be multiplied by $1/a$ so that the pivot element will be 1.

STEP 3 of the Simplex Algorithm Choose the pivot element.

basic simplex algorithm

STEP 4 of the Simplex Algorithm Use the pivot element and the pivot row to obtain all 0s in the pivot column except for the pivot element. This is done by using Gaussian elimination.

STEP 5 of the Simplex Algorithm Using the new simplex tableau obtained in Step 4, go to Step 2.

These five steps complete the *Basic Simplex Algorithm.* The process will terminate after a finite number of steps and will end at Step 2.

The five steps just described will now be applied to the initial tableau of system (1):

Step 1

$$\begin{array}{c} \begin{matrix} x & \;\; y & \;\; t & \;\; s & \;\; P & \end{matrix} \\ \left[\begin{array}{cc:ccc:c} -1 & 1 & 1 & 0 & 0 & 1 \\ 2 & 1 & 0 & 1 & 0 & 4 \\ \hdashline -2 & -3 & 0 & 0 & 1 & 0 \end{array}\right]. \\ \quad\;\;\uparrow \end{array}$$

Step 2. Choose the pivot column. The arrow indicates that -3 is the most negative number of the last row and hence column 2 is the pivot column.

Step 3. Choose the pivot element. To do this we must form quotients using elements in the pivot column. These are indicated below:

$$\begin{array}{c} \\ \end{array}\left[\begin{array}{cc:ccc:c} x & y & t & s & P & \\ -1 & \boxed{1} & 1 & 0 & 0 & \boxed{1} \\ 2 & \boxed{1} & 0 & 1 & 0 & \boxed{4} \\ \hdashline -2 & -3 & 0 & 0 & 1 & 0 \end{array}\right] \qquad \begin{array}{l} \underline{\text{Quotients}} \\ \dfrac{1}{1} = 1 \leftarrow \\ \dfrac{4}{1} = 4 \end{array}$$

Since 1 is the smallest positive quotient, row one is the pivot row. It is customary to circle the pivot element as shown:

$$\text{Pivot row} \rightarrow \left[\begin{array}{cc:ccc:c} x & y & t & s & P & \\ -1 & \textcircled{1} & 1 & 0 & 0 & 1 \\ 2 & 1 & 0 & 1 & 0 & 4 \\ \hdashline -2 & -3 & 0 & 0 & 1 & 0 \\ & \uparrow & & & & \\ & \text{Pivot column} & & & & \end{array}\right].$$

Step 4. Using Gaussian elimination, the pivot row, and the pivot element, we obtain 0s in the pivot column:

$$\left[\begin{array}{cc:ccc:c} x & y & t & s & P & \\ -1 & \textcircled{1} & 1 & 0 & 0 & 1 \\ 2 & 1 & 0 & 1 & 0 & 4 \\ \hdashline -2 & -3 & 0 & 0 & 1 & 0 \end{array}\right] \quad \begin{array}{l} R_2 \rightarrow -R_1 + R_2 \\ R_3 \rightarrow 3R_1 + R_3 \end{array}$$

$$\left[\begin{array}{cc:ccc:c} -1 & 1 & 1 & 0 & 0 & 1 \\ 3 & 0 & -1 & 1 & 0 & 3 \\ \hdashline -5 & 0 & 3 & 0 & 1 & 3 \end{array}\right]$$

Step 5. Go to Step 2.

Step 2. Use the new simplex tableau from Step 4 and look for negative numbers corresponding to variables in the last row. If there are any, choose the most negative one. There is one in this case:

$$\left[\begin{array}{cc:ccc:c} x & y & t & s & P & \\ -1 & 1 & 1 & 0 & 0 & 1 \\ 3 & 0 & -1 & 1 & 0 & 3 \\ \hdashline -5 & 0 & 3 & 0 & 1 & 3 \\ \uparrow & & & & & \end{array}\right].$$

The arrow shows that column one is the new pivot column.

Step 3. Choose the pivot element by forming the appropriate quotients:

$$\begin{array}{c} \begin{matrix} x & y & t & s & P & \end{matrix} \\ \left[\begin{array}{cc:ccc:c} -1 & 1 & 1 & 0 & 0 & 1 \\ \textcircled{3} & 0 & -1 & 1 & 0 & 3 \\ \hdashline -5 & 0 & 3 & 0 & 1 & 3 \end{array}\right]. \end{array} \qquad \begin{array}{l} \underline{\text{Quotients}} \\ \dfrac{1}{-1} = -1 \\ \dfrac{3}{3} = 1 \leftarrow \end{array}$$

There is only one positive quotient here and that is in row two. Thus row two is the pivot row, and 3 is the pivot element:

$$\begin{array}{c} \begin{matrix} x & y & t & s & P & \end{matrix} \\ \left[\begin{array}{cc:ccc:c} -1 & 1 & 1 & 0 & 0 & 1 \\ \textcircled{3} & 0 & -1 & 1 & 0 & 3 \\ \hdashline -5 & 0 & 3 & 0 & 1 & 3 \end{array}\right] \\ \left[\begin{array}{cc:ccc:c} -1 & 1 & 1 & 0 & 0 & 1 \\ \textcircled{1} & 0 & -\frac{1}{3} & \frac{1}{3} & 0 & 1 \\ \hdashline -5 & 0 & 3 & 0 & 1 & 3 \end{array}\right]. \end{array} \qquad R_2 \to \left(\frac{1}{3}\right)R_2$$

Step 4. Obtain 0s in the pivot column except for the pivot element:

$$\begin{array}{c} \begin{matrix} x & y & t & s & P & \end{matrix} \\ \left[\begin{array}{cc:ccc:c} -1 & 1 & 1 & 0 & 0 & 1 \\ \textcircled{1} & 0 & -\frac{1}{3} & \frac{1}{3} & 0 & 1 \\ \hdashline -5 & 0 & 3 & 0 & 1 & 3 \end{array}\right] \\ \left[\begin{array}{cc:ccc:c} 0 & 1 & \frac{2}{3} & \frac{1}{3} & 0 & 2 \\ 1 & 0 & -\frac{1}{3} & \frac{1}{3} & 0 & 1 \\ \hdashline 0 & 0 & \frac{4}{3} & \frac{5}{3} & 1 & 8 \end{array}\right]. \end{array} \qquad \begin{array}{l} R_1 \to R_2 + R_1 \\ R_3 \to 5R_2 + R_3 \end{array}$$

Step 5. Go to Step 2.

Step 2. In the new simplex tableau resulting from Step 4 there are no negative numbers in the last row. Hence, the process terminates and the solution may be read by using ROP:

$$x = 1, \quad y = 2, \quad t = 0, \quad s = 0, \quad P = 8.$$

Applying ROP to a tableau

EXAMPLE 2 Apply ROP to the following tableau.

$$\begin{array}{cccc|ccc|c} & x & y & z & t & s & P & \\ \hline & 5 & 0 & 1 & 1 & 0 & 0 & 10 \\ & 2 & 1 & 0 & 0 & -3 & 0 & 12 \\ \hline & -7 & 0 & -1 & 0 & 4 & 1 & 20 \end{array}.$$

Solution: The x-column contains more than one nonzero element, and by ROP $x = 0$. The y-column contains only one nonzero number which is 1 and therefore $y = 12$. Similarly, $z = 0$, $t = 10$, $s = 0$, $P = 20$.

Using the simplex algorithm

EXAMPLE 3 Solve the LP problem by using the simplex algorithm.

Maximize $P = 25x + 12y$
subject to the constraints

$$x \leq 3,$$
$$y \leq 4,$$
$$2x + y \leq 8,$$
$$x \geq 0, \quad y \geq 0.$$

Solution: To set up the initial tableau, write the problem as a system of equations. Three slack variables are needed. We choose t, s, u, and the equations are

$$\begin{aligned} x \quad\quad + t \quad\quad\quad\quad &= 3, \\ y \quad + s \quad\quad &= 4, \\ 2x + y \quad\quad\quad + u \quad &= 8, \\ -25x - 12y \quad\quad\quad\quad + P &= 0. \end{aligned}$$

The tableau for Step 1 is

$$\begin{array}{cc|cccc|c} x & y & t & s & u & P & \\ \hline 1 & 0 & 1 & 0 & 0 & 0 & 3 \\ 0 & 1 & 0 & 1 & 0 & 0 & 4 \\ 2 & 1 & 0 & 0 & 1 & 0 & 8 \\ \hline -25 & -12 & 0 & 0 & 0 & 1 & 0 \end{array}.$$

Step 2. Choose the pivot column. This is the x-column since -25 is the most negative number of the last row.

Step 3. Choose the pivot element. For this, the correct quotients must be computed:

$$\begin{array}{c}
\begin{array}{cc:cccc:c} x & y & t & s & u & P & \end{array} \\
\left[\begin{array}{cc:cccc:c}
\boxed{1} & 0 & 1 & 0 & 0 & 0 & \boxed{3} \\
0 & 1 & 0 & 1 & 0 & 0 & 4 \\
\boxed{2} & 1 & 0 & 0 & 1 & 0 & \boxed{8} \\
\hdashline
-25 & -12 & 0 & 0 & 0 & 1 & 0
\end{array}\right]
\end{array}
\qquad
\begin{array}{l}
\underline{\text{Quotients}} \\
\dfrac{3}{1} = 3 \leftarrow \\
\dfrac{8}{2} = 4
\end{array}$$

This shows that row one is the pivot row since 3 is the smallest quotient. The pivot element is then in the first row, first column.

Step 4. Use Gaussian elimination to obtain 0s in the pivot column:

$$\begin{array}{cc:cccc:c} x & y & t & s & u & P & \end{array}$$

$$\left[\begin{array}{cc:cccc:c}
\textcircled{1} & 0 & 1 & 0 & 0 & 0 & 3 \\
0 & 1 & 0 & 1 & 0 & 0 & 4 \\
2 & 1 & 0 & 0 & 1 & 0 & 8 \\
\hdashline
-25 & -12 & 0 & 0 & 0 & 1 & 0
\end{array}\right]
\quad
\begin{array}{l} R_3 \to -2R_1 + R_3 \\ R_4 \to 25R_1 + R_4 \end{array}$$

$$\left[\begin{array}{cc:cccc:c}
1 & 0 & 1 & 0 & 0 & 0 & 3 \\
0 & 1 & 0 & 1 & 0 & 0 & 4 \\
0 & 1 & -2 & 0 & 1 & 0 & 2 \\
\hdashline
0 & -12 & 25 & 0 & 0 & 1 & 75
\end{array}\right].$$

Step 5. Go to Step 2.

Step 2. Choose the pivot column. In the new tableau obtained in Step 4 the y-column is the only one containing a negative number in the last row. Thus the second column is the pivot column.

Step 3. Choose the pivot element.

$$\begin{array}{c}
\begin{array}{cc:cccc:c} x & y & t & s & u & P & \end{array} \\
\left[\begin{array}{cc:cccc:c}
1 & 0 & 1 & 0 & 0 & 0 & 3 \\
0 & \boxed{1} & 0 & 1 & 0 & 0 & \boxed{4} \\
0 & \boxed{1} & -2 & 0 & 1 & 0 & \boxed{2} \\
\hdashline
0 & -12 & 25 & 0 & 0 & 1 & 75
\end{array}\right].
\end{array}
\qquad
\begin{array}{l}
\underline{\text{Quotients}} \\
\dfrac{4}{1} = 4 \\
\dfrac{2}{1} = 2 \leftarrow
\end{array}$$

From the quotients, row 3 is the pivot row, and the pivot element is in row three, column two.

Step 4. Use Gaussian elimination to obtain 0s in the pivot column:

$$\begin{array}{cc|cccc|c} x & y & t & s & u & P & \\ 1 & 0 & 1 & 0 & 0 & 0 & 3 \\ 0 & 1 & 0 & 1 & 0 & 0 & 4 \\ 0 & \textcircled{1} & -2 & 0 & 1 & 0 & 2 \\ \hline 0 & -12 & 25 & 0 & 0 & 1 & 75 \end{array} \quad \begin{array}{l} R_2 \to -R_3 + R_2 \\ R_4 \to 12R_3 + R_4 \end{array}$$

$$\left[\begin{array}{cc|cccc|c} 1 & 0 & 1 & 0 & 0 & 0 & 3 \\ 0 & 0 & 2 & 1 & -1 & 0 & 2 \\ 0 & 1 & -2 & 0 & 1 & 0 & 2 \\ \hline 0 & 0 & 1 & 0 & 12 & 1 & 99 \end{array}\right].$$

Step 5. Go to Step 2.

Step 2. The algorithm terminates at this point since the new tableau of Step 4 contains no negative numbers in the last row. Therefore, by ROP we read the optimal solution:

$$x = 3, \quad y = 2, \quad P = 99.$$

---

basic simplex algorithm

A summary of the steps in the Simplex Algorithm is given below, but the student is reminded that we are making some special assumptions as listed at the beginning of this section, and the steps in the algorithm apply under these assumptions. The algorithm summarized here will be referred to as the *Basic Simplex Algorithm.*

SUMMARY OF STEPS IN THE BASIC SIMPLEX ALGORITHM

Step 1. Construct the initial Simplex tableau.

Step 2. Terminate the algorithm or choose the pivot column.

(a) If there are no negative numbers in the last row which correspond to variables, the problem is solved and the optimal solution is read by ROP.

(b) If there are negative numbers in the last row which correspond to variables, then the pivot column is that column containing the most negative of these numbers.

Step 3. Choose the pivot element. In the pivot column and column of constants, compute the appropriate quotients.

The pivot row is that row corresponding to the smallest positive quotient. The pivot element is the element in the pivot row and pivot column.

Step 4. Use Gaussian elimination and the pivot element to obtain 0s in the pivot column (except in the pivot row).

Step 5. Go to Step 2.

Using the simplex method

EXAMPLE 4 Use the simplex method to solve the following LP problem:

$$\text{Maximize } P = 10x + 12y + 8z$$
subject to the constraints
$$x + 2y + z \le 8,$$
$$x + y + 2z \le 10,$$
$$x \ge 0, \quad y \ge 0, \quad z \ge 0.$$

Solution: The letters t, s are used as slack variables to write the linear system

$$\begin{aligned} x + 2y + z + t \qquad &= 8, \\ x + y + 2z \quad + s \quad &= 10, \\ -10x - 12y - 8z \qquad + P &= 0. \end{aligned}$$

Then the steps in the simplex method are as follows:

$$\begin{array}{c} \begin{matrix} x & y & z & t & s & P & \end{matrix} \\ \left[\begin{array}{ccc:ccc:c} 1 & \textcircled{2} & 1 & 1 & 0 & 0 & 8 \\ 1 & 1 & 2 & 0 & 1 & 0 & 10 \\ \hdashline -10 & -12 & -8 & 0 & 0 & 1 & 0 \end{array}\right] \end{array} \qquad \text{Quotients: } \frac{8}{2} = 4 \leftarrow \quad \frac{10}{1} = 10$$

$R_1 \to 0.5R_1$

$$\left[\begin{array}{ccc:ccc:c} 0.5 & \textcircled{1} & 0.5 & 0.5 & 0 & 0 & 4 \\ 1 & 1 & 2 & 0 & 1 & 0 & 10 \\ \hdashline -10 & -12 & -8 & 0 & 0 & 1 & 0 \end{array}\right]$$

$R_2 \to -R_1 + R_2$
$R_3 \to 12R_1 + R_3$

$$\left[\begin{array}{ccc:ccc:c} \textcircled{0.5} & 1 & 0.5 & 0.5 & 0 & 0 & 4 \\ 0.5 & 0 & 1.5 & -0.5 & 1 & 0 & 6 \\ \hdashline -4 & 0 & -2 & 6 & 0 & 1 & 48 \end{array}\right] \qquad \frac{4}{0.5} = 8 \leftarrow \quad \frac{6}{0.5} = 12$$

$$\left[\begin{array}{ccc:ccc:c} ① & 2 & 1 & 1 & 0 & 0 & 8 \\ 1 & 0 & 3 & -1 & 2 & 0 & 12 \\ \hdashline -4 & 0 & -2 & 6 & 0 & 1 & 48 \end{array}\right] \begin{array}{l} R_1 \to 2R_1 \\ R_2 \to 2R_2 \end{array}$$

$$\begin{array}{l} R_2 \to -R_1 + R_2 \\ R_3 \to 4R_1 + R_3 \end{array} \left[\begin{array}{ccc:ccc:c} 1 & 2 & 1 & 1 & 0 & 0 & 8 \\ 0 & -2 & 2 & -2 & 2 & 0 & 4 \\ \hdashline 0 & 8 & 2 & 10 & 0 & 1 & 80 \end{array}\right].$$

The algorithm terminates since there are no more negative numbers in the last row, and the solution is

$$x = 8, \quad y = 0, \quad z = 0, \quad P = 80.$$

EXERCISES

Set A–Using Slack Variables and Constructing Simplex Tableaux

In Exercises 1 through 5, write the given LP problem as a system of linear equations by using slack variables. Also construct the initial simplex tableau.

1. Maximize $P = 25x - 13y$ subject to the constraints

$$2x + y \le 12,$$
$$x + y \le 9,$$
$$x \ge 0, \quad y \ge 0$$

2. Maximize $P = 10x + 5y$ subject to the constraints

$$x + 2y \le 10,$$
$$2x + y \le 8,$$
$$x \ge 0, \quad y \ge 0$$

3. Maximize $P = 2x - y$ subject to the constraints

$$2x + y \le 6,$$
$$-x + 2y \le 2,$$
$$x \ge 0, \quad y \ge 0$$

4. Maximize $P = -4x + 7y$ subject to the constraints

$$3x + 2y \le 24,$$
$$x + y \le 9,$$
$$x + 4y \le 24,$$
$$x \ge 0, \quad y \ge 0$$

5. Maximize $P = 2x + 3y + 4z$ subject to the constraints

$$5x - y + 2z \le 20,$$
$$2x + y + 3z \le 15,$$
$$x + 4y - z \le 18,$$
$$x \ge 0, \quad y \ge 0, \quad z \ge 0$$

Set B–Setting Up the Initial Simplex Tableau and Using ROP

In each of Exercises 6 through 9, set up the initial simplex tableau and apply ROP to obtain an initial solution for the system of equations.

6. Maximize $P = 30x - 12y$ subject to the constraints

$$3x - y \le 14,$$
$$x + y \le 8,$$
$$x \ge 0, \quad y \ge 0$$

7. Maximize $P = 11x + 6y$ subject to the constraints

$$2x + 3y \le 12,$$
$$2x + 5y \le 9,$$
$$x \ge 0, \quad y \ge 0$$

8. Maximize $P = 3x + 2y$ subject to the constraints

$$3x + 2y \le 7,$$
$$-x + 4y \le 2,$$
$$x \ge 0, \quad y \ge 0$$

9. Maximize $P = -4x + 5y$ subject to the constraints

$$4x + 3y \le 24,$$
$$2x + y \le 9,$$
$$2x + 4y \le 24,$$
$$x \ge 0, \quad y \ge 0$$

Set C–Solving LP problems by the simplex method

In Exercises 10 through 19, use the Simplex Algorithm to solve each of the LP problems given

10. Maximize $P = 2x + 4y$ subject to the constraints

$$3x + 2y \le 6,$$
$$x + y \le 2,$$
$$x \ge 0, \quad y \ge 0$$

11. Maximize $P = 8x + 6y$ subject to the constraints

$$3x + y \le 33,$$
$$x + 2y \le 10,$$
$$x \ge 0, \quad y \ge 0$$

12. Maximize $P = 8x + 6y$ subject to the constraints

$$2x + 3y \le 22,$$
$$2x + y \le 20,$$
$$x \ge 0, \quad y \ge 0$$

13. Maximize $P = 20x + 30y$ subject to the constraints

$$x + 2y \le 8,$$
$$x + 6y \le 30,$$
$$x \ge 0, \quad y \ge 0$$

14. Maximize $P = 12x + 10y$ subject to the constraints

$$\begin{aligned} 2x + 7y &\le 32, \\ 2x + 5y &\le 20, \\ 2x + y &\le 8, \\ x \ge 0, \quad y &\ge 0 \end{aligned}$$

15. Maximize $P = 6x + 8y$ subject to the constraints

$$\begin{aligned} x + 3y &\le 27, \\ x &\le 10, \\ y &\le 8, \\ x \ge 0, \quad y &\ge 0 \end{aligned}$$

16. Maximize $P = 6x + 2y + 8z$ subject to the constraints

$$\begin{aligned} 2x + z &\le 14, \\ y + z &\le 12, \\ x \ge 0, \quad y \ge 0, \quad z &\ge 0 \end{aligned}$$

17. Maximize $P = 6x + 4y + 2z$ subject to the constraints

$$\begin{aligned} x + 2z &\le 8, \\ x + 2y + z &\le 10, \\ x \ge 0, \quad y \ge 0, \quad z &\ge 0 \end{aligned}$$

18. Maximize $P = 6x - 2y + 8z$ subject to the constraints

$$\begin{aligned} 2x \quad + z &\le 14, \\ -2x + y + z &\le 12, \\ x - y + z &\le 10, \\ x \ge 0, \quad y \ge 0, \quad z &\ge 0 \end{aligned}$$

19. Maximize $P = 6x - 2y + 8z$ subject to the constraints

$$\begin{aligned} x - y + z &\le 2, \\ x + y + z &\le 4, \\ -x + y + z &\le 6, \\ x \quad - z &\le 8, \\ x \ge 0, \quad y \ge 0, \quad z &\ge 0 \end{aligned}$$

Set D–Applications of LP Problems

Maximizing profits

20. The Stillon Oil Company produces regular, high-test, and unleaded gasoline and makes a profit of \$1.00 per barrel of regular, \$1.20 per barrel of high-test, and \$1.40 per barrel of unleaded. The Stillon refinery can produce no more than 4 million barrels of regular per day, 5 million barrels of high-test per day, and 6 million barrels of unleaded per day. The total of all three fuels cannot exceed the plant capacity of 8 million barrels per day. How many barrels of each kind of fuel should Stillon produce per day for maximum profits?

Maximizing return on investment

21. Mr. Jones has \$100,000 which he plans to invest in stocks, bonds, and money-market certificates. Stocks return 20%,

bonds 15%, and certificates 16%. As guidelines for his investments Jones requires that

(a) \$2000 more than the combined investment in bonds and certificates is no less than the amount invested in stocks.

(b) the amount invested in bonds is no more than 25% of the combined investment in stocks and certificates.

How much should Jones invest in each asset in order to obtain maximum return?

Scheduling for maximum profit

22. The Strong Desk Company is currently making three models of desks: the Executive, the Secretary, and the Student. These desks must be processed in the cabinet, finishing, and crating departments, and the time requirements for each desk model in these departments are as follows:

	Exec.	Sec.	Stu.
Cabinet	2	1	1
Finishing	1	2	1
Crating	1	1	.5

The available times in these departments are 16 hours per day in the cabinet, 16 hours per day in the finishing, and 10 hours per day in the crating departments, respectively. If the Company makes \$150 profit on each Executive model, \$125 on each Secretary, and \$50 on each Student, how many desks of each model should be scheduled each day for maximum profits?

Maximizing profits

23. Given the same information as in Exercise 22 except the profit on each Student desk is now \$80.

(a) With this change, how many of each model should be made per day to maximize profit, and what is this profit?

(b) The results of Exercise 22 suggest that the Strong Desk Company should consider eliminating the Student model as long as profit from each of these is only \$50. What is the smallest amount of profit per desk that the Company needs to make on the Student desk so that it is reasonable for this line to be continued?

5.3 THE SIMPLEX ALGORITHM, PART II: MIXED CONSTRAINTS

The simplex method discussed in the previous section works for many maximum problems which satisfy the special restrictions stated in Section 5.2. The object of this section is to broaden the scope of the algorithm so that a wider range of problems can be solved. The nonnegativity requirement on variables is retained here but other requirements will be dropped. It will no longer be required that constraints have the form

$$L \leq c$$

where $c \geq 0$ (of course, L must still be a linear function). For example, the constraint

$$2x + y \geq 6$$

will be allowed in this section. Each $\geq$ constraint will be converted to a $\leq$ constraint by multiplying both sides of the inequality by -1. Thus $2x + y \geq 6$ is converted to

$$-2x - y \leq -6.$$

mixed constraints

When an LP problem has both $\leq$ and $\geq$ constraints (not counting the nonnegativity constraints), we say that the constraints are *mixed*. To see how a problem with mixed constraints is handled, consider the following:

Maximize $P = 10x + 5y$
subject to the constraints

$$2x + y \geq 6,$$
$$4x + y \leq 10,$$
$$x \geq 0, \quad y \geq 0.$$

Begin by changing the first constraint to the $\leq$-form. Then the problem is

Maximize $P = 10x + 5y$
subject to the constraints

$$-2x - y \leq -6,$$
$$4x + y \leq 10,$$
$$x \geq 0, \quad y \geq 0.$$

Proceeding in the usual way, choose slack variables t, s and express the problem as a system of equations:

$$\begin{aligned} -2x - y + t &= -6, \\ 4x + y + s &= 10, \\ -10x - 5y + P &= 0. \end{aligned} \qquad (1)$$

The initial tableau is

$$\begin{array}{cc|ccc|c} x & y & t & s & P & \\ -2 & -1 & 1 & 0 & 0 & -6 \\ 4 & 1 & 0 & 1 & 0 & 10 \\ \hline -10 & -5 & 0 & 0 & 1 & 0 \end{array}.$$

Using ROP, we have

$$x = 0, \quad y = 0, \quad \underbrace{t = -6,}_{\uparrow} \quad s = 10, \quad P = 0.$$

The difficulty here shows up immediately: the variable t is negative, and this violates the nonnegativity condition on variables. To see how this can be overcome, consider the equation

$$-2x - y + t = -6$$

of system (1). Solve for t:

$$t = -6 + 2x + y \tag{2}$$

Equation (2) shows that if either x or y has positive values, t will be greater than -6 and, of course, this is what we want. In fact, we want to increase t from -6 to 0. Therefore, it appears that we should increase x or y from 0 to some positive value. *This is reflected in the initial tableau by having negative numbers* (-2 and -1) *in the x-, y-columns of row one.* It is apparent that it might be helpful to eliminate x or y from all equations of system (1) except one. This is equivalent to using -2 or -1 of row one as a pivot element in the initial tableau. To see how this works, -1 is chosen as pivot below:

$$\begin{array}{cc|ccc|c} x & y & t & s & P & \\ -2 & \boxed{-1} & 1 & 0 & 0 & -6 \\ 4 & 1 & 0 & 1 & 0 & 10 \\ \hline -10 & -5 & 0 & 0 & 1 & 0 \end{array} \qquad \begin{array}{l} R_2 \to R_1 + R_2 \\ R_3 \to -5R_1 + R_3 \end{array}$$

$$\begin{array}{cc|ccc|c} -2 & -1 & 1 & 0 & 0 & -6 \\ 2 & 0 & 1 & 1 & 0 & 4 \\ \hline 0 & 0 & -5 & 0 & 1 & 30 \end{array}.$$

By ROP we have:

$$x = 0, \quad y = 6, \quad t = 0, \quad s = 4, \quad P = 30.$$

The difficulty has been cleared up, and we may now use the basic Simplex Algorithm to finish the problem:

$$\left[\begin{array}{cc:ccc:c} -2 & -1 & 1 & 0 & 0 & -6 \\ 2 & 0 & \textcircled{1} & 1 & 0 & 4 \\ \hdashline 0 & 0 & -5 & 0 & 1 & 30 \end{array}\right] \qquad \begin{array}{l} \underline{\text{Quotients}} \\ -\dfrac{6}{1} = -6 \\ \dfrac{4}{1} = 4 \leftarrow \end{array}$$

$$R_1 \to -R_2 + R_1$$
$$R_3 \to 5R_2 + R_3$$

$$\left[\begin{array}{cc:ccc:c} -4 & -1 & 0 & -1 & 0 & -10 \\ 2 & 0 & 1 & 1 & 0 & 4 \\ \hdashline 10 & 0 & 0 & 5 & 1 & 50 \end{array}\right].$$

There are no more negative numbers in the last row, so we are done, and the optimal solution is

$$x = 0, \quad y = 10, \quad P = 50.$$

These considerations lead us to the following rules:

RULE FOR DEALING WITH MIXED CONSTRAINTS If an LP problem contains a $\geq$ constraint (other than nonnegativity constraints), each of these is converted to a $\leq$ constraint by multiplying each side of the expression by -1.

RULE FOR DEALING WITH NEGATIVE VARIABLES If a variable has a negative value when ROP is applied, then in the row corresponding to the negative variable choose one of the negative elements* and use it as a pivot.

The rule for dealing with negative variables must sometimes be applied several times before all negatives are removed. Example 1 shows how this can happen.

---

* In a case such as this there will always be negative entries in the row containing the negative constant. If there were none, the row would correspond to a constraint such as $2x + 3y \leq -6$ which has no solution in nonnegative values of x and y. Since we are not considering problems without solutions, this situation will not arise.

Using simplex with mixed constraints

EXAMPLE 1 Solve the following LP problem.

Maximize $P = 2x + 3y$
subject to the constraints

$$y \le 1,$$
$$x + y \ge 1,$$
$$x - y \le 0,$$
$$x \ge 0, \quad y \ge 0.$$

Solution: The second constraint is in $\ge$ form and must be changed to the $\le$ form. Then the inequality is $-x - y \le -1$. Using slack variables t, s, u, we obtain the system of equations

$$\begin{aligned} y + t \qquad\qquad &= 1, \\ -x - y \quad + s \qquad &= -1, \\ x - y \qquad + u \quad &= 0, \\ -2x - 3y \qquad\quad + P &= 0. \end{aligned}$$

From the system the initial tableau is

$$\begin{array}{cc|cccc|c} x & y & t & s & u & P & \\ 0 & 1 & 1 & 0 & 0 & 0 & 1 \\ -1 & -1 & 0 & 1 & 0 & 0 & -1 \\ 1 & -1 & 0 & 0 & 1 & 0 & 0 \\ \hline -2 & -3 & 0 & 0 & 0 & 1 & 0 \end{array}.$$

Using ROP for this tableau gives

$$x = 0, \quad y = 0, \quad t = 1, \quad s = -1, \quad u = 0, \quad P = 0.$$

The equation $s = -1$ is a violation of the nonnegativity requirement for variables and occurs in row two of the tableau. Therefore, we choose a negative number as pivot in this row. There are two negative entries here and we may choose either one of them. In this case we shall use the -1 in column one.

$$\begin{array}{cc|cccc|c} x & y & t & s & u & P & \\ 0 & 1 & 1 & 0 & 0 & 0 & 1 \\ \circled{-1} & -1 & 0 & 1 & 0 & 0 & -1 \\ 1 & -1 & 0 & 0 & 1 & 0 & 0 \\ \hline -2 & -3 & 0 & 0 & 0 & 1 & 0 \end{array} \quad R_2 \to -R_2$$

$$\left[\begin{array}{rr|rrrr|r} 0 & 1 & 1 & 0 & 0 & 0 & 1 \\ \textcircled{1} & 1 & 0 & -1 & 0 & 0 & 1 \\ 1 & -1 & 0 & 0 & 1 & 0 & 0 \\ \hline -2 & -3 & 0 & 0 & 0 & 1 & 0 \end{array}\right] \quad \begin{array}{l} R_3 \to -R_2 + R_3 \\ R_4 \to 2R_2 + R_4 \end{array}$$

$$\left[\begin{array}{rr|rrrr|r} 0 & 1 & 1 & 0 & 0 & 0 & 1 \\ 1 & 1 & 0 & -1 & 0 & 0 & 1 \\ 0 & -2 & 0 & 1 & 1 & 0 & -1 \\ \hline 0 & -1 & 0 & -2 & 0 & 1 & 2 \end{array}\right].$$

Now use ROP again:

$$x = 1, \quad y = 0, \quad t = 1, \quad s = 0, \quad u = -1, \quad P = 2.$$

The presence of $u = -1$ indicates that we are still not out of the woods and must repeat the process again. This time row three is the pivot row and -2 is the pivot element:

$$\left[\begin{array}{rr|rrrr|r} 0 & 1 & 1 & 0 & 0 & 0 & 1 \\ 1 & 1 & 0 & -1 & 0 & 0 & 1 \\ 0 & \textcircled{-2} & 0 & 1 & 1 & 0 & -1 \\ \hline 0 & -1 & 0 & -2 & 0 & 1 & 2 \end{array}\right] \quad R_3 \to -0.5R_3$$

$$\left[\begin{array}{rr|rrrr|r} 0 & 1 & 1 & 0 & 0 & 0 & 1 \\ 1 & 1 & 0 & -1 & 0 & 0 & 1 \\ 0 & \textcircled{1} & 0 & -0.5 & -0.5 & 0 & 0.5 \\ \hline 0 & -1 & 0 & -2 & 0 & 1 & 2 \end{array}\right] \quad \begin{array}{l} R_1 \to -R_3 + R_1 \\ R_2 \to -R_3 + R_2 \\ R_4 \to R_3 + R_4 \end{array}$$

$$\left[\begin{array}{rr|rrrr|r} 0 & 0 & 1 & 0.5 & 0.5 & 0 & 0.5 \\ 1 & 0 & 0 & -0.5 & 0.5 & 0 & 0.5 \\ 0 & 1 & 0 & -0.5 & -0.5 & 0 & 0.5 \\ \hline 0 & 0 & 0 & -2.5 & -0.5 & 1 & 2.5 \end{array}\right].$$

By ROP

$$x = 0.5, \quad y = 0.5, \quad t = 0.5, \quad s = 0, \quad u = 0, \quad P = 2.5.$$

The solution here shows no negative variables and indicates that we may now begin the standard algorithm:

$$\left[\begin{array}{rr|rrrr|r} 0 & 0 & 1 & 0.5 & 0.5 & 0 & 0.5 \\ 1 & 0 & 0 & -0.5 & 0.5 & 0 & 0.5 \\ 0 & 1 & 0 & -0.5 & -0.5 & 0 & 0.5 \\ \hline 0 & 0 & 0 & -2.5 & -0.5 & 1 & 2.5 \end{array}\right]$$

$$\left[\begin{array}{cc:cccc:c} 0 & 0 & 2 & \textcircled{1} & 1 & 0 & 1 \\ 2 & 0 & 0 & -1 & 1 & 0 & 1 \\ 0 & 2 & 0 & -1 & -1 & 0 & 1 \\ \hdashline 0 & 0 & 0 & -5 & -1 & 2 & 5 \end{array}\right] \quad \begin{array}{l} R_1 \to 2R_1 \\ R_2 \to 2R_2 \\ R_3 \to 2R_3 \\ R_4 \to 2R_4 \end{array}$$

$$\begin{array}{l} R_2 \to R_1 + R_2 \\ R_3 \to R_1 + R_3 \\ R_4 \to 5R_1 + R_4 \end{array}$$

$$\left[\begin{array}{cc:cccc:c} 0 & 0 & 2 & 1 & 1 & 0 & 1 \\ 2 & 0 & 2 & 0 & 2 & 0 & 2 \\ 0 & 2 & 2 & 0 & 0 & 0 & 2 \\ \hdashline 0 & 0 & 10 & 0 & 4 & 2 & 10 \end{array}\right].$$

The algorithm terminates here since there are no negative numbers in the last row and the optimum solution is

$$x = 1, \quad y = 1, \quad P = 5.$$

In solving LP problems by the simplex method it is important that ROP be applied after each pivot operation in order to determine whether the basic algorithm applies. For the usual kind of situation the basic algorithm automatically calculates the objective function at each corner point of the set of feasible solutions until it reaches the optimal one. Some care must be exercised in the use of the algorithm because it is possible to begin the process in the set of feasible solutions but move outside this set as a result of pivot operations. This is illustrated in Example 2.

Using simplex with unusual conditions

EXAMPLE 2 Solve the following LP problem.

Maximize $P = 2x + y$
subject to the constraints

$$\begin{aligned} x + y &\le 3, \\ -2x + y &\le 0, \\ x - 2y &\le 0, \\ x \ge 0, \quad y &\ge 0. \end{aligned}$$

Solution: The initial tableau for this problem is

$$\begin{array}{cc:cccc:c} x & y & t & s & u & P & \\ \hline 1 & 1 & 1 & 0 & 0 & 0 & 3 \\ -2 & 1 & 0 & 1 & 0 & 0 & 0 \\ 1 & -2 & 0 & 0 & 1 & 0 & 0 \\ \hdashline -2 & -1 & 0 & 0 & 0 & 1 & 0 \end{array}$$

ROP:

$$x = 0, \quad y = 0, \quad t = 3, \quad s = 0, \quad u = 0, P = 0.$$

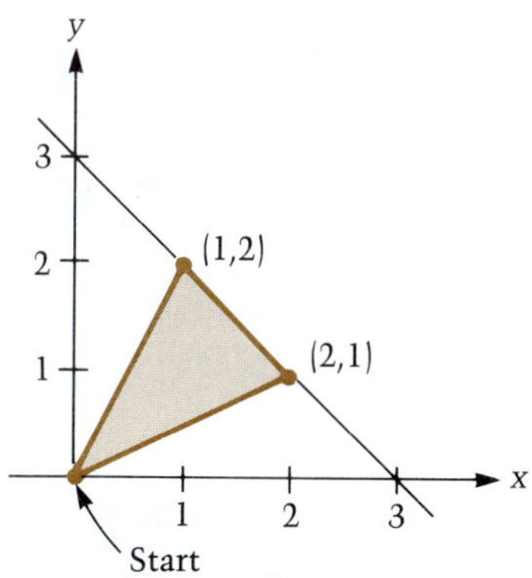

The values of the variables here indicate that this solution is within the feasible set. Thus the standard algorithm may be applied. The only positive quotient is in row one, so that the pivot element is in row one, column one:

$$\left[\begin{array}{cc:cccc:c} \textcircled{1} & 1 & 1 & 0 & 0 & 0 & 3 \\ -2 & 1 & 0 & 1 & 0 & 0 & 0 \\ 1 & -2 & 0 & 0 & 1 & 0 & 0 \\ \hdashline -2 & -1 & 0 & 0 & 0 & 1 & 0 \end{array}\right]$$
$\uparrow$

$$\begin{array}{l} R_2 \to 2R_1 + R_2 \\ R_3 \to -R_1 + R_3 \\ R_4 \to 2R_1 + R_4 \end{array}$$

$$\left[\begin{array}{cc:cccc:c} 1 & 1 & 1 & 0 & 0 & 0 & 3 \\ 0 & 3 & 2 & 1 & 0 & 0 & 6 \\ 0 & -3 & -1 & 0 & 1 & 0 & -3 \\ \hdashline 0 & 1 & 2 & 0 & 0 & 1 & 6 \end{array}\right].$$

At this stage there are no negative entries in the last row, and it is tempting to conclude that the problem has been solved. However, by ROP (see the figure)

$$x = 3, \quad y = 0, \quad t = 0, \quad s = 6, \quad \underbrace{u = -3}_{\uparrow}, \quad P = 6.$$

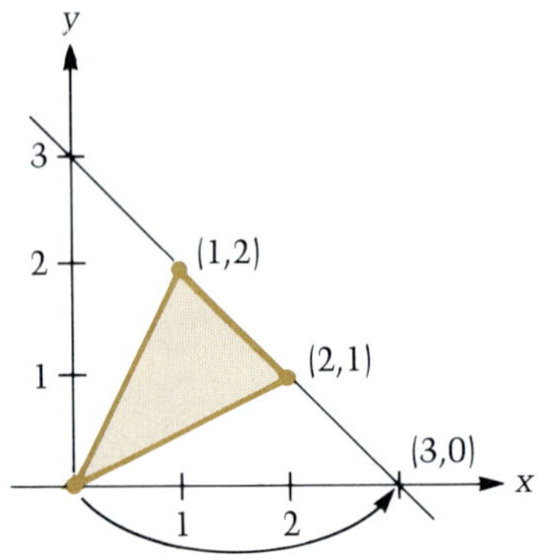

Thus a pivot operation must be used to remove the negative variable.

$$\left[\begin{array}{cc:cccc:c} 1 & 1 & 1 & 0 & 0 & 0 & 3 \\ 0 & 3 & 2 & 1 & 0 & 0 & 6 \\ 0 & -3 & \textcircled{-1} & 0 & 1 & 0 & -3 \\ \hdashline 0 & 1 & 2 & 0 & 0 & 1 & 6 \end{array}\right] \quad \begin{array}{l} R_1 \to R_3 + R_1 \\ R_2 \to 2R_3 + R_2 \\ R_4 \to 2R_3 + R_4 \end{array}$$

$$\left[\begin{array}{cc:cccc:c} 1 & -2 & 0 & 0 & 1 & 0 & 0 \\ 0 & -3 & 0 & 1 & 2 & 0 & 0 \\ 0 & -3 & -1 & 0 & 1 & 0 & -3 \\ \hdashline 0 & -5 & 0 & 0 & 2 & 1 & 0 \end{array}\right].$$

ROP:

$$x = 0, \quad y = 0, \quad t = 3, \quad s = 0, \quad u = 0, \quad P = 0.$$

(See figure.)

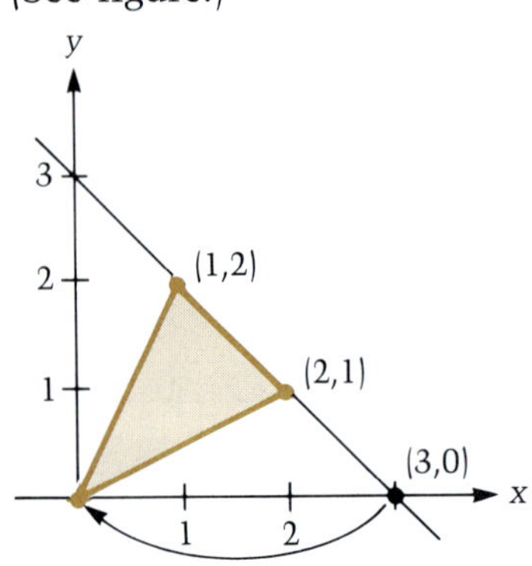

This solution is feasible and is the same as that obtained from the initial tableau, although it is derived from a *different* tableau. Now the standard algorithm may be applied again. The only positive quotient in column two is in row three:

$$\left[\begin{array}{cc|cccc|c} 1 & -2 & 0 & 0 & 1 & 0 & 0 \\ 0 & -3 & 0 & 1 & 2 & 0 & 0 \\ 0 & \textcircled{-3} & -1 & 0 & 1 & 0 & -3 \\ \hline 0 & -5 & 0 & 0 & 2 & 1 & 0 \end{array}\right] \quad R_3 \to \left(-\frac{1}{3}\right)R_3$$

$$\left[\begin{array}{cc|cccc|c} 1 & -2 & 0 & 0 & 1 & 0 & 0 \\ 0 & -3 & 0 & 1 & 2 & 0 & 0 \\ 0 & \textcircled{1} & \frac{1}{3} & 0 & -\frac{1}{3} & 0 & 1 \\ \hline 0 & -5 & 0 & 0 & 2 & 1 & 0 \end{array}\right] \quad \begin{array}{l} R_1 \to 2R_3 + R_1 \\ R_2 \to 3R_3 + R_2 \\ R_4 \to 5R_3 + R_4 \end{array}$$

$$\left[\begin{array}{cc|cccc|c} 1 & 0 & \frac{2}{3} & 0 & \frac{1}{3} & 0 & 2 \\ 0 & 0 & 1 & 1 & 1 & 0 & 3 \\ 0 & 1 & \frac{1}{3} & 0 & -\frac{1}{3} & 0 & 1 \\ \hline 0 & 0 & \frac{5}{3} & 0 & \frac{1}{3} & 1 & 5 \end{array}\right].$$

ROP:

$$x = 2, \quad y = 1, \quad t = 0, \quad s = 3, \quad u = 0, \quad P = 5.$$

(See figure.)

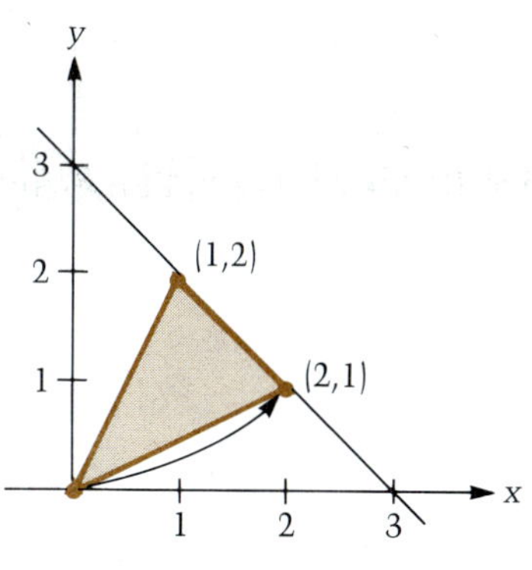

This solution is feasible, and there are no negative numbers in the last row. Therefore, the algorithm terminates with the optimal solution $x = 2$, $y = 1$, $P = 5$.

EXERCISES

Set A–Using the Simplex Method with Mixed Constraints

In Exercises 1 through 10, solve the problems using the Simplex Algorithm and in Exercises 1 through 7 check your results graphically.

1. (a) Maximize $P = 3x + 6y$ subject to the constraints

$$x + y \le 2,\quad x \ge 1,\quad x \ge 0,\quad y \ge 0$$

(b) Maximize $P = 6x + 3y$ subject to the constraints

$$x + y \le 2,\quad x \ge 1,\quad x \ge 0,\quad y \ge 0$$

2. Maximize $P = x - y$ subject to the constraints

$$x + y \ge 1,\quad x + 2y \le 2,\quad x \ge 0,\quad y \ge 0$$

3. Maximize $P = x + y$ subject to the constraints

$$x + 3y \ge 3,\quad 2x + 3y \le 8,\quad x \ge 0,\quad y \ge 0$$

4. (a) Maximize $P = 2x + y$ subject to the constraints

$$x + y \le 4,\quad 4x + y \ge 4,\quad x \ge 0,\quad y \ge 0$$

(b) Maximize $P = x + 2y$ subject to the constraints

$$x + y \le 4,\quad 4x + y \ge 4,\quad x \ge 0,\quad y \ge 0$$

5. (a) Maximize $P = x + 2y$ subject to the constraints

$$x + y \le 2,\quad x + y \ge 1,\quad x \ge 0,\quad y \ge 0$$

(b) Maximize $P = 2x + y$ subject to the constraints

$$x + y \le 2,\quad x + y \ge 1,\quad x \ge 0,\quad y \ge 0$$

6. (a) Maximize $P = 4x + 8y$ subject to the constraints

$$x - y \le 0,\quad x + y \le 4,\quad x \ge 1,\quad x \ge 0,\quad y \ge 0$$

(b) Maximize $P = 8x + 4y$ subject to the constraints

$$x - y \le 0,\quad x + y \le 4,\quad x \ge 1,\quad x \ge 0,\quad y \ge 0$$

7. Maximize $P = 3x + y$ subject to the constraints

$$\begin{aligned} x + y &\geq 1, \\ x + y &\leq 3, \\ -x + y &\leq 1, \\ x &\leq 2, \\ x \geq 0, \quad y &\geq 0 \end{aligned}$$

8. Maximize $P = x - y + z$ subject to the constraints

$$\begin{aligned} x - 2y + z &\geq 6, \\ x + y + z &\geq 10, \\ x \geq 0, \quad y \geq 0, \quad z &\geq 0 \end{aligned}$$

9. Maximize $P = x + 2y + 4z$ subject to the constraints

$$\begin{aligned} x + 2y + z &\leq 10, \\ 2x + y + 2z &\geq 12, \\ x \geq 0, \quad y \geq 0, \quad z &\geq 0 \end{aligned}$$

10. Maximize $P = x + 2y + 4z$ subject to the constraints

$$\begin{aligned} x + z &\leq 10, \\ x + y &\geq 4, \\ -y + z &\geq 4, \\ x \geq 0, \quad y \geq 0, \quad z &\geq 0 \end{aligned}$$

Set B–Applied Problems to Be Solved by the Simplex Method

Maximizing profits

11. The Stillon Oil Company produces gasoline and diesel fuel. Profit on a barrel of gasoline is \$2 and that on diesel is \$1. How many barrels of each kind of fuel should the Company produce each day for maximum profit if the following conditions must be followed?
    (a) The Company must produce at least 4 million barrels of diesel each day.
    (b) The amount of diesel must not be more than 80% of total production.
    (c) Total production cannot exceed 10 million barrels per day.

Maximizing return on investment

12. Mr. Jones has \$100,000 to invest. He plans to invest at least \$20,000 in stocks and bonds. The amount invested in bonds must be at least two-thirds the amount invested in stocks. For maximum return how much should Mr. Jones invest in each kind of security if
    (a) stocks pay 20% and bonds 15%?
    (b) stocks pay 15% and bonds 18%?

Maximizing return on investment

13. A banker is arranging an investment program for a customer who has a total of \$15,000 to invest in stocks and bonds. The customer insists that the banker adhere to the following rules:
    (a) The total investment in stocks and bonds must be at least \$2000.

(b) The amount invested in stocks must not be more than \$10,000.

(c) The amount invested in bonds must be at least 75% of \$6000 less than the amount invested in stocks.

If stocks pay 18% and bonds 15%, how much should be invested in each kind of security for maximum return?

5.4 THE SIMPLEX ALGORITHM, PART III: MINIMIZATION PROBLEMS

Up to this point the simplex method has been applied only to maximization LP problems. This section shows that the Simplex Algorithm can be applied to minimization problems as well. The basis here for solving minimization problems is the observation that each such problem can be converted to an LP maximum problem. The original problem is then solved by finding a solution to the maximum problem. The first task will be to explain the conversion from one kind of problem to the other.

The conversion process involves only the objective function of an LP problem and to show how it works, we will demonstrate with the function

$$C = 3x - y.$$

Suppose function C is to be minimized on the feasible set shown in the figure. In the table the values of C and also $-C$ have been calculated at the corner points of the feasible set. The values of C range, in order, from -3 to $+3$, and those of $-C$ from $+3$ to -3. We can see that as one of these functions increases, the other decreases and vice versa. This suggests that if we have found the

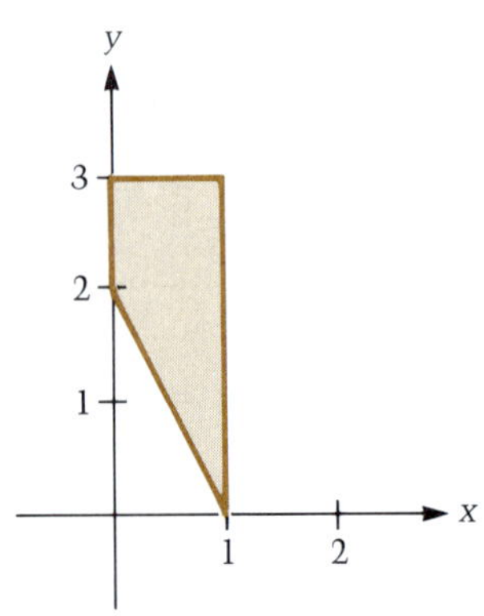

TABLE 5.1

Corner points	Values of C	Values of $-C$
(0, 3)	−3	+3
(0, 2)	−2	+2
(1, 3)	0	0
(1, 0)	+3	−3

maximum of $-C$ (which is +3 in this example) the minimum of C is found by changing the sign of the maximum of $-C$:

$$\text{Maximum of } -C \longrightarrow \text{Change sign} \longrightarrow \text{Minimum of } C$$

$$+3 \xrightarrow{\hspace{6cm}} -3$$

Now suppose we are given any LP problem in which the objective function C is to be minimized. The problem is solved by finding the maximum of $-C$ and changing sign.

STEPS FOR SOLVING LP MINIMUM PROBLEMS

1. Given an LP problem in which the objective function C is to be minimized. Convert this LP problem to a new problem having the same constraints but objective function $-C$. Find the optimal solution of the new problem in which $-C$ is to be maximized.
2. The optimal solution of the original minimum problem is the same as that of the maximum problem except that the signs of the values of the two objective functions are opposite (after the value of the objective function has been found in the maximum problem, change its sign to get the value of the function in the minimum problem).

A Min problem using simplex

EXAMPLE 1 Solve

Minimum $C = 32x + 12y$
subject to the constraints

$$2x + 5y \geq 20,$$

$$x + y \geq 7,$$

$$x \geq 0, \quad y \geq 0.$$

Solution: The first step is to convert to a maximum problem. Let $P = -C = -32x - 12y$. Then the new problem is

$$\begin{aligned}&\text{Maximize } P = -32x - 12y\\&\text{subject to the constraints}\\&\quad 2x + 5y \geq 20,\\&\quad x + y \geq 7,\\&\quad x \geq 0, \quad y \geq 0.\end{aligned}$$

Now the new problem is solved by the methods we already know. The first step is to write the problem with $\leq$-constraints:

$$\begin{aligned}&\text{Maximize } P = -32x - 12y\\&\text{subject to the constraints}\\&\quad -2x - 5y \leq -20,\\&\quad -x - y \leq -7,\\&\quad x \geq 0, \quad y \geq 0.\end{aligned}$$

Using slack variables t, s, we obtain the initial tableau

$$\begin{array}{c} \begin{array}{ccccccc} x & y & t & s & P & & \end{array} \\ \left[\begin{array}{cc|ccc|c} -1 & -1 & 1 & 0 & 0 & -7 \\ -2 & -5 & 0 & 1 & 0 & -20 \\ \hline 32 & 12 & 0 & 0 & 1 & 0 \end{array}\right]\end{array}$$

ROP:

$$x = 0, \quad y = 0, \quad t = -7, \quad s = -20, \quad P = 0.$$

Since $t = -7$, we will pivot on the element in row one, column one:

$$\left[\begin{array}{cc|ccc|c} \textcircled{-1} & -1 & 1 & 0 & 0 & -7 \\ -2 & -5 & 0 & 1 & 0 & -20 \\ \hline 32 & 12 & 0 & 0 & 1 & 0 \end{array}\right] \quad \begin{array}{l} R_2 \to -2R_1 + R_2 \\ R_3 \to 32R_1 + R_3 \end{array}$$

$$\left[\begin{array}{cc|ccc|c} -1 & -1 & 1 & 0 & 0 & -7 \\ 0 & -3 & -2 & 1 & 0 & -6 \\ \hline 0 & -20 & 32 & 0 & 1 & -224 \end{array}\right].$$

ROP:

$$x = 7, \quad y = 0, \quad t = 0, \quad s = -6, \quad P = -224.$$

A negative variable still occurs in row two, so we pivot on the element in row two, column two:

$$\left[\begin{array}{rr|rrr|r} -1 & -1 & 1 & 0 & 0 & -7 \\ 0 & \textcircled{-3} & -2 & 1 & 0 & -6 \\ \hline 0 & -20 & 32 & 0 & 1 & -224 \end{array}\right] \quad R_2 \to \left(-\frac{1}{3}\right)R_2$$

$$\left[\begin{array}{rr|rrr|r} -1 & -1 & 1 & 0 & 0 & -7 \\ 0 & \textcircled{1} & \frac{2}{3} & -\frac{1}{3} & 0 & 2 \\ \hline 0 & -20 & 32 & 0 & 1 & -224 \end{array}\right] \quad \begin{array}{l} R_1 \to R_2 + R_1 \\ R_3 \to 20R_2 + R_3 \end{array}$$

$$\left[\begin{array}{rr|rrr|r} -1 & 0 & \frac{5}{3} & -\frac{1}{3} & 0 & -5 \\ 0 & 1 & \frac{2}{3} & -\frac{1}{3} & 0 & 2 \\ \hline 0 & 0 & \frac{136}{3} & -\frac{20}{3} & 1 & -184 \end{array}\right].$$

ROP:

$$x = 5, \quad y = 2, \quad t = 0, \quad s = 0, \quad P = -184.$$

This solution is feasible, and the basic algorithm may be started:

$$\left[\begin{array}{rr|rrr|r} -1 & 0 & \frac{5}{3} & \textcircled{-\frac{1}{3}} & 0 & -5 \\ 0 & 1 & \frac{2}{3} & -\frac{1}{3} & 0 & 2 \\ \hline 0 & 0 & \frac{136}{3} & -\frac{20}{3} & 1 & -184 \end{array}\right] \quad \begin{array}{l} R_2 \to -R_1 + R_2 \\ R_3 \to -20R_1 + R_3 \end{array}$$

$$\left[\begin{array}{rr|rrr|r} -1 & 0 & \frac{5}{3} & -\frac{1}{3} & 0 & -5 \\ 1 & 1 & -1 & 0 & 0 & 7 \\ \hline 20 & 0 & 12 & 0 & 1 & -84 \end{array}\right].$$

ROP:

$$x = 0, \quad y = 7, \quad t = 0, \quad s = 15, \quad P = -84.$$

The solution is feasible, and there are no negative numbers in the last row (except −84, which does not count). This means that the solution

$$x = 0, \quad y = 7, \quad P = -84$$

is optimal for the maximum problem. Therefore, the optimal solution for the minimum problem is

$$x = 0, \quad y = 7, \quad C = 84.$$

One instance of what has come to be known as the transportation problem is discussed in Example 2.

Minimizing costs—the transportation problem

EXAMPLE 2 The Tumpkin Roller Bearing Company operates plants in Chicago and Cleveland. The Company supplies Atlanta, Baltimore, and Houston with its Type A bearing. Atlanta must receive 60 cartons, Baltimore 80 cartons, and Houston 60 cartons. The Cleveland plant has 100 cartons of Type A on hand and Chicago also has 100 cartons available. The costs of shipping from Cleveland are: \$2.00/carton to Atlanta, \$2.00/carton to Baltimore, \$2.00/carton to Houston. Shipping costs from Chicago are: \$2.00/carton to Atlanta, \$3.00/carton to Baltimore, \$4.00/carton to Houston. How should Tumpkin arrange shipments to keep costs as low as possible?

Solution: For this type of problem it is helpful to draw a diagram as shown:

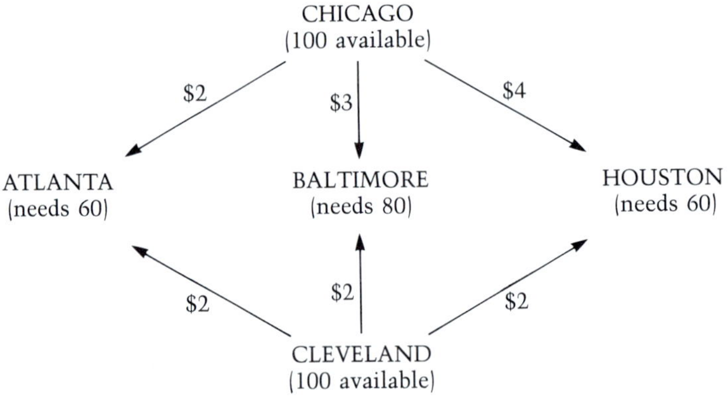

We do not know how many cartons from Chicago will be shipped to Atlanta, Baltimore, and Houston. Thus we let

x = the number of cartons to be shipped from Chicago to Atlanta;

y = the number of cartons to be shipped from Chicago to Baltimore;

z = the number of cartons to be shipped from Chicago to Houston.

Atlanta needs 60 cartons, and if x cartons come from Chicago, then the remainder must come from Cleveland. That is, $60 - x$ represents the number of cartons to come from Cleveland if x cartons come from Chicago. We may also represent the number of cartons to be shipped from Cleveland to Baltimore as $80 - y$ and the number of cartons to be shipped from Cleveland to Houston as $60 - z$. The diagram can now be improved with this information:

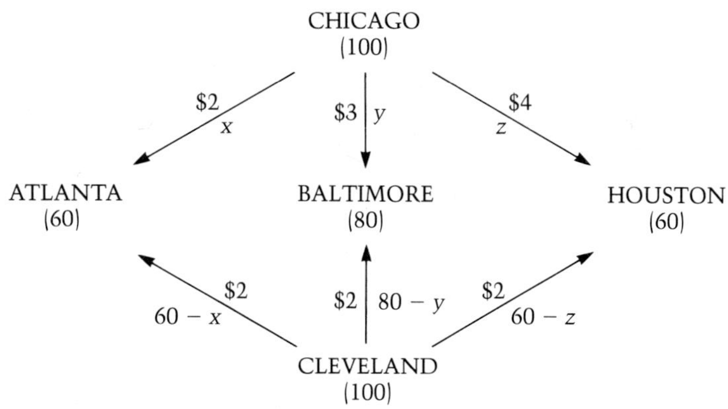

Now we are ready to write constraints:

$$\text{Atlanta needs 60 cartons: } x \leq 60,$$
$$\text{Baltimore needs 80 cartons: } y \leq 80,$$
$$\text{Houston needs 60 cartons: } z \leq 60,$$
$$\text{Chicago has available 100 cartons: } x + y + z \leq 100,$$
$$\text{Cleveland has available 100 cartons: } (60 - x) + (80 - y) + (60 - z) \leq 100.$$

The Cleveland inequality can be simplified as follows:

$$\begin{aligned} (60 - x) + (80 - y) + (60 - z) &\leq 100, \\ 200 - x - y - z &\leq 100, \\ -x - y - z &\leq -100, \\ x + y + z &\geq 100. \end{aligned}$$

The Chicago constraint is $x + y + z \le 100$, and the Cleveland constraint is $x + y + z \ge 100$. Since both of these must hold, we conclude that

$$x + y + z = 100.$$

Now the constraints are

$$\begin{aligned} &x + y + z = 100, \\ &x \le 60, \quad y \le 80, \quad z \le 60, \\ &x \ge 0, \quad y \ge 0, \quad z \ge 0. \end{aligned} \tag{1}$$

The objective function comes from the costs of shipping. From the diagram

$$2x + 3y + 4z \text{ is the cost of shipping from Chicago}$$

and

$$2(60 - x) + 2(80 - y) + 2(60 - z) = 400 - 2x - 2y - 2z$$

is the cost of shipping from Cleveland. Adding these costs results in total shipping costs:

$$R = y + 2z + 400.$$

This is the linear function to be minimized. It may be noted that this function has no x-term but a constant term of 400. We need to concentrate only on $y + 2z$ since if we can minimize this, our total costs, R, will be minimized. Therefore, we take

$$C = y + 2z$$

as our objective function.

From the first constraint of (1) we may solve for x in terms of y and z:

$$x = 100 - y - z, \tag{2}$$

and this equation allows us to express our problem in terms of y and z only. Using (2), we convert the constraints

$$x \ge 0 \quad \text{and} \quad x \le 60$$

into ones containing only y and z:

$$x \geq 0 \quad \text{becomes} \quad 100 - y - z \geq 0 \quad \text{or} \quad y + z \leq 100$$
$$x \leq 60 \quad \text{becomes} \quad 100 - y - z \leq 60 \quad \text{or} \quad y + z \geq 40$$

Now we have the following LP model:

$$\begin{aligned}&\text{Minimize } C = y + 2z\\&\text{subject to the constraints}\\&\quad y + z \leq 100,\\&\quad y + z \geq 40,\\&\quad y \leq 80,\\&\quad z \leq 60,\\&\quad y \geq 0, \quad z \geq 0.\end{aligned}$$

To solve this, convert to the maximum problem:

$$\begin{aligned}&\text{Maximize } P = -y - 2z\\&\text{subject to the constraints}\\&\quad y + z \leq 100,\\&\quad -y - z \leq -40,\\&\quad y \leq 80,\\&\quad z \leq 60,\\&\quad y \geq 0, \quad z \geq 0.\end{aligned}$$

Using slack variables t, s, u, v we obtain the following initial tableau:

$$\begin{array}{c} \begin{matrix} y & z & t & s & u & v & P & \end{matrix} \\ \left[\begin{array}{cc:ccccc:c} 1 & 1 & 1 & 0 & 0 & 0 & 0 & 100 \\ \textcircled{-1} & -1 & 0 & 1 & 0 & 0 & 0 & -40 \\ 1 & 0 & 0 & 0 & 1 & 0 & 0 & 80 \\ 0 & 1 & 0 & 0 & 0 & 1 & 0 & 60 \\ \hdashline 1 & 2 & 0 & 0 & 0 & 0 & 1 & 0 \end{array}\right]. \end{array}$$

ROP:

$$y = 0, \quad z = 0, \quad t = 100, \quad s = -40, \quad u = 80, \quad v = 60, \quad P = 0.$$

Since $s = -40$, we pivot on -1 in column 1 and obtain

$$\left[\begin{array}{rr|rrrrr|r} 0 & 0 & 1 & 1 & 0 & 0 & 0 & 60 \\ -1 & -1 & 0 & 1 & 0 & 0 & 0 & -40 \\ 0 & -1 & 0 & 1 & 1 & 0 & 0 & 40 \\ 0 & 1 & 0 & 0 & 0 & 1 & 0 & 60 \\ \hline 0 & 1 & 0 & 1 & 0 & 0 & 1 & -40 \end{array}\right].$$

ROP:

$$y = 40, \quad z = 0, \quad t = 60, \quad s = 0, \quad u = 40, \quad v = 60, \quad P = -40.$$

This is a feasible solution with no negative numbers in the bottom row. Therefore,

$$y = 40, \quad z = 0, \quad P = -40$$

is the optimal solution for the maximum problem, and the solution

$$y = 40, \quad z = 0, \quad C = 40$$

is optimal for the minimum problem. To interpret this in the original problem it is necessary to recall x:

$$\begin{aligned} x &= 100 - y - z \\ &= 100 - 40 - 0 \qquad \text{substituting } y = 40 \text{ and } z = 0 \\ &= 60. \end{aligned}$$

Also the total cost is

$$\begin{aligned} R &= y + 2z + 400 \\ &= 400 + C \\ &= 400 + 40 \qquad \text{substituting } C = 40 \\ &= 440. \end{aligned}$$

Therefore, $x = 60$, $y = 40$, $z = 0$, $R = 440$ is the optimal solution of the original problem. Going back to the diagram and using these values of x, y, z, we have the following shipping schedule

From Chicago:
- to Atlanta ship 60 cartons
- to Baltimore ship 40 cartons
- to Houston ship 0 cartons

From Cleveland: $\begin{cases} \text{to Atlanta ship 0 cartons} \\ \text{to Baltimore ship 40 cartons} \\ \text{to Houston ship 60 cartons} \end{cases}$

and the total shipping costs will be \$440.

EXERCISES

Set A–Using the Simplex Method to Solve Minimizing LP Problems

In Exercises 1 through 14, use the simplex method in all cases.

1. Minimize $C = x + 5y$ subject to the constraints

$$3x + y \geq 3$$
$$x + 3y \geq 3$$
$$x \geq 0, \quad y \geq 0$$

2. Minimize $C = 2x + 4y$ subject to the constraints

$$x + 2y \geq 3,$$
$$x + y \geq 3,$$
$$x \geq 0, \quad y \geq 0$$

3. Minimize $C = 10x + 5y$ subject to the constraints

$$2x + y \geq 20,$$
$$x + 5y \geq 55,$$
$$x \geq 0, \quad y \geq 0$$

4. Minimize $C = 3x + 3y$ subject to the constraints

$$2x + 3y \geq 48,$$
$$x + y \geq 20,$$
$$x \geq 0, \quad y \geq 0$$

5. Minimize $C = x + y$ subject to the constraints

$$3x + y \geq 3,$$
$$x + 3y \geq 3,$$
$$x \geq 0, \quad y \geq 0$$

6. Minimize $C = 3x + y$ subject to the constraints

$$x + 2y \geq 3,$$
$$x + y \geq 2,$$
$$x \geq 0, \quad y \geq 0$$

7. Minimize $C = 3x + 2y$ subject to the constraints

$$2x + y \geq 20,$$
$$x + 5y \geq 55,$$
$$x \geq 0, \quad y \geq 0$$

8. Minimize $C = x + 2y$ subject to the constraints

$$2x + 3y \geq 48,$$
$$x + y \geq 20,$$
$$x \geq 0, \quad y \geq 0$$

9. (a) Minimize $C = 9x + 8y$ subject to the constraints

$$\begin{aligned} x + y &\le 3, \\ x \quad &\ge 1, \\ y &\ge 1, \\ x \ge 0, \quad y &\ge 0 \end{aligned}$$

(b) Minimize $C = 9x - 8y$ subject to the constraints

$$\begin{aligned} x + y &\le 3, \\ x \quad &\ge 1, \\ y &\ge 1, \\ x \ge 0, \quad y &\ge 0 \end{aligned}$$

10. (a) Minimize $C = 10x + 11y$ subject to the constraints

$$\begin{aligned} x - y &\le 2, \\ x + y &\ge 4, \\ y &\le 2, \\ x \ge 0, \quad y &\ge 0 \end{aligned}$$

(b) Minimize $C = 11x + 10y$ subject to the constraints

$$\begin{aligned} x - y &\le 2, \\ x + y &\ge 4, \\ y &\le 2, \\ x \ge 0, \quad y &\ge 0 \end{aligned}$$

11. (a) Minimize $C = 2x + 3y + z$ subject to the constraints

$$\begin{aligned} x + 2y \qquad &\le 10, \\ 2x \qquad + z &\ge 15, \\ x \ge 0, \quad y \ge 0, \quad z &\ge 0 \end{aligned}$$

(b) Minimize $C = 2x - y + z$ subject to the constraints

$$\begin{aligned} x + 2y \qquad &\le 10, \\ 2x \qquad + z &\ge 15, \\ x \ge 0, \quad y \ge 0, \quad z &\ge 0 \end{aligned}$$

12. (a) Minimize $C = 4x + y + 2z$ subject to the constraints

$$\begin{aligned} x + 2y + 8z &\le 20, \\ 4x + y + 2z &\ge 8, \\ x \ge 0, \quad y \ge 0, \quad z &\ge 0 \end{aligned}$$

(b) Minimize $C = 10x + y + z$ subject to the constraints

$$\begin{aligned} x + 2y + 8z &\le 20, \\ 4x + y + 2z &\ge 8, \\ x \ge 0, \quad y \ge 0, \quad z &\ge 0 \end{aligned}$$

13. (a) Minimize
$C = x + 2y + 2z$
subject to the constraints

$$\begin{aligned} x + y + z &\geq 4, \\ x \quad\quad &\leq 20, \\ y \quad &\leq 20, \\ z &\leq 20, \\ x \geq 0, \quad y \geq 0, &\quad z \geq 0 \end{aligned}$$

(b) Minimize
$C = 2x + 2y + z$
subject to the constraints

$$\begin{aligned} x + y + z &\geq 4, \\ x \quad\quad &\leq 20, \\ y \quad &\leq 20, \\ z &\leq 20, \\ x \geq 0, \quad y \geq 0, &\quad z \geq 0 \end{aligned}$$

14. (a) Minimize
$C = x + 2y + z$
subject to the constraints

$$\begin{aligned} 2x + 4y + 6z &\leq 30, \\ x + y - z &\geq 12, \\ 4x - 2y + z &\geq 4, \\ x \geq 0, \quad y \geq 0, &\quad z \geq 0 \end{aligned}$$

(b) Minimize
$C = 2x + y + z$
subject to the constraints

$$\begin{aligned} 2x + 4y + 6z &\leq 30, \\ x + y - z &\geq 12, \\ 4x - 2y + z &\geq 4, \\ x \geq 0, \quad y \geq 0, &\quad z \geq 0 \end{aligned}$$

Set B–Applications of the Simplex Method to Minimizing LP Problems

Scheduling to minimize costs

15. The Thor Coal Company owns three mines: Abe I, Abe II, and Abe III. In one day's operation Abe I produces 1 ton each of hard, ordinary, and soft coal. A day's operation of Abe II produces 2 tons each of hard and ordinary coal and only 1 ton of soft coal. At Abe III in a day the production is 1 ton of hard coal and 2 tons each of ordinary and soft coal. Daily operating costs are \$250 for Abe I, \$310 for Abe II, and \$350 for Abe III. In order to fulfill contracts the Thor Co. must produce at least 10 tons of hard, 12 tons of ordinary, and 16 tons of soft coal. To meet these requirements at minimum cost, how many days should the three mines be operated?

Minimizing operation costs

16. The Stillon Oil Company operates three refineries, A, B, C, and each produces regular, high-test, and unleaded gasoline. It costs \$10,000/day to operate refinery A, \$12,000/day to

operate B, and \$14,000/day to operate C. The daily production at these refineries is given in the table below:

	A	B	C
Regular	1000	4000	2000
High-test	3000	5000	1000
Unleaded	2000	1000	2000

Stillon has orders from its own stations for 32,000 gallons of regular, 47,000 gallons of high-test, and 22,000 gallons of unleaded (any gasoline produced in excess of these amounts can be sold to independent dealers). How many days should each of the refineries operate to fill these orders and minimize costs?

Scheduling to minimize costs

17. Buildarama sells a small heater at two of its stores, one located at the North Mall and the other at the South Mall. The North Mall store needs 50 of these heaters, and the South Mall store needs 40. The heaters must come from two warehouses: I and II. It costs \$4 to ship a heater from I to North Mall and \$5 to ship from I to South Mall. Shipping a heater from warehouse II to the North or South Mall costs \$6. Warehouse I has 75 heaters in stock and warehouse II has 80. How should Buildarama schedule shipments so that shipping costs will be as low as possible?

Scheduling to minimize costs—transportation problem

18. The Jones Furniture Company has three retail outlets: A, B, C. Each of these outlets has ordered 20 student desks (made by the Strong Desk Company). The desks must be supplied by the company's two warehouses, I and II. Each warehouse has 30 of these desks in stock. Shipping a desk from warehouse I to outlet A or from I to B costs \$5; shipping a desk from II to B or from II to A costs \$4. Shipping a desk from I or II to C costs \$6. How should the Jones Company schedule shipments so as to keep costs as low as possible?

5.5 SUMMARY AND REVIEW EXERCISES

REVIEW OF CONCEPTS

I. The Read-Out Principle (ROP) for a Simplex Tableau

1. If a column for a variable has more than one nonzero entry, then that variable is set equal to zero.
2. If a column for a variable has exactly one nonzero entry and this is 1, then the variable is set equal to the constant in the row containing the 1.

II. Steps in the Basic Simplex Algorithm

1. Terminate the algorithm or choose the pivot column. If there are no negative numbers in the last row which correspond to variables, the problem is solved and the optimal solution is read by ROP. If there are negative numbers in the last row which correspond to variables, then the pivot column is that column containing the most negative of these numbers.
2. Choose the pivot element. In the pivot column and column of constants compute the appropriate quotients. The pivot row is that row corresponding to the smallest positive quotient. The pivot element is the element in the pivot row and pivot column.
3. Use Gaussian elimination and the pivot element to obtain 0s in the pivot column (except in the pivot row).
4. Go to Step 1.

III. Rule for Dealing with Mixed Constraints

If an LP problem contains mixed constraints, then each $\geq$-constraint is converted to a $\leq$-constraint by multiplying each side of the inequality by -1.

IV. Rule for Dealing with Negative Variables

If a variable has a negative value when ROP is applied, then in the row corresponding to the negative variable choose one of the negative elements and use it as a pivot.

V. Steps for Solving Minimum LP Problems

1. Convert the given LP problem to a new problem having the same constraints but objective function $-C$ (the negative of the original objective function). Find the optimal solution of the new problem in which $-C$ is maximized.

2. The optimal solution of the original minimum problem is the same as that of the maximum problem except that the signs of the values of the two objective functions are opposite (after the value of the objective function has been found in the maximum problem, change its sign to get the value of the function in the minimum problem).

REVIEW EXERCISES

1. Write the following LP problem as a system of equations by introducing slack variables. Then construct the initial simplex tableau.

   Maximize $P = 5x + 9 + z$
   subject to the constraints

$$3x + y + z \leq 15,$$
$$x + 2y + 3z \leq 10,$$
$$2x + 3y + 5z \leq 12,$$
$$x \geq 0, \quad y \geq 0, \quad z \geq 0$$

2. Apply ROP (the Read-Out Principle) to the initial simplex tableau obtained for Exercise 1.

3. Use the Simplex Algorithm to solve the following LP problem.

   Maximize $P = 6x + 4y + 2z$
   subject to the constraints

$$x + y \leq 8,$$
$$x + 2y + z \leq 10,$$
$$x \geq 0, \quad y \geq 0, \quad z \geq 0$$

4. Use the Simplex Algorithm to solve the following LP problem.

   Maximize $P = x + y$
   subject to the constraints

$$-x + y \leq 1,$$
$$2x + y \leq 7,$$
$$x + 2y \geq 5,$$
$$x \geq 0, \quad y \geq 0$$

5. Use the Simplex Algorithm to solve the following problem.

Minimize $P = x - y + 2z$
subject to the constraints

$$\begin{aligned} x + y + z &\geq 4, \\ x &\leq 20, \\ y &\leq 20, \\ z &\leq 20, \\ x \geq 0, \quad y \geq 0, \quad z &\geq 0 \end{aligned}$$

CHAPTER

6

COMBINATORICS

6

6.1 INTRODUCTION

This chapter deals briefly with some elementary but sophisticated methods of counting. The techniques discussed are important for direct real-world applications as well as in relation to other areas of mathematics. One important use of this topic for us is in probability theory—discussed in Chapter 7.

When the contents of this chapter have been mastered, the reader will be able to solve problems of the following kinds.

1. A personnel manager must fill three positions *A*, *B*, *C*. The manager has four employees all of whom are qualified for the three jobs. In how many ways can the manager fill the positions?

2. A computer represents a number by a sequence of eight lights:

Each light is either on or off, and each configuration of on-off lights represents a number. How many numbers can be represented in this way? ■

6.2 EXPERIMENTS AND THE COUNTING PRINCIPLE

Suppose we spin the arrows on the two spinners shown in Figure 6.1. How many possible letter positions can the two arrows assume? This question may be answered by listing all the positions and counting them. Figure 6.2 lists six positions, and this exhausts all possibilities.

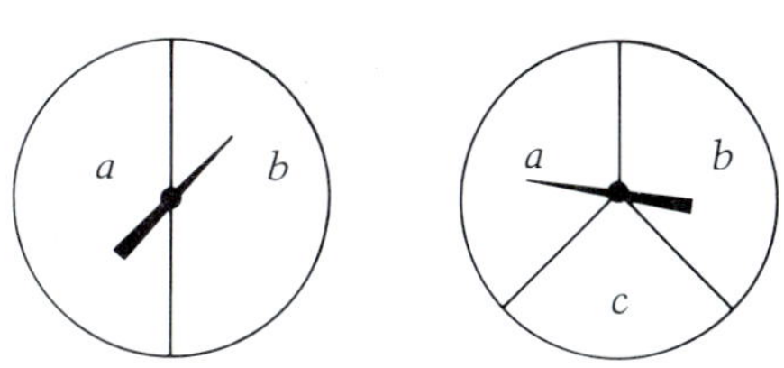

FIG. 6.1. Spinner 1 Spinner 2

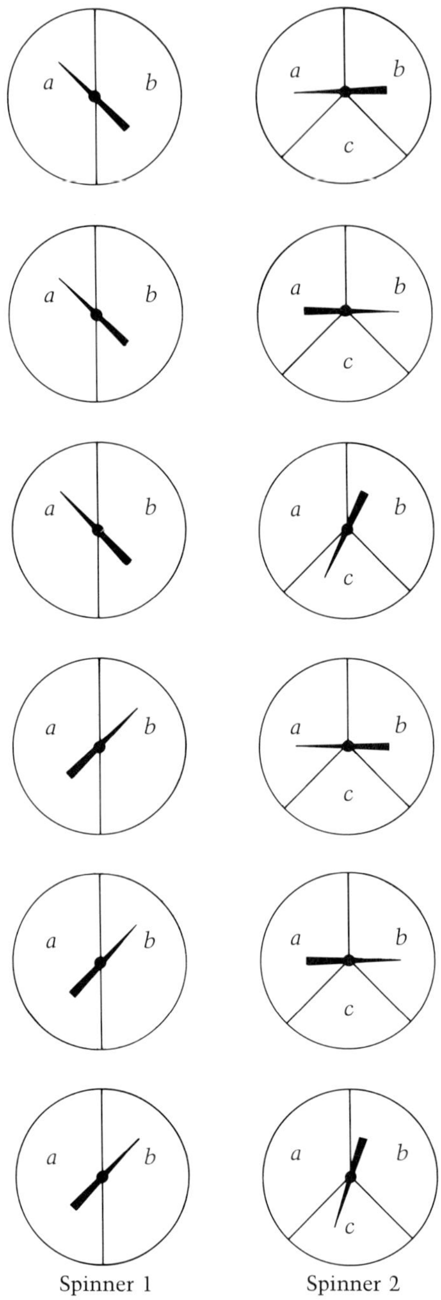

FIG. 6.2.

experiments
outcome
sample space

In mathematics such acts as spinning the arrows on these two spinners are often called *experiments,* and each possible position is called an *outcome* of the experiment. The collection of all outcomes is called the *sample space* of the experiment. Thus, we say that there are six possible outcomes of the experiment involving the spinners.

Consider the experiment of rolling a die (singular for dice) one time. (See Figure 6.3.) For this experiment there are six possible outcomes because a die has six faces. It is convenient to represent these six outcomes by the set

$$1, 2, 3, 4, 5, 6,$$

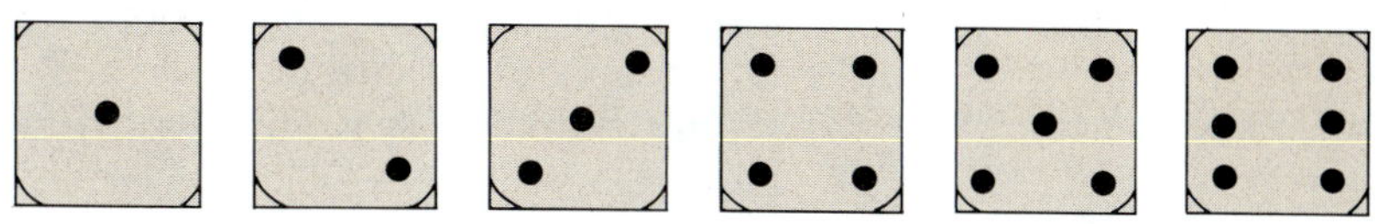

FIG. 6.3.

and this set is the sample space for the experiment of rolling the die one time. It is easy to see that rolling a die one time results in six possible outcomes because the experiment is so simple. However, the experiment which consists of spinning the arrows on the two spinners is more complicated because it has two parts. It is, in essence, an experiment which consists of two simple experiments, and we may carry it out by performing the first experiment (spinning the arrow on spinner 1) and then performing the second experiment (spinning the arrow on spinner 2). The arrow on spinner 1 will point to a or b, and the arrow on spinner 2 will point to a, b, or c. Thus, there are two outcomes on the first spin and three outcomes on the second. All the outcomes are conveniently represented as pairs:

$$(a, a), \quad (a, b), \quad (a, c), \quad (b, a), \quad (b, b), \quad (b, c).$$

The pair (b, a), for example, represents the fact that the arrow on spinner 1 points to b, and the arrow on spinner 2 points to a.

The diagram below shows a map of the roads which connect the towns of Smithfield, Dikemoore, and Brockton. Two roads,

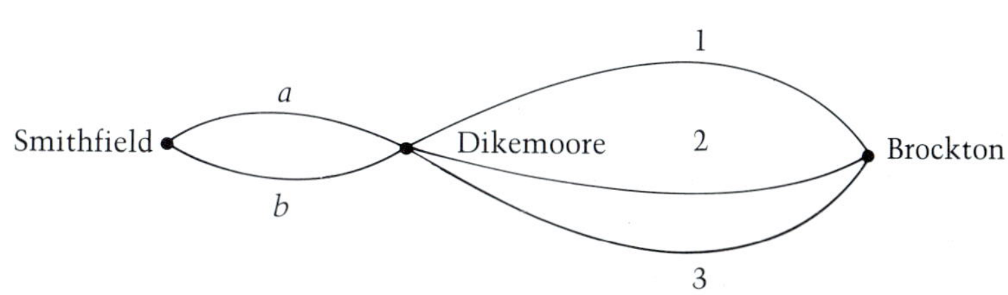

called a and b in the map, connect Smithfield to Dikemoore, and three roads, called 1, 2, and 3, connect Dikemoore to Brockton. We wish to consider the total number of routes one could take in traveling from Smithfield to Brockton. In this example such a trip is the "experiment," and any of the possible routes is an "outcome." The collection of all the routes is the sample space. The total number of routes may be determined as we have done in the previous examples by listing all the routes and counting. Here is the list of all possible routes:

$$(a, 1), \quad (a, 2), \quad (a, 3), \quad (b, 1), \quad (b, 2), \quad (b, 3).$$

The notation $(a, 3)$, for example, means that road a was taken from Smithfield to Dikemoore and road 3 was taken from Dikemoore to Brockton. Thus $(a, 3)$ is one of the possible routes. From the given list we may conclude that there are six routes in all.

tree diagram A device which will be used frequently in connection with experiments is the *tree diagram*. This is illustrated in Figure 6.4 for the two spinners. In the tree diagram two points are chosen on line 1 to represent the two outcomes for spinner 1; from each point on line 1 there are three lines to three points on line 2 to represent the three outcomes of spinner 2. This shows that if the

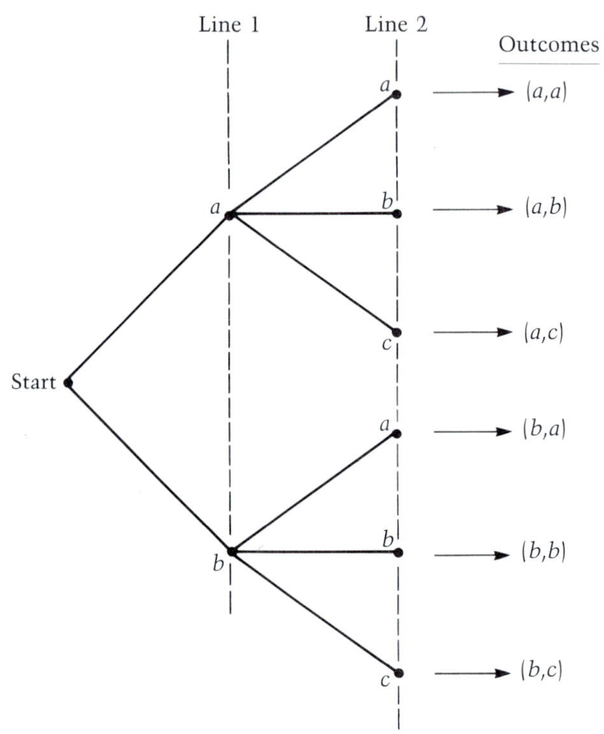

FIG. 6.4.

arrow on spinner 1 points to either a or b, then the arrow on spinner 2 can point to a, b, or c. Note that the outcomes may be listed by following each branch from the starting point to its end.

Tree diagram

EXAMPLE 1 Perform the experiment which consists of rolling a die once and then tossing a coin once. Construct a tree diagram, list the outcomes, and determine the sample space.

Solution: The outcomes on the die may be represented by

$$1, 2, 3, 4, 5, 6,$$

and the outcomes for the coin by

$$H, T.$$

If the die is rolled first, the tree diagram is constructed as shown in Figure 6.5. From the diagram we can see that the sample space

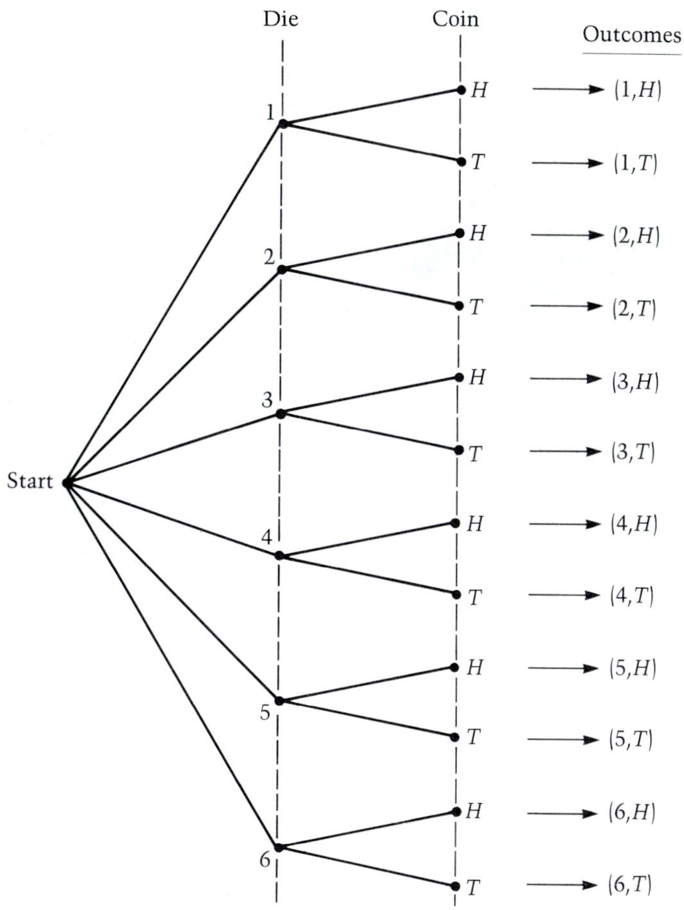

FIG. 6.5.

consists of the points

$$(1, H),\quad (1, T),\quad (2, H),\quad (2, T),\quad (3, H),\quad (3, T),$$
$$(4, H),\quad (4, T),\quad (5, H),\quad (5, T),\quad (6, H),\quad (6, T).$$

In Example 1 we see that the experiment has 12 outcomes because all the outcomes are easily listed from the tree diagram. Since it is not always easy or practical to list all the outcomes of an experiment, it is useful to have a method for counting them which does not depend upon a list. One such method is called the *Counting Principle.*

counting principle

THE COUNTING PRINCIPLE Suppose that an experiment E consists of two parts. There are m outcomes of the first part and n outcomes of the second part. If, after the first part has been performed, all n outcomes of the second part are possible, then there are $m \cdot n$ outcomes of the experiment E.

The experiment E which is performed by spinning each of the two spinners shown in Figure 6.6 once consists of two parts; the first part has 2 outcomes (see Figure 6.6), and the second 3. All 3 outcomes on spinner 2 are possible regardless of the outcome on spinner 1 (we say that the two parts are *independent*). Therefore, by the Counting Principle, there are $2 \cdot 3$, or 6, outcomes of this experiment.

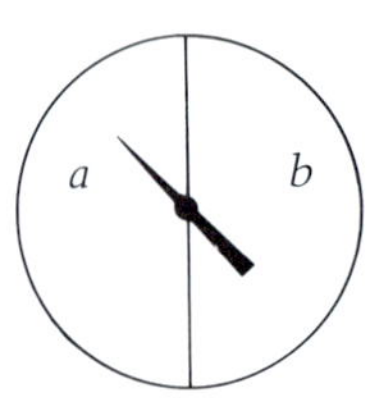

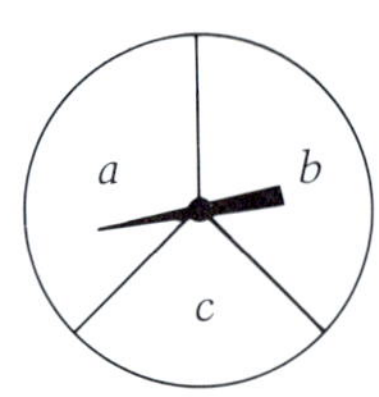

FIG. 6.6. Spinner 1 Spinner 2

In Example 1, the experiment E consists of rolling a die and then tossing a coin. There are 6 outcomes for the die and 2 for the coin. Since the two parts are independent, there are $6 \cdot 2$ $(= 12)$ outcomes of the experiment.

Using the counting principle

EXAMPLE 2 Record selections are made on a juke box by pressing one letter button and one number button on an array of buttons such as the one shown in the figure. How many selections are possible?

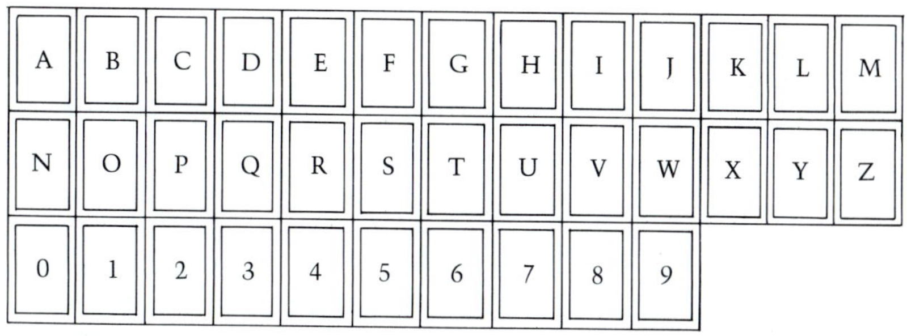

Solution: Making a selection may be thought of as an experiment consisting of two parts: selecting a letter and then selecting a number (the juke box is constructed so that these two parts are independent). There are 26 outcomes of the first part and 10 outcomes of the second. Therefore, there are $26 \cdot 10$ (= 260) outcomes of the experiment by the Counting Principle. Thus, 260 selections are possible.

PRACTICE PROBLEM Spin the arrows on the two spinners shown in Figure 6.7 one time. How many outcomes are possible?

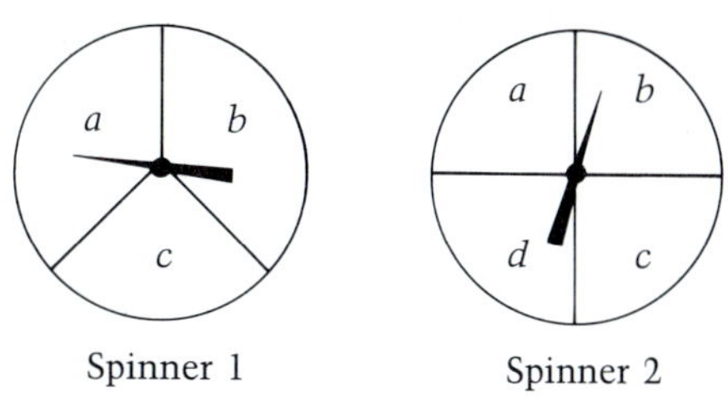

FIG. 6.7.

Answer: 12. ■

As the following example shows the Counting Principle easily extends to experiments having more than two parts.

The counting principle

EXAMPLE 3 Spin the arrows on the three spinners shown in Figure 6.8 one time. How many outcomes are possible?

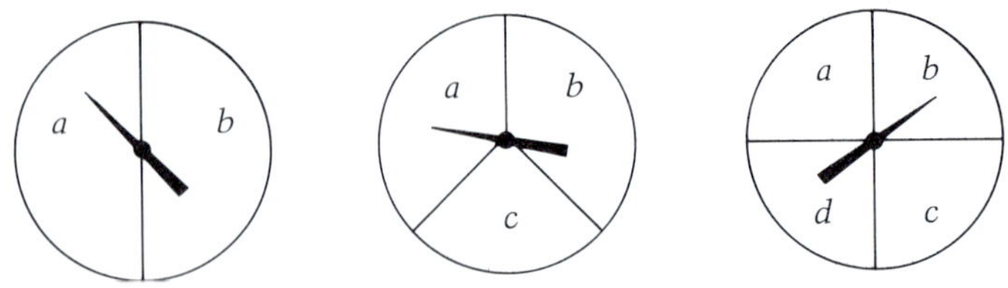

FIG. 6.8. Spinner 1 Spinner 2 Spinner 3

Solution: This experiment has three parts, but we may think of it as having only two by combining the first two parts into a single experiment. Using the Counting Principle, we see that there are $2 \cdot 3$, or 6, outcomes for spinners 1 and 2. Therefore, using the Counting Principle again, there are $6 \cdot 4$, or 24, outcomes for spinners 1, 2, and 3.

EXERCISES

Outcomes

1. An experiment consists of selecting one number from the set $\{0, 2, 4, 6, 8\}$. How many outcomes are possible?

Outcomes

2. An experiment consists of choosing an odd number between 0 and 10. How many outcomes are possible? Give the sample space.

Outcomes
Sample space

3. If we are allowed to choose one entree from the following menu, how many outcomes are possible? What is the sample space?

MENU

Meat loaf

Turkey

Rabbit stew

Liver and onions

Fish sticks

Duck pie

Outcomes
Sample space

4. An experiment consists of obtaining the remainders when each of the numbers 33, 47, 56, 95, 125 is divided by 4. How many outcomes are possible? Give the sample space.

Outcomes Sample space

5. A box contains 20 ping-pong balls which are numbered from 1 to 20. The balls with odd numbers are white, and the others are black. An experiment consists of choosing one of these balls from the box.
   (a) If we are interested in the number of the ball that is chosen, how many outcomes are possible?
   (b) If we are interested only in the color of the ball chosen, how many outcomes are possible?
   Give the sample space in each case.

Tree diagram

6. An experiment consists of spinning the arrows on two given spinners one time. Construct a tree diagram and list the outcomes as pairs [for example, one outcome is the pair $(b, 3)$].

Tree diagram

7. An experiment consists of tossing a coin twice. Construct a tree diagram and list the outcomes. What is the sample space?

Tree diagram

8. An experiment consists of tossing a coin three times. Construct a tree diagram and list the outcomes. What is the sample space?

Tree diagram

9. An experiment consists of selecting a number from the set $A = \{0, 1\}$ and then selecting a letter from the set $B = \{a, b, c\}$. Construct a tree diagram and list the outcomes. What is the sample space?

Tree diagram

10. Choose two numbers from the set $S = \{0, 1, 2\}$; a first number and then a second is chosen from those remaining after the first selection. Construct a tree diagram and list the possible outcomes. What is the sample space?

Tree diagram

11. Two numbers are to be chosen from the set $S = \{0, 1, 2, 3\}$; a first number is chosen and then a second number is chosen from those remaining after the first selection. Construct a tree diagram and list the outcomes. What is the sample space?

Tree diagram

12. Three numbers are to be chosen from the set $S = \{0, 1, 2, 3\}$; a first number is chosen and then a second number is chosen from those remaining after the first selection etc. Construct a tree diagram and list the outcomes. What is the sample space?

Counting principle

13. If A and B are the sets given below, in how many ways can we choose a pair of numbers, first one from A and then one from B?

$$A = \{1, 2, 3, 4, 5\}, \qquad B = \{10, 11, 12, 13\}$$

Number of outfits

14. A girl has 3 skirts and 4 blouses and each skirt will match

each blouse. How many different outfits does she have which consist of a skirt and a blouse?

Number of routes

15. Town A is connected to town B by 3 roads and B is connected to C by 4 roads. In how many ways is it possible to travel from A to C by passing through B?

Number of routes

16. There are 5 roads connecting town A to town B and 8 roads connecting town B to town C. How many different ways are there of traveling from A to C by going through B?

Number of routes

17. There are 6 roads connecting town A to town B, 9 roads connecting town B to town C, and 4 roads connecting town C to town D. How many different ways are there of going from A to D by passing through B and C?

Number of meals in a hospital

18. A hospital prepares menus by selecting one item each from four food groups: A, B, C, D. If these groups contains 6, 12, 5, and 9 items, respectively, how many different meals are possible?

Number of test completions

19. A test consists of 10 true-false questions. In how many ways can the test be completed?

Number of test completions

20. In how many ways can the test of Exercise 19 be answered so that all the odd questions are marked F?

Number of test completions

21. A test consists of 10 four-answer multiple choice questions. In how many ways can the test be completed?

Number of test completions

22. In how many ways can the test of Exercise 21 be completed so that all the odd questions are answered correctly and the even questions are answered incorrectly?

Number of schedules

23. A student who is registering for the fall term wishes to have a 9, a 10, and a 12 o'clock class. At 9 there are 6 courses available to her, at 10 there are 4 possible courses, and at 12 there are 5 courses. From how many different schedules can she choose?

Football contest

24. Two football teams (11 players on each team) decide to settle a game with a single arm-wrestling contest. Each team chooses one of its members to represent it in the contest. In how many ways can the contest take place?

Health spa

25. Each morning at breakfast a health spa furnishes its guests with free vitamins: multiple vitamins and vitamin C. On a table, dish M contains 30 Multitabs, and dish C contains 20 Rosehip Super C's.
    (a) In how many ways can Tom choose one vitamin, either from dish M or dish C?
    (b) In how many different ways can Tom choose one vitamin of each kind?

Radio station call letters

26. Call letters for some radio stations consist of either 3- or 4-letter Latin alphabet sequences; suppose each sequence begins with "W."

(a) How many 3-letter sequences begin with "W"?
(b) How many 4-letter sequences begin with "W"?
(c) How many possible radio stations can have call letters beginning with "W"?

Telephone numbers

27. Local telephone numbers have the pattern

XXX – XXXX

where each position must be a digit 0 through 9.

(a) If there are no restrictions on the digits in any of the positions, how many phone numbers are possible?
(b) If the first position cannot be a 0 or 9, how many numbers are possible?

Auto plates

28. An auto license plate has the following design:

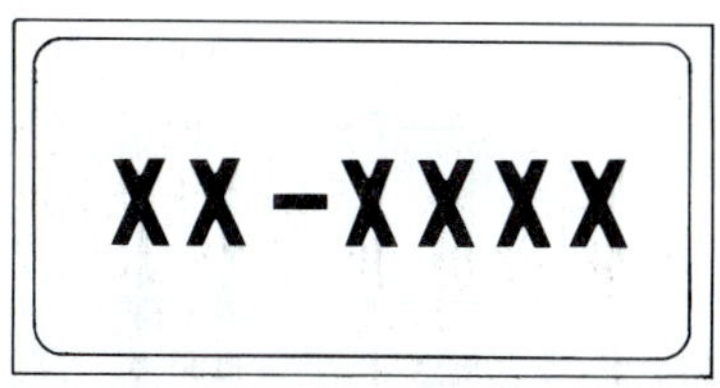

(a) If the first two positions can be occupied by any letters of the alphabet and the last four by any of the digits 0 through 9, how many plates of this design are possible?
(b) If the letter "O" cannot occur in the first two positions, and the number 0 cannot appear in position three, how many plates are possible?

Number of games to be played

29. There are 5 contestants in a chess tournament. If each contestant must play each of the other contestants exactly twice, how many games must be played?

Number of games to be played

30. If each contestant of Exercise 29 must play each of the other contestants exactly 3 times, how many games must be played?

k-state devices

31. An on-off relay is an example of a 2-state digital device. Such a device can be in only one of two possible states. For example, it may be in state 0 or 1. A 3-state device can be in state 0,

1, or 2. A k-state device is possible for any whole number k greater than 1. We can think of these devices as being a row of buttons such as those found on juke boxes:

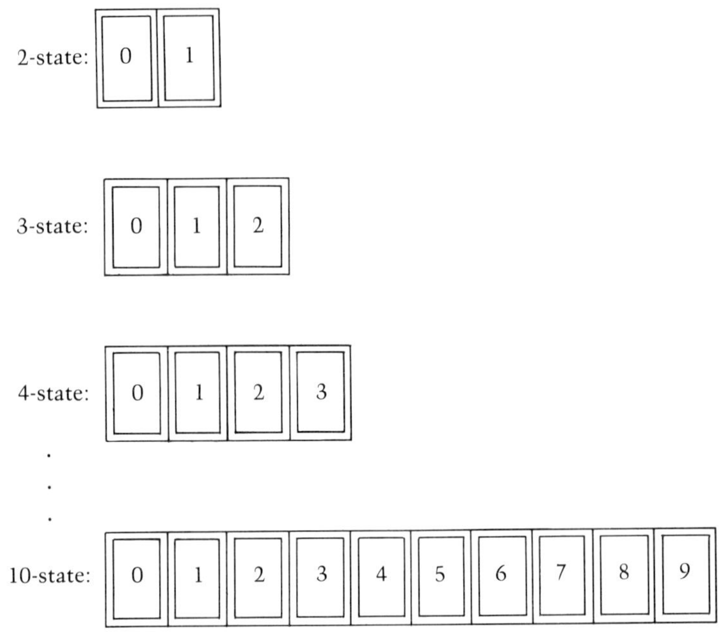

The state of the device is determined by pressing a button.

Suppose we combine two 2-state devices:

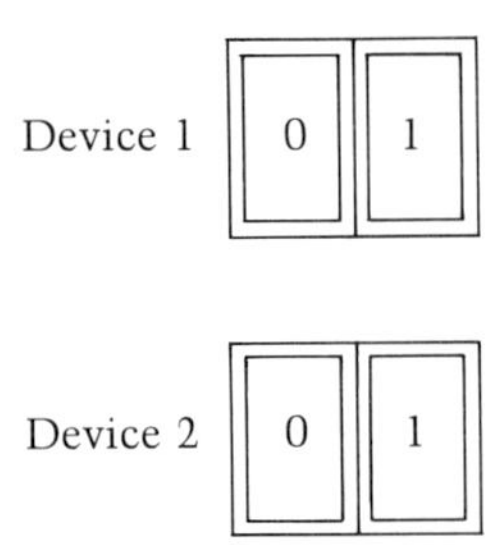

Messages may be transmitted by using the states associated with the devices. There are only 4 messages possible in the case of two 2-state devices. They are:

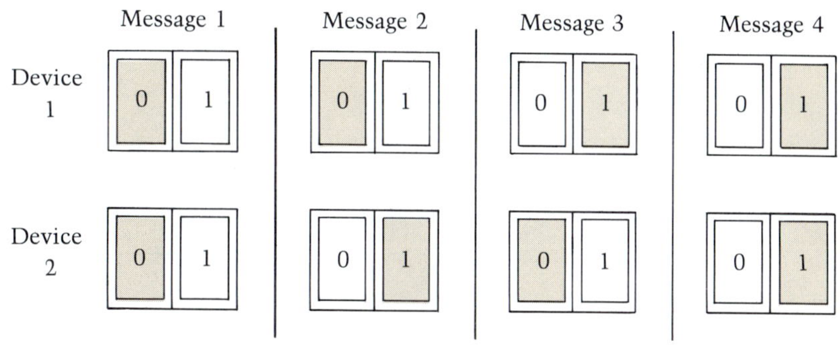

(a) How many messages are possible if six 2-state devices are used?
(b) How many messages are possible if two 12-state devices are used?
(c) How many messages are possible if eight 3-state devices are used?
(d) Suppose we have 24 buttons to arrange in rows of k-state devices (each row must have k states for some fixed k). For example, we could make two rows of 12-state devices. How many rows should be used to maximize the number of messages?

6.3 PERMUTATIONS

Suppose we have 3 identical ping-pong balls and on each one we paint one of the letters A, B, C as shown in Figure 6.9. Now the balls are placed in a box and an experiment will be performed which consists of drawing them out of the box, one at a time.

drawing with replacement

Drawing with replacement: We will draw the 3 ping-pong balls from the box, one at a time, and after each draw the letter on the ball will be noted and then the ball replaced in the box. This is called *drawing with replacement.* We wish to determine the sam-

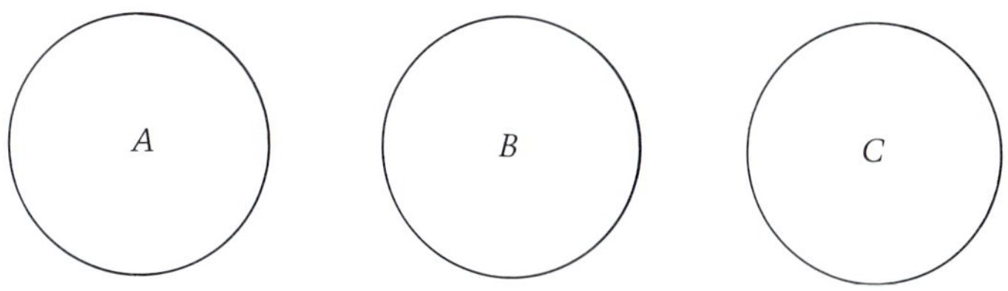

FIG. 6.9.

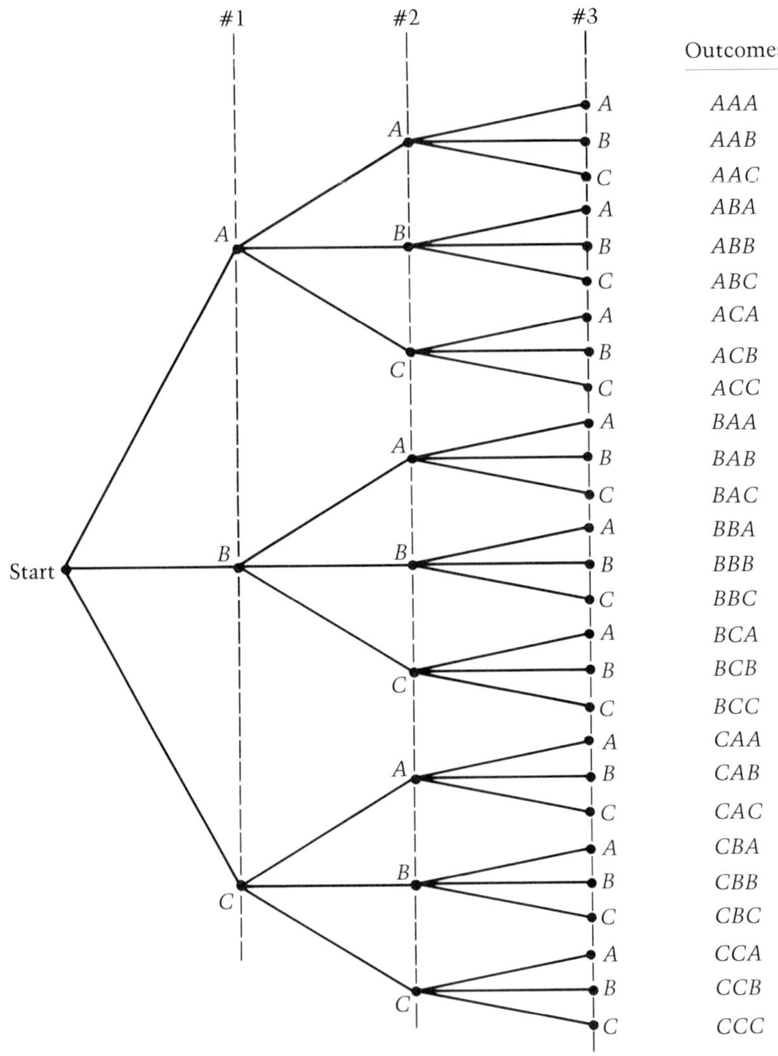

FIG. 6.10.

ple space of this experiment. The tree diagram is shown in Figure 6.10. This experiment leads to 27 outcomes as the diagram shows. These may be counted by using the Counting Principle in the following way. An outcome may be obtained by filling the blanks below:

$$(\_\_\_, \_\_\_, \_\_\_)$$

In the first blank we place the letter which results from the first draw. The second and third blanks are filled in a similar way.

Since each blank can be filled in three ways (remember the drawing is with replacement) there are

$$3 \cdot 3 \cdot 3 = 27$$

ways of filling the 3 blanks. Hence, this experiment has 27 outcomes.

Drawing with replacement

EXAMPLE 1 How many two-digit numbers can be formed by using only the numerals 3, 4, 5? It is understood that repetitions are allowed.

Solution: We may think of forming a two-digit number by filling two blanks:

$$(\_\_, \_\_)$$

Each blank must be filled with one of the numerals 3, 4, 5 and, since repetitions are allowed (this means the same as replacement), each blank may be filled in 3 ways. By the Counting Principle there are $3 \cdot 3 = 9$ ways of filling the blanks. Hence, there are 9 two-digit numbers which involve only the digits 3, 4, 5. These are listed below:

33, 43, 53,
34, 44, 54,
35, 45, 55.

PRACTICE PROBLEM How many three-digit numbers can be formed by using only the numbers 3, 4, 5 if repetitions are allowed?

Answer: 27. ■

drawing without replacement

Drawing without replacement: We return now to the box containing the three ping-pong balls marked A, B, C. As before, we draw three balls from the box, one after the other. In this experiment, however, we do not replace a ball which has been drawn. This is called *drawing without replacement.* To determine the sample space for this experiment, the tree diagram shown in Figure 6.11 is constructed.

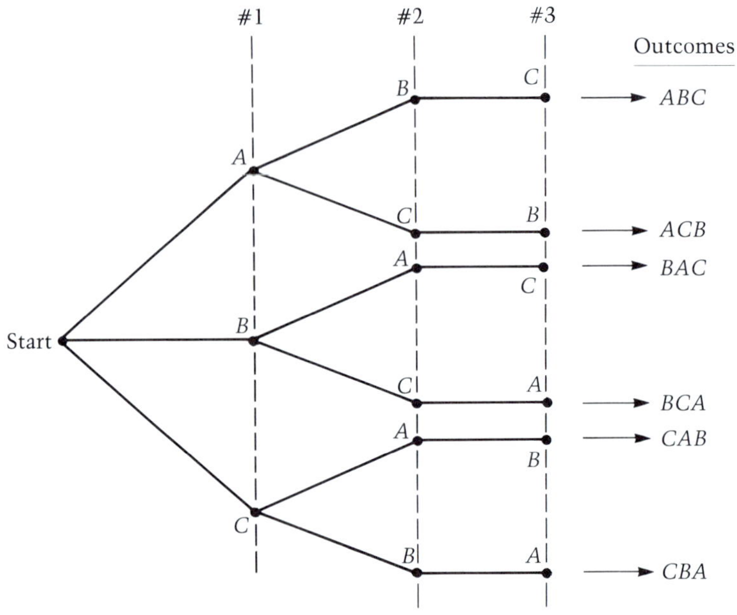

FIG. 6.11.

Note in the tree diagram of Figure 6.11 that if A is obtained, for example, on the first draw, then it is only possible to choose B or C on the second draw, since A is not replaced in the box after the first draw. This experiment results in only 6 outcomes. These may also be counted using the Counting Principle but since the drawings are not independent, our reasoning must be modified somewhat. Each outcome may be obtained by filling in three blanks as before:

$$(\_\_\_, \_\_\_, \_\_\_)$$

On the first draw there are 3 outcomes so that the first blank can be filled in 3 ways. After the first draw has been made, there are only 2 possible outcomes for the second draw, since the ball obtained on the first draw was not replaced. Therefore, the second blank can be filled in only 2 ways. In a similar manner, after the first and second draws, there is only one possible outcome for the third draw. Thus, the third blank can be filled in one way only after the first and second blanks have been filled. By the Counting Principle, the three blanks may be filled in

$$3 \cdot 2 \cdot 1 = 6$$

ways.

Drawing without replacement

EXAMPLE 2 The proud parents of a baby girl wish to choose their baby's first and second names from the following list:

April	Anne
Joy	Paige
Carol	Joan

If we assume that the first and second names will be different, how many choices are possible?

Solution: Since the two names must be different, this is a problem of drawing without replacement: After a choice has been made for the first name, any of the remaining names can be used for the second. The first name may be chosen in 6 ways (since there are 6 names to choose from), and then the second may be chosen in 5 ways (since the second name must be different from the first). Therefore, there are $6 \cdot 5 = 30$ possible choices.

PRACTICE PROBLEM In Example 2 suppose that the name Lynn is added to the list. How many choices are possible?

Answer: 42. ■

We have been considering experiments in which the parts are performed in a definite order: a first part, a second part, etc. This, in turn, leads to outcomes in which order is important. For example, in drawing the three ping-pong balls without replacement, we obtained the following sample space:

$$(A, B, C),$$
$$(A, C, B),$$
$$(B, A, C),$$
$$(B, C, A),$$
$$(C, A, B),$$
$$(C, B, A).$$

This sample space consists of all possible ways of arranging the letters A, B, C in a definite order. Every arrangement is called a *permutation* of the letters A, B, C.

permutation

DEFINITION OF PERMUTATION A **permutation** of a set of objects is an arrangement of those objects.

This definition will usually be applied here to sets of *distinct* objects such as x, y, z (see Example 3), but in some cases the objects will not be distinct (see Example 4) such as x, x, y.

Permutations

EXAMPLE 3 List all the permutations of the letters a, b.

Solution: Here we have two distinct objects and, by the definition, we should list all possible arrangements of the letters a, b. There are only two:

$$a,\ b,$$
$$b,\ a.$$

PRACTICE PROBLEM List all the permutations of the letters a, b, c.

Answer

$a, b, c,\quad a, c, b,\quad b, a, c,\quad b, c, a,\quad c, a, b,\quad c, b, a.$ ■

Permutations

EXAMPLE 4 List all the permutations of the letters a, a, b, b.

Solution: This is a case of 4 nondistinct objects, and it will be helpful to use a scheme which will tell us when the problem has been completed. One such scheme in this example is to list all those permutations which begin with the letter a and then all those which begin with b:

$$\text{All these begin with } a \begin{cases} \mathbf{a}, a, b, b \\ \mathbf{a}, b, a, b \\ \mathbf{a}, b, b, a \end{cases} \qquad \text{All these begin with } b \begin{cases} \mathbf{b}, a, a, b \\ \mathbf{b}, a, b, a \\ \mathbf{b}, b, a, a \end{cases}$$

PRACTICE PROBLEM List all the permutations of the letters a, a, b.

Answer

$$a, a, b, \qquad a, b, a, \qquad b, a, a. \blacksquare$$

Permutations

EXAMPLE 5 If 4 contestants are in a race, in how many ways can the race be completed if there are no ties?

Solution: Let us call the contestants a, b, c, d. Then a completion of the race is a permutation of the letters. For example, the permutation *bacd* means that b came in first, a second, c third, and d fourth. Therefore, the problem will be answered by knowing the number of permutations of a, b, c, d. To determine this we consider each permutation as being constructed by filling in four blanks:

$$(\_\_\_, \_\_\_, \_\_\_, \_\_\_)$$

The first blank may be filled in 4 ways with any one of the letters a, b, c, d. After the first blank is filled, the second can be filled in only 3 ways (since the letter in the first blank cannot be used again). Similarly, the third and fourth blanks can be filled in 2 ways and 1 way, respectively. Hence, by the Counting Principle there are

$$4 \cdot 3 \cdot 2 \cdot 1 = 24$$

ways of filling the 4 blanks and thus 24 permutations of the letters a, b, c, d.

PRACTICE PROBLEM What is the answer to Example 5 if there are 5 contestants instead of 4?

Answer: 120. ■

Permutations

EXAMPLE 6 In how many of the permutations of the letters a, b, c, d are the letters a and d adjacent?

Solution: We may reason as follows: if a, d are adjacent to each other, they are in positions 1 and 2, 2 and 3, or 3 and 4:

Positions 1 and 2	Positions 2 and 3	Positions 3 and 4
***a*, *d*,** *b*, *c*	*b*, ***a*, *d*,** *c*	*b*, *c*, ***a*, *d***
***a*, *d*,** *c*, *b*	*c*, ***a*, *d*,** *b*	*c*, *b*, ***a*, *d***
***d*, *a*,** *b*, *c*	*b*, ***d*, *a*,** *c*	*b*, *c*, ***d*, *a***
***d*, *a*,** *c*, *b*	*c*, ***d*, *a*,** *b*	*c*, *b*, ***d*, *a***

Since there are 4 permutations in each group, we see that there are

$$4 + 4 + 4 = 12$$

permutations of a, b, c, d in which a, d are adjacent.

PRACTICE PROBLEM How many of the permutations of a, b, c, d begin with the letter d?

Answer: 6. ■

EXERCISES

Tree diagram

1. Two ping-pong balls labeled a, b are placed in a box. Draw the two balls from the box, one at a time, without replacement.
   (a) Construct a tree diagram of this experiment.
   (b) What is the sample space?
   (c) How many outcomes have different letters?

Tree diagram

2. Two ping-pong balls labeled a, b are placed in a box. Draw the two balls from the box, one at a time, with replacement.
   (a) Construct a tree diagram for this experiment.
   (b) What is the sample space?

Tree diagram

3. We write one of the numbers 1, 2, 3 on each of three slips of paper, a different number on each slip. The slips are placed in a box and two are drawn out one after the other, with replacement.
   (a) Construct a tree diagram.
   (b) What is the sample space?
   (c) On how many of the outcomes will the sum of the numbers be less than 6?
   (d) On how many of the outcomes will the sum of the numbers be less than 5?

Tree diagram

4. We write one of the numbers 1, 2, 3 on each of three slips of paper, a different number on each slip. The slips of paper are

placed in a box and two are drawn out, one after the other, without replacement.

(a) Construct a tree diagram.
(b) What is the sample space?
(c) On how many of the outcomes will the sum of the numbers be at least 5?

Tree diagram

5. One white and two black marbles are placed in a box. Three marbles are drawn from the box, one at a time, without replacement.
(a) Construct a tree diagram.
(b) What is the sample space?

Counting principle

6. On each of four slips of paper one of the numbers 1, 2, 3, 4 is written, a different number on each slip. The slips of paper are put into a box and four are drawn out, one at a time, with replacement.
(a) Use the Counting Principle to determine the number of outcomes.
(b) How many outcomes begin with 1?
(c) How many outcomes begin with an even number?
(d) How many outcomes begin and end with an even number?
(e) On how many outcomes is the sum of the numbers even?

Counting principle

7. On each of four slips of paper one of the numbers 1, 2, 3, 4 is written, a different number on each slip. The slips of paper are put into a box and four are drawn out, one at a time, without replacement.
(a) Use the Counting Principle to determine the number of outcomes.
(b) How many outcomes begin with 1?
(c) How many outcomes begin with an even number?
(d) How many outcomes begin and end with an even number?
(e) On how many outcomes is the sum of the numbers even?

Counting principle

8. An experiment consists of tossing a coin four times.
(a) Use the Counting Principle to determine the number of outcomes.
(b) How many outcomes begin with heads?
(c) How many outcomes begin and end with heads?
(d) How many outcomes have two heads and two tails?
(e) How many outcomes have at least two heads?
(f) How many outcomes have at least one head and at least one tail?

Permutations

9. Using the numbers 1, 2, 3, 4, 5, we wish to form three-digit numbers.

(a) If repetitions are allowed, how many such numbers are possible?
(b) If repetitions are not allowed, how many such numbers are possible?

Permutations

10. Using the numbers 0, 1, 2, 3, 4, 5, 6, 7, 8, 9, determine how many two-digit numbers there are if repetitions are allowed. [*Hint:* 0 is not allowed in the first position. For example, 01 is considered a one-digit number.]

Permutations

11. Using the numbers 0, 1, 2, 3, 4, 5, 6, 7, 8, 9 with repetitions allowed, determine how many
(a) three-digit numbers there are.
(b) three-digit even numbers there are.
(c) three-digit numbers there are which are divisible by 5. [*Hint:* A whole number is divisible by 5 if its last digit is 0 or 5.]

Permutations

12. How many whole numbers are there greater than 99 but less than 1000?

Permutations

13. How many whole numbers are there greater than 99 but less than 10,000?

Permutations

14. Write down all the permutations of the letters a, a, b, c.

Permutations

15. Write down all the permutations of the letters a, a, b, b, c.

Permutations

16. Write down all the permutations of the letters a, a, b, b, b.

Permutations

17. Write down all the permutations of the letters a, b, c, d which begin with (a) the letter a, (b) the letter b, (c) the letter c, (d) the letter d.

Permutations

18. Use Exercise 17 to determine the total number of permutations of the letters a, b, c, d.

Permutations

19. Use Exercise 18 to determine the total number of permutations of any four distinct objects.

Counting principle

20. Use the Counting Principle to determine the number of permutations of four distinct objects.

Counting principle

21. Use the Counting Principle to determine the number of permutations of the letters u, v, w, x, y.

Permutations

22. How many of the permutations of the letters u, v, w, x, y have the letters u, v in adjacent positions?

Permutations

23. How many of the permutations of the letters u, v, w, x, y have the letter w in the third position?

Permutations

24. Use the Counting Principle to determine the number of permutations of 6 distinct objects.

Hexadecimal numbers

25. The hexadecimal number system uses the "digits"

0, 1, 2, 3 ,4, 5, 6, 7, 8, 9, A, B, C, D, E, F.

If 0 is allowed to be a first digit, how many two-digit hexadecimal numbers are there?

Computer chips

26. Five computer chips must be placed on a board whose diagram is shown in the figure. Using the five chips, $X1$, $X2$, $X3$, $Y1$, $Y2$, determine in how many ways these can be placed on the board if the Xs must go in I and the Ys in II.

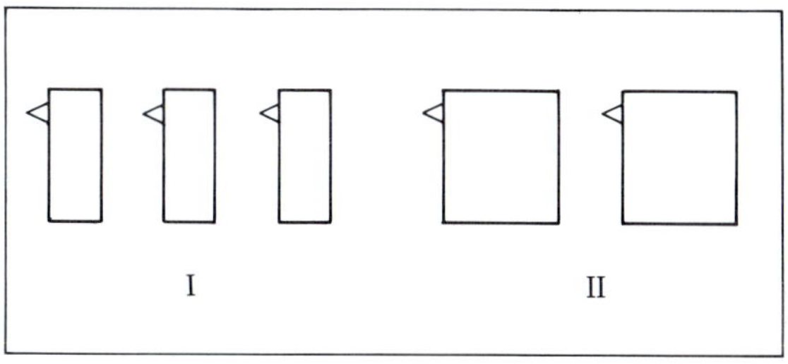

Group portrait

27. In how many ways can 5 people line up for a group portrait?

Seating arrangements

28. In how many ways can 4 boys and 4 girls be seated in a row of 8 chairs if
    (a) the seats are not assigned?
    (b) the first and last chairs are to be occupied by boys only?
    (c) boys and girls are seated alternately?

Seating arrangements

29. Eight invited guests are seated in a row. How many different seating arrangements are possible if 2 guests refuse to sit next to each other?

Seating arrangements

30. Four couples are to sit in a row. How many different seating arrangements are possible if all the men sit together and all the women sit together?

6.4. PERMUTATIONS (CONTINUED)

The previous section was devoted to problems which are related to permutations. For example, we considered problems such as determining the number of permutations of the letters a, b, c, d, e. This means, of course, that the total number of ways of arranging the letters a, b, c, d, e in order must be counted. This is easily done by thinking of the problem as an experiment which consists of 5 parts. Each of the permutations may be constructed by filling 5 blanks with a, b, c, d, e:

$$\_\_\_, \_\_\_, \_\_\_, \_\_\_, \_\_\_$$

We reason that the first blank can be filled in 5 ways, the second in 4 ways, the third in 3 ways, the fourth in 2 ways, and the fifth in 1 way. Hence, by the Counting Principle, all 5 blanks can be filled in

$$5 \cdot 4 \cdot 3 \cdot 2 \cdot 1 = 120$$

ways. Thus there are 120 permutations of the letters a, b, c, d, e.

factorial

Expressions such as $5 \cdot 4 \cdot 3 \cdot 2 \cdot 1$ are conveniently expressed in *factorial* notation. To do this the symbol

$$5!$$

(read as "five factorial") is used as a short way of writing $1 \cdot 2 \cdot 3 \cdot 4 \cdot 5$.

DEFINITION OF FACTORIAL If n is a whole number and $n > 0$, then $n!$ is defined by the equation

$$n! = 1 \cdot 2 \cdot 3 \cdot 4 \cdots (n-1) \cdot n$$

Also 0! is defined to be 1, that is,

$$0! = 1.$$

ILLUSTRATION By the definition of factorial we have

$$0! = 1$$
$$1! = 1$$
$$2! = 1 \cdot 2 = 2$$
$$3! = 1 \cdot 2 \cdot 3 = 6$$
$$4! = 1 \cdot 2 \cdot 3 \cdot 4 = 24$$

Factorials

EXAMPLE 1 Calculate the following. (a) 7! (b) 8! (c) $\frac{10!}{6!}$ (d) $\frac{120!}{118!}$

Solution

(a) $$\begin{aligned}7! &= 1\cdot 2\cdot 3\cdot 4\cdot 5\cdot 6\cdot 7\\ &= 6\cdot 4\cdot 5\cdot 6\cdot 7\\ &= 24\cdot 5\cdot 6\cdot 7\\ &= 120\cdot 6\cdot 7\\ &= 720\cdot 7\\ &= 5040.\end{aligned}$$

(b) $$\begin{aligned}8! &= 1\cdot 2\cdot 3\cdot 4\cdot 5\cdot 6\cdot 7\cdot 8\\ &= (1\cdot 2\cdot 3\cdot 4\cdot 5\cdot 6\cdot 7)\cdot 8\\ &= (7!)\cdot 8\\ &= (5040)\cdot 8 \qquad \text{from (a) } 7! = 5040\\ &= 40{,}320.\end{aligned}$$

(c) $$\begin{aligned}\frac{10!}{6!} &= \frac{1\cdot 2\cdot 3\cdot 4\cdot 5\cdot 6\cdot 7\cdot 8\cdot 9\cdot 10}{1\cdot 2\cdot 3\cdot 4\cdot 5\cdot 6}\\ &= \frac{(1\cdot 2\cdot 3\cdot 4\cdot 5\cdot 6)\cdot 7\cdot 8\cdot 9\cdot 10}{1\cdot 2\cdot 3\cdot 4\cdot 5\cdot 6}\\ &= 7\cdot 8\cdot 9\cdot 10\\ &= 5040.\end{aligned}$$

(d) $$\begin{aligned}\frac{120!}{118!} &= \frac{1\cdot 2\cdot 3\cdot 4\cdot \cdots \cdot 118\cdot 119\cdot 120}{1\cdot 2\cdot 3\cdot 4\cdot \cdots \cdot 118}\\ &= \frac{(1\cdot 2\cdot 3\cdot 4\cdot \cdots \cdot 118)\cdot 119\cdot 120}{1\cdot 2\cdot 3\cdot 4\cdot \cdots \cdot 118}\\ &= 119\cdot 120\\ &= 14{,}280.\end{aligned}$$

PRACTICE PROBLEM Calculate the following: (a) 9! (b) $\dfrac{27!}{20!}$.

Answers: (a) 362,880, (b) 4,475,671,200. ■

Consider again the permutations of *a*, *b*, *c*, *d*, *e*. Suppose we choose 2 of these letters and form all possible permutations of them. For example, we might choose b, e. There are only 2 permutations of these letters:

$$b, e \qquad \text{and} \qquad e, b.$$

These two arrangements are examples of *permutations* of a, b, c, d, e *taken 2 at a time.* Now it is important to know how many permutations there are of a, b, c, d, e taken 2 at a time. This is easy to determine. As before we may determine the number of ways of filling two blanks:

$$\_\_\_, \_\_\_$$

with the letters a, b, c, d, e. There are 5 ways of filling the first blank since we may choose any one of the 5 letters a, b, c, d, e: once the first blank has been filled, the second one may be filled in 4 ways. Therefore, the 2 blanks may be filled in

$$5 \cdot 4 = 20$$

ways. We conclude that there are 20 permutations of the 5 letters a, b, c, d, e taken 2 at a time. The symbol $P(5, 2)$ is used to denote the number of permutations of 5 objects taken 2 at a time. Thus we have determined that

$$P(5, 2) = 20.$$

$P(n, k)$

DEFINITION OF $P(n, k)$ If n is a positive whole number and k is a whole number no larger than n, then the symbol $P(n, k)$ denotes the number of permutations of n objects taken k at a time.

Computing $P(n, k)$

EXAMPLE 2 Compute each of the following. (a) $P(5, 3)$, (b) $P(5, 5)$.

Solution

(a) To compute $P(5, 3)$ we must determine the number of permutations of 5 things taken 3 at a time. For this we may use the 5 letters a, b, c, d, e and imagine filling 3 blanks:

$$\_\_\_, \_\_\_, \_\_\_$$

Using the Counting Principle, we find that there are

$$5 \cdot 4 \cdot 3 = 60$$

ways of filling these 3 blanks. Hence,

$$P(5, 3) = 60.$$

(b) To determine $P(5, 5)$ we may imagine filling 5 blanks:

$$\_\_\_, \_\_\_, \_\_\_, \_\_\_, \_\_\_$$

Therefore,

$$P(5, 5) = 5 \cdot 4 \cdot 3 \cdot 2 \cdot 1 = 120.$$

Note in this case that $P(5, 5) = 5!$

PRACTICE PROBLEM Compute $P(6, 4)$.

Answer: 360. ■

Permutations

EXAMPLE 3 There are 20 contestants in a race. In how many ways is it possible for the first, second, and third places to be taken?

Solution: The question is equivalent to asking for the number of permutations of 20 things taken 3 at a time, and this number is

$$P(20, 3) = 20 \cdot 19 \cdot 18 = 6840.$$

PRACTICE PROBLEM Fifteen contestants enter a checkers tournament. In how many ways is it possible for first and second place to be taken?

Answer: $P(15, 2) = 210$. ■

By using reasoning similar to that of Example 2 the following formula for $P(n, k)$ may be obtained.

FIRST FORMULA OF PERMUTATIONS

$$P(n, k) = n \cdot (n - 1) \cdot (n - 2) \cdot (n - 3) \cdot \cdots \cdot (n - k + 1).$$

We may state this formula in words by saying that *the number of permutations of n objects taken k at a time is obtained by multiplying k factors, the first of which is n and each succeeding factor is 1 less than the previous one.*

Consider, for example, the number $P(10, 6)$. In this case, there will be 6 factors which begin with 10 and decrease by 1 until all factors are obtained:

Begin here . . . decrease by 1 until 6 factors are obtained

$\to 10, \to 9, \to 8, \to 7, \to 6, \to 5$

Therefore,

$$P(10, 6) = 10 \cdot 9 \cdot 8 \cdot 7 \cdot 6 \cdot 5 = 151{,}200.$$

The number

$$10 \cdot 9 \cdot 8 \cdot 7 \cdot 6 \cdot 5$$

which we have just been considering may be written like this

$$\begin{aligned} 10 \cdot 9 \cdot 8 \cdot 7 \cdot 6 \cdot 5 &= \frac{(10 \cdot 9 \cdot 8 \cdot 7 \cdot 6 \cdot 5) \cdot (4 \cdot 3 \cdot 2 \cdot 1)}{(4 \cdot 3 \cdot 2 \cdot 1)} \\ &= \frac{10!}{4!}. \end{aligned}$$

Therefore,

$$P(10, 6) = \frac{10!}{4!} \qquad \text{or} \qquad P(10, 6) = \frac{10!}{(10 - 6)!}.$$

In this same way, it is possible to obtain another formula for $P(n, k)$.

SECOND FORMULA FOR PERMUTATIONS $P(n, k) = \dfrac{n!}{(n - k)!}$ if $0 \le k < n$.

ILLUSTRATION Using the second formula for permutations, we obtain

$$\begin{aligned}P(10, 3) &= \frac{10!}{(10-3)!}\\ &= \frac{10!}{7!}\\ &= \frac{(1\cdot 2\cdot 3\cdot 4\cdot 5\cdot 6\cdot 7)\cdot 8\cdot 9\cdot 10}{1\cdot 2\cdot 3\cdot 4\cdot 5\cdot 6\cdot 7}\\ &= 8\cdot 9\cdot 10\\ &= 720.\end{aligned}$$

Permutations

EXAMPLE 4 We are going to consider permutations of the letters a, b, c, A, B, C. How many permutations are there of these letters taken

(a) all at a time?

(b) two at a time?

(c) two at a time in which both are lowercase letters or both are uppercase letters?

(d) two at a time in which one letter is lowercase and the other is uppercase?

Solution

(a) $P(6, 6) = 6! = 720$.

(b) $P(6, 2) = 6 \cdot 5 = 30$.

(c) Some of the permutations in this question involve only lowercase letters and the others involve only uppercase letters. We may calculate separately those of each kind and add these numbers to obtain the result. The number of permutations of the lowercase letters taken 2 at a time is

$$P(3, 2) = 6.$$

This is also the number of permutations of the uppercase letters taken 2 at a time. Thus the answer is

$$6 + 6 = 12.$$

These permutations are as follows:

lowercase only			uppercase only		
ab	*ac*	*bc*	*AB*	*AC*	*BC*
ba	*ca*	*cb*	*BA*	*CA*	*CB*

(d) For this question we may first determine those permutations, taken 2 at a time, which begin with a lowercase letter. Next, we shall determine the number of permutations which begin with an uppercase letter. The answer to the question will then be the sum of these two numbers. We will imagine filling two blanks:

$$\underline{\quad}, \underline{\quad}$$

By the Counting Principle, if the first blank is filled with a lowercase letter and the second by an uppercase letter, there are

$$3 \cdot 3 = 9$$

permutations taken 2 at a time which have a lowercase letter first and an uppercase letter second. Since 9 is also the number of permutations taken 2 at a time in which an uppercase letter comes first and a lowercase letter second, the answer to the original question is

$$9 + 9 = 18.$$

Here is a list of these permutations:

lowercase first			uppercase first		
aA	*aB*	*aC*	*Aa*	*Ab*	*Ac*
bA	*bB*	*bC*	*Ba*	*Bb*	*Bc*
cA	*cB*	*cC*	*Ca*	*Cb*	*Cc*

PRACTICE PROBLEM How many permutations are there of the letters a, b, A, B, C taken 2 at a time in which (a) the lowercase letters are together? (b) one letter is lowercase and the other uppercase?

Answer: (a) 2. (b) 12. ■

Permutations

EXAMPLE 5 On a single shelf we are to arrange 4 history books and 5 mathematics books. In how many ways can this be done if the history books are to be grouped together and the mathematics books are also to be together?

Solution: We can see that this problem is about arrangements and hence about permutations. But we must think about permutations of the two sets of books together. The question may be analyzed as an experiment consisting of three parts:

Part 1: Choose the order of the subjects (either history or mathematics).

Part 2: Choose the order of the history books.

Part 3: Choose the order of the mathematics books.

Each of these parts involves permutations and the number of ways of performing each part may be calculated by use of one of the permutation formulas:

Part 1 may be performed in $P(2, 2)$ ways,

Part 2 may be performed in $P(4, 4)$ ways,

Part 3 may be performed in $P(5, 5)$ ways.

Now by the Counting Principle, the number of ways of arranging the books on the shelf with the history books together and the mathematics books together is

$$\begin{aligned} P(2, 2) \cdot P(4, 4) \cdot P(5, 5) &= (2!) \cdot (4!) \cdot (5!) \\ &= 2 \cdot 24 \cdot 120 \\ &= 5760. \end{aligned}$$

PRACTICE PROBLEM In Example 5, what is the answer to the question if the history books must come before the mathematics books?

Answer: 2880. ■

EXERCISES

Set A–Using Formulas for Permutations and Factorial Notation

Calculate each of the expressions in Exercises 1 through 20.

1. $P(4, 2)$ **2.** $P(6, 3)$ **3.** $P(5, 1)$ **4.** $P(12, 2)$

5. $P(8, 4)$ **6.** $P(n, 1)$, where n is a natural number

7. $\dfrac{P(8, 4)}{P(4, 4)}$ **8.** $\dfrac{P(10, 5)}{P(5, 5)}$ **9.** $\dfrac{P(12, 3)}{P(3, 3)}$ **10.** $5!$

11. (a) $2! + 4!$ (b) $(2 + 4)!$

12. (a) $2! + 5!$ (b) $(2 + 5)!$

13. (a) $(2!) \cdot (3!)$ (b) $(2 \cdot 3)!$

14. (a) $(3!) \cdot (3!)$ (b) $(3 \cdot 3)!$

15. (a) $2! \cdot (2 + 3)!$ (b) $2! \cdot (2! + 3!)$

16. (a) $2! \cdot (5 - 2)!$ (b) $2! \cdot (5! - 2!)$

17. $\dfrac{8!}{2!(8 - 2)!}$

18. $\dfrac{10!}{5!(10 - 5)!}$

19. $\dfrac{30!}{(30 - 3)!3!}$

20. $\dfrac{100!}{2!(100 - 2)!}$

Set B–Questions About Permutations

21. Find the number of permutations of the objects a, b, c, d, e, f taken
(a) 1 at a time (b) 2 at a time (c) 3 at a time
(d) 4 at a time (e) 5 at a time (f) 6 at a time

22. How many permutations are there of the numbers 1, 2, 3, 4, 5, 6, 7 taken
(a) 1 at a time (b) 2 at a time (c) 3 at a time
(d) 4 at a time (e) 5 at a time (f) 6 at a time
(g) 7 at a time

23. Seven ping-pong balls marked 1, 2, 3, 4, 5, 6, 7 are placed in a box and four are drawn out, one after the other, without replacement.
(a) How many outcomes are there for this experiment?
(b) How many outcomes involve only 4, 5, 6, 7?
(c) In how many outcomes does 7 occur second?
(d) In how many outcomes does 7 occur first or last?
(e) In how many outcomes does 7 occur, but not 6?
(f) In how many outcomes does 7 occur first and 1 last?
(g) In how many outcomes does 7 not occur?

24. We will form permutations of the letters a, b, A, B, C. How many permutations are there of these letters taken
(a) all at a time? (b) 2 at a time?
(c) 2 at a time and involving only lower-case letters?
(d) 2 at a time and involving only upper-case letters?
(e) 2 at a time and involving one lower- and one upper-case letter?
(f) all at a time with the lower-case letters together and the upper-case letters together?

Winners in a contest

25. There are 10 contestants in a beauty contest. In how many ways is it possible to award first, second, and third place?

Club officers

26. A club having 15 members is electing officers: President,

Vice-president, Secretary, Treasurer. In how many ways can these officers be elected? (No member can hold more than one office.)

Computer chips

27. From the 7 computer chips $X1, X2, X3, \ldots, X7$, 3 must be chosen and placed on a board of the design shown in the figure. In how many ways can this be done?

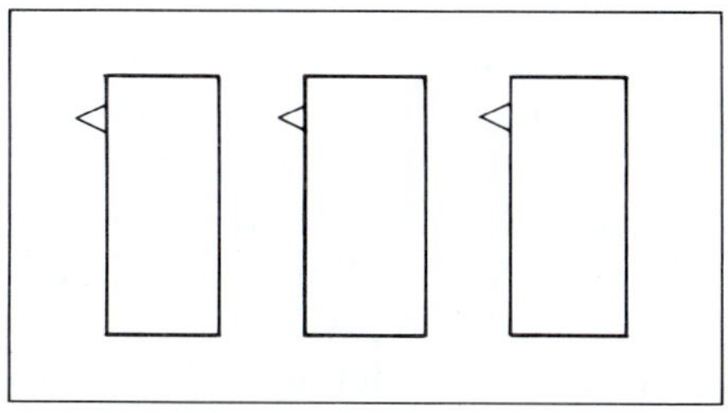

Ship flags

28. A ship has 12 signal flags. Each flag displays one of the four symbols $+$, $>$, $<$, $<>$, and each symbol is contained on 3 different-colored flags. How many signals can be sent by means of

(a) 1 flag? (b) 2 flags? (c) 3 flags? (d) 1, 2, or 3 flags?

Poetry reading

29. At each meeting of the Yoga Poetry Society 4 members are selected to give readings. If its March meeting is attended by 10 members, in how many ways is it possible for the readings to be given?

Psychology test

30. In a particular psychological test a subject is asked to separate the objects $+$, $<$, $>$ into two sets of at least one object each and then order the objects in each set. In how many ways can this be done?

6.5 COMBINATIONS

Suppose we have a dollar, a half-dollar, a quarter, a nickel, and a penny. Denote these coins by D, H, Q, N, P, respectively. We wish to determine the number of ways of choosing 2 of these coins. In this case, we are interested only in the 2 coins chosen, not the order in which they may be chosen. The problem is easily solved by first considering the permutations of these 5 coins taken 2 at a time. Using the first formula for permutations, we find that

$$P(5, 2) = 5 \cdot 4 = 20.$$

These 20 permutations are listed below:

DH	*DQ*	*DN*	*DP*	*HQ*
HD	*QD*	*ND*	*PD*	*QH*
HN	*HP*	*QN*	*QP*	*NP*
NH	*PH*	*NQ*	*PQ*	*PN*

Since we are interested only in the value of the coins and not in the order of choosing them, we do not wish to distinguish between *DH* and *HD*, for example. To indicate that we are not interested in order, we use the notation $\{D, H\}$ as opposed to the notation *DH* in which order is important. The reader will recognize $\{D, H\}$ as simply a set of two elements—also called a "combination of two elements." Thus the two permutations *DH* and *HD* lead to the same combination, namely, $\{D, H\}$.

combination

DEFINITION OF COMBINATION A *combination* of objects is the set containing those objects.

Table 6.1 shows the relation between 2-element permutations and 2-element sets. The pattern which should be observed in this table is that there are twice as many permutations as combinations because of the fact that each combination of two elements leads to 2! permutations. Thus we see in this example that the following equations holds:

The number of permutations = 2!(the number of combinations)

The symbol

$$\binom{5}{2}$$

is used to denote the number of combinations of 5 objects taken 2 at a time [another standard notation for this is $C(5, 2)$]. Using this notation, we may express the above equation as follows:

$$P(5, 2) = 2!\binom{5}{2}.$$

This equation may now be used to express the number of combi-

TABLE 6.1

	Permutations	Combinations
1.	DH, HD →	$\{D, H\}$
2.	DQ, QD →	$\{D, Q\}$
3.	DN, ND →	$\{D, N\}$
4.	DP, PD →	$\{D, P\}$
5.	HQ, QH →	$\{H, Q\}$
6.	HN, NH →	$\{H, N\}$
7.	HP, PH →	$\{H, P\}$
8.	QN, NQ →	$\{Q, N\}$
9.	QP, PQ →	$\{Q, P\}$
10.	NP, PN →	$\{N, P\}$

nations in terms of the number of permutations. By solving the equation for

$$\binom{5}{2},$$

we have

$$\begin{aligned}\binom{5}{2} &= \frac{P(5, 2)}{2!}\\ &= \frac{20}{2}\\ &= 10.\end{aligned}$$

DEFINITION If n is a positive integer and k is a nonnegative integer no larger than n, then the symbol

$$\binom{n}{k}$$

denotes the number of combinations of n objects taken k at a time.

In the example of the coins, in which 2 were chosen from 5, the following pattern emerged:

The number of permutations = 2!(The number of combinations)

$$P(5, 2) = 2!\binom{5}{2}.$$

The kind of argument employed in this example may be used to obtain a more general pattern which deals with k objects chosen from n.

$$\underbrace{\text{The number of permutations}} = k!\underbrace{(\text{The number of combinations})}$$

$$\longrightarrow P(n, k) = k!\binom{n}{k}. \longleftarrow$$

From this last equation comes the first formula for combinations obtained by solving for

$$\binom{n}{k}.$$

FIRST FORMULA FOR COMBINATIONS

$$\binom{n}{k} = \frac{P(n, k)}{k!}.$$

Recall that the second formula for permutations is

$$P(n, k) = \frac{n!}{(n-k)!}.$$

By substituting this into the first formula for combinations we obtain the following formula for combinations.

SECOND FORMULA FOR COMBINATIONS

$$\binom{n}{k} = \frac{n!}{k!(n-k)!}.$$

Combinations

EXAMPLE 1 A school club has 22 members. The club wishes to choose 3 of its members as a committee to plan a club party. In how many ways can the committee be chosen?

Solution: In this case, order is not important (only the members of the committee are important and not the order in which they are chosen), so the problem is equivalent to the number of combinations of 22 objects taken 3 at a time. Therefore, the solution is

$$\begin{aligned}\binom{22}{3} &= \frac{22!}{3!19!} \\ &= \frac{(1 \cdot 2 \cdot 3 \cdots 19) \cdot 20 \cdot 21 \cdot 22}{(1 \cdot 2 \cdot 3) \cdot (1 \cdot 2 \cdot 3 \cdots 19)} \\ &= 1540.\end{aligned}$$

PRACTICE PROBLEM In Example 1, what is the answer if the club has only 20 members?

Answer: 1140. ■

SOME IMPORTANT POINTS TO REMEMBER ABOUT PERMUTATIONS AND COMBINATIONS

1. Some questions about permutations omit the phrase which indicates that so many objects are taken at a time. In cases like this it is understood that *all* the objects are to be taken at a time. For example, the question

   How many permutations are there of 5 objects

   means

   How many permutations are there of 5 objects *taken 5 at a time*?

2. For questions involving order, permutations are usually indicated. For example, the phrase

   President, Vice-president, Secretary

   indicates order. It is not enough to know that Jones, Smith, and Green hold these three offices; we must know which office is held by each individual.

3. For questions in which order is not important, combinations are usually called for. For example, if the question should concern a committee as in Example 1, order is not important, so we use combinations. A committee is simply a collection of individuals without order or status (unless such is specified). Typical questions which require combinations for their solution are those which ask for *the number of ways of choosing a given number of objects* (without regard to order) *from a given set.* For example,

   In how many ways can we choose a set of 2 numbers from the set $\{1, 2, 3, 4, 5\}$?

   The first use of the word "set" in this case indicates that order is not to be considered, and so the answer is

$$\binom{5}{2} = 10.$$

4. For problems which involve permutations, it is very important to know whether or not all the objects are distinct (we do not consider combinations of nondistinct objects). The First and Second Formulas for Permutations apply to *distinct objects only.* For example, a question about the objects

$$x, x, y, y, y$$

involves nondistinct letters. See Example 5 below for a question of this type.

5. Do not fall into the trap of trying to use formulas at all cost! They don't always work.

Combinations

EXAMPLE 2 We wish to choose elements from the set $\{x, y, z, w, u, v\}$. In how many ways can we choose a set

(a) of 3 elements?

(b) of 3 elements, one of which is u?

(c) of 3 elements, one of which is u and another w?

(d) of 3 elements which is $\{x, y, z\}$?

(e) of 3 elements which does not contain u or v?

Solution

(a)
$$\binom{6}{3} = \frac{6!}{3!3!} = 20.$$

(b) If one of the 3 elements must be u, we may choose 2 elements from $\{x, y, z, w, v\}$—the set not containing u—to go with u. Therefore, the question is equivalent to asking in how many ways we can choose 2 elements out of 5:

$$\binom{5}{2} = \frac{5!}{2!3!} = 10.$$

(c) This is like part (b) except that now 2 elements are already determined: u, w. We may pick any one of the remaining elements x, y, z, v to go with u, w. The number of possibilities is

$$\binom{4}{1} = \frac{4!}{1!3!} = 4.$$

(d) There is only one way to choose the 3-element set $\{x, y, z\}$.

(e) If the 3 elements chosen must not contain u or v, then they must be picked from $\{x, y, z, w\}$. Hence, we must choose 3 elements out of 4:

$$\binom{4}{3} = \frac{4!}{3!1!} = 4.$$

PRACTICE PROBLEM In Example 2, in how many ways is it possible to choose a set of 4 elements, one of which is u?

Answer: 10. ■

Combinations

EXAMPLE 3 We are going to consider combinations of the letters a, b, c, A, B, C. How many combinations are there of these letters taken

(a) all at a time?

(b) 2 at a time?

(c) 2 at a time in which no lowercase and uppercase letters are mixed?

(d) 2 at a time in which one letter is lowercase and the other is uppercase?

Solution

(a) There is only one combination of all these letters:

$$\binom{6}{6} = \frac{6!}{6!(6-6)!} = \frac{6!}{6!0!} = 1.$$

(b)
$$\binom{6}{2} = \frac{6!}{2!4!}$$
$$= 15.$$

(c) Since there are 3 lowercase letters to choose from, the number of combinations involving 2 of these letters is

$$\binom{3}{2} = \frac{3!}{2!1!}$$
$$= 3.$$

Also the number of combinations involving 2 uppercase letters is 3. Therefore, the total number of combinations involving either two lowercase letters or two uppercase letters is

$$3 + 3 = 6.$$

These combinations are

Lowercase letters	Uppercase letters
$\{a, b\}, \{a, c\}, \{b, c\}$	$\{A, B\}, \{A, C\}, \{B, C\}.$

(d) Although order is not important here, we may agree to make these 2-element sets by choosing first the lowercase letter and then choosing the uppercase one. Each one of these choices may be made in 3 ways. Thus by the Counting Principle, we may choose one of each in

$$3 \cdot 3 = 9$$

ways. We conclude that there are 9 combinations of the desired type, and they are

$$\{a, A\}, \{a, B\}, \{a, C\},$$
$$\{b, A\}, \{b, B\}, \{b, C\},$$
$$\{c, A\}, \{c, B\}, \{c, C\}.$$

PRACTICE PROBLEM In Example 3, how many combinations are there which contain 3 letters, exactly one of which is lowercase.

Answer: 18. ■

Committees

EXAMPLE 4 The House and Senate have both established *ad hoc* committees to study the question of legislative reform. The House Committee contains 7 members and the Senate Committee 6. Two members must be chosen from each of these committees to form a conference committee. If the chairman from neither committee is to be chosen, in how many ways can the conference committee be formed?

Solution: We look upon this problem as an experiment consisting of these two parts:

Part 1: Choose the 2 members of the House committee (do not choose the chairman); thus we choose 2 members from 6, and this can be done in

$$\binom{6}{2}$$

ways.

Part 2: Choose the 2 members of the Senate committee (do not choose the chairman); this can be done in

$$\binom{5}{2}$$

ways.

Then by the Counting Principle the new committee can be formed in

$$\begin{aligned}\binom{6}{2}\cdot\binom{5}{2} &= \frac{6!}{2!4!}\cdot\frac{5!}{2!3!}\\ &= \frac{6!}{2!4!}\cdot\frac{5!}{2!3!}\\ &= \frac{5\cdot 6}{1\cdot 2}\cdot\frac{4\cdot 5}{1\cdot 2}\\ &= 15\cdot 10\\ &= 150\end{aligned}$$

ways.

PRACTICE PROBLEM In Example 4, how many ways can the conference committee be formed if the chairmen can be included?

Answer: 315. ■

Permutations of nondistinct objects

EXAMPLE 5 How many permutations of the letters x, x, y, y, y are there?

Solution: Here we have 5 objects, but that are not distinct: 2 are alike and 3 are alike. This implies that a simple use of the permutation formulas will not solve the problem. Starting at the most basic level, we imagine filling 5 blanks:

$$\_\_\_, \_\_\_, \_\_\_, \_\_\_, \_\_\_$$

Each of these permutations may be constructed by choosing the 2 blanks for the xs. For once the xs are placed, the ys must be placed in the remaining blanks. Therefore, we need to know the number of ways of choosing two blanks for the xs. But there is no order involved in placing the two xs since they are indistinguishable. Thus, we may choose 2 places for the xs in

$$\binom{5}{2} = \frac{5!}{2!3!} = 10$$

ways. We conclude that there are 10 permutations of x, x, y, y, y. It may be helpful to list them:

xxyyy	*yxxyy*	*yyxxy*	*yyyxx*	*xyxyy*
yxyxy	*yyxyx*	*xyyxy*	*yxyyx*	*xyyyx*

It should be noted that this problem is about permutations but a combination formula was used to solve it.

PRACTICE PROBLEM How many permutations of the letters x, x, y are there?

Answer: 3. ■

EXERCISES

Set A–Using Formulas for Combinations

Evaluate the expressions in Exercises 1 through 10.

1. $\binom{8}{2}$ 2. $\binom{10}{3}$ 3. $\binom{7}{4}$ 4. $\binom{12}{10}$ 5. $\binom{100}{2}$ 6. $\binom{98}{3}$

7. $\binom{8}{k}$ for $k = 5$

8. $\binom{10}{10 - k}$ for $k = 2$

9. $\binom{8 + k}{8 - k}$ for $k = 3$

10. $\binom{3k + 1}{2k - 1}$ for $k = 3$

Set B–Questions About Combinations

In Exercises 11 through 14, how many combinations are there

11. of 5 objects taken 3 at a time?
12. of 13 objects taken 2 at a time?
13. of 20 objects taken 4 at a time?
14. of 20 objects taken 6 at a time?
15. How many combinations are there of 9 objects taken
    (a) none at a time? (b) 1 at a time?
    (c) 2 at a time (d) 3 at a time?
    (e) 4 at a time? (f) 5 at a time?
    (g) 6 at a time? (h) 7 at a time?
    (i) 8 at a time? (j) 9 at a time?

In Exercises 16 through 20, in how many ways is it possible to choose

16. 12 objects out of 13?
17. 3 objects out of 16?
18. 5 objects out of 20?
19. 22 objects out of 25?
20. 4 objects out of 30?

Set C–Problems to be Solved by Permutations or Combinations

21. Four ping-pong balls marked x, y, z, w are placed in a box and a set of 2 are drawn out (no order).
    (a) How many outcomes are there of this experiment?
    (b) In how many outcomes does x appear?
    (c) In how many outcomes does x not appear?
22. Six ping-pong balls marked 1, 2, 3, 4, 5, 6 are placed in a box, and 4 are withdrawn simultaneously (no order). How many outcomes
    (a) are there of this experiment? (b) will have a 6?
    (c) involve a 1 and a 2? (d) will have a 1 but not a 2?
    (e) involve a 1 or a 2 but not both?
    (f) involve a 1 and/or a 2?

23. We wish to choose elements from the set $\{a, b, c, d, e\}$. In how many ways can we choose a set of
   (a) 4 elements? (b) 4 elements one of which is a?
   (c) 4 elements one of which is a and another b?
   (d) 4 elements 3 of which are c, d, e?
   (e) 4 elements which does not contain c?

24. We will form combinations of the letters a, b, A, B, C. How many combinations are there of these letters taken
   (a) all at a time? (b) 2 at a time?
   (c) 2 at a time and involving only lowercase letters?
   (d) 2 at a time and involving only uppercase letters?
   (e) 2 at a time and involving one lowercase and one uppercase letter?

25. In English 101 each student must read 4 books from the following list:

   The Adventures of Tom Sawyer

   The Heart Is a Lonely Hunter

   The Scarlet Letter

   The Rise of Silas Lapham

   The Life of Johnson

   Crime and Punishment

   The Caine Mutiny

   In how many ways can this assignment be completed?

26. In the Clarkston General Post Office there are 10 mail boxes. If we wish to mail 5 pieces of identical mail in 5 different boxes, in how many ways can this be done?

27. In Exercise 26 suppose that every two pieces of mail are different.
   (a) In how many ways can the mail be posted if it must go into 5 different boxes?
   (b) In how many ways can the mail be posted if any letter can go into any box?

28. In a school club there are 6 males and 8 females. A committee of 4, consisting of 2 males and 2 females, is to be chosen to represent the club at a national convention. In how many ways can this committee be formed?

29. (a) George has 23 stamps and Sue has 18. In how many ways can George trade 1 of his stamps for 1 of Sue's.
   (b) In how many ways can Sue trade 2 of her stamps for 2 of George's?

30. We wish to make a gift package which contains 5 stamps and 7 coins. If there are 10 stamps and 12 coins to choose from, in how many ways can the package be made up?

31. Mrs. Jones has 12 books which she wishes to give to her children Connie and Clyde. In how many ways can she make up the 2 book packages if Connie is to receive 7 and Clyde 5 books?

32. A guerilla leader has 20 assignments which he wishes to parcel out among 3 of his men: John, Teddy, and Ronnie. In how many ways can he give out these assignments if John is to get 9, Teddy 7, and Ronnie 4?

33. On July 7, 1982, it was reported that the manager of the Cincinnati Reds (a baseball team) had used 66 lineups in 78 games. If we assume that a lineup is an assignment of 9 player positions to 9 different batting positions, how many lineups are possible?

34. A basket contains 15 apples and 8 oranges. In how many ways can Tom choose
(a) one apple or one orange?
(b) one apple and one orange?
(c) two apples and two oranges?

35. A publisher wishes to produce a series of dictionaries which will allow translation between any 2 of the following languages:

Greek, Latin, English, French, German, Russian.

How many volumes must be published if each volume will contain no more than 2 languages?

36. A military company contains 2 officers, 4 sergeants, and 50 privates. In how many ways can a detachment be formed consisting of 1 officer, 2 sergeants, and 25 privates?

37. A precinct captain has 3 detectives, 2 sergeants, and 4 patrolmen to assign to 12 city blocks. In how many ways can this be done
(a) if no block is assigned more than 1 policeman?
(b) if no block is assigned more than 1 policeman of a given rank?

38. Box A contains 10 pennies and box B contains 10 nickels. In how many ways can 10 coins be chosen from the boxes?

39. A chess board has 8 rows and 8 columns, 64 squares in all. Squares alternate between the colors red and black. In how many ways can

(a) 2 squares be chosen?
(b) 2 squares be chosen from different rows and columns?
(c) two squares of different colors be chosen from different rows and different columns?

40. A border patrol commander has 6 new patrolmen. Each day for 20 days he takes a different group of 3 new men with him on patrol. In how many ways can this be done?

41. How many permutations are there of x, x, y, y?

42. How many permutations are there of x, x, x, y, y, y?

43. How many permutations are there of x, y, y, z?

44. How many permutations are there of x, x, y, y, z?

45. How many permutations are there of the letters $x, x, y, y, y, y, z, z, z$?

Set D–Calculator Exercises

Use a hand-held calculator to compute each of the following.

46. $\binom{12}{5}$ 47. $\binom{15}{8}$ 48. $\binom{18}{10}$ 49. $\binom{21}{12}$ 50. $\binom{23}{13}$ 51. $\binom{26}{18}$

6.6 SUMMARY AND REVIEW EXERCISES

REVIEW OF CONCEPTS

I. The Counting Principle

Suppose that an experiment E consists of 2 parts. There are m outcomes of the first part and n outcomes of the second part. If, after the first part has been performed, all n of the outcomes of the second part are possible, then there are $m \cdot n$ outcomes of the experiment E.

II. Factorial Notation

$n! = n \cdot (n-1) \cdot (n-2) \cdot (n-3) \cdots 3 \cdot 2 \cdot 1.$

III. Formulas for Permutations

The number of permutations of n objects taken k at a time, denoted by $P(n, k)$, is given by the formulas:

1. $P(n, k) = n \cdot (n-1) \cdot (n-2) \cdot (n-3) \cdots (n-k+1).$

2. $P(n, k) = \dfrac{n!}{(n-k)!}.$

IV. Formulas for Combinations

The number of combinations of n objects taken k at a time, denoted by

$$\binom{n}{k},$$

is given by the formulas:

$$\text{1. } \binom{n}{k} = \frac{P(n, k)}{k!} \qquad \text{2. } \binom{n}{k} = \frac{n!}{k!(n-k)!}$$

REVIEW EXERCISES

1. An experiment consists of tossing a coin and then rolling a die. How many outcomes are possible?
2. A test contains 10 questions, each of which is a 3-answer multiple choice question. If each question must be answered, in how many ways can the test be completed?
3. In how many ways is it possible to complete the test of Exercise 2 so that each question is answered incorrectly?
4. Compute the following:

   (a) $\dfrac{10!}{8}$ (b) $P(10, 4)$ (c) $\dbinom{13}{10}$
5. How many permutations are there of 8 objects taken 4 at a time?
6. How many combinations are there of 10 objects taken 3 at a time?
7. In how many ways is it possible to choose 5 objects from a set of 12?
8. A student club contains 20 members: 12 females and 8 males. The club must form a committee containing 2 females and 2 males. In how many ways can this be done?
9. How many permutations are there of the letters x, x, y, a, a, b, b?
10. How many permutations are there of the letters x, x, y, A, A, B, B in which the lowercase letters come first?
11. How many permutations are there of the letters of Exercise 10 in which the lowercase letters all come first or last?
12. A personnel manager must fill three positions A, B, C. The manager has 4 employees all of whom are qualified for the three jobs. In how many ways can the manager fill the positions?

CHAPTER

7

PROBABILITY

7

7.1 INTRODUCTION

Some of the material of Chapter 6 involved experiments, and our point of focus there was mainly on determining the number of possible outcomes which can result as a consequence of one performance of an experiment. In this chapter, we are again concerned with experiments, but now the interest is not only with a single performance but also with results of a large number of repetitions of the experiments. Every repetition of an experiment is called a *trial* and so we are going to consider the long-range effects of many trials.

trial

The experiments which are considered here will sometimes seem trivial and unimportant because of their simplicity. But these examples are important precisely because of their simplicity. Perhaps the best-known experiments of a simple nature are those related to games of chance, and these will furnish a basis for understanding the subject of this chapter—probability.

The student should expect to be able to solve the following kinds of problems when the material of this chapter has been mastered.

1. A student takes a true-false test of 10 questions on a subject he knows nothing about. To choose his answers he flips a coin and marks the question T if heads turn up and F otherwise. If, in order to pass, he must get at least 6 correct answers, what are his chances of getting a passing grade?
2. A pharmaceutical company claims that their brand of aspirin for children is preferred 3 to 2 by those physicians who recommend aspirin. If this claim is correct, what is the probability that in a sample of 20 physicians who recommend aspirin at least 15 prefer this company's aspirin? ■

7.2 WHAT IS PROBABILITY?

Probability is concerned with those experiments which are not predictable in the short run but which are predictable in the long run. As an example of this phenomenon consider the toss of a coin. It is very hard (if not impossible) to predict the result of a single toss of an ordinary coin, but if it is tossed repeatedly, it is possible to predict, with a certain degree of accuracy, the proportion of times that heads and tails will appear. Table 7.1 shows the results of simulating a coin toss 10,000 times on a computer.

Note in the table that as the number of trials (denoted by n) increases, the number of heads and tails tend to be equal. This is reflected in columns 2, 3, and 4. In the table, n is the number of trials, $\#(H)$ is the number of heads, and $\#(T)$ is the number of tails.

TABLE 7.1

N	$\#(H)$	$\#(T)$	$\#(H)/\#(T)$	$\#(H)/n$
1,000	496	504	0.9841	0.4960
2,000	1029	971	1.0597	0.5145
3,000	1520	1480	1.0270	0.5067
4,000	2011	1989	1.0111	0.5028
5,000	2496	2504	0.9968	0.4992
6,000	2982	3018	0.9881	0.4970
7,000	3480	3520	0.9886	0.4971
8,000	3980	4020	0.9900	0.4975
9,000	4473	4527	0.9881	0.4970
10,000	4969	5031	0.9877	0.4969

A little reflection on the nature of this experiment will reveal that these results should not be surprising. When a coin is tossed, there are only two possible *outcomes*: heads (H) or tails (T). Neither of these outcomes seems dominant in ordinary coins and, therefore, in a large number of trials each will claim its rightful share of results. That is, we expect heads and tails each to occur about half the time. We describe this situation by saying that on one toss of an ordinary coin the *probability* of getting heads is $\frac{1}{2}$. Although this is a statement about a single toss of the coin, it should be interpreted as information about the results of *many* trials of the experiment. This meaning is called the *frequency interpretation of probability.*

outcomes

probability

frequency interpretation of probability

Suppose that a given probability is, for example, 0.635. According to the frequency interpretation, this probability has a long-range meaning which we may explain in the following way: If the experiment which produced this probability is carried out many, many times (always under the same conditions), then we could expect this outcome, on the average, about 635 times out of every 1000 trials.

Frequency interpretation

EXAMPLE 1 A given probability is 0.438. Give the frequency interpretation of this number.

Solution: Suppose A is the outcome which has probability 0.438. Then, if the experiment is carried out many times, the outcome A will occur (on the average) about 438 times out of every 1000 trials.

PRACTICE PROBLEM Given that the probability of an outcome is 0.287. If the experiment is carried out 100 times, about how many times would you expect this outcome to occur?

Answer: About 29. ■

equally likely
fair

Let us consider the tree diagram for the coin-toss experiment as shown in Figure 7.1. There are two outcomes in this case, H and T, which lead to the sample space $\{H, T\}$. For ordinary coins (i.e. coins which are not loaded), we say that the two outcomes are *equally likely* and that the coin is *fair.* In this case "equally likely" and "fair" are synonymous. (There are experiments in which outcomes are not equally likely, but they will not be considered in this section.) Because of the frequency interpretation of probability, we say that the probability of getting each of these two outcomes is $\frac{1}{2}$ and we write

$$P(H) = 0.5 \qquad \text{and} \qquad P(T) = 0.5.$$

It is convenient to place these probabilities on the branches of the tree diagram as shown in Figure 7.1.

equally likely model

The experiment which consists of one toss of a fair coin is sometimes called the *equally likely model* of a coin toss. Each

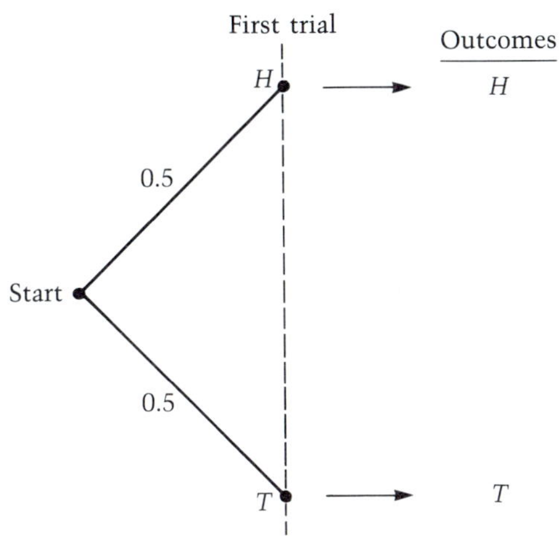

FIG. 7.1. Tree diagram of 1 coin toss.

outcome in the model is assigned a probability which reflects the long-range outcome of many trials, but, as we have seen, the probability also turns out to be simply 1 divided by the number of outcomes. This is the basis for the following definition.

DEFINITION Suppose that an experiment E has n outcomes. **The equally likely model** of this experiment assigns probability $1/n$ to each outcome.

Consider a fair die which is rolled 1 time. In this case, there are 6 outcomes as shown on the tree diagram of Figure 7.2. Since the die is fair, this is the equally likely model, and each outcome is assigned probability $\frac{1}{6}$. For example, on one roll, the probability of getting a 3 is $\frac{1}{6}$. The frequency interpretation here is that in a large number of trials, 3 will occur about $\frac{1}{6}$ of the time. In Table 7.2 the results of rolling a die 6000 times have been recorded. Note that the number of times 3 occurs is 1028, giving a ratio of

$$\frac{1028}{6000} = 0.1713 \quad \text{(to four places)}$$

This compares very well to $\frac{1}{6} = 0.1667$ (also to four places), since

$$0.1713 - 0.1667 = 0.0046,$$

and this is less than a 3% error.

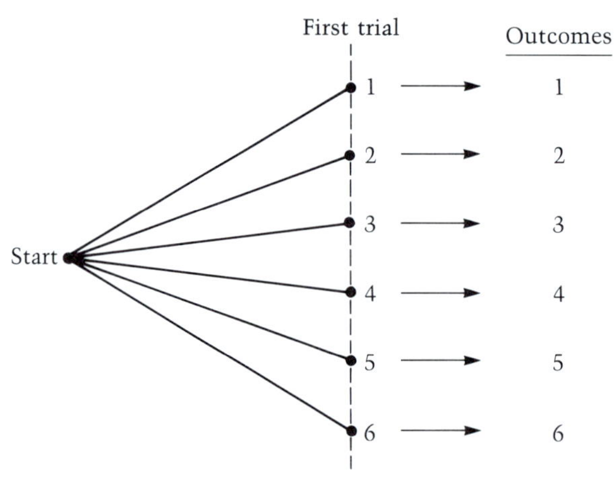

FIG. 7.2. Tree diagram of 1 roll of a die.

TABLE 7.2 **6000 Rolls of a Fair Die**

Outcomes	Number of occurrences	Frequency of outcome
1	1055	0.1758
2	941	0.1568
3	1028	0.1713
4	1026	0.1710
5	954	0.1590
6	996	0.1660
Totals	6000	0.9999

random

In many situations the word *random* is used to describe an experiment. This is usually a signal that the equally likely model is intended. We may speak of drawing a card at random from a well-shuffled deck of 52 cards. This means, for example, that the probability of drawing the ten of hearts is $\frac{1}{52}$ since there are 52 equally likely outcomes, and the ten of hearts is one of them.

chance

Sometimes the word *chance* is used to describe a probability. In drawing one card at random from a well-shuffled deck of cards the probability of drawing the ten of hearts may be expressed as *1 chance in* 52. Using this same terminology, we may ask

What are the chances of drawing a ten?

This question means the same as

What is the probability of drawing a ten?

Using the frequency interpretation of probability, we may reason as follows: there are 4 tens in a deck of 52 cards, and in the long run a ten will be drawn about $\frac{4}{52}$ of the time. That is, there are 4 chances out of 52 of drawing a ten or the probability of drawing a ten is $\frac{4}{52}$ or $\frac{1}{13}$.

events

In probability theory it is common practice to speak of *events*. In the experiment of drawing 1 card at random from a well-shuffled deck of cards, we asked for the probability of drawing a ten. This may also be expressed by referring to the act of drawing a ten as an "event." Then we may ask

occurs

What is the probability that the event *occurs*?

Note that the event in question is identified with a subset of the sample space: the sample space consists of the 52 possible outcomes of which there are 4 tens. Thus, the event which is the act of drawing a ten is associated with the subset of the sample space which consists of the 4 tens. The event *occurs* if a ten is drawn.

simple events

compound events

The single elements of a sample space are called *simple events* and the subsets consisting of more than a single element are called *compound events.* The 4 tens in a deck of cards is a compound event if the experiment is to draw a single card from the deck.

In Figure 7.3 we have drawn the tree diagram for 3 tosses of a fair coin. In the diagram, there are 8 outcomes which are equally likely since the coin is fair. Thus each simple outcome has probability $\frac{1}{8}$. Some of the events associated with this experiment have been listed in Table 7.3.

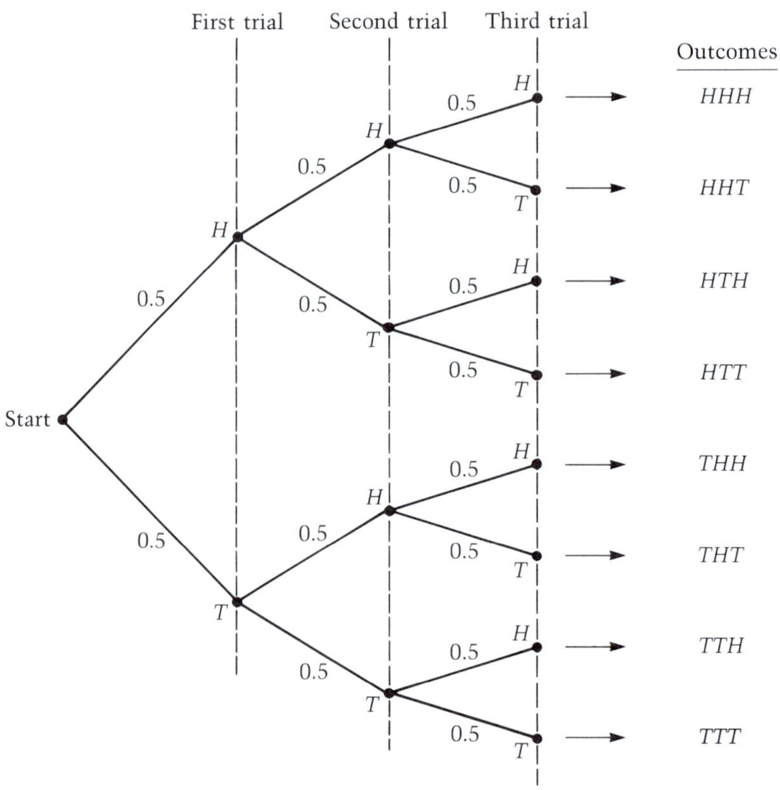

FIG. 7.3. Tree diagram of 3 tosses of a fair coin.

TABLE 7.3

Simple events	Simple and compound events	Probabilities
HHH	Getting exactly 3 heads	$\frac{1}{8} = 0.125$
HHT *HTH* *THH*	Getting exactly 2 heads	$\frac{3}{8} = 0.375$
HTT *THT* *TTH*	Getting exactly 1 head	$\frac{3}{8} = 0.375$
TTT	Getting no heads	$\frac{1}{8} = 0.125$

Consider the event

$$A = \text{getting the same result on the first two tosses.}$$

If we look at the tree diagram of Figure 7.3 or Table 7.3, we can see that A consists of the following 4 outcomes out of the total of 8:

$$HHH,\ HHT,\ TTH,\ TTT$$

We can conclude from this that if this experiment is repeated many times, we will have the same result on the first two tosses about half of the time. Therefore,

$$P(A) = 0.5.$$

DEFINITION Suppose that E is an experiment which has n outcomes and that A is an event associated with E which has k elements. The **probability that A occurs,** in the equally likely model, is k/n, that is,

$$P(A) = k/n.$$

As a practical matter, this definition may be approached in the following way. We are given an experiment E and an event A associated with E. Among all the elements in the sample space, count those associated with the event A and divide this number by the total number of outcomes in the sample space. The resulting ratio is the probability that A occurs.

Probability definition

EXAMPLE 2 A box contains three ping-pong balls numbered 1, 2, 3. We draw 2 balls from the box at random, one after the other, with replacement.

(a) Draw a tree diagram for this experiment and place the appropriate probabilities on the branches.

(b) Determine the total number of outcomes and the probability of each.

(c) Determine the probability that the sum of the numbers on the two balls is 4.

Solution

(a) The tree diagram is shown in Figure 7.4.

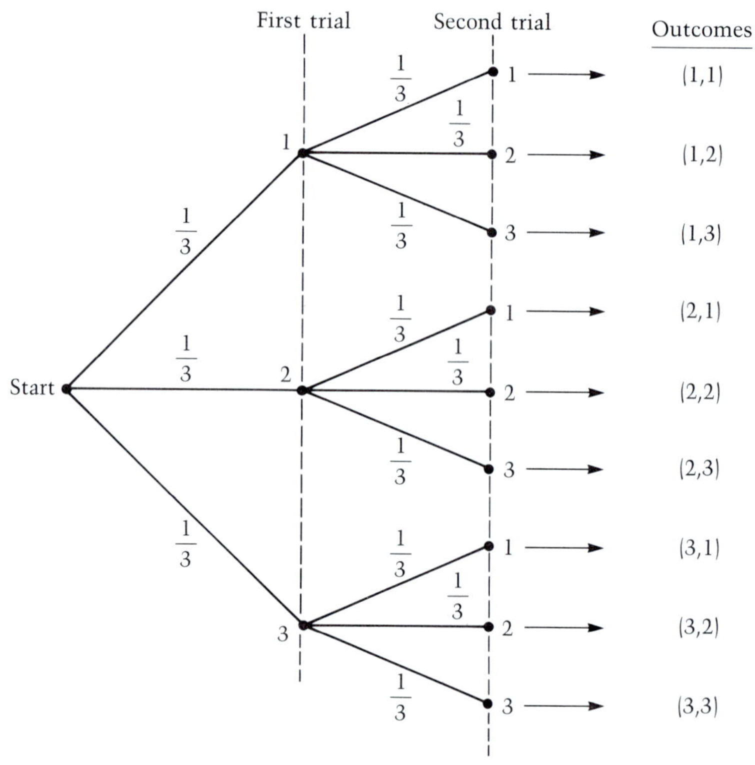

FIG. 7.4.

(b) As can be seen from the tree diagram, there are 9 outcomes, and since the drawing is random, each of these has probability $\frac{1}{9}$.

(c) Using the tree diagram again, we see that the outcomes which show a sum of 4 are (1, 3), (2, 2), and (3, 1). Since there are 3 of these, the probability that the two balls have a sum of 4 is $\frac{3}{9}$ or $\frac{1}{3}$.

PRACTICE PROBLEM In this example determine the probability that the sum of the numbers on the balls is less than 4.

Answer: $\frac{1}{3}$. ■

Probability definition

EXAMPLE 3 Form all the permutations of the letters x, x, y, y, y and choose one of these permutations at random.

(a) List the elements in the sample space and determine the probability of each.

(b) What is the probability that in the permutation chosen the two x's are adjacent to each other?

(c) What is the probability that in the permutation chosen the two x's are separated by at least one y?

Solution

(a) All the permutations of these letters were determined in Example 5 of Section 6.5. As that example showed, there are 10 of these permutations:

1. *xxyyy*	**2.** *xyxyy*
3. *xyyxy*	**4.** *xyyyx*
5. *yxxyy*	**6.** *yxyxy*
7. *yxyyx*	**8.** *yyxxy*
9. *yyxyx*	**10.** *yyyxx*

Since we are to choose one permutation at random from this collection, this is the equally likely model, and the probability of each outcome is 0.1.

(b) Among the 10 permutations, only in the following ones are the two x's adjacent:

1. *xxyyy*

5. *yxxyy*

8. *yyxxy*

10. *yyyxx*

Therefore, since there are 4 of these, we conclude that the probability is 0.4 that in the permutation drawn the two x's will be adjacent to each other.

(c) The permutations in which the x's are separated are

2. *xyxyy*	**3.** *xyyxy*
4. *xyyyx*	**6.** *yxyxy*
7. *yxyyx*	**9.** *yyxyx*

Thus, the probability is 0.6 in this case.

PRACTICE PROBLEM In this example determine the probability that at least two y's come first in the permutation chosen.

Answer: 0.3. ■

Probability definition

EXAMPLE 4 From the set $\{A, B, a, b, c\}$ we are going to select at random a 2-element subset.

(a) Determine the number of outcomes of this experiment and give the probability of each.

(b) What is the probability that the 2 letters chosen are both uppercase?

(c) What is the probability that the 2 letters chosen are both lowercase?

(d) What is the probability that of the 2 letters, 1 is uppercase and 1 is lowercase?

Solution

(a) From the set $\{A, B, a, b, c\}$ we are to choose 2 letters, i.e., we are to choose 2 elements out of 5. This can be done in

$$\binom{5}{2} = 10$$

ways. Therefore, there are 10 outcomes, and since the selection is random, the probability of each outcome is 0.1.

(b) There is only 1 way to choose 2 uppercase letters, and so there is 1 chance in 10 that the letters chosen will both be uppercase.

(c) In the set $\{A, B, a, b, c\}$ there are 3 lowercase letters, and so we may choose 2 of them in

$$\binom{3}{2} = 3$$

ways. Therefore, there are 3 chances in 10 that the 2 letters selected will both be lowercase. Thus, the probability is 0.3 that 2 lowercase letters will be chosen.

(d) There are 2 ways of choosing an uppercase letter and 3 ways of choosing a lowercase letter. Then, by the Counting Principle, there are $2 \cdot 3 = 6$ ways of choosing one uppercase and one lowercase letter. Hence, the probability is 0.6 that of the 2 letters 1 will be uppercase and 1 will be lowercase.

PRACTICE PROBLEM In this example, what is the probability that the 2 letters drawn are both a's or both b's (either lowercase or uppercase)?

Answer: 0.2. ■

Families having three children

EXAMPLE 5 A sociologist is studying families having exactly 3 children. The main emphasis of the study is the sex of the children. Such a family is therefore reported by giving the number of a child and its sex. For example, gbg means

First child is female (g),

Second child is male (b),

Third child is female (g).

Assume that a family is chosen at random and that outcomes are equally likely.

(a) Construct a tree diagram and determine the sample space.

(b) Find the probability of a family having all girls.

(c) Find the probability of a family having 2 girls and 1 boy in that order.

(d) Find the probability of a family having exactly 2 girls.

(e) Find the probability of a family having at least 2 girls.

Solution

(a) The tree diagram is shown in the figure. From the diagram, the sample space is

$$\{ggg, ggb, gbg, gbb, bgg, bgb, bbg, bbb\}.$$

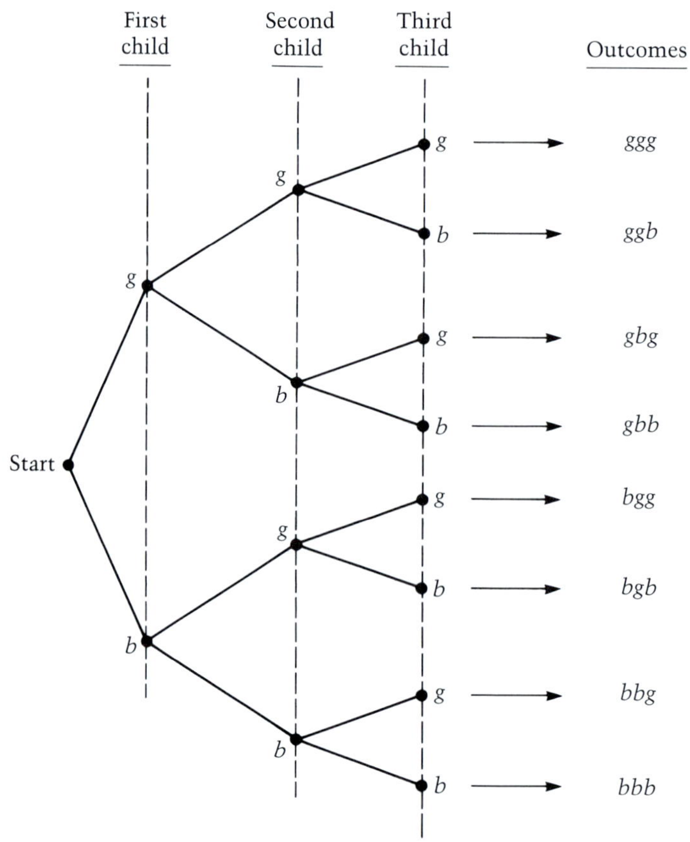

(b) Since outcomes are equally likely and the sample space has 8 elements, $\frac{1}{8}$ is the probability of each outcome. The only outcome showing all girls is *ggg*, and thus the probability of a family having all girls is $\frac{1}{8}$.

(c) The only outcome having 2 girls and 1 boy in that order is ggb. Then

$$P(ggb) = \frac{1}{8}.$$

(d) The elements of the sample space which have 2 g's are ggb, gbg, bgg. Since each of these has probability $\frac{1}{8}$, we have by definition

$$\begin{aligned} P(\text{two girls}) &= P(ggb) + P(gbg) + P(bgg) \\ &= \frac{1}{8} + \frac{1}{8} + \frac{1}{8} \\ &= \frac{3}{8}. \end{aligned}$$

(e) If a family has at least 2 girls, then it has 2 or 3 girls. The subset of the sample space associated with the event "at least 2 g's" is

$$\{ggb, gbg, bgg, ggg\}.$$

Then

$$\begin{aligned} P(\text{at least two } g\text{'s}) &= \frac{1}{8} + \frac{1}{8} + \frac{1}{8} + \frac{1}{8} \\ &= \frac{1}{2}. \end{aligned}$$

PRACTICE PROBLEM In Example 5, what is the probability that a family has all girls or all boys?

Answer: $\frac{1}{4}$. ■

EXERCISES

Set A–The Frequency Interpretation of Probability

1. The probability of a certain event is given to be 0.390. If this experiment is carried out in 100 trials, about how many times would you expect the given event to occur?

2. The probability of a certain event is given to be 0.584. The experiment is carried out 100 times. About how many times would you expect the event to occur?
3. The probability for an event A is given to be 0.481. If this experiment is carried out in 1000 trials, about how many times would you expect the event to occur?
4. The probability for an event A is given to be 0.269. If the experiment is carried out 1000 times, about how many times would you expect the event A to occur?
5. The probability of an event X is given to be 0.805. If the experiment is carried out in 200 trials, about how many times would you expect the event to occur?
6. The probability of an event Y is given to be 0.741. If the experiment is carried out in 2000 trials, about how many times would you expect the event Y to occur?
7. Assume that a poker hand is a set of 5 cards drawn randomly from a deck of 52 cards. The probability of obtaining a full house (one pair and one triple of the same face value) is 0.0014. If a person plays poker on a regular basis, about how often should he or she expect a full house?
8. The probability of obtaining four of a kind (4 cards of the same face value) in poker is 0.00024. If a person plays poker on a regular basis, about how often should they expect to obtain four of a kind?
9. In the game of poker, the probability of obtaining a royal flush (ten, Jack, Queen, King, Ace in a single suit) is 0.0000015. If a person plays poker on a regular basis, about how often should he or she expect to be dealt a royal flush?
10. In the game of poker, the probability of obtaining a straight flush (five in a sequence in a single suit, but not a royal flush) is 0.000014. If a person plays poker on a regular basis, about how often should he or she expect to be dealt a straight flush?

Set B–Computing Probabilities in the Equally Likely Model

11. Choose one of the numbers 0, 1, 2, 3, 4, 5, 6, 7, 8, 9 at random. What is the probability that
    (a) 5 is chosen?
    (b) an even number is chosen?
    (c) a multiple of 3 is chosen?
    (d) a number less than 4 is chosen?
    (e) 3 or 7 is chosen?

12. One card is selected at random from a well-shuffled deck of 52 cards. What is the probability that
    (a) the ace of spades is chosen?
    (b) a black ace is drawn?
    (c) an ace is drawn?
    (d) a face card is drawn (Jack, Queen, King)?
    (e) a red card is drawn?
    (f) a red card or a face card is drawn?
13. Choose one of the numbers, 21, 31, 47, 52, 75, 80, 96, 97, 98, 99 at random. What is the probability that the number chosen
    (a) is less than 80?
    (b) is even?
    (c) is larger than 50?
    (d) is at least 80?
    (e) will leave a remainder of 1 if divided by 5?
14. A class of 30 students consists of 10 boys and 20 girls. If 1 student is chosen at random, what is the probability that
    (a) a girl is chosen?
    (b) a boy is chosen?
15. A box contains 12 ping-pong balls: 4 white ones and 8 black ones. If 1 ball is drawn at random, what is the probability that the ball drawn is
    (a) white (b) black
16. A box contains 15 ping-pong balls: 4 white, 5 black, and 6 red ones. If 1 ball is drawn at random, what is the probability that the ball drawn is
    (a) white? (b) black? (c) red?
    (d) black or white? (e) black or red?
    (f) white or red? (g) black or white or red?
17. If a letter is chosen at random from the word "introduction," what is the probability that the letter chosen is
    (a) r? (b) d? (c) i? (d) n? (e) t?
    (f) o? (g) u? (h) a vowel? (i) a consonant?
18. An experiment consists of tossing a fair coin and then rolling a fair die. Draw a tree diagram and list the elements in the sample space. What is the probability
    (a) of each simple event?
    (b) that the die shows an even number?
    (c) that the coin turns up heads?
    (d) that the number on the die is less than 3?
19. An experiment consists of randomly choosing 2 numbers from the set {1, 2, 3, 4} (no order is assumed). List all possible

outcomes and give the probability of each. What is the probability that
(a) both numbers are even?
(b) at least one number is even?
(c) the sum of the numbers is 4?
(d) the sum of the numbers is 7?
(e) the sum of the numbers is even?

20. Choose at random two numbers from the set {1, 2, 3, 4}, one after the other, with replacement. Draw a tree diagram and list the outcomes. What is the probability
(a) of each simple event?
(b) that a 4 is drawn first?
(c) that the sum of the two numbers is odd?
(d) that the sum of the two numbers is less than 7?
(e) that the sum of the two numbers is at least 4?

21. Choose at random two numbers from the set {1, 2, 3, 4}, one after the other, without replacement. Draw a tree diagram and list all the outcomes. What is the probability
(a) of each simple event?
(b) that a 4 is drawn first?
(c) that the sum of the two numbers is odd?
(d) that the sum of the two numbers is less than 7?
(e) that the sum of the two numbers is at least 4?

22. Form all the permutations of the letters x, x, y, y and choose one of these at random. List the elements of the sample space and determine the probability that in the permutation chosen
(a) x is the first letter.
(b) the x's and y's alternate (starting with either).
(c) the two x's are adjacent.
(d) at least one x is to the left of at least one y.

23. Form all the permutations of the letters x, x, y, y, z and choose one of these at random. List the elements in the sample space and determine the probability of each simple event. What is the probability that in the permutation chosen
(a) x is the first letter?
(b) the x's and y's alternate (starting with either)?
(c) the two x's are adjacent?
(d) at least one x is to the left of at least one y?
(e) z does not come first or last?

Set C–Some Probability Word Problems

24. In Wadloux, Wyoming, it is possible to get TV reception from 4 commercial and 2 educational stations. Charles Lamb, who

is one of Wadloux's leading exponents of the arts, has a black and white TV set featuring 10 channels. If Charles selects a channel at random, what is the probability that
(a) he will get a commercial station?
(b) he will get an educational station?
(c) he will get no station?
(d) he will get a picture (assuming his set is working properly)?

25. Joe Swain is dressing for a date when the lights in his room go out. Joe finishes dressing in the dark and is concerned that his socks do not match. He owns only three pairs of socks: red, blue, and brown. If Joe keeps his socks loosely strewn in his dresser drawer and if he chooses two socks (not 2 pairs) at random, what is the probability that he gets a match?

26. Two boys and 2 girls are arranged randomly in a row for a panel discussion. What is the probability that the boys and the girls alternate in a row?

27. Three couples are attending a dance and the 3 ladies choose their partners randomly from the 3 men. What is the probability that no gentleman is dancing with his date?

Tax returns

28. Among 25 income tax returns, 6 contain errors favoring the taxpayer. What is the probability that among 4 randomly chosen returns 2 will contain an error?

Lottery

29. A lottery sells 25 tickets and gives 5 prizes. If one buys 3 tickets, what is the probability of winning a prize?

Law enforcement

30. A game warden inspects a fisherman's catch by examining 2 fish which he selects at random. What is the probability that the fisherman will be arrested if his catch of 10 includes 3 undersized fish?

Quality control

31. A company tests its product by examining 6 randomly selected items from each batch. If a defective item is found, the batch is rejected. What is the probability that a batch of 1000 items will be rejected if it contains 10 defective items.

7.3 SOME PRINCIPLES OF PROBABILITY

Suppose we consider an experiment E such as drawing a card at random from a deck of 52 cards. If A is an event associated with E and $\#(A)$ is the number of elements in A, then

$$\frac{\#(A)}{n}$$

is the probability that A occurs, where n is the total number of outcomes of the experiment E. That is

$$P(A) = \frac{\#(A)}{n}. \tag{1}$$

Since A is a subset of all the outcomes of E, it follows that

$$0 \le \#(A) \le n. \tag{2}$$

If we divide the preceding inequality by n, we obtain

$$0 \le \frac{\#(A)}{n} \le 1. \tag{3}$$

Using (1) and (3) together, we obtain

$$0 \le P(A) \le 1. \tag{4}$$

Statement (4) asserts that *no probability is smaller than 0 or larger than 1.* The two extremes of 0 and 1 have the following meanings: *an event which has probability* 0 *will not occur* and *an event which has probability* 1 *will occur without fail.* This statement is illustrated on the number line below.

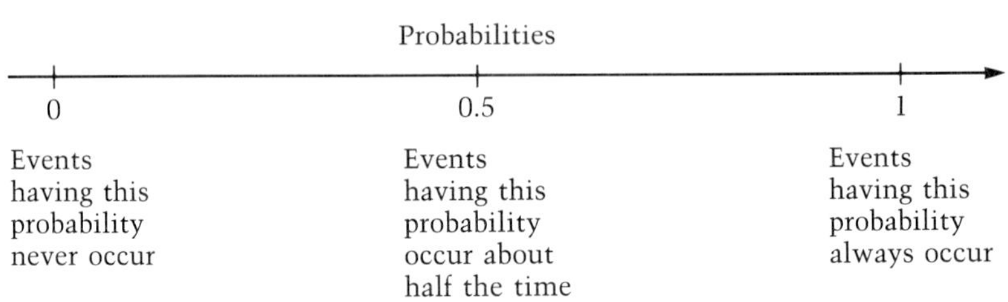

The equally likely model of an experiment E assigns probability $1/n$ to each outcome, where n is the total number of outcomes on E. Therefore, if we add the probabilities of all the outcomes, the result is 1:

$$\underbrace{\frac{1}{n} + \frac{1}{n} + \frac{1}{n} + \cdots + \frac{1}{n}}_{\text{There are } n \text{ terms here.}} = \frac{n}{n} = 1.$$

This result is basic to all probabilities even if the model is not the equally likely one (an example of this is considered below). If we let $e_1, e_2, e_3, \ldots, e_n$ be all the outcomes of E, this result may be expressed as

$$P(e_1) + P(e_2) + P(e_3) + \cdots + P(e_n) = 1. \qquad (5)$$

Equation (5) may be stated in words by saying that *the sum of the probabilities of all the simple events of an experiment is 1.*

To illustrate this with an experiment we have considered before, let E be the act of rolling a fair die 1 time. The simple outcomes and their probabilities are as follows:

$$\begin{array}{cccccc l} 1 & 2 & 3 & 4 & 5 & 6 & \text{Outcomes} \\ \downarrow & \downarrow & \downarrow & \downarrow & \downarrow & \downarrow & \\ \frac{1}{6} & \frac{1}{6} & \frac{1}{6} & \frac{1}{6} & \frac{1}{6} & \frac{1}{6} & \text{Probabilities} \end{array}$$

We see that the sum of these probabilities is 1:

$$\left(\frac{1}{6}\right) + \left(\frac{1}{6}\right) + \left(\frac{1}{6}\right) + \left(\frac{1}{6}\right) + \left(\frac{1}{6}\right) + \left(\frac{1}{6}\right) = 1.$$

Example 1 shows how this property can be used in conjunction with other information to assign probabilities to outcomes in the nonequally likely case.

Assigning probabilities to outcomes

EXAMPLE 1 Given an experiment with the simple outcomes a, b, c; that is, the sample space for this experiment is $\{a, b, c\}$. If the outcome a is twice as likely as b, and c is three times as likely as b, what are the probabilities $P(a)$, $P(b)$, $P(c)$?

Solution: From the given information we have

$$P(a) = 2P(b) \qquad \text{and} \qquad P(c) = 3P(b).$$

Now substitute this information into the equation

$$\begin{aligned} P(a) + P(b) + P(c) &= 1, \\ 2P(b) + P(b) + 3P(b) &= 1, \\ 6P(b) &= 1. \end{aligned}$$

Therefore,

$$P(b) = \frac{1}{6}.$$

Then

$$P(a) = 2P(b)$$
$$= \frac{1}{3},$$
$$P(c) = 3P(b)$$
$$= \frac{1}{2}.$$

PRACTICE PROBLEM In Example 1, suppose that outcome c is just as likely as a and that outcome b is four times as likely as a. Find $P(a)$, $P(b)$, $P(c)$.

Answer: $P(a) = \frac{1}{6}$, $P(b) = \frac{2}{3}$, $P(c) = \frac{1}{6}$. ■

The definition of probability in the equally likely model has been given in the previous section but a definition has not yet been stated in the nonequally likely case. Such a definition requires that we set forth the rules for computing $P(A)$, where A is any event associated with an experiment.

DEFINITION OF PROBABILITY Suppose that S is the sample space associated with some experiment E and that the probability of each simple outcome is known. Then, if A is any event associated with E, that is, A is a subset of E, the **probability of A,** $P(A)$, is the sum of the probabilities of the simple outcomes in A.

Probability definition

EXAMPLE 2 Voters in a certain locality are classified according to sex and political party. The accompanying table shows the percentage of the voters in various categories. An experiment will consist of randomly choosing a voter.

Voters

	Democrat (D)	Republican (R)	Independent (I)
Female (F)	27%	23%	5%
Male (M)	22%	21%	2%

(a) Construct an appropriate sample space.

(b) Determine the probability that the chosen voter is female.

(c) Determine the probability that the chosen voter is independent.

Solution

(a) Any voter chosen will fall into one of the following categories:

C_1 = Female-Democrat	C_4 = Male-Democrat
C_2 = Female-Republican	C_5 = Male-Republican
C_3 = Female-Independent	C_6 = Male-Independent

Therefore, an appropriate sample space is

$$S = \{C_1, C_2, C_3, C_4, C_5, C_6\}.$$

(b) The event F (female) is the following subset of the sample space:

$$F = \{C_1, C_2, C_3\},$$

and according to the given table

$$P(C_1) = 0.27, P(C_2) = 0.23, P(C_3) = 0.05.$$

Then, by the definition of probability,

$$\begin{aligned} P(F) &= P(C_1) + P(C_2) + P(C_3) \\ &= 0.27 + 0.23 + 0.05 \\ &= 0.55. \end{aligned}$$

(c) Since I (independent) is the following subset of the sample space

$$I = \{C_3, C_6\},$$

we have

$$\begin{aligned} P(I) &= P(C_3) + P(C_6) \\ &= 0.05 + 0.02 \\ &= 0.07. \end{aligned}$$

PRACTICE PROBLEM In Example 2 determine the probability that the chosen voter is Republican.

Answer: 0.44. ■

Consider again the experiment E which consists of drawing one card at random from a deck of 52 cards. We wish to determine the probability of drawing a black card or a face card. For this purpose, let

$$A = \text{drawing a black card}$$
$$B = \text{drawing a face card.}$$

Using this notation, we may express the probability to be determined as

$$P(A \text{ or } B).$$

Thus, we need to find

$$\#(A \text{ or } B),$$

that is, the number of cards which are either black or face cards. To obtain this number, one might first think of adding $\#(A)$ and $\#(B)$, but this is not the correct result, since there are black cards which are also face cards: Jack, Queen, King of clubs and spades. Thus if we add $\#(A)$ and $\#(B)$, the black face cards are counted twice. To correct for this overcount, we subtract $\#(A \text{ and } B)$. Therefore,

$$\#(A \text{ or } B) = \#(A) + \#(B) - \#(A \text{ and } B). \quad (6)$$

Since

$$\#(A) = 26, \qquad \#(B) = 12, \qquad \#(A \text{ and } B) = 6,$$

we have

$$\begin{aligned}\#(A \text{ or } B) &= 26 + 12 - 6\\ &= 32.\end{aligned}$$

Then the probability of drawing a black card or a face card is

$$\frac{32}{52} \text{ or } \frac{8}{13}$$

Equation (6) is correct for any experiment E and any two events A, B. Therefore, if n is the number of outcomes of E, we have [by dividing (6) by n]

$$\frac{\#(A \text{ or } B)}{n} = \frac{\#(A)}{n} + \frac{\#(B)}{n} - \frac{\#(A \text{ and } B)}{n}.$$

Using the definition of probability for the equally likely model we find that this equation gives us the following formula:

addition formula

THE ADDITION FORMULA FOR PROBABILITY

$$P(A \text{ or } B) = P(A) + P(B) - P(A \text{ and } B).$$

The Addition Formula has been derived in the equally likely model only, but it is true in *all cases.* It is common practice in probability to use the symbols ∪ and ∩ from set theory to denote "or" and "and," respectively. Then the Addition Formula is expressed as

$$P(A \cup B) = P(A) + P(B) - P(A \cap B).$$

Addition formula

EXAMPLE 3 Choose one of the integers 0, 1, 2, 3, 4, 5, 6, 7, 8, 9 at random. What is the probability that the number chosen is less than 5 or a multiple of 3?

Solution: Let

$$A = \text{the number chosen is less than 5,}$$
$$B = \text{the number chosen is a multiple of 3.}$$

Then the problem is to find $P(A \text{ or } B)$ and, by the Addition Formula,

$$P(A \text{ or } B) = P(A) + P(B) - P(A \text{ and } B).$$

Hence, we need to find $P(A)$, $P(B)$, and $P(A \text{ and } B)$. This is easily done by underlining numbers less than 5 and placing a bar above those which are multiples of 3:

$$\underline{\overline{0}}\ \underline{1}\ \underline{2}\ \underline{\overline{3}}\ \underline{4}\ 5\ \overline{6}\ 7\ 8\ \overline{9}.$$

Counting the numbers which are underlined gives

$$P(A) = \frac{5}{10} = 0.5,$$

and counting those with bars on top yields

$$P(B) = \frac{4}{10} = 0.4.$$

Finally, $P(A \text{ and } B)$ is obtained by counting the numbers with underlines and bars:

$$P(A \text{ and } B) = \frac{2}{10} = 0.2.$$

Then

$$\begin{aligned} P(A \text{ or } B) &= P(A) + P(B) - P(A \text{ and } B) \\ &= 0.5 + 0.4 - 0.2 \\ &= 0.7. \end{aligned}$$

Using the frequency interpretation of probability, we conclude that if this experiment is carried out many times, about 7 out of every 10 trials, on the average, will produce either a number less than 5 or a multiple of 3.

PRACTICE PROBLEM Using the experiment of Example 3, find the probability that the number drawn is even or greater than 5.

Answer: 0.7. ■

If we draw a single card at random from a deck of cards, it is not possible to obtain both a red and black card. Let

$$\begin{aligned} A &= \text{drawing a red card}, \\ B &= \text{drawing a black card}. \end{aligned}$$

We say that the events A and B are *exclusive;* i.e. they cannot both occur on the same draw. This means that the event

$$A \text{ and } B$$

cannot occur, and hence

$$P(A \text{ and } B) = 0.$$

exclusive events

DEFINITION OF EXCLUSIVE EVENTS Events A, B are said to be **exclusive** if

$$P(A \text{ and } B) = 0.$$

Using the addition formula and the definition of exclusive events A, B, we can derive the following important result.

THE ADDITION FORMULA FOR EXCLUSIVE EVENTS If A, B are exclusive events, then

$$P(A \text{ or } B) = P(A) + P(B).$$

The meaning of exclusive for two events may be stated by saying that the *events cannot occur simultaneously* or that the *occurrence of one of them precludes the occurrence of the other.*

Exclusive events

EXAMPLE 4 Roll a die one time. Determine which of the possible pairs of the following events are exclusive:

$$A = \text{rolling a number less than 4,}$$
$$B = \text{rolling a number at least 5,}$$
$$C = \text{rolling an odd number.}$$

Solution: The sample space for this experiment is

$$\{1, 2, 3, 4, 5, 6\},$$

and

$$A = \{1, 2, 3\}, \qquad B = \{5, 6\}, \qquad C = \{1, 3, 5\}.$$

The possible pairs to consider are

$$A, B \quad A, C \quad B, C.$$

Pair A, B: This pair is exclusive since if one of the events occurs, then the other one does not.

Pair A, C: This pair is not exclusive since both occur if the die shows a 1.

Pair B, C: This pair is not exclusive since they both occur if the die shows a 5.

PRACTICE PROBLEM For the experiment in Example 4, let D be the event that the die shows a number less than 4. If A, B, C are the events given in the example, which of the following pairs are exclusive?

$$A, D, \quad B, D, \quad C, D.$$

Answer: A, D are not exclusive; B, D are exclusive; C, D are not exclusive. ■

complement

If A is an event, then A' is the *complement* of A, that is, A' is the set of all outcomes in the sample space which are not in A. In other words, A and A' account for all the outcomes of the experiment. For example, consider a single roll of a die. If A is the event "getting an even number," then

$$A = \{2, 4, 6\},$$

and A' is the event "not getting an even number" or

$$A' = \{1, 3, 5\}.$$

On a single roll of the die we are sure to get an outcome in either A or A', that is, the event A or A' is sure to occur. In the same way, we can conclude that the event A or A' will always occur no matter what the experiment or the event A. But an event that is a certainty has probability 1:

$$P(A \text{ or } A') = 1.$$

The definition of A' implies that the events A and A' are exclusive or that the event A and A' cannot occur:

$$P(A \text{ and } A') = 0.$$

Now apply the Addition Formula for Exclusive Events to A or A':

$$P(A \text{ or } A') = P(A) + P(A').$$

But we have just seen that $P(A \text{ or } A') = 1$ and hence that

$$1 = P(A) + P(A').$$

This last equation may be expressed as

$$P(A') = 1 - P(A).$$

complementary events

FORMULA FOR PROBABILITY OF COMPLEMENTARY EVENTS
If A is an event, then

$$P(A') = 1 - P(A).$$

The usefulness of this formula lies in the fact that we can always compute $P(A')$ from a knowledge of $P(A)$. [*Note:* We read $P(A')$ as "the probability that *A does not occur.*"]

Complementary event

EXAMPLE 5 Draw a card at random from a deck of 52 cards. What is the probability of not getting a face card?

Solution: To solve the problem, we may let

$$A = \text{drawing a face card}$$

and then use the formula for the complement of A since the probability of not drawing a face card is $P(A')$. We have already seen that the probability of drawing a face card is $\frac{12}{52}$ since there are 12 face cards in the deck and outcomes are equally likely. Then

$$\begin{aligned} P(A') &= 1 - P(A) \\ &= 1 - \frac{12}{52} \\ &= \frac{40}{52} \\ &= \frac{10}{13}. \end{aligned}$$

Therefore, the probability of not drawing a face card is $\frac{10}{13}$.

PRACTICE PROBLEM For the experiment of Example 5 find the probability of not drawing a 4.

Answer: $\frac{12}{13}$. ■

Transportation

EXAMPLE 6 A vice-president of a company visits New York often. The probabilities concerning her modes of transportation are:

Mode of transportation	Probability
Her own car	0.24
Company car	0.27
Train	0.18
Bus	0.19
Plane	0.12

Find the probability that

(a) she will not catch a plane;

(b) she will go by car;

(c) she will go by car or bus.

Solution

(a) Let A be the event "she will go by plane." Then we need to compute $P(A')$:

$$\begin{aligned} P(\text{she will not catch a plane}) &= P(A') \\ &= 1 - P(A) \\ &= 1 - 0.12 \\ &= 0.88. \end{aligned}$$

(b) Let X and Y be the events "she will go by her own car" and "she will go by company car," respectively. Then the event "she will go by car" is

$$X \quad \text{or} \quad Y.$$

It is clear that she can use only one mode of transportation per trip and that the events X and Y are exclusive. Therefore,

$$\begin{aligned} P(\text{she will go by car}) &= P(X \text{ or } Y) \\ &= P(X) + P(Y) \\ &= 0.24 + 0.27 \\ &= 0.51. \end{aligned}$$

(c) We have already determined the probability that she will go by car in part (b). Then

$$\begin{aligned} &P(\text{she will go by car or bus}) \\ &\qquad = P(\text{she will go by car}) + P(\text{she will go by bus}) \\ &\qquad = 0.51 + 0.19 \\ &\qquad = 0.70. \end{aligned}$$

PRACTICE PROBLEM In Example 6 determine the probability that she will go by car or train.

Answer: 0.69. ■

EXERCISES

Set A–Exercises Using Formulas and Definitions

1. Given events A and B with $P(A) = 0.15$, $P(B) = 0.40$, and $P(A \text{ and } B) = 0.06$, find $P(A \text{ or } B)$.
2. Given events A and B with $P(A) = 0.2$, $P(B) = 0.08$, and $P(A \text{ and } B) = 0.01$, find $P(A \text{ or } B)$.
3. Given events A and B with $P(A) = 0.44$, $P(B) = 0.93$, and $P(A \text{ and } B) = 0.40$, find $P(A \text{ or } B)$.
4. Given events A and B with $P(A) = 0.34$, $P(B) = 0.80$, and $P(A \text{ and } B) = 0.21$, find $P(A \text{ or } B)$.
5. Given events A and B which are exclusive and $P(A) = 0.65$, $P(B) = 0.25$. Find
   (a) $P(A \text{ and } B)$. (b) $P(A \text{ or } B)$.
6. Given events A and B which are exclusive and $P(A) = 0.86$, $P(B) = 0.13$. Find
   (a) $P(A \text{ and } B)$. (b) $P(A \text{ or } B)$.

7. Given events A and B which are exclusive and $P(A) = 0.56$, $P(B) = 0.37$. Find
   (a) $P(A \text{ and } B)$. (b) $P(A \text{ or } B)$.
8. Given events A and B which are exclusive and $P(A) = 0.88$, $P(B) = 0.11$. Find
   (a) $P(A \text{ and } B)$. (b) $P(A \text{ or } B)$.
9. Suppose the events A and B are exclusive and that $P(A) = 0.15$, $P(B) = 0.35$. Find
   (a) $P(A \text{ and } B)$. (b) $P(A')$. (c) $P(B')$. (d) $P(A \text{ or } B)$.
10. Suppose that X, Y are exclusive events and that $P(X) = 0.12$, $P(Y) = 0.30$. Find
    (a) $P(X \text{ and } Y)$. (b) $P(X')$. (c) $P(Y')$. (d) $P(X \text{ and } Y)$.
11. Is it possible to have two exclusive events M, N such that $P(M) = 0.5$ and $P(N) = 0.5$? If your answer is "no," give a reason, and if "yes," give an example.
12. Is it possible to have two exclusive events M, N such that $P(M) = 0.55$ and $P(N) = 0.48$? Give a reason for your answer.
13. Suppose an experiment has exactly two outcomes, A and B, and that A is twice as likely to occur as B. Find $P(A)$ and $P(B)$.
14. If an experiment has exactly two outcomes, A and B, what is $P(A \text{ and } B)$?
15. An experiment has exactly three outcomes, A, B, C. If $P(A \text{ or } B) = 0.65$ and $P(A \text{ and } B) = 0.15$, what is $P(C)$?
16. Suppose an experiment has exactly three outcomes A, B, C and that $P(B \text{ or } C) = 0.48$ and $P(B \text{ and } C) = 0.23$. Find $P(A)$.
17. Suppose an experiment has exactly three outcomes, A, B, C. Also A, B are exclusive and A, C are exclusive. If $P(A \text{ or } B) = 0.4$ and $P(A \text{ or } C) = 0.65$, what are $P(A)$, $P(B)$, and $P(C)$.

Set B–Some Probability Word Problems

Some of the exercises below involve playing cards. Assume that the ace is always high and never 1.

18. Choose a card at random from a deck of 52 cards. Determine the probability of
    (a) drawing a King or a Queen.
    (b) drawing a King or a red card.
    (c) drawing an Ace, King or Queen.
    (d) drawing a red card or a card with a number less than 10.
19. Roll a pair of fair dice one time. What is the probability that the dice show
    (a) the same number? (b) the sum of 6? (c) a sum larger

than 6? (d) an even sum? (e) an odd sum? (f) an even sum or a sum larger than 6?

20. A box contains 2 white and 2 black ping-pong balls. Two balls are drawn at random from the box, one after the other, without replacement.
 (a) Construct a tree diagram and determine the probability of each simple event.
 (b) What is the probability the first ball drawn is black and the second is white?
 (c) What is the probability the 2 balls drawn have different colors?
 (d) What is the probability that the 2 balls drawn have the same color?

21. Four black and 3 white marbles are placed in a box, and 2 are drawn at random, one after the other, with replacement. What is the probability that the marbles drawn are
 (a) both black? (b) both white? (c) of different color?

22. Perform the same experiment as in Exercise 21 except without replacement. Then answer the questions of that exercise.

Weather forecast

23. Suppose the weather forecaster says that the probability of rain today is $\frac{1}{3}$, and that the probability of rain tomorrow is $\frac{1}{4}$.
 (a) What is the probability of rain today or tomorrow (assume exclusive events)?
 (b) What is the probability that it rains on neither day?

Course enrollment

24. Of 300 college students, 115 are enrolled in economics, 245 are enrolled in mathematics, and 85 are enrolled in both. What is the probability that a randomly selected student is enrolled in
 (a) either economics or mathematics?
 (b) neither of these subjects?

Voting patterns

25. For married couples living in a certain suburb, the probability that a husband will vote is 0.38, the probability that a wife will vote is 0.27, and the probability that both will vote is 0.21. What is the probability that
 (a) the husband will not vote?
 (b) the wife will not vote?
 (c) at least one member of a couple will vote?
 (d) neither husband nor wife will vote?

Traffic control

26. An engineer who is studying traffic control has determined that if a traffic light turns red while an auto is in the zones given on the diagram, the probabilities that the vehicle will stop are as shown.

What is the probability that such a vehicle
(a) will not stop if the light turns red in one of these zones?
(b) will stop if the light turns red before zone C?
(c) will stop if the light turns red after zone A?

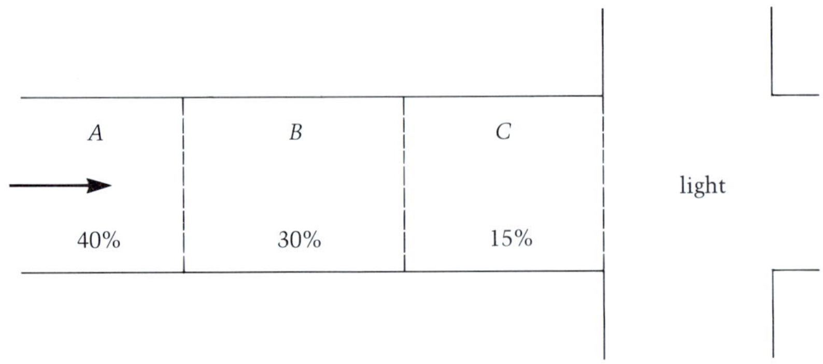

7.4 CONDITIONAL PROBABILITY AND INDEPENDENT EVENTS

A box contains 2 white and 3 black ping-pong balls. Choose two balls at random, one after the other, without replacement. This experiment consists of two trials: a first draw and then a second draw (see the tree diagram of Figure 7.5).

The tree diagram of Figure 7.5 shows the contents of the box

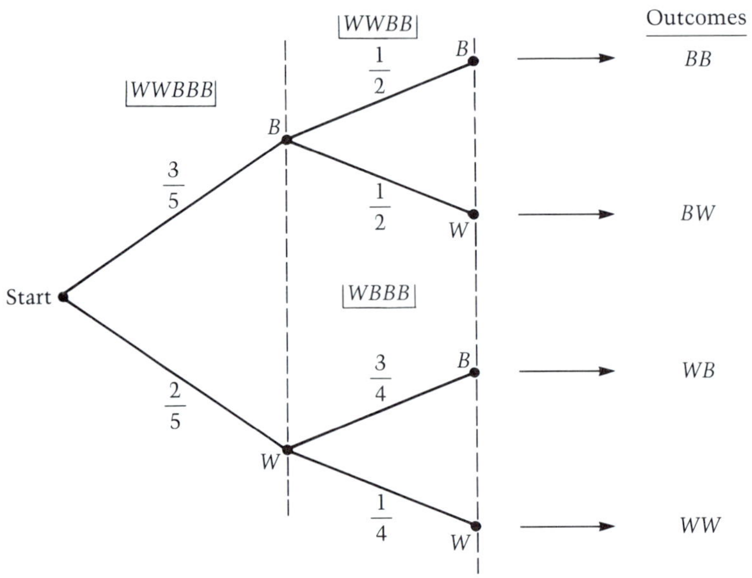

FIG. 7.5.

before the first draw and also after the first draw. On the first draw the probability of getting W is $\frac{2}{5}$. But after the first draw the chances of getting W on the second draw are either 1 in 4 or 2 in 4, depending upon what color was drawn first. The diagram shows these two possible states of the box after the first draw. This is an example of *dependent trials.* If the drawing had been

independent trials

with replacement, the trials would have been *independent* because the chances of getting W on the first and second draw would have been the same. Tossing a coin repeatedly is also an example of an experiment which consists of independent trials.

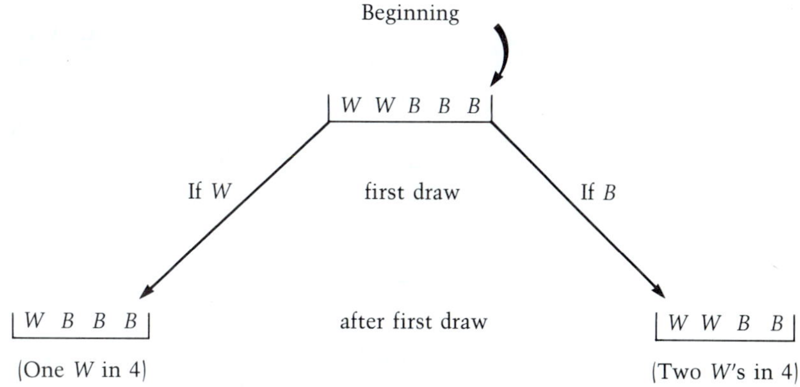

Continuing with the experiment shown in Figure 7.5, let

$$W1 = \text{drawing white on the first draw,}$$
$$W2 = \text{drawing white on the second draw,}$$
$$B1 = \text{drawing black on the first draw,}$$
$$B2 = \text{drawing black on the second draw.}$$

The notation

$$P(W2 \mid W1)$$

is read as

> the probability of getting white on the second draw, given that white is drawn on the first draw.

From the tree diagram of Figure 7.5 we can see that

$$P(W2 \mid W1) = \frac{1}{4}.$$

This is an example of *conditional probability.* Some other conditional probabilities in this experiment are (see Figure 7.5)

$$P(W2 \mid B1) = \frac{1}{2},$$

$$P(B2 \mid W1) = \frac{3}{4},$$

$$P(B2 \mid B1) = \frac{1}{2}.$$

For a conditional probability $P(B \mid A)$—the probability that B occurs, given that A occurs—we are asked to compute the probability that B occurs, given that A has *already* occurred. This implies that we must look for the event B among that part of the sample space which contains A, that is, we must look for the event (A and B) among the event A. This suggests that

$$P(B \mid A) = \frac{\#(A \text{ and } B)}{\#(A)}. \tag{1}$$

Formula (1) is correct for the equally likely model and if we divide numerator and denominator by n, the number of outcomes of the experiment, we obtain the following formula which holds in all cases.

$$P(B \mid A) = \frac{P(A \text{ and } B)}{P(A)}. \tag{2}$$

Equation (2) is the basis for the following definition.

conditional probability

DEFINITION OF CONDITIONAL PROBABILITY If A and B are events and $P(A)$ is not 0, then the **conditional probability,** $P(B \mid A)$, is defined by the following equation:

$$P(B \mid A) = \frac{P(A \text{ and } B)}{P(A)}. \tag{3}$$

Conditional probability

EXAMPLE 1 A fair die is rolled one time. What is the probability that an even number is obtained, given that a number less than 4 is obtained?

Solution: Let

$$B = \text{an even number is obtained,}$$
$$A = \text{a number less than 4 is obtained.}$$

Then the question may be expressed as $P(B \mid A)$. By the definition, we must compute $P(A \text{ and } B)$ and $P(A)$. The sample space is

$$1 \quad 2 \quad 3 \quad 4 \quad 5 \quad 6$$

If we place a bar above the numeral for B and underline for A, we have

$$A \longrightarrow \quad \underline{1} \quad \underline{\bar{2}} \quad \underline{3} \quad \bar{4} \quad 5 \quad \bar{6} \quad \longleftarrow B$$

Then,

$$P(A) = \frac{3}{6} = \frac{1}{2}$$

and

$$P(A \text{ and } B) = \frac{1}{6}.$$

Hence

$$\begin{aligned} P(B \mid A) &= \frac{P(A \text{ and } B)}{P(A)} \\ &= \frac{\frac{1}{6}}{\frac{1}{2}} \\ &= \left(\frac{1}{6}\right)\left(\frac{2}{1}\right) \\ &= \frac{1}{3}. \end{aligned}$$

PRACTICE PROBLEM For the experiment of Example 1 compute the probability that a number less than 5 is obtained, given that the number obtained is odd.

Answer: $\frac{2}{3}$. ■

Homes for senior citizens

EXAMPLE 2 In a certain state, homes for senior citizens are classified as Rest Homes and Nursing Homes. Each of these facilities is inspected periodically and determined to be "satisfactory," "unsatisfactory," or "marginal." The table shows the results of one inspection.

Homes for Senior Citizens

	Rest Homes (R)	Nursing Homes (N)	Totals
Satisfactory (S)	105	45	150
Unsatisfactory (U)	5	10	15
Marginal (M)	40	19	59
Totals	150	74	224

If a home is chosen at random, what is the probability that

(a) the home is unsatisfactory, given that it is a nursing home?

(b) the home is a rest home, given that it is marginal?

Solution

(a) Using the notation given in the table, we need to compute

$$P(U \mid N).$$

Using the definition of conditional probability gives

$$\begin{aligned} P(U \mid N) &= \frac{P(U \text{ and } N)}{P(N)} \\ &= \frac{\frac{10}{224}}{\frac{74}{224}} \\ &= \frac{10}{74} \\ &= 0.135. \end{aligned}$$

(b) Again we use the notation of the table and the definition of conditional probability:

$$
\begin{aligned}
P(R \mid M) &= \frac{P(R \text{ and } M)}{P(M)} \\
&= \frac{\frac{40}{224}}{\frac{59}{224}} \\
&= \frac{40}{59} \\
&= 0.678
\end{aligned}
$$

PRACTICE PROBLEM In Example 2 compute the probability that the home chosen is marginal, given that it is a nursing home.

Answer: 0.257. ■

From the definition of $P(B \mid A)$ we may derive the following formula which is known as the *Multiplication Formula for Probability.* The formula is derived from Eq. (3) by multiplying both sides by $P(A)$.

multiplication formula

MULTIPLICATION FORMULA FOR PROBABILITY

$$P(A \text{ and } B) = P(A)P(B \mid A).$$

Return for a moment to the tree diagram of Figure 7.5. Consider the event $(W1 \text{ and } W2)$, getting white on both the first and second draw. On the tree diagram $(W1 \text{ and } W2)$ is shown as WW, that is,

$$(W1 \text{ and } W2) = WW.$$

Using the Multiplication Formula for the event $(W1 \text{ and } W2)$ gives

$$
\begin{aligned}
P(W1 \text{ and } W2) &= P(W1)P(W2 \mid W1) \\
&= \left(\frac{2}{5}\right)\left(\frac{1}{4}\right) \\
&= \frac{1}{10}.
\end{aligned}
$$

Therefore, the simple event WW has probability 0.1. This suggests that *we are able to determine the probability of the simple events from a tree diagram by multiplying the probabilities on the branches which lead to the simple event.* On the tree diagram of Figure 7.5 the probabilities of the simple events have been obtained in this way. It will be noted that the simple events in this experiment are not equally likely since, for example,

$$P(WW) = \frac{1}{10} \qquad \text{and} \qquad P(WB) = \frac{3}{10}.$$

But the probabilities of the simple events still add up to 1 as they should:

$$\begin{aligned} P(WW) + P(WB) + P(BW) + P(BB) &= \frac{1}{10} + \frac{3}{10} + \frac{3}{10} + \frac{3}{10} \\ &= 1. \end{aligned}$$

Probability in nonequally likely model

EXAMPLE 3 In the experiment for Figure 7.5 determine the probability of getting 2 ping-pong balls of different colors.

Solution: Let

$$A = \text{getting 2 ping-pong balls of different colors.}$$

From Figure 7.5 we see that

$$A = \{WB, BW\}.$$

But this is not the equally likely model, so that the formula

$$\frac{\#(A)}{4}$$

cannot be used. If this formula were correct, the probability would be $\frac{1}{2}$. To compute the correct probability, it may be reasoned that to obtain different-colored ping-pong balls we must obtain WB or BW. This suggests that the Addition Formula be used, and since WB, BW are exclusive,

$$
\begin{aligned}
P(WB \text{ or } BW) &= P(WB) + P(BW) \\
&= \frac{3}{10} + \frac{3}{10} \\
&= \frac{6}{10} \\
&= \frac{3}{5}.
\end{aligned}
$$

PRACTICE PROBLEM For the experiment of Example 3 determine the probability of drawing white on the second draw.

Answer: $\frac{2}{5}$. ■

The definition of probability which has often been stressed is only for the equally likely model and, as Example 3 shows, this cannot be used in all cases. But this is no great problem so long as the probabilities of simple events are known. When the probabilities of simple events are known, we may always use the general definition of probability or the Addition Formula to compute probabilities of other events as was done in Example 3.

independent events

DEFINITION OF INDEPENDENT EVENTS Two events A and B are said to be **independent** if

$$P(B \mid A) = P(B).$$

This definition may be paraphrased in the following way: Two events A and B are independent if the *occurrence of A does not affect the probability that B occurs.*

Independent events

EXAMPLE 4 Draw a single card from a deck of 52 cards. Let

$$A = \text{drawing a face card,}$$

$$B = \text{drawing a red card.}$$

Show that the events A and B are independent.

Solution: We must compute $P(B \mid A)$, $P(B)$ and show that these two probabilities are equal. There are 26 red cards in the deck so that

$$P(B) = \frac{26}{52} = \frac{1}{2}.$$

To compute $P(B \mid A)$ we need $P(A \text{ and } B)$ and $P(A)$:

$$P(A) = \frac{12}{52} = \frac{3}{13} \quad \text{(there are 12 face cards)},$$

and

$$P(A \text{ and } B) = \frac{6}{52} = \frac{3}{26} \quad \text{(there are 6 red face cards)}.$$

Then

$$\begin{aligned} P(B \mid A) &= \frac{P(A \text{ and } B)}{P(A)} \\ &= \frac{\frac{3}{26}}{\frac{3}{13}} \\ &= \left(\frac{3}{26}\right)\left(\frac{13}{3}\right) \\ &= \frac{1}{2}. \end{aligned}$$

Therefore, $P(B \mid A) = P(B)$, and by definition, A, B are independent events.

PRACTICE PROBLEM Use the experiment of Example 4 and let

A = drawing a card having an even number,

B = drawing a black card.

Show that A and B are independent events.

Answer: $P(B \mid A) = \frac{1}{2}$ and $P(B) = \frac{1}{2}$. ■

If A, B are independent events, we may substitute $P(B)$ for $P(B|A)$ in the multiplication formula and obtain the following result.

RESULT FOR INDEPENDENT EVENTS If A and B are independent events, then

$$P(A \text{ and } B) = P(A)P(B).$$

Independent events

EXAMPLE 5 Suppose X and Y are independent events and that $P(X) = 0.12$ and $P(Y) = 0.44$. Compute $P(X \text{ and } Y)$.

Solution: Since X and Y are independent events, we may use the result just obtained and write

$$\begin{aligned} P(X \text{ and } Y) &= P(X)P(Y) \\ &= (0.12)(0.44) \\ &= 0.0528. \end{aligned}$$

PRACTICE PROBLEM Assume that U and V are independent events and $P(U) = 0.37$, $P(V) = 0.23$. Find $P(U \text{ and } V)$.

Answer: 0.0851. ■

EXERCISES

Set A–Assigning Probabilities on Tree Diagrams

Draw a tree diagram for each of Exercises 1 through 6 and assign probabilities to each branch of the diagram. Then find the probability of each simple event.

1. A box contains 4 marbles: 1 white and 3 black. Draw 2 marbles from the box, one after the other, without replacement. What is the probability that the 2 marbles drawn are
   (a) both white? (b) both black?
   (c) the same color? (d) of different colors?
2. The same as Exercise 1 except that the box contains 2 marbles of each color.
3. A box contains 4 marbles: 1 white and 3 black. Draw 3 marbles from the box, one after the other, without replacement.

What is the probability that of the 3 marbles drawn
(a) all are white? (b) all are black?
(c) all are the same color? (d) exactly 2 are white?
(e) at least 2 are white?

4. The same as Exercise 3 except that the box contains 2 marbles of each color.

5. Consider the following experiment which uses the spinner and two boxes containing black and white marbles as shown in the figure. The experiment consists of two parts: first a spin of the spinner and then a draw from box X or Y. If the spinner points to A, a marble is drawn at random from X, otherwise a marble is drawn from Y.

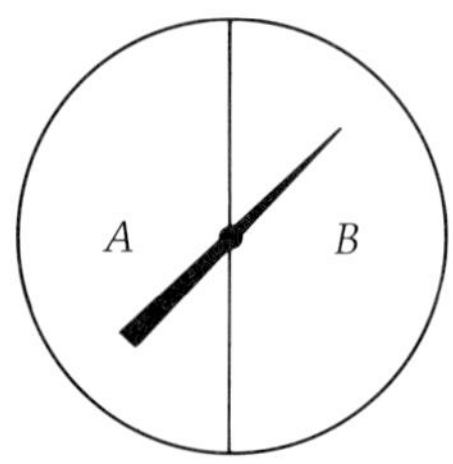

X: |W W B B|

Y: |W B B B|

What is the probability that
(a) the spinner points to A, and W is drawn?
(b) the spinner points to A, and B is drawn?
(c) the spinner points to B, and W is drawn?
(d) the spinner points to B, and B is drawn?

6. An experiment consists of drawing 3 marbles from the following box

|W W B B B|

If the marble drawn on any trial is W, it is replaced before the next draw. If the marble drawn is B, it is not replaced. What is the probability that of the 3 marbles drawn
(a) all have the same color?
(b) 2 are white and 1 is black?
(c) at least 1 is white?

Set B–Computing Conditional Probabilities from the Definition

7. A fair die is rolled one time. What is the probability a number less than 4 is obtained, given that an even number is obtained?

8. A card is selected at random from a deck of 52 cards. What is the probability the card is an ace, given that it is red?
9. A card is selected at random from a deck of 52 cards. What is the probability that the card is red, given that it is a face card?
10. A card is selected at random from a deck of 52 cards. What is the probability that the card is a club, given that it is a face card?
11. Given the set $A = \{1, 2, 3, 4, 5\}$. Choose a set of two elements from A at random.
    (a) How many outcomes are possible?
    (b) List the outcomes.
    (c) What is the probability that one of the two elements chosen is 3?
    (d) What is the probability that one of the elements chosen is 3, given that the other is 4?
    (e) What is the probability that one of the elements chosen is 3, given that the other is less than 5?
    (f) What is the probability that one of the elements chosen is 3, given that the other is 1 or 2?
12. A fair coin is tossed twice. What is the probability that
    (a) H is obtained on both tosses, given that H is obtained on the first toss?
    (b) H is obtained on both tosses, given that H is obtained on at least one toss?
13. A fair coin is tossed 3 times. What is the probability that
    (a) T is obtained 3 times, given that T is obtained on the first and second toss?
    (b) T is obtained on all 3 tosses, given that T is obtained at least twice?
    (c) T is obtained on all 3 tosses, given that T is obtained at least once?
14. Roll a fair pair of dice one time. One die is red and one is blue. What is the probability that the sum of the numbers on the dice is
    (a) four, given that a 3 turns up on the red die?
    (b) seven, given that a 3 turns up on the red die?
    (c) eight, given that the sum is at least 5?
    (d) six, given that the number on the blue die is no more than 4?
15. Two cards are selected at random from a deck of 52 cards, one at a time, without replacement. What is the probability that
    (a) the second card is an ace, given that the first card is an ace?

(b) the second card is a face card, given that the first card is a face card?
(c) the second card is a face card, given that the first card is red?

Set C–Some Probability Word Problems

16. Two boxes each contain 10 marbles. The first box contains 3 white and 7 black marbles, and the second box contains 3 black and 7 white marbles. A box is chosen at random, and then a marble is drawn from that box. What is the probability that
(a) a black marble is drawn? (b) a white marble is drawn?

17. Two boxes each contain 10 black and white marbles which are distributed as in Exercise 16. A biased coin with $P(H) = 0.6$ is flipped one time. If H turns up on the coin, we choose one of the boxes at random and draw a marble from it. If T turns up, we place one of the black marbles of the first box in the second box and then, after choosing one of the boxes at random, we draw a marble. What is the probability that
(a) a black marble is drawn? (b) a white marble is drawn?

18. A box contains 1 white and 2 black marbles. A single marble is chosen from the box and replaced. If the marble drawn is white, an additional white marble is added to the box. If this experiment is repeated 3 times,
(a) what is the probability that at least one white marble is drawn?
(b) what is the probability that a black marble is chosen on the third draw?

Quality control

19. A carton contains 100 light bulbs, 15 defectives and 85 nondefectives. If 2 bulbs are selected at random, what is the probability that
(a) both are defective?
(b) both are defective, given that the first is defective?
(c) both are defective, given that at least 1 is defective?

Selecting committees

20. A group of 10 teachers which includes Jones, Smith, and Green wishes to select a committee consisting of 3 members. Assume that the committee is selected at random. What is the probability that
(a) the members selected are Jones, Smith, and Green?
(b) the committee selected includes Jones and Smith?
(c) the committee selected includes Jones and Smith, given that Green is 1 of its members?
(d) the committee selected includes Jones, given that Smith and Green are 2 of its members?

(e) the committee selected has Smith as 1 of its members, given that Green is not a member?

Student body composition

21. The table contains information about the student body at Midtown College.

	Freshmen	Sophomores	Juniors	Seniors	Totals
Male	215	186	175	168	744
Female	236	201	198	187	822
Totals	451	387	373	355	1566

Assume that a student is selected at random. What is the probability that the student selected is

(a) male? (b) female? (c) a freshman? (d) a senior?
(e) a male, given that he/she is a freshman?
(f) a sophomore, given that she is female?
(g) a female, given that the student is a junior or senior?

Coffee composition

22. Two brands of coffee are each composed of Columbian, Brazilian, and African beans. The composition of the brands is given in the table.

	Columbian	Brazilian	African
Brand 1	50%	40%	10%
Brand 2	70%	25%	5%

A 100-pound bag of Brand 1 and a 100-pound bag of Brand 2 are available. One of these bags is chosen at random and then 2 beans are randomly chosen. If we assume that the first bean is not replaced before the second is drawn but that the composition is not changed by its removal, what is the probability that

(a) both beans are Columbian?
(b) 1 bean is Columbian and 1 is African?
(c) at least 1 is African?

Army personnel

23. A group of 300 enlisted personnel in the U.S. Army are classified according to sex and rank. The result is provided in the table.

	Pvt (P)	Cpl (C)	Sgt (S)	Totals
Male (M)	121	70	9	200
Female (F)	55	30	15	100
Totals	176	100	24	300

Assume that a soldier is randomly chosen from this group. Compute the following probabilities.

(a) The probability that the soldier chosen is female.
(b) The probability that the soldier chosen is female, given the soldier is a private.
(c) Use the results of (a) and (b) to determine whether F and P are independent events.
(d) The probability that the soldier chosen is a sergeant.
(e) The probability that the soldier chosen is a sergeant, given the soldier is a female.
(f) Use the results of (d) and (e) to determine whether S and F are independent.

Disease among ethnic groups

24. A Public Health Office classified a group of 500 individuals as black or white and as having disease X or not having this disease. The table gives the results of this classification.

	Have disease X (H)	**Do not have disease X**	**Totals**
Black (B)	40	60	100
White (W)	160	240	400
Totals	200	300	500

If an individual is selected at random from this group, determine the following probabilities.

(a) The probability the individual selected has the disease.
(b) The probability the individual selected has the disease, given he is black.
(c) Use (a) and (b) to determine whether the events H and B are independent.
(d) The probability the individual chosen is white.
(e) The probability the individual chosen is white, given he does not have the disease.
(f) Use (d) and (e) to determine whether the events W and H are independent.

Maze problem

25. A rat is placed in a box as shown in Figure 7.6. The rat will randomly try each door in the box until it reaches the food; it will not try the same door twice. Construct an appropriate tree diagram with probabilities. What is the probability that the rat tries doors A and B before it gets the food?

Maze problem

26. A rat is to run the maze shown in Figure 7.7. If the rat goes straight ahead at any stage, it is eaten by a cat (X); if it turns

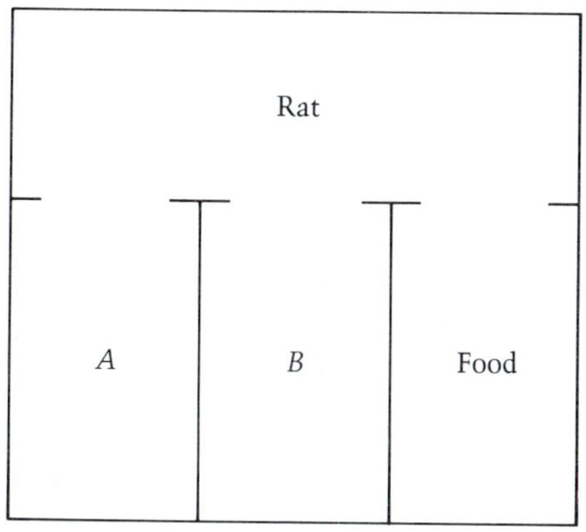

FIG. 7.6. From Hardy, F. Lane, *Finite Mathematics,* Harper & Row, 1977.

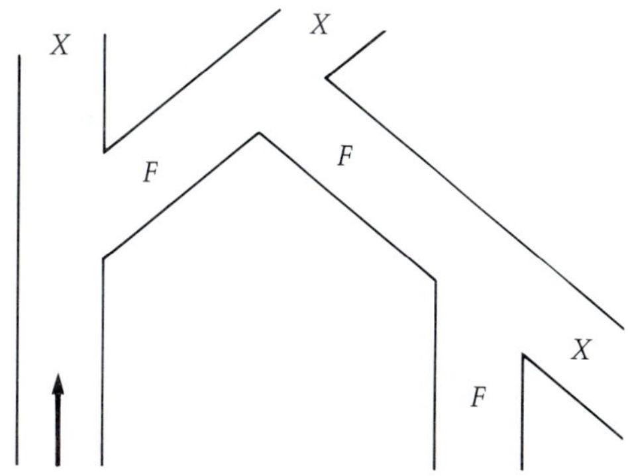

FIG. 7.7. From Hardy, F. Lane, *Finite Mathematics,* Harper & Row, 1977.

to the right, it obtains food (F). At the first stage, the probability that the rat goes straight ahead is 0.5. If it turns to the right at any stage, this action is reinforced by the food and increases the probability that it will turn right at the next stage by 0.1. Draw a tree diagram and determine the probabilities for the simple events. What is the probability that the rat

(a) gets exactly one meal?
(b) gets exactly two meals?
(c) does not get eaten by the cat?

Human populations

27. Assume that the births of males and females in human populations are equally likely. In families having 2 children, what is the probability that both children are male? What is the probability that both children are female, given that the first born is female?

Computer chip failure

28. It has been determined that the chances are 1 in 10 that the model XZ computer chip will fail in the first 100 hours of use if the chip is nondefective. Suppose we have a group of 5 XZ chips, all nondefective except 1. Also suppose that the chances are 3 in 5 that the defective chip will fail in the first 100 hours of use. If we do not know which chip is defective and we choose one at random and it fails in less than 100 hours, what is the probability that the defective chip was chosen?

Tennis match

29. Hall and Fulwood are to play a 2-out-of-3 set tennis match. The probability that Hall wins the first set is 0.7 but drops to

0.5 in the second set. His probability of winning the third set (if it is played) is 0.4. What is the probability that
(a) Fulwood wins the match?
(b) the match ends in 2 sets?
(c) Hall wins the match in two sets?

Tennis match

30. Two tennis players A and B are evenly matched. What is the probability that a winner will be determined in
(a) exactly 2 sets if the players play a 2-out-of-3 match?
(b) exactly 3 sets in a 2-out-of-3 set match?

Tennis match

31. Two evenly matched tennis players play a 3-out-of-5 set match. What is the probability that the match will end in exactly
(a) 3 sets? (b) 4 sets? (c) 5 sets?

Tennis match

32. Two tennis players A and B are not evenly matched. The probability that A wins any set with B is 0.6. If a 2-out-of-3 set match is played, what is the probability that the match will end in exactly
(a) 2 sets? (b) 3 sets?

Tennis match

33. Two tennis players A and B are evenly matched in the first set. For the player who wins a set, the probability of winning the next set increases by 0.1. In a 3-out-of-5 match, what is the probability that the match ends in exactly
(a) 3 sets? (b) 4 sets? (c) 5 sets?

World series

34. In the World Series the two teams play a 4-out-of-7 series. If the teams are evenly matched, what is the probability that the series ends in exactly
(a) 4 games? (b) 5 games? (c) 6 games? (d) 7 games?

7.5 BINOMIAL EXPERIMENTS

binomial experiments

We have frequently considered experiments which involve tossing coins. Such experiments are examples of *binomial experiments;* that is, we are dealing with a sequence of independent trials for each of which there are two outcomes. If we repeatedly toss a coin, the probability of obtaining heads does not change from one toss to the next (this means the trials are independent), and there are always the two possible outcomes "heads" and "tails."

success
failure

When discussing binomial experiments, we customarily label one of the two outcomes S (for success) and the other F (for failure). These choices are usually quite arbitrary, and for this reason we should not conclude that an outcome which has been called a "success" is in some way desirable. In tossing a coin, we may decide to refer to H as success and T as failure. If we draw a single marble from a container containing red and blue marbles, we could describe success as obtaining a blue marble.

Before discussing other examples we summarize below the important characteristics of binomial experiments.

CHARACTERISTICS OF BINOMIAL EXPERIMENTS

1. There is a fixed number of trials denoted by n:

$$n = \text{the number of trials.}$$

2. There are always two outcomes, S and F, at each trial.
3. Trials are independent (probabilities for S and F remain the same for all trials).

NOTATION FOR BINOMIAL EXPERIMENTS

1. The number of trials:

$$n = \text{the number of trials.}$$

2. The probability of success and failure:

$$p = \text{the probability of success} = P(S),$$
$$q = \text{the probability of failure} = P(F).$$

3. The number of successes in n trials:

$$k = \text{the number of } S\text{'s (successes) in } n \text{ trials}$$

Each trial of a binomial experiment results in either success or failure, i.e., the event S or F is sure to occur. Therefore,

$$P(S \text{ or } F) = 1.$$

But the events S and F are exclusive (they cannot both occur on the same trial) and this allows us to conclude that

$$P(S \text{ or } F) = P(S) + P(F)$$

or

$$p + q = 1.$$

This equation may also be expressed as

$$q = 1 - p.$$

The equation just obtained is very important since it allows us to easily calculate the value of q when the value of p is known.

Suppose a coin is biased with $P(H) = 0.7$. If we consider H as success, then

$$p = 0.7 \qquad \text{and} \qquad q = 0.3.$$

One trial may be represented by a tree diagram as shown in Figure 7.8. Figures 7.9 and 7.10 show 2 and 3 trials of the same experiment.

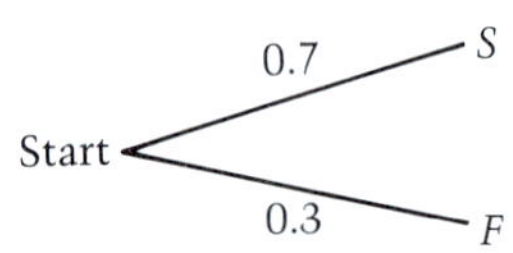

FIG. 7.8.

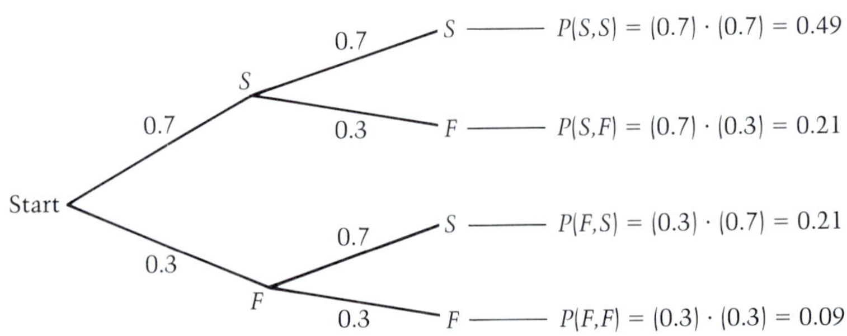

FIG. 7.9. From Hardy, F. Lane, *Finite Mathematics*, Harper & Row, 1977.

It is instructive to observe that each outcome in the tossing of a coin 3 times may be represented as a triple (x, y, z), where x, y, z are each either S or F. Thus x is the result of the first toss, y the result of the second, and z the result of the third. In the example of Figure 7.10 we have $P(S) = p = 0.7$ and $P(F) = q = 0.3$. Thus the probability of the simple event SFS, for example, is obtained as follows:

$$\begin{array}{ccc} S & F & S \\ \downarrow & \downarrow & \downarrow \\ 0.7 & 0.3 & 0.7 \end{array}$$

$$(0.7)(0.3)(0.7) = 0.147.$$

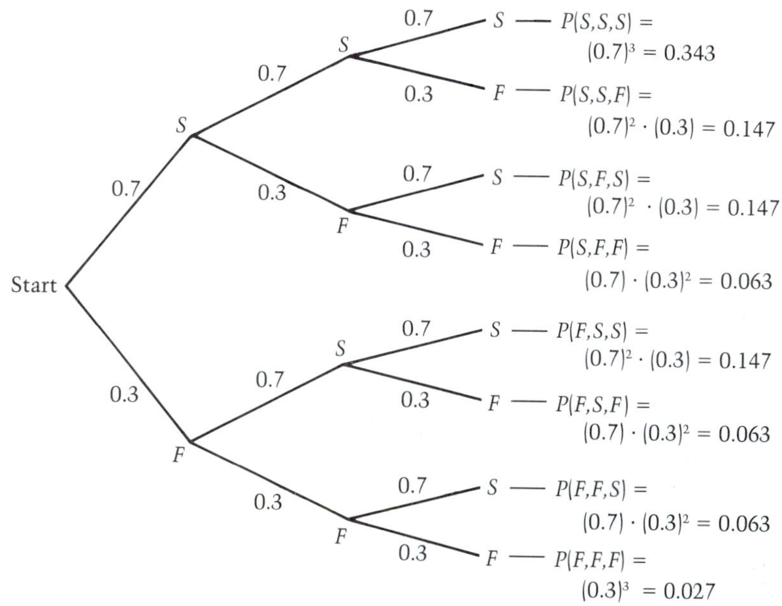

FIG. 7.10. From Hardy, F. Lane, *Finite Mathematics,* Harper & Row, 1977.

For this event we have multiplied 0.7, 0.3, 0.7 (since these are the probabilities on the branches of the tree diagram). The number 0.7 occurs twice because of the two *S*'s, and 0.3 occurs only once because of the single *F*. Observe that it is not important that one *S* is first and the other last—only that two *S*'s occur. Thus, all the simple events *SFS*, *FSS*, *SSF* have the same probability, namely, (0.7) (0.3) (0.7). Then we conclude, by the Addition Formula, that the probability of obtaining exactly 2 successes in 3 trials is

$$3(0.7)(0.3)(0.7).$$

The factor 3 occurs in this case because there are 3 simple events involved and all 3 have the same probability. The reason that we are dealing here with 3 simple events is that there are 3 ways of having exactly 2 successes in 3 trials, for each simple event is a triple (x, y, z), and in this case we must choose 2 of the 3 letters x, y, z to be *S*'s and the number of ways of choosing 2 items out of 3 is 3.

Now consider a binomial experiment of 3 trials in which the probability of success is unspecified. Then the tree diagram will look like the one in Figure 7.11. If we are interested in the probability of exactly 2 successes in 3 trials, the simple events involved

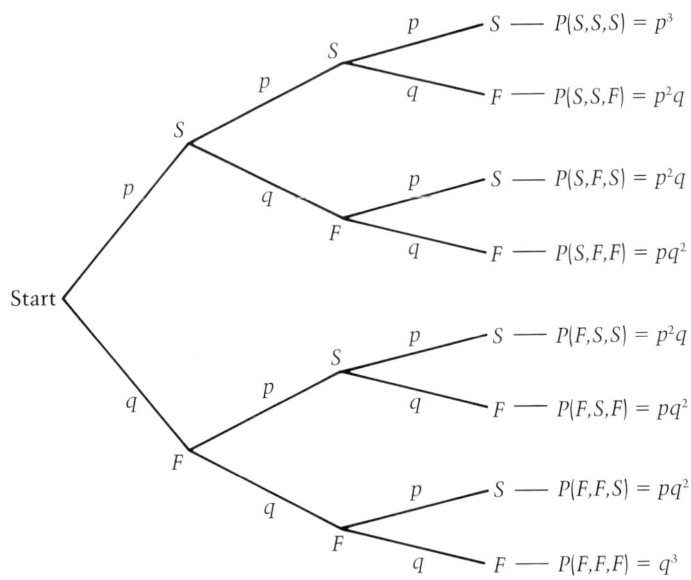

FIG. 7.11.

are just as before: *SSF*, *SFS*, *FSS*. Therefore, the probability of exactly 2 successes in 3 trials is

$$\begin{aligned} P(\text{SSF or } SFS \text{ or } FSS) &= P(SSF) + P(SFS) + P(FSS) \\ &= ppq + ppq + ppq \\ &= p^2q + p^2q + p^2q \\ &= 3p^2q. \end{aligned}$$

Let us imagine performing the same experiment except that the number of trials is increased to 10. It is no longer practical to draw a tree diagram since such a diagram would have 1024 endpoints. However, the probabilities may be computed without difficulty. For instance, we may compute the probability of obtaining exactly 4 successes ($k = 4$) in 10 trials as follows. Since there are 10 trials, we may think of each simple event as being constructed by filling 10 blanks with either *S* or *F*. In this case, we are interested in 4 *S*'s and 6 *F*'s. For example, one of the sequences we must consider is

$$F\ \underline{S}\ \underline{S}\ F\ F\ \underline{S}\ F\ \underline{S}\ F\ F$$

The probability of this simple event is

$$p^4q^6,$$

and, of course, *every* simple event having 4 S's and 6 F's has this same probability. Therefore, the probability of exactly 4 successes will be the sum of all these numbers, that is,

$$p^4q^6$$

will occur as a summand one time for each simple event containing exactly 4 S's. Hence, we need to know the number of simple events which contain exactly 4 S's. But this is the number of ways of choosing 4 places out of 10 for the 4 S's. That is, the number of simple events in question is

$$\binom{10}{4},$$

and we conclude that the probability of exactly 4 successes in 10 trials is

$$\binom{10}{4}p^4q^6.$$

The notation

$$b(k; n, p)$$

represents *the probability of exactly k successes in n trials of a binomial experiment in which the probability of success on any single trial is p.* By carrying out reasoning similar to that used in the above examples, we can show that the following formula is correct for binomial probabilities.

FORMULA FOR BINOMIAL PROBABILITIES

$$b(k; n, p) = \binom{n}{k}p^kq^{n-k}.$$

It may be of interest to those readers familiar with the Binomial Theorem that the probabilities $b(k; n, p)$ are the terms in the expansion of the expression

$$(p + q)^n.$$

Binomial experiment

EXAMPLE 1 Suppose a coin will come up heads with probability 0.8 on any toss. If this coin is tossed 10 times, what is the probability that exactly 7 heads will be obtained?

Solution: Interpret heads as "success." Then

$$P(S) = p = 0.8 \qquad \text{and} \qquad P(F) = q = 1 - p = 0.2.$$

Also $n = 10$ and $k = 7$. Using the Formula for Binomial Probabilities, we obtain

$$\begin{aligned} b(7;\, 10,\, 0.8) &= \binom{10}{7}(0.8)^7(0.2)^3 \\ &= \frac{10!}{7!3!}(0.8)^7(0.2)^3 \\ &= \frac{8 \cdot 9 \cdot 10}{1 \cdot 2 \cdot 3}(0.2097)(0.0080) \\ &= 0.2013. \end{aligned}$$

PRACTICE PROBLEM In Example 1, compute the probability of getting exactly 6 heads in 10 tosses.

Answer: 0.0881. ■

Binomial experiment

EXAMPLE 2 A fair die is rolled 5 times. What is the probability that on exactly 2 of the 5 rolls the number obtained on the face of the die is a multiple of 3?

Solution: Success may be interpreted here as getting a multiple of 3 (i.e. a 3 or 6) and failure as not getting a multiple of 3. Since the probability of getting a 3 or 6 on one roll of a fair die is $\frac{1}{3}$, we see that $p = \frac{1}{3}$ and $q = \frac{2}{3}$. Then, with $n = 5$ and $k = 2$,

$$\begin{aligned} b\left(2;\, 5,\, \frac{1}{3}\right) &= \binom{5}{2}\left(\frac{1}{3}\right)^2\left(\frac{2}{3}\right)^3 \\ &= \frac{5!}{2!3!}\left(\frac{1}{3}\right)^2\left(\frac{2}{3}\right)^3 \\ &= 0.3292. \end{aligned}$$

Therefore, we see that if a fair die is rolled 5 times, the probability of obtaining a 3 or a 6 twice is less than $\frac{1}{3}$.

PRACTICE PROBLEM In Example 2, compute the probability of getting a multiple of 3 exactly three times on 5 rolls.

Answer: 0.1646. ■

Binomial experiment

EXAMPLE 3 A fair coin is tossed 5 times. What is the probability that no more than 2 heads are obtained?

Solution: This question may be viewed as a question about 2 or fewer successes, that is, about 2, 1 or 0 successes. In other words, we wish to know the probability of having 0 or 1 or 2 successes in 5 trials. Therefore, the event in question is made up of 3 simple events which are exclusive; for example, we cannot have exactly 1 success and also exactly 2 successes. Hence, the probability of obtaining no more than 2 heads in 5 tosses is (by the Addition Formula)

$$\begin{aligned} b(0; 5, 0.5) + b(1; 5, 0.5) + b(2; 5, 0.5) \\ &= \binom{5}{0}(0.5)^0(0.5)^5 + \binom{5}{1}(0.5)^1(0.5)^4 + \binom{5}{2}(0.5)^2(0.5)^3 \\ &= 0.03125 + 0.15625 + 0.3125 \\ &= 0.5. \end{aligned}$$

PRACTICE PROBLEM In Example 3 compute the probability of obtaining at least 2 heads in 5 tosses.

Answer: 0.8125. ■

Table 5 of Appendix A contains a listing of some useful binomial probabilities. These probabilities are not exact but have been given to three places only. For example, the problem of Example 3 may be solved by using Table 5 with $n = 5$, $p = 0.5$ and $k = 0, 1, 2$:

$$0.031 + 0.156 + 0.312 = 0.499.$$

Note that the result obtained from Table 5 is in error by 0.001 (0.500 − 0.499). The error results, of course, from the fact that the entries in the table have been rounded to three places.

Binomial experiment

EXAMPLE 4 If Dale Murphy of the Atlanta Braves has a batting average of 0.300, what is the probability that he will get at least 3 hits in his next 7 times at bat?

Solution: In a problem of this type, we take 0.300 to be the probability that Murphy will get a hit at any time at bat. Then this may be considered a binomial experiment in which

$$S = \text{getting a hit},$$

$$F = \text{not getting a hit},$$

$$p = 0.300,$$

$$n = 7.$$

Since the question is about at least 3 hits, we wish to know about at least 3 successes in 7 trials. Using Table 5 with $n = 7$ and $k = 3, 4, 5, 6, 7$, and $p = 0.3$, we have

$$0.227 + 0.097 + 0.025 + 0.004 + 0.000 = 0.353.$$

PRACTICE PROBLEM Referring to Example 4, compute the probability that Murphy does not get more than 3 hits in his next 7 times at bat.

Answer: 0.8750. ■

Medical research

EXAMPLE 5 Several medical research workers are studying the common cold and part of their research involves effective procedures for infecting healthy individuals with a cold virus. The researchers have determined that Procedure X is effective in 40% of the cases. If 25 subjects are chosen at random and Procedure X is administered to each, what is the probability that at least 10 of the 25 will come down with a cold?

Solution: Let S = subject is infected with cold virus after Procedure X is followed. Then

$$p = P(S) = 0.40,$$

and

$$n = 25.$$

We need to determine the probability of 10 or more successes and for this Table 5 of Appendix A may be used. We find

k	10	11	12	13	14	15	16	17	18
Probability	0.161	0.147	0.114	0.076	0.043	0.021	0.009	0.003	0.001

Table 5 shows that all probabilities $b(k; 25, 0.4)$ for $k > 18$ are 0.000. Therefore, we total the probabilities in the above table and find that 0.575 is the probability we are seeking.

PRACTICE PROBLEM In Example 5, determine the probability that fewer than 6 subjects out of 25 will be infected after being treated with Procedure X.

Answer: 0.029. ■

EXERCISES

Set A–Calculating Binomial Probabilities by Using the Formula

Some of Exercises 1 through 12 will require a calculator. Use the formula; do *not* use Table 5.

1. $b(1; 2, 0.5)$ **2.** $b(2; 3, 0.5)$ **3.** $b(1; 4, 0.4)$
4. $b(2; 4, 0.4)$ **5.** $b(3; 4, 0.4)$ **6.** $b(1; 5, 0.2)$
7. $b(0; 5, 0.2)$ **8.** $b(5; 5, 0.2)$ **9.** $b(5; 10, 0.6)$
10. $b(8; 12, 0.6)$ **11.** $b(10; 12, 0.7)$ **12.** $b(12; 15, 0.45)$

Set B–Using Table 5 of Appendix A to Determine Binomial Probabilities

Use Table 5 of Appendix A to determine the probabilities given in Exercises 13 through 24.

13. $b(3; 5, 0.2)$ **14.** $b(2; 5, 0.2)$ **15.** $b(4; 6, 0.5)$
16. $b(3; 6, 0.5)$ **17.** $b(9; 11, 0.4)$ **18.** $b(8; 11, 0.4)$
19. $b(10; 15, 0.3)$ **20.** $b(15; 18, 0.8)$ **21.** $b(17; 20, 0.6)$
22. $b(18; 20, 0.5)$ **23.** $b(18; 22, 0.4)$ **24.** $b(20; 25, 0.5)$

Set C–Some Word Problems About Binomial Experiments

You may use Table 5 of Appendix A for Exercises 25 through 35.

25. What is the probability of exactly 11 successes in 20 trials of a binomial experiment when the probability of success on any single trial is 0.3?

26. What is the probability of exactly 5 or 6 successes in 10 trials of a binomial experiment when the probability of success on any single trial is 0.2?

27. What is the probability of having at least 5 successes in 10 trials of a binomial experiment if $p = 0.3$?

28. What is the probability of at least 12 successes in 15 trials of a binomial experiment if $p = 0.5$?

29. What is the probability of at least 8 successes in 11 trials of a binomial experiment if the probability of success on each trial is 0.2?

30. What is the probability of no more than 5 successes in 12 trials of a binomial experiment if the probability of success on each trial is 0.2?

31. What is the probability of no more than 6 successes in 13 trials of a binomial experiment if the probability of success on each trial is 0.3?

32. What is the probability of at most 12 failures in 15 trials of a binomial experiment if $p = 0.5$?

33. What is the probability of at most 15 failures in 20 trials of a binomial experiment if $p = 0.6$?

34. What is the probability of at least one success and at least one failure in 8 trials of a binomial experiment if $p = 0.5$?

35. What is the probability of at least 3 successes and at least 3 failures in 12 trials of a binomial experiment if $p = 0.5$?

Set C–Models of Binomial Experiments

Construct a binomial model for Exercises 36 through 42 and solve.

True-false test

36. A student takes a true-false test of 10 questions on a subject he knows nothing about. To choose his answers he flips a coin and marks "true" as his answer if heads turns up and otherwise marks "false." If, in order to pass, he must get at least 6 correct answers, what is the probability that he will receive a passing grade?

Quality control

37. The Apex Company receives mirrors in 100-lot cartons. Apex has determined that about 1% of the mirrors received are broken. If 25 mirrors are chosen at random from a carton, what is the probability that at least 1 mirror is broken?

Human births

38. Assume that in human families the probability of a male

birth is 0.5. What is the probability that in a family of 4 children there will be
(a) all males? (b) 2 males and 2 females? (c) 3 females, 1 male?

Sales contacts

39. Mike Starr is a salesman for an auto company. He estimates that he sells cars to about 10% of his contacts. What is the probability that Mike will make at least 2 sales in his next 10 contacts?

Effects of coffee

40. Nancy Neumark claims that coffee keeps her awake at night. In fact, she estimates that coffee is 90% effective in keeping her awake. If she drinks coffee for the next 12 nights, what is the probability she will get at least one good night's sleep during this period?

Tennis

41. Suppose that in a tennis match Bitsy Grant is getting in 70% of his first serves. What is the probability that of his next 15 serves at least 10 first serves go in?

Use of aspirin

42. A pharmaceutical company claims that its brand of aspirin for children is preferred 3 to 2 by those physicians who recommend aspirin for children. If this claim is correct, what is the probability that in a random sample of 20 physicians who recommend aspirin for children, at least 15 will prefer this company's aspirin?

7.6 BAYES' THEOREM (OPTIONAL)

This section contains information about alternative methods for computing conditional probabilities. In one case, the technique involves redrawing a tree diagram, and the other uses a method known as Bayes' theorem. Tree diagrams will be studied first.

Consider the following problem. We are given 3 boxes containing black and white marbles, and these are distributed among the boxes as shown in the table below:

NUMBER OF MARBLES

Box	White	Black	Totals
I	4	1	5
II	3	2	5
III	2	6	8
Totals	9	9	18

One of the boxes is chosen at random and a single marble is then drawn at random from the box. If the marble drawn is black, what is the probability that box III was chosen? A tree diagram for this experiment is shown in Figure 7.12. The probability which we wish to determine is the conditional probability $P(\text{III} \mid B)$. This may be determined by reversing the order on the tree diagram of Figure 7.12. This produces a new diagram as shown in Figure 7.13. Here the color of the marbles comes first and the boxes second. In order to compute $P(\text{III} \mid B)$ we only have to determine the probabilities on the branches of the new diagram. We know that the probabilities at the ends of the branches are the same as in the original diagram since these are the probabilities of the simple events and, for example, $P(\text{I and } B)$ is the same as $P(B \text{ and I})$. Next we need the probability on the branch to B and also to W. But these can be calculated from the diagram in Figure 7.12. Thus,

$$P(B) = \frac{1}{15} + \frac{2}{15} + \frac{1}{4}$$
$$= \frac{9}{20}.$$

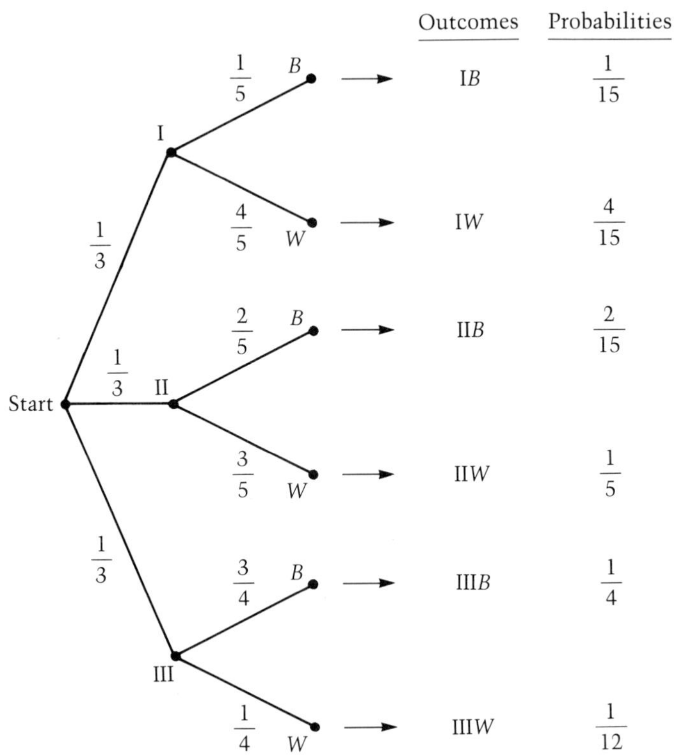

FIG. 7.12.

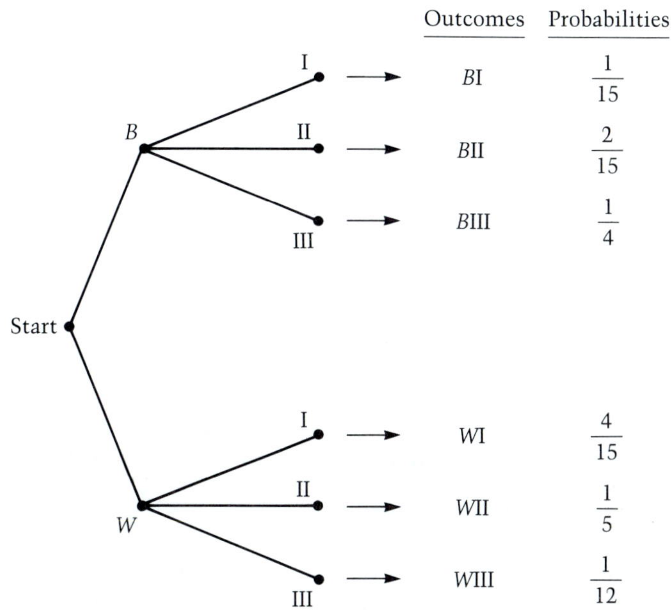

FIG. 7.13.

From this we may conclude that $P(W) = \frac{11}{20}$. Inserting these probabilities into the diagram of Figure 7.13 we have the diagram of Figure 7.14. This diagram is complete except for the probabilities on the second set of branches. These probabilities are now deter-

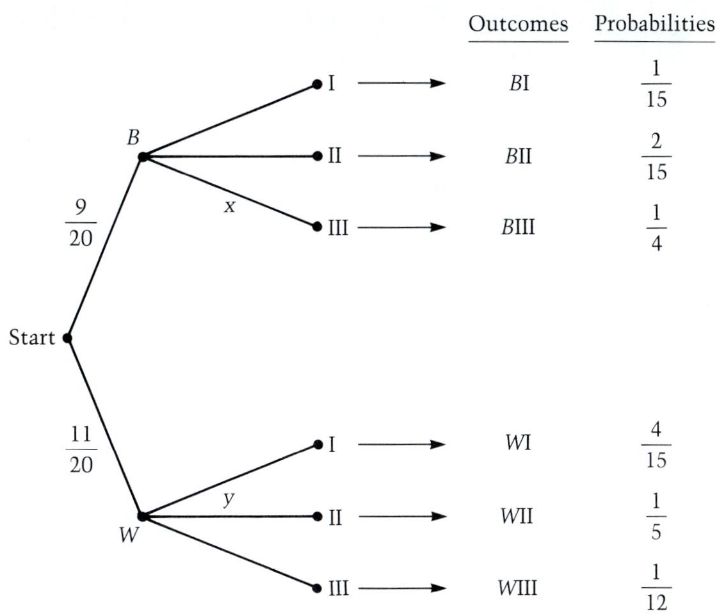

FIG. 7.14.

mined from the fact that the probabilities of simple events are found by multiplying along the branches. Therefore, if x is the probability on the branch from B to III as shown in Figure 7.14, we have

$$\left(\frac{9}{20}\right)x = \frac{1}{4}.$$

Solving this equation for x gives

$$x = \left(\frac{1}{4}\right)\left(\frac{20}{9}\right) = \frac{5}{9}.$$

But this is the probability we were looking for, that is,

$$P(\text{III} \mid B) = \frac{5}{9}.$$

Conditional probability

EXAMPLE 1 Using the information given in Figure 7.12, compute the probability for the branch from W to II in Figure 7.14, that is $P(\text{II} \mid W)$.

Solution: Let y be the probability which we are to compute. Since we already know that $P(W) = \frac{11}{20}$ and that the probability at the end of this branch is $\frac{1}{5}$, we have

$$\left(\frac{11}{20}\right)y = \frac{1}{5}.$$

Therefore,

$$y = \left(\frac{1}{5}\right)\left(\frac{20}{11}\right) = \frac{4}{11}.$$

PRACTICE PROBLEM Use Figure 7.12 to compute $P(\text{I} \mid B)$; that is, compute the probability in Figure 7.14 on the branch from B to I.

Answer: $\frac{4}{27}$. ■

The experiment illustrated in Figure 7.12 requires that one of the three boxes I, II, III be chosen at random and that next a marble be drawn. Hence, if a black marble is drawn, it must have come from I, II, or III. This implies that the event

$$B = \text{drawing a black marble}$$

is the same as the event

$$(\text{I and } B) \quad \text{or} \quad (\text{II and } B) \quad \text{or} \quad (\text{III and } B).$$

Therefore,

$$P(B) = P(\text{I and } B) + P(\text{II and } B) + P(\text{III and } B). \tag{1}$$

From the definition of conditional probability, we have

$$P(B \mid \text{I}) = \frac{P(B \text{ and I})}{P(\text{I})}$$

or

$$P(B \text{ and I}) = P(\text{I})P(B \mid \text{I}).$$

But since $P(B \text{ and I}) = P(\text{I and } B)$, it follows that

$$P(\text{I and } B) = P(\text{I})P(B \mid \text{I}). \tag{2}$$

By the same argument, we have

$$P(\text{II and } B) = P(\text{II})P(B \mid \text{II}), \tag{3}$$

$$P(\text{III and } B) = P(\text{III})P(B \mid \text{III}). \tag{4}$$

Substituting Eqs. (2), (3), and (4) into (1) gives

$$P(B) = P(\text{I})P(B \mid \text{I}) + P(\text{II})P(B \mid \text{II}) + P(\text{III})P(B \mid \text{III}). \tag{5}$$

If we wish to calculate $P(\text{III} \mid B)$, we may begin with the definition of conditional probability:

$$P(\text{III} \mid B) = \frac{P(\text{III and } B)}{P(B)}. \tag{6}$$

Now substitute from (4) and (5) into (6):

$$P(\text{III} \mid B) = \frac{P(\text{III})P(B \mid \text{III})}{P(\text{I})P(B \mid \text{I}) + P(\text{II})P(B \mid \text{II}) + P(\text{III})P(B \mid \text{III})}. \tag{7}$$

The probabilities needed in (7) may be read from Figure 7.12:

$$
\begin{aligned}
P(\text{III} \mid B) &= \frac{\left(\frac{1}{3}\right)\left(\frac{3}{4}\right)}{\left(\frac{1}{3}\right)\left(\frac{1}{5}\right) + \left(\frac{1}{3}\right)\left(\frac{2}{5}\right) + \left(\frac{1}{3}\right)\left(\frac{3}{4}\right)} \\
&= \frac{\frac{3}{4}}{\frac{1}{5} + \frac{2}{5} + \frac{3}{4}} \\
&= \frac{15}{4 + 8 + 15} \\
&= \frac{15}{27} \\
&= \frac{5}{9}.
\end{aligned}
$$

Equation (7) is useful because the probabilities on the right-hand side are easily computed. Equation (7) is an instance of Bayes' theorem. The key to the result in this example is the set of events I, II, III. If this experiment is carried out, one and only one of the events I, II, III will occur. Such events as I, II, III are called a *complete set of alternatives.* Conditional probabilities may be computed by Bayes' theorem in the presence of a complete set of alternatives. The statement of Bayes' theorem given below is for four alternatives only, but the result is true for any number so long as the alternatives form a complete set.

Bayes' theorem

STATEMENT OF BAYES' THEOREM Let E be an experiment and let X be an event associated with E. If A, B, C, D are a complete set of alternatives for E, then

$$P(A \mid X) = \frac{P(A)P(X \mid A)}{P(A)P(X \mid A) + P(B)P(X \mid B) + P(C)P(X \mid C) + P(D)P(X \mid D)}. \quad (8)$$

This statement of Bayes' theorem is incomplete. In addition to Eq. (8), there are three other equations which involve $P(B \mid X)$, $P(C \mid X)$ and $P(D \mid X)$. To obtain the equation for $P(B \mid X)$, for example, make the following changes in (8): Replace $P(A \mid X)$ by $P(B \mid X)$

on the left and replace $P(A)P(X \mid A)$ by $P(B)P(X \mid B)$ in the numerator on the right.

Quality control

EXAMPLE 2 In manufacturing metal office equipment, the *XYZ* Company uses nuts supplied by companies *A, B, C*. Company *A* supplies 40% of the nuts used by *XYZ* while *B* and *C* supply 35% and 25%, respectively. It is known that 0.5% of the nuts from *A* are defective; the corresponding figures from *B* and *C* are 1% and 1.2%. If a nut is selected at random and found to be defective, which of the companies is most likely to have supplied it?

Solution: Let the events *A, B, C, X* be defined as follows:

A—the nut was supplied by *A*,

B—the nut was supplied by *B*,

C—the nut was supplied by *C*,

X—the nut was defective.

The problem may be solved by computing the following conditional probabilities:

$$P(A \mid X), \quad P(B \mid X), \quad P(C \mid X).$$

The events *A, B, C* form a complete set of alternatives and Bayes' theorem is applicable:

$$\begin{aligned} P(A \mid X) &= \frac{P(A)P(X \mid A)}{P(A)P(X \mid A) + P(B)P(X \mid B) + P(C)P(X \mid C)} \\ &= \frac{(0.4)(0.005)}{(0.4)(0.005) + (0.35)(0.01) + (0.25)(0.012)} \\ &= 0.2353 \quad \text{(rounded to four places).} \end{aligned}$$

A similar calculation shows that $P(B \mid X) = 0.4118$. The solution will be complete when $P(C \mid X)$ has been computed.

PRACTICE PROBLEM Finish Example 2 by computing $P(C \mid X)$.

Answer: $P(C \mid X) = 0.3529$. Company *B* is most likely to have supplied the defective nut. ■

Medical test

EXAMPLE 3 A team of medical researchers has produced a test for a disease found in rats. In attempting to determine the effectiveness of the test, rats are classified into groups I, II, III as follows: group I contains rats that have the disease, group II contains only rats that do not have the disease but are ill from other causes, group III consists of healthy animals. The researchers have determined that their test is 99% accurate for group I; that is, when rats from group I are tested, the test indicates that the rats have the disease in 99% of the cases. For group II, the test indicates the presence of the disease in 8% of the animals (in this case the test is in error), and for group III, the test indicates the disease in 3% of the cases (also in error). In a particular group of 100 rats it has been established that 5 belong to group I, 30 belong to group II, and 65 are in group III. If a rat is chosen at random from this group of 100, and the test indicates that the disease is present, what is the probability that the animal actually has the disease?

Solution: Define events A, B, C, X as follows:

A—the rat chosen is from group I,

B—the rat chosen is from group II,

C—the rat chosen is from group III,

X—the test indicates the rat has the disease.

The probability we seek in this case is $P(A \mid X)$ and again we may use Bayes' theorem:

$$\begin{aligned} P(A \mid X) &= \frac{P(A)P(X \mid A)}{P(A)P(X \mid A) + P(B)P(X \mid B) + P(C)P(X \mid C)} \\ &= \frac{(0.05)(0.99)}{(0.05)(0.99) + (0.30)(0.08) + (0.65)(0.03)} \\ &= 0.5323. \end{aligned}$$

This result indicates that the test is not as reliable as might be desired.

PRACTICE PROBLEM In Example 3 compute the probability that if the test indicates that the disease is present, the rat is from group II.

Answer: 0.2581. ■

EXERCISES

Set A–Tree Diagrams

In Exercises 1 through 4, supply the missing probabilities on the tree diagrams given and compute the requested probabilities.

1. (a) $P(B \mid \text{I})$ (b) $P(A \mid \text{II})$ (c) $P(A)$
   (d) $P(B)$

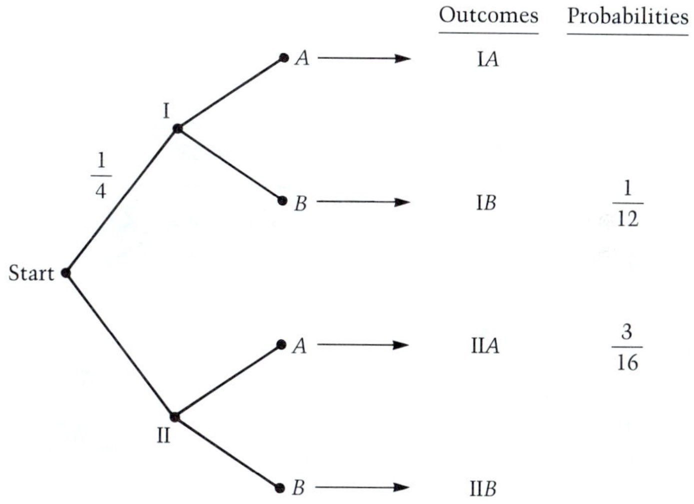

Tree Diagram for Exercise 1.

2. (a) $P(A \mid \text{I})$ (b) $P(A \mid \text{II})$ (c) $P(A \mid \text{III})$
   (d) $P(A)$ (e) $P(B)$

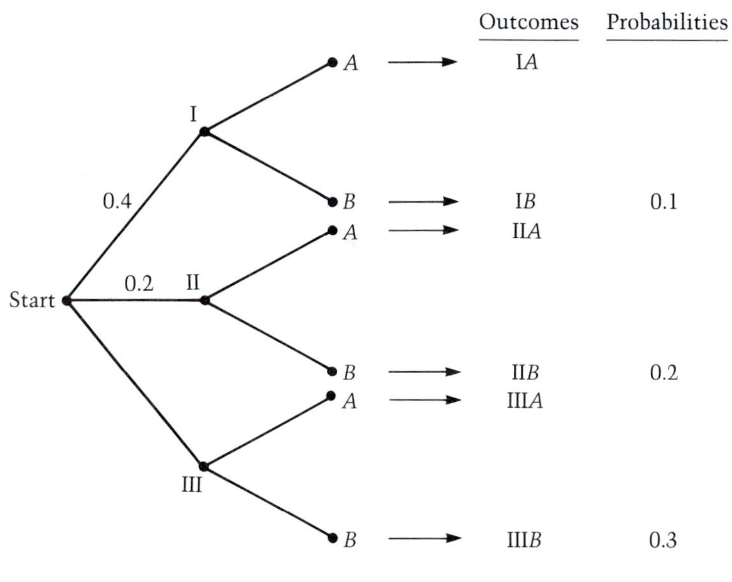

Tree Diagram for Exercise 2.

3. (a) $P(A \mid \text{I})$ (b) $P(A \mid \text{II})$ (c) $P(A \mid \text{III})$
(d) $P(A)$ (e) $P(B)$ (f) $P(C)$

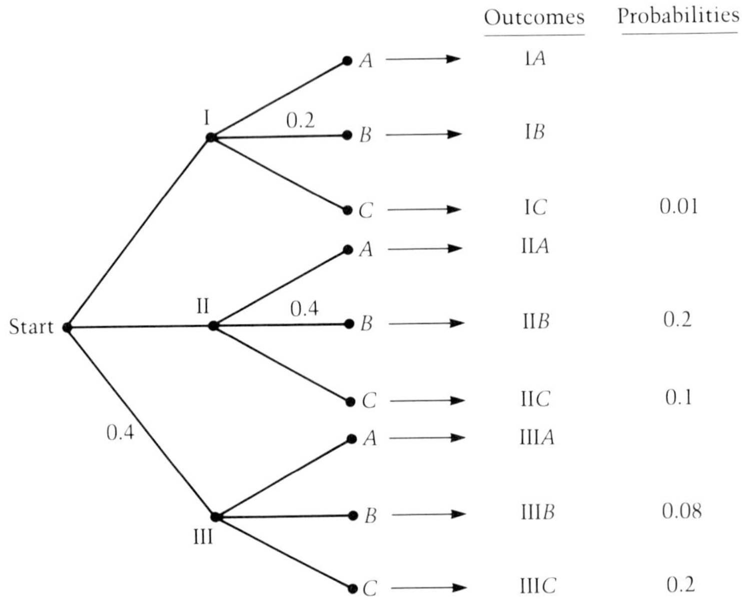

Tree Diagram for Exercise 3.

4. (a) $P(B \mid \text{I})$ (b) $P(A \mid \text{II})$ (c) $P(\text{III and } B)$
(d) $P(B \mid \text{IV})$ (e) $P(A)$ (f) $P(B)$

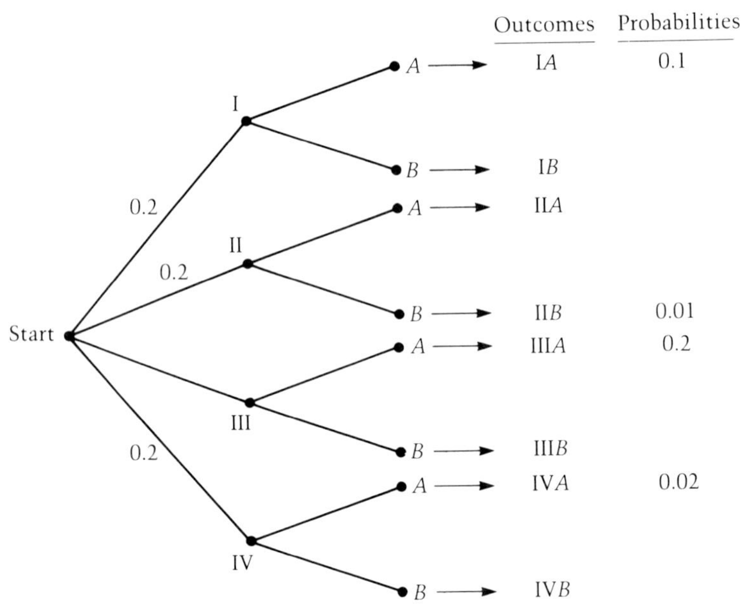

Tree Diagram for Exercise 4.

Set B–Tree Diagrams to Compute Conditional Probabilities

5. Draw a tree diagram which is in reverse order of that given in Exercise 1 and compute the following.
(a) $P(\text{I}\,|\,A)$ (b) $P(\text{II}\,|\,A)$ (c) $P(\text{I}\,|\,B)$ (d) $P(\text{II}\,|\,B)$

6. Draw a tree diagram which is in reverse order of that given in Exercise 2 and compute the following.
(a) $P(\text{I}\,|\,A)$ (b) $P(\text{I}\,|\,B)$ (c) $P(\text{II}\,|\,A)$
(d) $P(\text{II}\,|\,B)$ (e) $P(\text{III}\,|\,A)$ (f) $P(\text{III}\,|\,B)$

7. Draw a tree diagram which is in reverse order of that given in Exercise 3 and compute the following.
(a) $P(\text{I}\,|\,A)$ (b) $P(\text{I}\,|\,B)$ (c) $P(\text{I}\,|\,C)$
(d) $P(\text{II}\,|\,A)$ (e) $P(\text{II}\,|\,B)$ (f) $P(\text{II}\,|\,C)$
(g) $P(\text{III}\,|\,A)$ (h) $P(\text{III}\,|\,B)$ (i) $P(\text{III}\,|\,C)$

8. Draw a tree diagram which is in reverse order of that given in Exercise 4 and compute the following.
(a) $P(\text{I}\,|\,A)$ (b) $P(\text{I}\,|\,B)$ (c) $P(\text{II}\,|\,A)$
(d) $P(\text{II}\,|\,B)$ (e) $P(\text{III}\,|\,A)$ (f) $P(\text{III}\,|\,B)$
(g) $P(\text{IV}\,|\,A)$ (h) $P(\text{IV}\,|\,B)$

Set C–Tree Diagrams and Conditional Probabilities

Draw tree diagrams to answer the questions of each of Exercises 9 through 14.

9. Two boxes each contain 10 marbles. Box I contains 3 white and 7 black marbles and box II contains 3 black and 7 white marbles. A box is chosen at random and then a marble is drawn from that box. If the marble drawn is white, what is the probability that it came from box I?

10. Two boxes each contain 10 marbles. Box I contains 3 white and 7 black marbles and box II contains 1 black and 9 white marbles. A box is chosen at random and a marble is drawn at random from that box. If the marble drawn is black, what is the probability that it came from box II?

11. Two boxes each contain 10 black and white marbles which are distributed as in Exercise 9. A biased coin with $P(H) = 0.6$ is tossed one time. If the coin turns up heads, we choose one of the boxes at random and draw a marble from it. If tails turns up, we place one of the black marbles of box I in box II and then, after choosing one of the boxes at random, we draw a marble.
(a) If the marble drawn is black, what is the probability that it came from box I?
(b) If the marble drawn is white, what is the probability that it came from box II?

12. Box I contains 2 white and 4 black marbles and box II contains 3 white and 3 black marbles. A biased coin with $P(H) = 0.6$ is tossed and if heads turns up, box I is selected and one marble is drawn. If tails is obtained on the coin, box II is selected and one marble is drawn.
   (a) Show that if the marble drawn is white, it is just as likely to have come from box I as from box II.
   (b) Show that if the marble drawn is black, it is more likely to have come from box I.

13. Three boxes A, B, and C contain black and white marbles as shown:

A		B		C	
White	Black	White	Black	White	Black
2	0	1	1	0	2

   A box is chosen at random and from the box a marble is selected at random. If the marble drawn is black, what is the probability that it was drawn from box
   (a) A? (b) B? (c) C?

14. A box contains 3 white and 5 black marbles. One marble is selected randomly from the box, and this drawing is observed by Richard and Jerry. Richard reports that the marble drawn is white and Jerry reports that it is black. If Richard tells the truth 3 times out of 4 and Jerry 4 out of 5, what is the probability that the marble drawn is
   (a) white? (b) black?

Set D–Word Problems to be Solved by Bayes' Theorem

15. (a) Given 2 coins, 1 fair and 1 biased with $P(H) = 0.9$. One of the coins is chosen at random and tossed 1 time. If heads turns up, what is the probability that the biased coin was chosen?
   (b) Given 3 coins, 2 fair and 1 biased with $P(H) = 0.9$. One of the coins is chosen at random and tossed 1 time. If heads turns up, what is the probability that the biased coin was chosen?
   (c) Given 4 coins, 3 fair and 1 biased with $P(H) = 0.9$. One of the coins is chosen at random and tossed 1 time. If heads turns up, what is the probability that the biased coin was chosen?
   (d) Given 20 coins, 19 fair and 1 biased with $P(H) = 0.9$. One

of the coins is chosen at random and tossed 1 time. If heads turns up, what is the probability that the biased coin was chosen?

16. (a) Given 2 coins, 1 fair and 1 biased with $P(H) = 0.9$. One of the coins is chosen at random and flipped twice. If heads appears both times, what is the probability that the biased coin was chosen?
(b) Given 3 coins, 2 fair and 1 biased with $P(H) = 0.9$. One of the coins is chosen at random and tossed twice. If heads appears both times, what is the probability that the biased coin was chosen?
(c) Given 4 coins, 3 fair and 1 biased with $P(H) = 0.9$. One of the coins is chosen at random and tossed twice. If heads appears both times, what is the probability that the biased coin was the one chosen?
(d) Given 20 coins, 19 fair and 1 biased with $P(H) = 0.9$. One of the coins is chosen at random and tossed twice. If heads appears both times, what is the probability that the biased coin was the one chosen?

17. Out of 20 coins, 19 are fair and 1 is biased with $P(H) = 0.9$. If one of the coins is chosen at random, tossed 5 times and H turns up each time, what is the probability that the biased coin was not chosen?

18. Suppose that the probability that a male college freshman will choose mathematics as a major is $\frac{1}{8}$ and that a female freshman chooses mathematics as a major with probability $\frac{1}{12}$. Assume that the freshman class of DeKalb College is 60% female. If a student selected at random from the freshman class is a mathematics major, what is the probability that the student is female?

Quality control

19. The Apex Tool Company which manufactures auto radiator caps owns 3 machines which turn out all the company's production of safety caps. Machine I turns out 40% of this item, and machines II and III turn out 35% and 25%, respectively. When in good running order, machine I turns out 1% of defective caps and machines II and III produce 0.9% and 0.8% of defective caps. At the end of one day's production, a cap is chosen at random and found to be defective. What is the probability that the cap was produced by
(a) machine I? (b) machine II?
(c) machine III?

Failure rates

20. Instructors A, B, C, D teach all the sections of the introductory statistics course at Middling State College. The table shows the proportion of students taught and the failure rate of each instructor. If a student is picked at random and found to have failed statistics, what is the probability that he/she was in a class taught by C?

Instructor	Proportion	Failure Rate
A	24%	49%
B	54%	12%
C	17%	36%
D	5%	58%

Quality control

21. An electronics manufacturing firm produces printed circuits on 5 different machines labeled I through V. The table shows the proportion of circuits produced by each machine and the proportion of defective items turned out by each. If a printed circuit is chosen at random and found to be defective, what is the probability that it was produced by machine V?

Machine	Proportion	Rate of Defectives
I	22%	2%
II	24%	6%
III	13%	3%
IV	34%	5%
V	7%	7%

Quality control

22. Soviet scientists have developed a test for determining whether a citizen has capitalist tendencies. In order to evaluate the effectiveness of the test a group of 1000 Soviet citizens are classified as follows:
Group A–400 pure Marxists with no capitalist tendencies;
Group B–400 suffering from revisionist ideas but showing no capitalist tendencies;
Group C–200 have definite capitalist tendencies.
This classification is certified correct by the KGB. It was shown that the test is 98% effective in group C but indicates (incorrectly) capitalist tendencies in 3% of groups A and B. If a citizen is chosen at random and on the basis of the test is sent to Siberia, what is the probability he/she was a pure Marxist?

7.7 MARKOV CHAINS (OPTIONAL)

Markov chains

We have previously studied experiments which consisted of independent trials. Now we are going to consider experiments in which probabilities for outcomes on a particular trial are dependent on the outcome of the immediately preceding trial. Such experiments are known as *Markov chains.*

As an example of a Markov chain consider the following simple experiment. Two boxes each contain 2 marbles: box A contains 2 white marbles and box B contains 2 black marbles. A trial will be carried out as follows:

Trial

First: Draw a marble from each box at random.
Second: Place the marble from A into B and the marble from B into A.

state

As a result of each trial, box A will contain a number of black marbles, and this number can be only 0, 1, 2 since there are only 2 black marbles in all. The composition of the two boxes at the end of each trial is called the *state* of the system. Since we can determine the composition of the two boxes just by knowing the number of black marbles in box A, we can see that the system can be in three possible states: 0, 1, 2. These three states are shown in the table.

TABLE

State	A	B	Color	Box A	Box B
#1	*ww*	*bb*	White	2	0
(0 black in box A)			Black	0	2
#2	*wb*	*bw*	White	1	1
(1 black in box A)			Black	1	1
#3	*bb*	*ww*	White	0	2
(2 black in box A)			Black	2	0

Now suppose the system is in state #1 (there are no black marbles in box A). If one trial is carried out, the table shows that a white marble will be drawn from A and a black one from B. After the marbles are switched in the boxes, the system will be in state #2. This shows that if the system is in state #1 and one trial is performed, it will go into state #2. From state #1 the system cannot go into state #1 or #3. We therefore have the following conditional probabilities:

$$P(\#1 \mid \#1) = 0, \qquad P(\#2 \mid \#1) = 1, \qquad P(\#3 \mid \#1) = 0.$$

That is, for example, $P(\#1 \mid \#1) = 0$ means that the probability is 0 that the system will go to state #1, given that it begins in state #1. It is also easy to see that if the system is in state #3, it must go into state #2, and this produces these probabilities:

$$P(\#1 \mid \#3) = 0, \qquad P(\#2 \mid \#3) = 1, \qquad P(\#3 \mid \#3) = 0.$$

What happens if the system is in state #2? A draw from each box will result in one of the following possibilities:

(a) white from A and white from B,

(b) white from A and black from B,

(c) black from A and white from B,

(d) black from A and black from B.

If the result is (a), the system will remain in state #2 since the 2 white marbles will simply be interchanged. Similarly, if the result is (d), the system remains in state #2. If the result is (b), a white marble will be added to B and a black marble will be added to A; this will put the system in state #3. Finally, if the result is (c), the system will pass to state #1. These possibilities have been summarized in Figure 7.15. This tree diagram indicates that if the system is in state #2, it can pass into any one of the states #1, #2, #3. The symbols p_{21}, p_{22}, p_{23} represent the appropriate probabilities. For example, p_{21} is the probability that the system moves from state #2 to state #1. This is a short way of expressing the conditional probability $P(\#1 \mid \#2)$. Note the change in the order of

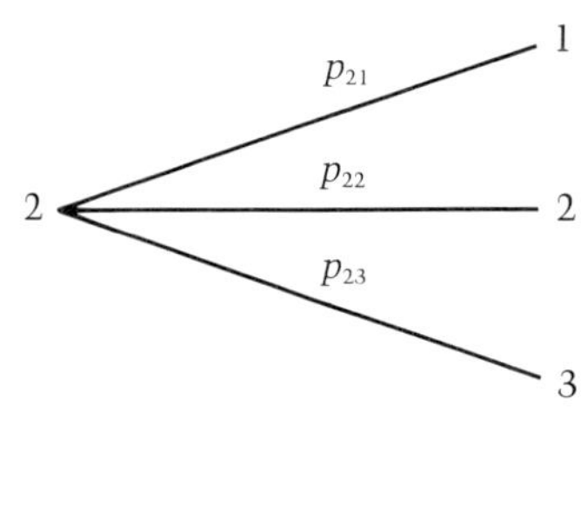

FIG. 7.15.

#1 and #2 in these two notations. Thus, for example,

$$P(\#1 \mid \#2) = p_{21}.$$

The probabilities p_{21}, p_{22}, p_{23} will now be determined. Figure 7.16 illustrates the fact that p_{21} is the probability that the system moves from state #2 to #1. This transition can only be made as a result of drawing black from A and white from B. Therefore, p_{21} is the probability that black is drawn from A and white is drawn from B. Hence, we see that

$$p_{21} = (0.5)(0.5) = 0.25.$$

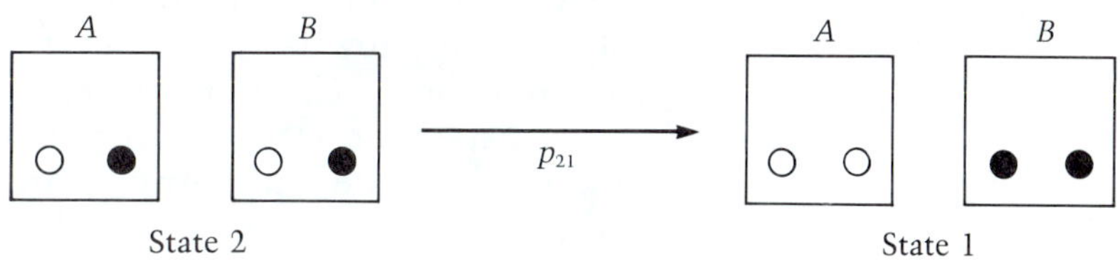

FIG. 7.16.

Now consider p_{22}, the probability that the system passes to state #2, given that it is in state #2. (See Figure 7.17.) This transition can be made in two ways: either by interchanging the whites or by interchanging the blacks. Thus, in this case,

$$p_{22} = 0.25 + 0.25 = 0.5.$$

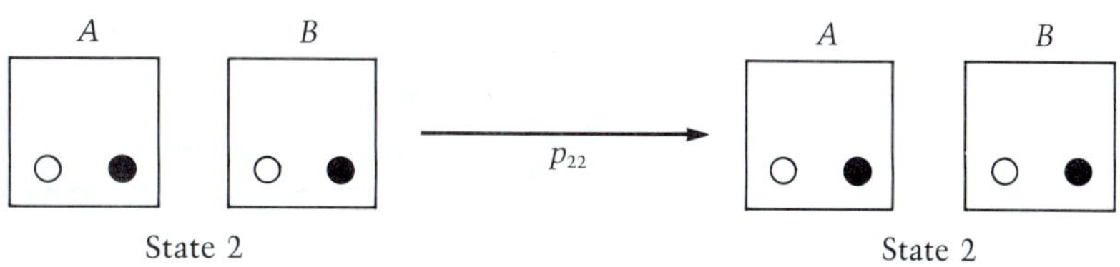

FIG. 7.17.

PRACTICE PROBLEM For this example compute p_{23}.

Answer: 0.25. ■

The probabilities p_{21}, p_{22}, p_{23} are examples of *transition probabilities.* For this example, there are 9 transition probabilities: a transition probability for passing from any one of the 3 states to any of the 3 states. It is convenient to express transition probabilities in a *transition matrix:*

transition matrix

$$T = \begin{bmatrix} p_{11} & p_{12} & p_{13} \\ p_{21} & p_{22} & p_{23} \\ p_{31} & p_{32} & p_{33} \end{bmatrix}.$$

In the present example, the transition matrix is

$$T = \begin{bmatrix} 0 & 1 & 0 \\ 0.25 & 0.50 & 0.25 \\ 0 & 1 & 0 \end{bmatrix}.$$

Transition matrices have the property that the sum of the entries of any row is 1; as an example, see the matrix T of the present problem. This property reflects the fact that if the system is in any state and a trial is carried out, then the system must pass to *some* one of the states. Any square matrix having this property is the transition matrix of some Markov chain.

If we are given the transition matrix T for a Markov chain, we can determine the probability that the system moves from any state to any state in one trial. But what about the probabilities after two trials or more? It turns out that this information is conveniently contained in the powers of the matrix T. For instance T^2 contains the probabilities needed for two trials. For the example we have been considering, T^2 is given as follows:

$$\begin{aligned} T^2 &= \begin{bmatrix} 0 & 1 & 0 \\ 0.25 & 0.50 & 0.25 \\ 0 & 1 & 0 \end{bmatrix}\begin{bmatrix} 0 & 1 & 0 \\ 0.25 & 0.50 & 0.25 \\ 0 & 1 & 0 \end{bmatrix} \\ &= \begin{bmatrix} 0.25 & 0.50 & 0.25 \\ 0.125 & 0.750 & 0.125 \\ 0.25 & 0.50 & 0.25 \end{bmatrix}. \end{aligned}$$

Suppose now that we wish to determine the *probability that the system is in state #2 after two trials, given that it started in state #2.* This is answered from the matrix T^2 as follows:

$$\begin{array}{ccc} & & \text{Ending states} \\ & & \begin{array}{ccc} \#1 & \#2 & \#3 \end{array} \\ \text{Beginning states} \longrightarrow & \begin{array}{c} \#1 \\ \#2 \\ \#3 \end{array} & \begin{bmatrix} 0.25 & 0.50 & 0.25 \\ 0.125 \longrightarrow & 0.750 & 0.125 \\ 0.25 & 0.50 & 0.25 \end{bmatrix}. \end{array}$$

The element in the *second row, second column* of T^2 is the probability that in two trials the system is in state #2, given that it began in state #2. From T^2 we can read that this probability is 0.750. We may verify that these are the correct probabilities by using tree diagrams.

Markov chain probabilities

EXAMPLE 1 Using the matrix T^2 given above determine the following probabilities.

(a) The probability that the system is in state #1 after two trials, given that it began in state #3.

(b) The probability that the system is in state #3 after two trials, given that it began in state #1.

Solution

(a) We read the element in the third row, first column of T^2: 0.25.

(b) The answer is the element in the first row, third column of T^2: 0.25.

PRACTICE PROBLEM Use T^2 to determine the probability that the system is in state #3 after two trials, given that it began in state #2.

Answer: 0.125. ■

We consider now the question of getting a Markov chain process started. This may be accomplished by using a chance device of some kind. For example, if a chain has two states, a coin could be tossed to determine the beginning state. In any case, there is a certain probability that the process begins in a given state. If a fair coin is used for a 2-state chain, these probabilities may be expressed by the vector $\langle 0.5 \quad 0.5 \rangle$, which means that the process begins in state #1 with probability 0.5 and the same is true for state #2. Such a vector is called the *initial state vector.*

state vector

Suppose we have a 2-state Markov chain with transition matrix

$$T = \begin{bmatrix} 0.5 & 0.5 \\ 0.4 & 0.6 \end{bmatrix}$$

and initial state vector $v = \langle 0.5 \quad 0.5 \rangle$. We have already seen that the powers of the matrix T,

$$T, T^2, T^3, T^4, T^5, \ldots$$

contain conditional probabilities for the Markov chain after numbers of trials. Now the vectors

$$vT,\ vT^2,\ vT^3,\ vT^4,\ vT^5,\ \ldots$$

also contain probabilities for the chain after numbers of trials. For example, it is possible to show, by using tree diagrams, that the vector

$$vT^4$$

contains the probabilities that the system is in the two states after four trials. These probabilities are determined by making the following computations:

$$vT = \langle 0.5 \quad 0.5 \rangle \begin{bmatrix} 0.5 & 0.5 \\ 0.4 & 0.6 \end{bmatrix} = \langle 0.450 \quad 0.550 \rangle,$$

$$vT^2 = (vT)T = \langle 0.450 \quad 0.550 \rangle \begin{bmatrix} 0.5 & 0.5 \\ 0.4 & 0.6 \end{bmatrix} = \langle 0.445 \quad 0.555 \rangle,$$

$$vT^3 = (vT^2)T = \langle 0.445 \quad 0.555 \rangle \begin{bmatrix} 0.5 & 0.5 \\ 0.4 & 0.6 \end{bmatrix} = \langle 0.444 \quad 0.556 \rangle,$$

$$vT^4 = (vT^3)T = \langle 0.444 \quad 0.556 \rangle \begin{bmatrix} 0.5 & 0.5 \\ 0.4 & 0.6 \end{bmatrix} = \langle 0.444 \quad 0.556 \rangle.$$

From the fact that $vT^4 = \langle 0.444 \quad 0.556 \rangle$, we can say that after four trials the system will be in state #1 with probability 0.444 and in state #2 with probability 0.556.

The sequence of vectors $vT, vT^2, vT^3, vT^4, \ldots$ are called *state vectors.* Note that the sum of the entries in the state vectors is always 1 since the system must be in some state after each trial. Vectors which have entries that add up to 1 are called *probability vectors.* We can see from the above calculations that the state vectors after three and four trials are the same (actually they are only approximately the same since the probabilities are rounded to three places). The system has reached a *steady state* when the state vectors become equal. It is a characteristic of Markov chains that they reach a steady state if some power of the transition matrix T has only positive entries.

probability vectors

Markov chain states

EXAMPLE 2 Suppose a 2-state Markov chain has the following initial state vector and transition matrix:

$$v = \langle 0.3 \quad 0.7 \rangle, \qquad T = \begin{bmatrix} 0.2 & 0.8 \\ 0.6 & 0.4 \end{bmatrix}.$$

Compute the state vectors after one and two trials.

Solution: The state vectors after one and two trials are vT and vT^2.

$$vT = \langle 0.3 \quad 0.7\rangle \begin{bmatrix} 0.2 & 0.8 \\ 0.6 & 0.4 \end{bmatrix} = \langle 0.480 \quad 0.520\rangle,$$

$$vT^2 = (vT)T = \langle 0.480 \quad 0.520\rangle T = \langle 0.408 \quad 0.592\rangle.$$

Therefore, the state vector after one trial is $\langle 0.480 \quad 0.520\rangle$ and after two trials is $\langle 0.408 \quad 0.592\rangle$.

PRACTICE PROBLEM For the Markov chain given in Example 2 compute the state vector after three trials.

Answer: $\langle 0.437 \quad 0.563\rangle$. ■

If a Markov chain reaches a steady state, then all subsequent state vectors will be equal. For instance, we may have that

$$vT^7 = vT^8$$

or

$$\begin{aligned} vT^7 &= vT^8 \\ &= (vT^7)T. \end{aligned} \tag{1}$$

If we let $m = vT^7$, then Eq. (1) may be written as follows:

$$mT = m. \tag{2}$$

fixed-point vector

The vector m is called a *fixed-point vector* for the matrix T because of Eq. (2). Our calculations indicate that we may determine the steady state of a Markov chain if we can find a fixed-point vector for the transition matrix.

Fixed-point vector

EXAMPLE 3 Determine the fixed-point probability vector for the transition matrix given below:

$$T = \begin{bmatrix} 0.25 & 0.75 \\ 0 & 1 \end{bmatrix}.$$

Solution: Suppose that $m = \langle a \quad b\rangle$ is the fixed-point probability vector for T. Then we have, by definition of fixed-point vector,

$$mT = m.$$

Calculating mT gives

$$\begin{aligned} mT &= \langle a \quad b\rangle \begin{bmatrix} 0.25 & 0.75 \\ 0 & 1 \end{bmatrix} \\ &= \langle 0.25a + 0 \quad 0.75a + b\rangle \\ &= \langle 0.25a \quad 0.75a + b\rangle. \end{aligned}$$

But, since $mT = m$, we have

$$\langle 0.25a \quad 0.75a + b\rangle = \langle a \quad b\rangle.$$

From this equation, it follows that

$$0.25a = a \qquad \text{and} \qquad 0.75a + b = b.$$

Solving the equation $0.75a + b = b$ results in $a = 0$, and since $a + b = 1$, $b = 1$. It may be concluded that $m = \langle 0 \quad 1\rangle$ is the fixed-point probability vector for the given transition matrix.

PRACTICE PROBLEM Verify, by calculating mT, that $m = \langle 0 \quad 1\rangle$ is the fixed-point vector for the transition matrix T of Example 3.

The fixed-point probability vector of a Markov chain (if it has one) has a greater significance than has been indicated so far. It is shown in more advanced courses in probability theory that if some power of the transition matrix T has only positive entries, then the Markov chain has a unique fixed-point probability vector, and the state vectors

$$vT, vT^2, vT^3, vT^4, vT^5, \ldots$$

approach closer and closer to the fixed-point vector for all possible initial state vectors v. This result has the following very important practical implication: no matter what the initial state of a system, if we know the fixed point of the chain, we can predict the state of the system in the long run.

Market management

EXAMPLE 4 The mathematics department of a college may be considered (each year) to be in one of two states with respect to a calculus textbook: the department either uses the text (state A) or it does not (state A'). The publisher of a text called *Calculus for the Many* (CM) has made a market survey and believes the following information to be accurate.

(a) 75% of the departments using CM in a given year will continue to use CM the following year and 25% will choose another book.

(b) 10% of the departments not using CM in a given year will change to CM next year and 90% will continue not to use CM in the next year.

i. Set up a transition matrix for this information.
ii. If a department is using CM this year, what is the probability that the department will be using CM in three years?
iii. Find the fixed-point vector for the transition matrix.
iv. If the market survey information remains correct, what is the probability a department will be using CM for a long period of time?

Solutions

(i) From the market survey information the transition matrix is

$$\text{This year} \quad \begin{array}{c} \\ A \\ A' \end{array}\!\!\begin{array}{c} \text{Next year} \\ \begin{array}{cc} A & A' \end{array} \\ \begin{bmatrix} 0.75 & 0.25 \\ 0.10 & 0.90 \end{bmatrix} \end{array} = T$$

(ii) If a department is using CM this year, the probability that the department is in state A is 1. Thus the initial state vector for such a department is

$$v = \langle 1 \quad 0 \rangle.$$

Then

$$\begin{aligned} vT &= \text{state vector after one year,} \\ vT^2 &= \text{state vector after two years,} \\ vT^3 &= \text{state vector after three years.} \end{aligned}$$

Therefore, we may calculate these vectors to obtain our answer:

$$vT = \langle 1 \quad 0 \rangle \begin{bmatrix} 0.75 & 0.25 \\ 0.10 & 0.90 \end{bmatrix} = \langle 0.75 \quad 0.25 \rangle,$$

$$vT^2 = (vT)T = \langle 0.75 \quad 0.25 \rangle \begin{bmatrix} 0.75 & 0.25 \\ 0.10 & 0.90 \end{bmatrix} = \langle 0.588 \quad 0.412 \rangle,$$

$$vT^3 = (vT^2)T = \langle 0.588 \quad 0.412 \rangle \begin{bmatrix} 0.75 & 0.25 \\ 0.10 & 0.90 \end{bmatrix} = \langle 0.482 \quad 0.518 \rangle.$$

Therefore, we see that for a department now using CM the probability that CM will be used 3 years hence (i.e. be in state A) is 0.482.

(iii) To find the fixed-point vector $m = \langle a \quad b\rangle$ for T, we solve the equation

$$mT = m,$$

$$\langle a \quad b\rangle \begin{bmatrix} 0.75 & 0.25 \\ 0.10 & 0.90 \end{bmatrix} = \langle a \quad b\rangle,$$

$$\langle 0.75a + 0.10b \qquad 0.25a + 0.90b\rangle = \langle a \quad b\rangle.$$

Then

$$0.75a + 0.10b = a \quad \text{or} \quad 0.10b = 0.25a,$$
$$0.25a + 0.90b = b \quad \text{or} \quad 0.25a = 0.10b.$$

The equations obtained are the same and may be written as

$$b = 2.5a. \tag{3}$$

Since $a + b = 1$, $b = 1 - a$, and substitution of $1 - a$ for b into (3) gives

$$1 - a = 2.5a,$$
$$1 = 3.5a,$$
$$a = \frac{1}{3.5},$$
$$a = 0.286 \quad \text{(to three places)}.$$

Thus $b = 1 - a = 0.714$, and the fixed-point vector for T is

$$m = \langle 0.286 \quad 0.714\rangle.$$

(iv) From the fixed-point vector of part (iii) we may conclude that the probability is 0.286 for a department to be using CM over a long period of time.

PRACTICE PROBLEM Referring to Example 4, determine the probability that a department will be using CM four years hence, assuming that it is using CM this year.

Answer: 0.413. ■

EXERCISES

Set A–Transition Matrices for Markov Chains

In Exercises 1 through 4 determine the probabilities requested by using the given transition matrices.

1. $$T = \begin{array}{c} \\ \#1 \\ \#2 \end{array}\!\!\begin{array}{c} \begin{array}{cc} \#1 & \#2 \end{array} \\ \begin{bmatrix} 0.5 & 0.5 \\ 0.3 & 0.7 \end{bmatrix} \end{array}$$

(a) After one trial what is the probability that the system is in state #1, given that it started in state #1?
(b) After one trial what is the probability that the system is in state #2, given that it started in state #2?
(c) After two trials what is the probability that the system is in state #1, given that it started in state #2?
(d) After two trials what is the probability that the system is in state #2, given that it started in state #1?

2. $$T = \begin{array}{c} \\ \#1 \\ \#2 \end{array}\!\!\begin{array}{c} \begin{array}{cc} \#1 & \#2 \end{array} \\ \begin{bmatrix} 0.88 & 0.12 \\ 0.36 & 0.64 \end{bmatrix} \end{array}$$

(a) After one trial what is the probability that the system is in state #1, given that it started in state #2?
(b) After one trial what is the probability that the system is in state #2, given that it started in state #1?
(c) After two trials what is the probability that the system is in state #1, given that it started in state #1?
(d) After two trials what is the probability that the system is in state #2, given that it started in state #2?

3. $$T = \begin{array}{c} \\ \#1 \\ \#2 \\ \#3 \end{array}\!\!\begin{array}{c} \begin{array}{ccc} \#1 & \#2 & \#3 \end{array} \\ \begin{bmatrix} 0.40 & 0.40 & 0.20 \\ 0.70 & 0.20 & 0.10 \\ 0.10 & 0.30 & 0.60 \end{bmatrix} \end{array}$$

(a) After one trial what is the probability that the system is in state #2, given that it started in state #3?
(b) After one trial what is the probability that the system is in state #3, given that it started in state #2?
(c) After one trial what is the probability that the system is in state #1, given that it started in state #2?
(d) After two trials what is the probability that the system is in state #1, given that it started in state #3?

(e) After two trials what is the probability that the system is in state #2, given that it started in state #2?
(f) After two trials what is the probability that the system is in state #3, given that it started in state #3?

4.
$$T = \begin{array}{c} \\ \#1 \\ \#2 \\ \#3 \end{array}\begin{array}{c} \begin{array}{ccc} \#1 & \#2 & \#3 \end{array} \\ \begin{bmatrix} 0.31 & 0.53 & 0.16 \\ 0.22 & 0.11 & 0.67 \\ 0.21 & 0.19 & 0.60 \end{bmatrix} \end{array}$$

(a) After one trial what is the probability that the system is in state #3, given that it started in state #1?
(b) After one trial what is the probability that the system is in state #2, given that it started in state #3?
(c) After one trial what is the probability that the system is in state #2, given that it started in state #2?
(d) After two trials what is the probability that the system is in state #3, given that it started in state #3?
(e) After two trials what is the probability that the system is in state #1, given that it started in state #2?
(f) After two trials what is the probability that the system is in state #3, given that it started in state #1?

Set B–Calculating State Vectors

5. Given the transition matrix

$$T = \begin{array}{c} \\ \#1 \\ \#2 \end{array}\begin{array}{c} \begin{array}{cc} \#1 & \#2 \end{array} \\ \begin{bmatrix} 0.4 & 0.6 \\ 1 & 0 \end{bmatrix} \end{array}$$

and the initial state vector $v = \langle 0.2 \quad 0.8 \rangle$, what is the probability that the system is in
(a) state #1 after two trials? (b) state #2 after two trials?
(c) state #1 after three trials? (d) state #2 after three trials?

6. Given the transition matrix

$$T = \begin{array}{c} \\ \#1 \\ \#2 \end{array}\begin{array}{c} \begin{array}{cc} \#1 & \#2 \end{array} \\ \begin{bmatrix} 0.3 & 0.7 \\ 0.8 & 0.2 \end{bmatrix} \end{array}$$

and the initial state vector $v = \langle 0.7 \quad 0.3 \rangle$, what is the probability that the system is in
(a) state #1 after two trials? (b) state #2 after two trials?
(c) state #1 after three trials? (d) state #2 after three trials?

7. Given the transition matrix

$$T = \begin{array}{c} \\ \#1 \\ \#2 \\ \#3 \end{array} \begin{array}{c} \begin{array}{ccc} \#1 & \#2 & \#3 \end{array} \\ \begin{bmatrix} 0.5 & 0.3 & 0.2 \\ 0.8 & 0.1 & 0.1 \\ 0.1 & 0.6 & 0.3 \end{bmatrix} \end{array}$$

and the initial state vector $\langle 0.1 \quad 0.4 \quad 0.5 \rangle$, what is the probability that the system is in
(a) state #1 after two trials? (b) state #2 after two trials?
(c) state #3 after two trials? (d) state #1 after three trials?
(e) state #2 after three trials? (f) state #3 after three trials?

8. Given the transition matrix

$$T = \begin{array}{c} \\ \#1 \\ \#2 \\ \#3 \end{array} \begin{array}{c} \begin{array}{ccc} \#1 & \#2 & \#3 \end{array} \\ \begin{bmatrix} 0.70 & 0.25 & 0.05 \\ 0.12 & 0.61 & 0.27 \\ 0.57 & 0.32 & 0.11 \end{bmatrix} \end{array}$$

and the initial state vector $v = \langle 0.26 \quad 0.30 \quad 0.44 \rangle$, what is the probability that the system is in
(a) state #1 after two trials? (b) state #2 after two trials?
(c) state #3 after two trials? (d) state #1 after three trials?
(e) state #2 after three trials? (f) state #3 after three trials?

Set C–Finding Fixed-Point Vectors

In Exercises 9 through 12 determine the fixed-point probability vector for the transition matrix given.

9. $T = \begin{bmatrix} 0.5 & 0.5 \\ 0.5 & 0.5 \end{bmatrix}$

10. $T = \begin{bmatrix} 0.30 & 0.70 \\ 0.75 & 0.25 \end{bmatrix}$

11. $T = \begin{bmatrix} 0.5 & 0 & 0.5 \\ 0 & 0.5 & 0.5 \\ 0.5 & 0.5 & 0 \end{bmatrix}$

12. $T = \begin{bmatrix} 0.5 & 0.3 & 0.2 \\ 0.8 & 0.1 & 0.1 \\ 0.1 & 0.6 & 0.3 \end{bmatrix}$

Set D–Markov Chain Word Problems

13. Box A contains 1 white marble and box B contains 1 white and 1 black marble. A trial consists of drawing randomly a marble from each box, placing the marble from box A into B and placing the marble from B into A. Take as a state the number of black marbles in box A (for example, state #1 means there are no black marbles in box A).
(a) Construct the transition matrix.
(b) What is the initial state vector?

(c) If the system is in state #2, what is the probability that the system will be in state #1 after two trials?
(d) After two trials what is the probability that the system will be in state #1?
(e) After three trials what is the probability that the system will be in state #2?
(f) Determine the fixed-point probability vector for the transition matrix.
(g) After 100 trials what is the probability that the system is in state #1?
(h) In the long run, what is the most likely state of this system?

14. Box A contains 2 white marbles and box B contains 3 black marbles. A trial is the same as that of Exercise 13. Take states as described below.

State #1: 0 black marbles in Box A.

State #2: 1 black marble in Box A.

State #3: 2 black marbles in Box A.

(a) Construct the transition matrix.
(b) What is the initial state matrix?
(c) If the system is in state #3, what is the probability that the system is in state #1 after two trials?
(d) If the system is in state #2, what is the probability that the system is in state #3 after three trials?
(e) After three trials what is the probability that the system is in state #1?
(f) After three trials what is the probability that the system is in state #2?
(g) After three trials what is the probability that the system is in state #3?
(h) Find the fixed-point probability vector for the transition matrix.
(i) In the long run, what is the most likely state for this system?

Maze problem

15. A rat is placed in a maze as shown in the figure. During a fixed time interval, the rat randomly chooses one of the doors available to it (depending upon which room it is in) and moves through that door to the next room—it does not remain in the room it occupies. Each movement of the rat is taken as a trial in a Markov chain in which a state is identified with the room the rat is in.

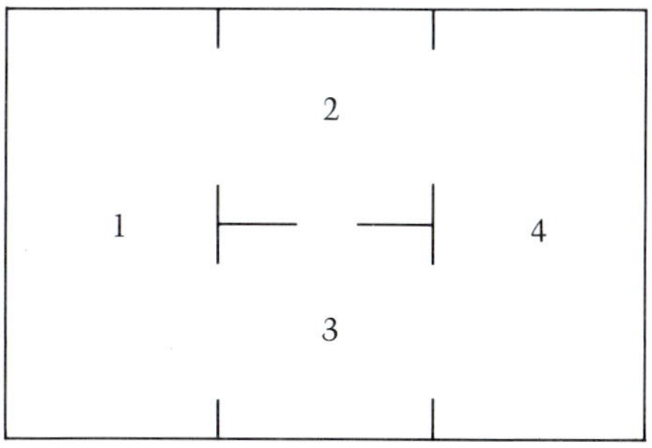

Maze for Exercise 15.

(a) Construct the transition matrix.
(b) If the rat is in room 1, what is the probability that it is in room 3 after two trials?
(c) Determine the fixed-point probability vector for the transition matrix.
(d) After a large number of trials, what is the probability that the rat is in room 4?
(e) In the long run, what is the probability that the rat is in room 2 or 3?

Maze problem

16. The details of this exercise are the same as those of Exercise 15 except that the maze is different. Answer the same questions as those asked in Exercise 15.

Maze problem

17. The details of this exercise are the same as those of Exercise 15 except that the maze is different. Answer the same questions as those asked in Exercise 15.

18. Imagine a particle positioned at one of the four points on the line in the figure shown. A fair coin is tossed: if heads is obtained and the particle is at 1, 2, or 3, it moves one position to the right, and if it is at position 4, it remains there. If tails

Maze for Exercise 16.

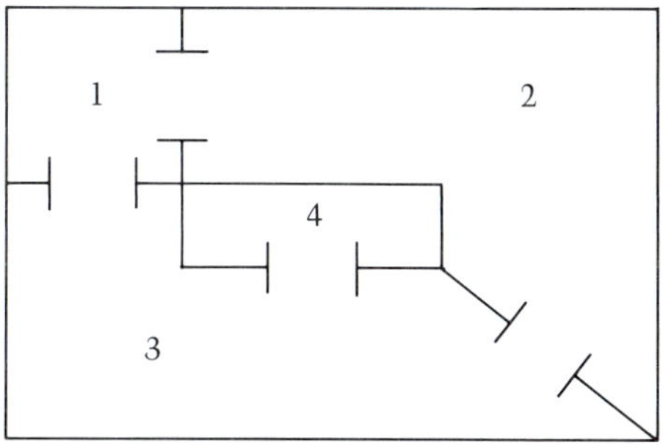

Maze for Exercise 17.

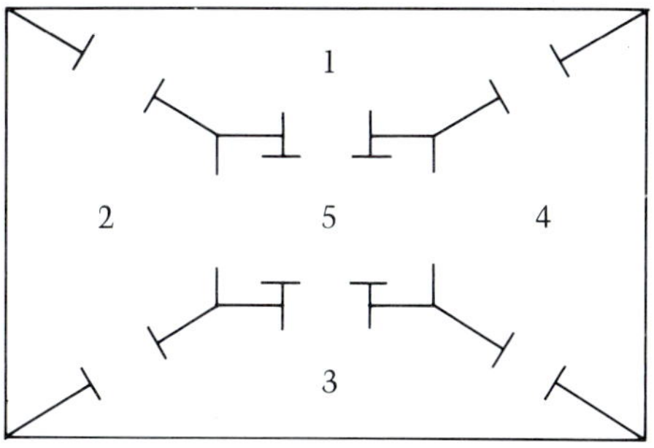

comes up and the particle is at 2, 3, or 4, it moves one position to the left, and if it is at position 1, it remains there. Let the states of this experiment be identified with the four positions. This is an example of a Markov chain known as a random walk.

(a) Determine the transition matrix T.
(b) Find the fixed-point vector.
(c) Can you draw any conclusions about the long-range probabilities?

Figure for Exercise 18.

19. The same as Exercise 18 except that the particle moves around the circle shown in the figure. The particle always moves one position in the clockwise direction if heads is obtained and one position counterclockwise if tails is obtained.

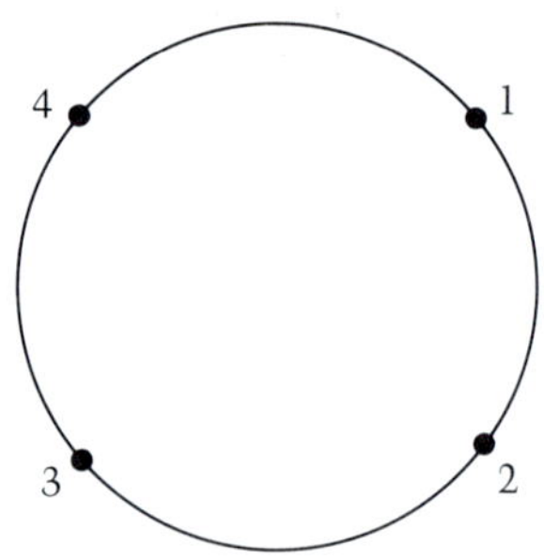

Figure for Exercise 19.

7.8 SUMMARY AND REVIEW EXERCISES

REVIEW OF CONCEPTS

1. *Equally Likely Model.* Let E be an experiment having n outcomes and let A be an event associated with E. If A can occur in k ways, the equally likely model assigns probability k/n to event A, that is,

$$P(A) = \frac{k}{n}.$$

2. *General Definition of Probability.* Suppose that S is the sample space associated with some experiment E and that the probability of each of the elements of S is known. Then, if A is any event, that is, a subset of S, $P(A)$ is the sum of the probabilities of the elements in the set A.

3. *The Addition Formula.* If A, B are events, then

$$P(A \text{ or } B) = P(A) + P(B) - P(A \text{ and } B).$$

If A and B are exclusive, then

$$P(A \text{ or } B) = P(A) + P(B).$$

4. *Formula for Probability of Complementary Event.* If A is an event, then

$$P(A') = 1 - P(A).$$

5. *Conditional Probability.* If A and B are events and $P(B) \neq 0$, then

$$P(A \mid B) = \frac{P(A \text{ and } B)}{P(B)}.$$

6. *Multiplication Formula.* If A and B are events, then

$$P(A \text{ and } B) = P(A)P(B \mid A).$$

7. *Independent Events.* If A and B are independent, then

$$P(A \mid B) = P(A)$$

and also

$$P(A \text{ and } B) = P(A)P(B).$$

8. *Bayes' Theorem.* Let E be an experiment and let X be an event associated with E. If A, B, C, D are a complete set of alternatives for E, then

$$P(A \mid X) = \frac{P(A)P(X \mid A)}{P(A)P(X \mid A) + P(B)P(X \mid B) + P(C)P(X \mid C) + P(D)P(X \mid D)}.$$

9. *Markov Chains.*
   (i) If T is the transition matrix for a Markov chain, then the powers

$$T, T^2, T^3, T^4, \ldots$$

contain conditional probabilities after 1, 2, 3, 4, . . . trials.

(ii) If v is an initial state vector, then the vectors

$$vT, vT^2, vT^3, vT^4, \ldots$$

are the state vectors after 1, 2, 3, 4, . . . trials.

(iii) If m is the fixed-point vector for a transition matrix T, then the entries of m give the probabilities for the system to be in the various states in the long run.

REVIEW EXERCISES

1. The probability of an event A is given to be 0.135. By the frequency interpretation of probability, about how many times, on the average, would A occur in 1000 trials?
2. A letter is chosen at random from the alphabet. What is the probability that a vowel is obtained?
3. Given $P(A) = 0.331$, $P(B) = 0.272$, and $P(A \text{ and } B) = 0.145$. What is $P(A \text{ or } B)$?
4. A school club has 20 members. Twelve of the members are female and 8 are males. If 2 club members are chosen at random (no order), what is the probability that one will be female and one male?
5. A number is chosen at random from the set $\{0, 1, 2, \ldots, 18, 19\}$. What is the probability of obtaining an even number or a multiple of 3?
6. A card is selected at random from a deck of 52. What is the probability that it is a black card, given it is a Jack or Queen?
7. Given three boxes A, B, C. Box A contains 2 black marbles, B contains 1 black and 1 white marble, and C contains 2 white marbles. A box is chosen at random and a marble is drawn at random. If the marble drawn is black, what is the probability that the other marble is also black?
8. A fair coin is tossed 5 times. What is the probability of getting 2 heads and 3 tails?
9. Given a binomial experiment with $p = 0.4$ and $n = 6$. What is the probability of obtaining 4 or 5 successes?
10. Given 5 coins, 4 fair and 1 with $P(T) = 0.2$. A coin is selected at random and tossed twice. If two heads are obtained, what is the probability that the coin chosen was the biased one?

CHAPTER

8

SEQUENCES AND MATHEMATICS OF FINANCE

8

8.1 INTRODUCTION

In this chapter we consider two special types of sequences of numbers and investigate their uses in practical problems. Some of the problems we shall learn how to solve are:

1. Suppose $4000 is deposited for ten years at 6% interest compounded quarterly. What would be the total amount on deposit after 10 years?
2. Suppose a person decides to pay off a loan of $6000 over a 3-year period by making equal monthly payments. If the annual interest rate on the loan is 12% compounded monthly, what would be the monthly payments?
3. Suppose $150 is deposited in a savings account at the beginning of each month, and the annual interest rate is 8% compounded monthly. What would be the total amount on deposit at the end of 10 years? At the end of 10 years, for how many months could $300 a month be withdrawn from the account provided the amount left on deposit continued to draw the same rate of interest?
4. Suppose a person borrows $6000 to buy a car and plans to pay off the loan in 48 equal monthly installments. The lender states that there is an *add-on interest rate* of 8.5% per year. What would be the monthly payments? What is the *annual percentage rate* for this loan? (The answer to the second question is 15.17%.)
5. How does one compute the monthly mortgage payments in the accompanying table which appeared in the business sec-

How Much a $50,000 30-Year Mortgage Costs You

Interest rate, %	Monthly payment, $	Total payout, $
8	366.89	132,080.40
9	402.32	144,835.20
10	438.79	157,964.40
11	476.16	171,417.60
12	514.31	185,151.60
13	553.10	199,116.00
14	592.44	213,278.40
15	632.22	227,559.20
16	672.38	242,056.80
17	712.84	256,622.40
18	753.54	271,274.40

tion of a newspaper? How would the monthly payments change if the 30-year mortgage were replaced by (a) a 20-year mortgage, (b) a 25-year mortgage?

6. Suppose a 30-year-old individual can invest money at an interest rate of 12% compounded monthly for an indefinite period of time. How much must he or she invest each month to be a millionaire by the age of 60? (Make a guess.) ■

8.2 ARITHMETIC SEQUENCES

Consider the sequence of odd positive integers: 1, 3, 5, 7, 9, 11, 13, ... What is the 50th term of this sequence? To answer this question, note that the second term can be obtained by adding two to the first term; the third by adding 2 twos to the first term; the fourth by adding 3 twos to the first term; and the 50th term can be obtained by adding 49 twos to the first term. Therefore, the 50th odd positive integer is

$$1 + (49)(2) = 99.$$

Furthermore, the nth term of this sequence is

$$1 + (n - 1)(2) = 2n - 1.$$

The sum S of the first 50 odd positive integers can be found in several ways. One is to list them all from 1 to 99 in a column and then add. Aside from the risk of making a mistake in addition, just writing down the integers would be a long and boring task.

A second and more efficient way to find the sum of the first 50 odd positive integers uses the fact that the order in which numbers are added does not affect the sum. If

$$S = 1 + 3 + 5 + \cdots + 95 + 97 + 99, \tag{1}$$

then

$$S = 99 + 97 + 95 + \cdots + 5 + 3 + 1. \tag{2}$$

Adding Eqs. (1) and (2) yields

$$2S = 100 + 100 + 100 + \cdots + 100 + 100 + 100.$$

Since the preceding sum contains fifty terms each of which is 100, it follows that

$$2S = 50(100)$$

and

$$S = 25(100) = 2500.$$

Suppose we want to find the sums of given sequences. For what kinds of sequences could we use the above technique? A careful investigation shows that if for a given sequence the difference between each pair of consecutive terms is the same, then the above technique will always produce a sum with identical terms and the sum can be found as in the example. Let us now consider such sequences.

arithmetic sequence

common difference

A sequence of numbers $a_1, a_2, a_3, \ldots$ in which the difference between each pair of consecutive terms is the same is called an **arithmetic sequence** (or *arithmetic progression*). In an arithmetic sequence with a_1 as its **first term,** the second term a_2 is $a_1 + d$ where d is the **common difference** between the terms. The third term a_3 is $a_1 + 2d$, the fourth term a_4 is $a_1 + 3d$, the fifth term a_5 is $a_1 + 4d$, and the nth term a_n of an arithmetic sequence is given by

nth term of arithmetic sequence

$$a_n = a_1 + (n - 1)d.$$

For example, 4, 7, 10, 13, 16, . . . is an arithmetic sequence with first term $a_1 = 4$ and common difference $d = 3$; the 14th term, for example, is

$$a_{14} = 4 + (13)(3) = 4 + 39 = 43.$$

Arithmetic sequence

EXAMPLE 1 If the first term of an arithmetic sequence is 11 and its common difference is −3, find its 31st term.

Solution: $a_{31} = a_1 + (30)d = 11 + (30)(-3) = -79.$

Arithmetic sequence

EXAMPLE 2 If the first term of an arithmetic sequence is 4 and the 23rd term is 50, what is the common difference?

Solution: Since $a_{23} = a_1 + (22d)$,

$$50 = 4 + (22)d,$$
$$22d = 46,$$
$$d = \frac{23}{11}.$$

Arithmetic sequence

EXAMPLE 3 For an arithmetic sequence with first term 2, common difference 3, and nth term 137, find the number of terms in the sequence.

Solution: Since $a_n = a_1 + (n - 1)d$, it follows that

$$137 = 2 + (n - 1)3,$$
$$3n - 3 = 135,$$
$$3n = 138,$$
$$n = 46.$$

PRACTICE PROBLEM

(a) In an arithmetic sequence, $a_1 = 3$ and $d = \frac{3}{2}$; find a_{31}.

(b) In an arithmetic sequence, $a_1 = 21$ and $a_{20} = 2$; find d.

(c) In an arithmetic sequence, $a_1 = 6$, $d = \frac{5}{8}$, and $a_n = 36$; find n.

Answer: (a) 48. (b) −1. (c) 49. ■

Savings plan

EXAMPLE 4 Suppose a person decides to save \$3.00 on Jan. 1 and to increase the amount saved each day thereafter by 75 cents each day. How much will he or she save on Jan. 21 of that year?

Solution: The savings pattern is an arithmetic sequence with $a_1 = 3$, $d = 0.75$, and $n = 21$. Consequently,

$$a_{21} = 3 + (20)(0.75) = 3 + 15 = 18.$$

Hence, \$18.00 would be saved on Jan. 21.

PRACTICE PROBLEM Following the savings pattern described in Example 4, how much would the person save on Jan. 30?

Answer: \$24.75. ■

The technique used to find the sum of the first 50 odd positive integers can be used to derive the formula for the sum of any arithmetic sequence of n terms. The sum s_n of n terms of an arithmetic sequence with first term a_1 and common difference d is given by

$$s_n = a_1 + [a_1 + d] + \cdots + [a_1 + (n-2)d] + [a_1 + (n-1)d].$$

Writing in terms of the preceding sum in reverse order, we get

$$s_n = [a_1 + (n-1)d] + [a_1 + (n-2)d] + \cdots + [a_1 + d] + a_1.$$

Adding the two above equalities yields

$$2s_n = [2a_1 + (n-1)d] + [2a_1 + (n-1)d] + \cdots + [2a_1 + (n-1)d].$$

Since this sum contains n terms of the form $[2a_1 + (n-1)d]$, we have

$$2s_n = n[2a_1 + (n-1)d]$$

and conclude that

sum of an arithmetic sequence

$$s_n = \frac{n}{2}[2a_1 + (n-1)d]$$

is the formula for the **sum** *of n terms of any arithmetic sequence* with a_1 as *first term* and d as *common difference.*

Total savings

EXAMPLE 5 Find the total amount saved in 21 days by the person whose savings plan is described in Example 4.

Solution: In Example 4, $n = 21$, $a_1 = 3$, and $d = 0.75$. Therefore,

$$s_{21} = \frac{21}{2}[2(3) + (20)(0.75)]$$
$$= 10.5[6 + 15] = 220.5.$$

Hence the person would save \$220.50.

PRACTICE PROBLEM Find the total amount saved in 30 days by the person whose savings plan is described in Example 4.

Answer: \$416.25. ■

Gross income

EXAMPLE 6 Suppose an editor secures a job with a starting salary of \$16,000 a year. She receives an \$800 raise each year thereafter for 20 years. Calculate her total gross income from the job for the 20-year period.

Solution: The salary schedule is an arithmetic sequence with $a_1 = 16{,}000$, $d = 800$, and $n = 20$. Therefore,

$$\begin{aligned} s_{20} &= \frac{20}{2}[32{,}000 + (19)(800)] \\ &= 10[32{,}000 + 15{,}200] \\ &= 472{,}000. \end{aligned}$$

The total gross income for the 20-year period would be \$472,000.

Since the nth term (last term) of an arithmetic sequence is given by

$$a_n = a_1 + (n - 1)d,$$

another important formula for the sum of an arithmetic sequence is obtained from the fact that

$$2a_1 + (n - 1)d = a_1 + [a_1 + (n - 1)d] = a_1 + a_n.$$

It is

sum of an arithmetic sequence

$$s_n = \frac{n}{2}(a_1 + a_n).$$

From the preceding formula, we see that *the sum of an arithmetic sequence with n terms is the number of terms times the arithmetic mean (average) of the first and last terms.*

PRACTICE PROBLEM Find a formula for $1 + 2 + 3 + \cdots + n$, the sum of the first n positive integers.

Answer: $n(n + 1)/2$. ■

Savings plan

EXAMPLE 7 In a given year a child saves 1 cent the first day, 2 cents the second, 3 cents the third, 4 cents the fourth, etc. How many days will it take the child to save a total of $40.95?

Solution: Since $40.95 is 4095 cents, we need to find the integer n, if it exists, such that

$$1 + 2 + 3 + \cdots + n = 4095.$$

Using the formula from the preceding practice problem yields

$$\begin{aligned} \frac{n(n+1)}{2} &= 4095 \\ n^2 + n &= 8190 \\ n^2 + n - 8190 &= 0. \end{aligned}$$

The quadratic equation can be solved either by the quadratic formula or by factoring. By factoring we obtain

$$(n + 91)(n - 90) = 0.$$

Although the quadratic equation has $n = -91$ and $n = 90$ as its two solutions, the positive solution is the only tenable one for this problem. Hence, 90 days are required to save $40.95. *Note:* If the goal were to save $41.00, an extra day would be needed. In this case, the resulting quadratic equation would not have an integral solution; the answer would be the smallest integer greater than its positive solution.

PRACTICE PROBLEM For the savings plan described in Example 7, how long would it take the child to save $100?

Answer: 141 days. ■

Finance charges

EXAMPLE 8 A person plans to pay off a $1200 debt by paying $100 at the end of each month plus a finance charge of 1.5% of the balance due before the payment is deducted. What will be the total finance charges?

Solution

*n*th month	Amount owed at the end of *n*th month	Finance charges paid at the end of the *n*th month
1	1200	0.015(1200) = 18.00
2	1100	0.015(1100) = 16.50
3	1000	0.015(1000) = 15.00
⋮	⋮	⋮
11	200	0.015(200) = 3.00
12	100	0.015(100) = 1.50

The sequence of finance charges is an arithmetic sequence containing 12 terms with first term 18.00 and last term 1.50. (*Note:* $d = -1.50$.) Hence the sum S is

$$\begin{aligned} S &= \frac{12}{2}(18.00 + 1.50) \\ &= 6(19.50) \\ &= \$117. \end{aligned}$$

(If only the finance charges were paid and the debt was not reduced, the total finance charges for the year would be

$$0.015(1200)(12) = \$216.$$

Since $\frac{216}{1200} = 0.18$, the person pays an 18% annual interest rate.)

PRACTICE PROBLEM In Example 8, suppose the person had decided to pay $200 plus a finance charge at the end of each month. What would have been the finance charges?

Answer: $63. ■

EXERCISES

Set A–Arithmetic Sequences

For each of the arithmetic sequences in Exercises 1 through 10, (a) find the common difference, (b) list the next 3 terms, and (c) find the sum of the first 21 terms of the sequence.

1. 8, 11, 14, . . .
2. 2, 4, 6, . . .
3. 1, 5, 9, . . .
4. 1.5, 3.0, 4.5, . . .
5. 2.3, 3.4, 4.5, . . .
6. −4, 1, 6, . . .
7. 3, −2, −7, . . .
8. 1, −3, −7, . . .
9. $\frac{2}{3}, \frac{5}{6}, 1, \ldots$
10. $\frac{2}{9}, \frac{1}{2}, \frac{7}{9}, \ldots$
11. An arithmetic sequence has 4 as its first term and 5 as its common difference.
    (a) What is the 50th term?
    (b) What is the sum of the first 50 terms of this sequence.
12. An arithmetic sequence has 3 as its first term and $\frac{4}{3}$ as the common difference.
    (a) What is the 50th term?
    (b) What is the sum of the first 50 terms?
13. In an arithmetic sequence having 5 as its first term and 98 as its 31st term, what is the common difference?
14. An arithmetic sequence has first term 6 and common difference $-\frac{2}{3}$.
    (a) What is the 20th term?
    (b) What is the sum of the first 20 terms?
15. Given: 5, 8, 11, . . . , 65 is an arithmetic sequence.
    (a) What is the common difference?
    (b) How many terms are there in the sequence if 65 is the last term?
    (c) What is the sum of the terms of the sequence?
16. Given: $5, \frac{11}{3}, \ldots, -11$ is an arithmetic sequence.
    (a) What is the common difference?
    (b) How many terms are there in the sequence if −11 is the last term?
    (c) Find the sum of the terms of the sequence.
17. What is the sum of the first 30 multiples of 3?
18. For the arithmetic sequence 8, s, t, u, 18, . . . , answer the following questions.
    (a) What is s?
    (b) What is t?
    (c) What is u?
    (d) What is the sum of the first 20 terms?

Set B–Applications of Arithmetic Sequences

Gross income

19. Suppose that a bookkeeper obtains a job with a starting salary of $15,000 and that he receives an $800 raise each year thereafter.
    (a) What would his salary be during the 15th year?

(b) What would his total gross income be for the 15-year period?

Gross income

20. Suppose that a college professor obtains a job with a starting salary of \$16,000 and that she receives a \$900 raise each year thereafter.
   (a) How much would her salary be during the 10th year?
   (b) What would her total gross income be for the 10-year period?

Salary comparison

21. Which of the following salary schedules would produce the greater total gross income over a 20-year period?
   (a) A starting salary of \$12,000 with a raise of \$900 per year, or
   (b) A starting salary of \$15,000 with a raise of \$600 per year.

Daily savings

22. A student begins the year by saving 5 cents the first day, 10 cents the second, 15 cents the third, 20 cents the fourth, etc. How many days will it take to save a total of \$91.50?

Daily savings

23. If the savings pattern described in Exercise 22 could be continued for one year (365 days), how much would the student save?

Monthly savings

24. A person starts the year by saving \$5.00 the first month, \$10.00 the second month, \$15.00 the third month, etc.
   (a) How many months will it take for the monthly savings to reach \$300?
   (b) At the end of that time, what would be the total amount of savings? (*Note:* We do not assume that the money is deposited in an interest-bearing account; such problems are discussed in Section 8.5.)

Grocery display

25. A grocery store manager has 25 cases of canned peaches, each containing 12 cans. He decides to display the cans by stacking them in a "pyramid" where each row after the bottom row contains one less can. Is it possible for him to use all the cans and end up with a top row having only one can? If so, how many cans should be put in the bottom row?

Grocery display

26. A grocery store manager has 5 cases of canned pears, each containing 24 cans. She decides to display the cans by stacking them in a "pyramid" where each row after the bottom row contains one less can. Is it possible for her to use all the cans and end up with a top row having only one can? If so, how many cans should be put in the bottom row?

Dental payments

27. The recipient of a dental bill for \$250 is offered the option of paying \$12.50 at the end of each month plus a finance charge of $1\frac{1}{2}\%$ on the balance due before the payment is deducted. If she accepts this option, what will be the total amount paid?

Furniture payments

28. A young couple receives a furniture bill for $1400 and is offered the option of paying $200 down and $100 at the end of each month plus a finance charge of $1\frac{1}{2}\%$ on the balance due before the payment is deducted. If this option is accepted, what will the total amount paid be?

Television payments

29. A television set is on sale for $900. It can be bought by paying $200 down and then paying $50 at the end of each month plus a finance charge of $1\frac{1}{2}\%$ on the balance due before the payment is deducted. If this payment plan is used, what will be the total cost of the set?

8.3 GEOMETRIC SEQUENCES

Consider the following salary problem. If Smith were paid a salary of $10,000 a day for 31 days and if Jones were paid 1 cent the first day, 2 cents the second day, 4 cents the third day, 8 cents the fourth day, 16 cents the fifth day, 32 cents the sixth day, and 2^{n-1} cents the nth day, which 31-day total salary is greater? Since we bother to ask the question, one may suspect that Jones has the better salary. How much do you think Smith would need to make each day in order to get the same total salary as Jones? (Make a guess.) To find out, let us determine Jones's total salary for 31 days.

If S is Jones's salary in cents for the 31 days, then

$$S = 1 + 2 + 2^2 + 2^3 + \cdots + 2^{29} + 2^{30}. \tag{1}$$

Although this sum could be found by adding the 31 given numbers, it is much easier to use the following procedure:

First multiply each side of Eq. (1) by 2 to obtain

$$2S = 2 + 2^2 + 2^3 + 2^4 + \cdots + 2^{30} + 2^{31}. \tag{2}$$

Next subtract Eq. (1) from Eq. (2) to get

$$S = 2^{31} - 1;$$

that is, $S = 2{,}147{,}483{,}647$ cents. Therefore,

$$S = \$21{,}474{,}836.47$$

is Jones's total salary in dollars. Since $21,474,836 divided by 31 is approximately $696,737, Smith would have to make more than

$\frac{2}{3}$ of a million dollars a day for his 31-day total salary to equal that of Jones.

The salary schedule for Jones forms what is called a *geometric sequence.* In general, if after the first term of a sequence each term can be obtained by multiplying the preceding term by the same number r, then the sequence is called a **geometric sequence;** the number r is called the **common ratio.** In a geometric sequence, if a is the first term and r is the common ratio, then the second term a_2 is ar, the third term a_3 is ar^2, the fourth term a_4 is ar^3, and the nth term a_n is given by

geometric sequence

common ratio

nth term of a geometric sequence

$$a_n = ar^{n-1}.$$

For example,

$$1, 2, 2^2, 2^3, \ldots, 2^{29}, 2^{30}$$

is a geometric sequence where $a = 1$, $r = 2$ and $n = 31$. Note that for any geometric sequence the ratio r can be found by dividing any term after the first by the preceding term in the sequence; that is,

$$r = \frac{a_n}{a_{n-1}}.$$

Geometric sequence

EXAMPLE 1

(a) If $6, 2, \frac{2}{3}, \ldots$ is a geometric sequence, find its common ratio.

(b) Find the seventh term of the geometric sequence $18, -12, 8, \ldots$.

Solution

(a) Since in any geometric sequence the common ratio r can be found by dividing any term after the first by the preceding term,

$$r = 2 \div 6 = \frac{1}{3}.$$

(b) Since $a = 18$ and $r = -12 \div 18 = -\frac{2}{3}$,

$$a_7 = 18\left(-\frac{2}{3}\right)^6 = 18\left(\frac{64}{729}\right) = \frac{128}{81}.$$

The procedure used to find the sum of the geometric sequence with $a = 1$, $r = 2$, and $n = 31$ can be employed to find the sum of any geometric sequence containing a finite number of terms. Let us derive the general formula. Consider the geometric sequence

$$a, ar, ar^2, ar^3, \ldots, ar^{n-2}, ar^{n-1}$$

with n terms and let

$$s_n = a + ar + ar^2 + ar^3 + \cdots + ar^{n-2} + ar^{n-1}. \tag{3}$$

Multiplying each side of Eq. (3) by the common ratio r, we obtain

$$rs_n = ar + ar^2 + ar^3 + ar^4 + \cdots + ar^{n-1} + ar^n. \tag{4}$$

By subtracting Eq. (3) from Eq. (4), all the terms except the first of Eq. (3) and the last of Eq. (4) drop out and we get

$$rs_n - s_n = ar^n - a,$$

$$s_n(r - 1) = ar^n - a.$$

Therefore,

sum of a geometric sequence

$$s_n = \frac{ar^n - a}{r - 1} \quad \text{provided} \quad r \neq 1.$$

Notes

(1) If $r = 1$, each term of the sequence is a and the sum is the product na.

(2) If the numerator and denominator of the fraction in the formula are multiplied by -1, then s_n is also given by

sum of a geometric sequence

$$s_n = \frac{a - ar^n}{1 - r} = a\left(\frac{1 - r^n}{1 - r}\right).$$

Geometric sequence

EXAMPLE 2 Let $a = 64$ be the first term of a geometric sequence and let $r = -\frac{1}{2}$ be the common ratio.

(a) Find the eighth term.

(b) Find the sum of the first eight terms.

Solution

(a) The eighth term is $64(-\frac{1}{2})^7 = -\frac{1}{2}$.

(b) $$s_8 = \frac{(64)(-\frac{1}{2})^8 - 64}{(-\frac{1}{2}) - 1} = \frac{\frac{1}{4} - 64}{-\frac{3}{2}} = \frac{85}{2}.$$

PRACTICE PROBLEM

(a) For the geometric sequence with 54 as first term and $-\frac{1}{3}$ as common ratio, find the sixth term.

(b) Find the sum of the first six terms.

Answer: (a) $a_6 = -\frac{2}{9}$. (b) $s_6 = \frac{364}{9}$. ■

Geometric sequence

EXAMPLE 3 If 27, u, v, 8 are the first four terms of a geometric sequence, what are the common ratio r and the terms u and v?

Solution: Since the fourth term of a geometric sequence is ar^3.

$$8 = 27r^3,$$

$$r^3 = \frac{8}{27},$$

$$r = \frac{2}{3}.$$

Hence, $u = (27)(\frac{2}{3}) = 18$ and $v = (18)(\frac{2}{3}) = 12$.

PRACTICE PROBLEM Let 125, u, v, and 8 be the first four terms of a geometric sequence. Find the common ratio r and the terms u and v.

Answer: $r = \frac{2}{5}$, $u = 50$, and $v = 20$. ■

Gross income

EXAMPLE 4 Suppose that a physician assistant's starting salary is \$15,000 a year and that he receives an 8% yearly salary increase for 10 years.

(a) What salary would he earn in the 10th year?

(b) What would be his gross income for the 10-year period?

Solution

(a) The salary for each year after the first would be the preceding year's salary plus an 8% increase. Thus

Second year salary: $\$15{,}000 + 0.08(\$15{,}000)$
$= \mathbf{\$15{,}000(1.08)}$

Third year salary: $\$15{,}000(1.08) + 0.08[\$15{,}000(1.08)]$
$= \$15{,}000(1.08)[1 + 0.08]$
$= \mathbf{\$15{,}000(1.08)^2}$

Fourth year salary: $\$15{,}000(1.08)^2 + 0.08[\$15{,}000(1.08)^2]$
$= \$15{,}000(1.08)^2[1 + 0.08]$
$= \mathbf{\$15{,}000(1.08)^3}$

⋮ ⋮ ⋮

Tenth year salary: $\mathbf{\$15{,}000(1.08)^9}$

Note that the annual salaries form a geometric sequence. By using a calculator, logarithms, or Table 1 in Appendix A, we find that the salary for the 10th year is

$$\$15{,}000(1.08)^9 \doteq \$15{,}000(1.999) = \$29{,}985.$$

Note: The annual salary would nearly double in 10 years.

(b) The total 10-year salary S_{10} is the sum of the geometric sequence where $a = 15{,}000$, $r = 1.08$, and $n = 10$. Hence

$$\begin{aligned} S_{10} &= 15{,}000\frac{(1.08)^{10} - 1}{1.08 - 1} \\ &= \frac{15{,}000}{0.08}[(1.08)^{10} - 1] \\ &= 187{,}500[2.158925 - 1] \\ &= 217{,}298.44. \end{aligned}$$

Consequently, the total salary (before taxes, etc.) would be \$217,298.44.

The solution to Example 4 can be used as a pattern to develop the following two general formulas. If B represents the starting (beginning) salary and i represents the percent of yearly increase, then the **salary** a_n for the nth year is given by

salary for nth year

$$a_n = B(1 + i)^{n-1}$$

Furthermore, the total **gross income** S_n for n years is given by

gross income for n years

$$S_n = \frac{B}{i}[(1 + i)^n - 1].$$

EXERCISES

Set A–Geometric Sequences

For each of the geometric sequences in Exercises 1 through 9, find the common ratio and the seventh term.

1. 3, −3, 3, . . .

2. 2, 6, 18, . . .

3. 4, −2, 1, . . .

4. 3, 6, 12, . . .

5. 81, −27, 9, . . .

6. 60, 30, 15, . . .

7. $\frac{2}{3}, \frac{1}{3}, \frac{1}{6}, \ldots$

8. $8, -2, \frac{1}{2}, \ldots$

9. 3, 3.6, 4.32, . . .

Find the sum of the first 6 terms of the geometric sequences in Exercises 10, 11, and 12.

10. 5, −5, 5, . . .

11. $8, -2, \frac{1}{2}, \ldots$

12. 81, −27, 9, . . .

13. For each of the geometric sequences, answer the given questions.
(a) 2, x, 9, . . . What is x if x is positive?
(b) 16, x, y, 54, . . . What are x and y?

14. For each of the geometric sequences, answer the given questions.
(a) 49, x, 16, . . . What is x if x is positive?
(b) 125, x, y, 8, . . . What are x and y?

Set B–Annual Salary and Gross Income

Gross income

15. Consider the following salary schedule: a $13,200 starting salary with a 5% yearly increase.
(a) What would be the salary during the 10th year?
(b) What would be the gross income during the first 10 years?

Gross income

16. Consider the following salary schedule: $12,000 starting salary with a 7% yearly increase.
(a) What would be the salary during the 10th year?
(b) What would be the gross income during the first ten years?

Gross income

17. A job pays a starting salary of $20,000 and calls for a 6% yearly increase thereafter.
(a) What would it pay in the 10th year?
(b) What would be the total gross income for the first 10 years?

Gross income

18. A job pays a starting salary of $18,000 and calls for an 8% yearly increase thereafter.
(a) What would be the salary for the 15th year?
(b) What would be the total gross income for the first 15 years?

Salary comparison

19. Consider the following salary schedules: (i) a $15,000 salary the first year and a $500 annual raise thereafter; (ii) a $12,000 salary the first year and a 6% annual raise thereafter.
(a) In each case, what would be the salary for the 10th year?
(b) In the first 10 years, which salary would provide the greater gross income? How much greater?
(c) In the first 20 years, which salary would provide the greater gross income? How much greater?

Effect of beginning salary

20. For a given position a person is offered a starting salary of $15,000 and an 8% yearly raise thereafter. However, the person insists on a starting salary of $15,800 and gets it. In 20 years, what will be the difference in the gross incomes obtained from the two starting salaries?

Effect of beginning salary

21. For a given position a person is offered a starting salary of $14,000 and a 6% yearly raise thereafter. However, the person insists on a starting salary of $14,800 and gets it. In 20 years, what will be the total difference in the gross incomes obtained from the two starting salaries?

Beginning salary

22. A manager's salary for her 15th year on a job is $36,715. If she received an 8% raise each year, what was her starting salary to the nearest dollar?

8.4 INTEREST: SIMPLE AND COMPOUND

interest

As we know, money paid for the use of money is called **interest.** If a sum of money, called **principal,** is invested and interest is earned, then the amount of interest received depends upon the **interest rate,** a number expressed as a percentage for a given time period. For example, a principal of \$100 invested at an annual interest rate of 6% earns

interest rate

$$0.06 \times 100 = \$6.00$$

interest in one year (that is, 6 cents on each dollar invested).

Simple interest is computed on principal only. For example, if \$100 were invested for two years at an annual simple interest rate of 6%, then the interest earned would be \$6.00 each year, a total of \$12.00 in interest. In general, if P represents principal, r denotes rate of interest per unit of time, t denotes the number of time units in which the interest rate is expressed, and I represents interest, then

simple interest

$$I = Prt$$

is the formula for *simple interest.* Since almost nobody invests money at simple interest (banks are forbidden by law to pay simple interest), our attention will be focused on what is called *compound interest.*

compound interest

Compound interest is interest paid not only on principal but also on interest. For example, if \$100 is invested for two years at a 6% annual interest rate compounded yearly, then the amount (balance) on deposit at the end of the first year would be the initial principal plus the first year's interest; that is,

$$P_1 = \$100 + (0.06)(\$100) = \$106,$$

where P_1 is the amount at the end of the first year. Although this is the same amount that would result from simple interest, things change in the second year. During the second year, interest is paid on \$106 and the second year's interest is

$$(0.06)(\$106) = \$6.36.$$

Therefore, the amount P_2 on deposit at the end of two years is

$$P_2 = \$106 + \$6.36 = \$112.36.$$

Hence, the compounding of interest yields \$12.36 in interest for the two years as opposed to \$12 if interest were not compounded. The difference in the amounts of interest may not seem very much in this case, but over longer periods of time it can be dramatic. For example, it would take more than 16 years to earn \$100 interest on \$100 invested at 6% simple interest while the time required to earn \$100 interest on \$100 invested at 6% interest compounded annually is less than 12 years. (We shall prove this fact soon.)

Suppose a person deposits P dollars where the annual interest rate i is 6% compounded annually. At the end of 1 year the total amount P_1 on deposit is the initial principal P plus the interest earned; hence,

$$P_1 = P + 0.06P = (1.06)P.$$

If the money is left on deposit, then at the end of the second year the amount of money P_2 on deposit is the sum of the amount on deposit at the end of the first year plus the second year's interest; that is,

$$\begin{aligned} P_2 &= (1.06)P + (0.06)[(1.06)P] \\ &= (1.06)P(1 + 0.06) \\ &= P(1.06)^2. \end{aligned}$$

In a similar way we find that the amount P_3 at the end of 3 years is

$$P_3 = (1.06)^2P + (0.06)[(1.06)^2P] = P(1.06)^3.$$

Continuing, we find that $P_1, P_2, P_3, P_4, \ldots$ is a geometric sequence, and the amount P_n on deposit at the end of n years starting from an initial deposit P at a 6% interest rate compounded annually is

$$P_n = P(1.06)^n.$$

(Note the similarity to the solution of Example 4, Section 8.3.)

In general, if P dollars is the initial principal invested at an interest rate of *i percent* **compounded annually,** then the amount P_n at the end of n years is given by

annual compound interest

$$P_n = P(1 + i)^n.$$

PRACTICE PROBLEM Suppose $1000 is deposited in a savings institution which pays 8% interest compounded annually. How much would be on deposit at the end of 4 years?

Answer: $1360.49. ■

To find how long it takes to double an original deposit P at 6% interest compounded annually involves finding the number n such that

$$2P = P(1.06)^n$$

or

$$2 = (1.06)^n.$$

One way to find n is by using interest tables. (See Appendix A, Table 1.) We find that

$$(1.06)^{12} = 2.01219647.$$

Since $(1.06)^{11} < 2$, it would take 12 years to double the amount.

Another way to solve the equation $(1.06)^n = 2$ for n is by using logarithms. If

$$(1.06)^n = 2,$$

then

$$\log_{10} (1.06)^n = \log_{10} 2,$$

and

$$n(\log_{10} 1.06) = \log_{10} 2.$$

Using logarithm tables (see Appendix A, Table 4), we obtain

$$0.0253n = 0.3010$$

$$n = \frac{3010}{253} \doteq 11.9.$$

Consequently, 12 years would more than double the original deposit.

Doubling a deposit at 7%

EXAMPLE 1 How many years does it take to double a given amount placed on deposit at 7% compounded annually?

Solution: We need to solve $(1 + 0.07)^n = 2$ for n. Using logarithms, interest tables, or a calculator, we find that if $n = 11$, then

$$(1.07)^{11} = 2.1048.$$

Furthermore, since $(1.07)^{10} < 2$, 11 years are required.

Suppose \$100 is deposited at an annual interest rate of 6%, and that the interest is compounded quarterly. How much will be on deposit at the end of the second quarter? At the end of the first quarter, the interest earned will be

$$\frac{1}{4}(0.06)(\$100) = (0.015)(\$100);$$

that is, \$1.50, which is one-fourth the yearly interest at 6%. The *amount* A_1 at the end of the first quarter on which interest will be computed during the second quarter is

$$A_1 = \$100 + (0.015)(\$100) = \$100(1.015).$$

At the end of the second quarter the amount A_2 will be \$100(1.015) plus the interest on it; that is,

$$\begin{aligned} A_2 &= \$100(1.015) + \$100(1.015)\frac{0.06}{4} \\ &= \$100(1.015)(1 + 0.015) \\ &= \$100(1.015)^2. \end{aligned}$$

Since $(1.015)^2 = 1.030\,225$, the amount at the end of two quarters will be \$103.02.

Let P be the principal invested and let *i percent be the annual interest rate.* If the interest is compounded four times per year (quarterly), then the amount on deposit at the end of the first quarter is

$$A_1 = P + P\left(\frac{i}{4}\right) = P\left(1 + \frac{i}{4}\right).$$

The amount at the end of the second quarter is

$$A_2 = P\left(1 + \frac{i}{4}\right) + P\left(1 + \frac{i}{4}\right)\left(\frac{i}{4}\right) = P\left(1 + \frac{i}{4}\right)^2,$$

and the amount at the end of the third quarter is

$$A_3 = P\left(1 + \frac{i}{4}\right)^2 + P\left(1 + \frac{i}{4}\right)^2\left(\frac{i}{4}\right) = P\left(1 + \frac{i}{4}\right)^3.$$

We see that $A_1, A_2, A_3, \ldots$ is a geometric sequence, and the amount at the end of m quarters is

$$A_m = P\left(1 + \frac{i}{4}\right)^m.$$

In fact, if there were q compoundings in a year, the amount A_m at the end of m compounding periods would be

$$A_m = P\left(1 + \frac{i}{q}\right)^m.$$

Note: In n years, $m = nq$.

interest rate per period

If we let $j = i/q$ be the **interest rate per period** and let m be the **number of compoundings,** we have the following general formula where P is the initial principal and A_m is the future value of the initial principal after m periods:

compound interest

$$A_m = P(1 + j)^m.$$

Quarterly compounding

EXAMPLE 2 Suppose \$100 is deposited in a savings institution where the annual interest rate is 6% compounded quarterly. How much would be on deposit at the end of two years?

Solution: Using the formula $A_m = P(1 + j)^m$ where $P = \$100$, $j = i/q = 0.06/4 = 0.015$, $m = nq = 8$, we get

$$\$100(1.015)^8 \doteq \$100(1.126\,49) \doteq \$112.64.$$

Important Notes

(1) The amount really is \$112.65 to the *nearest cent,* but most institutions ignore tenths of a cent when "rounding off" since this "omission" is in their favor.

(2) The phrase "i percent interest" means an annual interest rate of i percent. Since

$$\frac{0.06}{4} = 0.015,$$

6% interest is equivalent to a *quarterly interest* rate of 1.5%.

PRACTICE PROBLEM Suppose $500 is deposited in a savings institution where the annual interest rate is 6% compounded monthly. How much will be on deposit at the end of two years?

Answer: $563.58. ■

Future value of a principal

EXAMPLE 3 If $5000 is deposited at 8% interest compounded quarterly, what will be the amount on deposit after 10 years?

Solution: Since the interest is compounded quarterly for 10 years,

$$m = qn = 4(10) = 40.$$

Since $i = 0.08$, $j = i/q = 0.08/4 = 0.02$ is the interest rate per period. Hence, we need to find A_{40} where $P = \$5000$. Using the formula and a calculator (or Table 1, Appendix A), we find

$$\begin{aligned} A_{40} &= \$5000(1 + 0.02)^{40} \\ &= \$5000(1.02)^{40} \\ &= \$5000(2.208\ 039\ 66) \\ &= \$11{,}040.19. \end{aligned}$$

The future value of $5000 deposited for 10 years at 8% compounded quarterly is $11,040.19.

The amount on deposit after 1 year starting with a principal of 1 dollar invested, for example, at an interest rate of 6% compounded quarterly is

$$\left(1 + \frac{0.06}{4}\right)^4 = (1.015)^4 \doteq 1.061\ 36.$$

The interest rate which would yield the same amount in one year without compounding (simple interest) is

$$(1.06136 - 1)(100) = 6.136 \text{ percent};$$

this is called the *effective interest rate* of 6% interest compounded quarterly.

When interest is compounded daily, it is generally assumed that the year contains 360 days. The *effective interest rate for daily compounding* of 6% interest is

$$\left[\left(1 + \frac{0.06}{360}\right)^{360} - 1\right]100\%.$$

which, with the aid of a calculator, we find to be 6.18312% correct to five decimal places. In general, if i is the annual interest rate and the compounding is done q times per year then the **effective interest rate** E_q is given by

effective interest rate

$$E_q = \left[\left(1 + \frac{i}{q}\right)^{q} - 1\right] \times 100.$$

Effective interest rate

EXAMPLE 4 What is the effective interest rate of 8% compounded monthly?

Solution

$$\begin{aligned} E_{12} &= \left[\left(1 + \frac{0.08}{12}\right)^{12} - 1\right]100 \\ &\doteq [(1.066\,666\,6)^{12} - 1]100 \\ &\doteq [1.082\,999 - 1]100 \\ &= 8.2999\%. \end{aligned}$$

Thus, it is 8.3% to the nearest hundredth.

PRACTICE PROBLEM What is the effective interest rate of 9% compounded monthly?

Answer: 9.38%. ■

Present value

EXAMPLE 5 How much money should be deposited at an interest rate of 8% compounded quarterly for 4 years so that the amount on deposit at the end of that time is \$5000? (The problem is to find the present value of \$5000 at the given interest rate for the given time period.)

Solution: Since $q = 4$, $i = 0.08$, and $n = 4$, $m = nq = 16$ and $j = i/q = 0.08/4 = 0.02$. We want to find P when $A_{16} = \$5000$. Since

$$A_{16} = P(1 + j)^{16},$$

it follows that

$$\$5000 = P(1.02)^{16},$$

$$P = \frac{\$5000}{1.372\,7857}$$

$$= \$3642.23.$$

Therefore the present value of \$5000 at 8% compounded quarterly for 4 years is \$3642.23; it is the principal needed to yield \$5000.

Note: Since $P = \$5000(1.02)^{-16}$, if a calculator is not available, we can use Table 2, Appendix A, to avoid dividing by a decimal.

PRACTICE PROBLEM How much money should be deposited at 12% interest compounded quarterly for 4 years so that the amount on deposit at the end of that time is \$5000?

Answer: \$3115.84. ■

EXERCISES

Set A–Interest Rate per Period

In Exercises 1 through 6, find the interest rate per period j and the number of periods m where the annual interest rate is i, the number of compoundings per year is q, and the number of years is n.

1. $i = 6\%, q = 12, n = 3$
2. $i = 5\%, q = 4, n = 5$
3. $i = 10\%, q = 4, n = 2$
4. $i = 12\%, q = 12, n = 4$
5. $i = 12\%, q = 2, n = 5$
6. $i = 16\%, q = 4, n = 5$

Set B–Compound Interest Problems

Future value
7. If $1000 is invested at 5% interest compounded quarterly, what is the amount of the investment at the end of 1 year?

Future value
8. If $3000 is invested at 6% interest compounded monthly, what is the amount of the investment at the end of 3 years?

Future value
9. What is the future value of $3000 deposited for 3 years at 8% compounded quarterly?

Future value
10. What is the future value of $10,000 deposited for 4 years at 6% compounded monthly?

Effective interest rate
11. What is the effective interest rate of 5% interest compounded quarterly?

Effective interest rate
12. What is the effective interest rate of 8% interest compounded quarterly?

Effective interest rate
13. What is the effective interest rate of 12% interest compounded monthly?

Effective interest rate
14. What is the effective interest rate of 16% interest compounded daily?

Interest
15. What is the interest obtained from $10,000 deposited for one year at 6% compounded daily? (Use 360 days.)

Interest
16. What is the interest obtained from $8000 deposited for one year at 8% compounded daily?

Doubling amount
17. How long will it take to double an amount left on deposit at 6% compounded quarterly?

Doubling amount
18. How long will it take to double an amount left on deposit at 5% compounded quarterly?

Doubling amount
19. How long will it take to double an amount left on deposit at 12% compounded monthly?

Doubling amount
20. How long will it take to double an amount left on deposit at 16% compounded daily?

Present value
21. What is the present value of $5000 if the interest rate is 8% compounded quarterly for 4 years?

Present value
22. What is the present value of $4000 if the interest rate is 10% compounded quarterly for 4 years?

Present value
23. What is the present value of $10,000 if the interest rate is 12% compounded quarterly for 4 years?

Present value
24. How much should be placed in a 4-year savings certificate earning 8% compounded quarterly in order to have $20,000 on deposit at maturity of the certificate?

Present value
25. How much should be placed in a 4-year savings certificate earning 12% compounded monthly in order to have $20,000 on deposit at maturity of the certificate?

8.5 CONTINUOUS INTEREST

Let us now reconsider the geometric sequence

$$P\left(1 + \frac{i}{q}\right)^q,$$

which gives the amount obtained in one year from an initial deposit P where the annual interest rate is i percent compounded q times in one year. If we let $u = i/q$, then $q = i/u$ and

$$P\left(1 + \frac{i}{q}\right)^q = P(1 + u)^{i/u} = P[(1 + u)^{1/u}]^i.$$

Since $u = i/q$, it follows for fixed i that if q increases, then u approaches 0, and

$$(1 + u)^{1/u}$$

approaches $e \doteq 2.718\ 28$. (See Section 2.2.) As the number of compoundings q increases, u increases, and

$$[(1 + u)^{1/u}]^i \qquad \text{approaches} \qquad e^i \doteq (2.718\ 28)^i.$$

With an initial principal P, the amount obtained in one year from what is *defined* as i percent *annual interest with* **continuous compounding** is denoted by A_C and given by the formula

Continuous interest

$$A_C = Pe^i.$$

Continuous interest

EXAMPLE 1 If \$5000 is put into an account drawing 8% annual interest with continuous compounding, what is the amount on deposit at the end of 1 year?

Solution: Using the formula $A_C = Pe^i$ where $P = \$5000$ and $i = 0.08$, we get

$$A_C = \$5000e^{0.08}.$$

From Table 3 in Appendix A (or by using the e^x key on a calculator), we find that

$$\begin{aligned} A_C &= \$5000(1.083\ 287) \\ &= \$5416.43. \end{aligned}$$

Note: The interest from the deposit is \$416.43.

PRACTICE PROBLEM If \$4000 is deposited into an account earning 5% annual interest with continuous compounding, what is the amount on deposit at the end of one year?

Answer: \$4205.08. ■

If we let $P = \$1.00$ and $i = 0.06$, the amount obtained from 6% interest with continuous compounding is

$$A_C = e^{0.06} \doteq 1.0618.$$

effective interest rate

Consequently, the interest on \$1.00 in one year at 6% interest with continuous compounding is 6.18 cents. Since this is equivalent to the interest on \$1.00 at 6.18% simple interest, 6.18% is the *effective interest rate of 6% interest with continuous compounding.* In general, the **effective interest rate** for an annual interest rate of *i percent compounded continuously* is given by

$$E_C = (e^i - 1)(100).$$

Effective interest rate

EXAMPLE 2 What is the effective interest rate for an annual interest rate of 8% with continuous compounding?

Solution: Using the formula $E_C = (e^i - 1)(100)$ where $i = 0.08$, we get

$$E_C = (e^{0.08} - 1)(100) \doteq 8.328\ 7.$$

Therefore, the effective interest rate is approximately 8.33%.

Daily and continuous interest

EXAMPLE 3 What would be the difference in the amount of interest obtained from \$8000 deposited for 1 year at 6% interest compounded daily and 6% compounded continuously?

Solution: Using the formula derived in the last section, we find that the amount obtained from 6% interest compounded daily is

$$A_{360} = \$8000\left(1 + \frac{0.06}{360}\right)^{360} \doteq \$8000(1.061\,831\,2) \doteq \$8494.64.$$

The amount obtained from continuous compounding at 6% interest is

$$A_C = \$8000e^{0.06} \doteq \$8000(1.061\,836\,5) \doteq \$8494.69.$$

Therefore, the difference in the amount of interest for 1 year is 5 cents.

Continuous compounding

EXAMPLE 4 What is the present value of a deposit invested at an interest rate of 7.5% compounded continuously for 1 year when the amount in the account at the end of the year is \$4000?

Solution: In the equation $A_C = Pe^i$, the problem is to find P if $A_C = \$4000$ and $i = 0.075$.

$$\$4000 = Pe^{0.075},$$
$$(1.077\,88)P = \$4000,$$
$$P = \$3710.97.$$

Hence, at 7.5% interest compounded continuously, your deposit must be \$3710.97 in order for you to have \$4000 at the end of 1 year.

PRACTICE PROBLEM Find the value of the deposit which, if the interest rate is 5% compounded continuously for one year grows to \$2000.

Answer: \$1902.46. ■

Annual interest rate

EXAMPLE 5 Suppose the interest earned in one year on a deposit of \$4000 at a given interest rate compounded continuously is \$298.62. What is the annual interest rate?

Solution: In the equation $A_C = Pe^i$, the problem is to find i if $A_C = \$4298.62$ and $P = \$4000$.

$$\$4298.62 = \$4000e^i,$$
$$e^i = 1.074\,655.$$

From Table 3, Appendix A (or by taking the logarithm of each side in order to solve the exponential equation), we find that

$$i = 0.072; \qquad \text{that is,} \quad i = 7.2\%.$$

In order to obtain the formula for the future value of a principal P deposited at an interest rate of i percent compounded continuously for n years, we use the fact that as q increases,

$$P\left(1 + \frac{i}{q}\right)^{qn} = P\left[\left(1 + \frac{i}{q}\right)^{q}\right]^{n} \qquad \text{approaches} \qquad P(e^i)^n.$$

We conclude that the future value after n years of continuous compounding at i percent is given by

$$Pe^{in}.$$

Continuous compounding

EXAMPLE 6 Find the amount resulting from a principal of \$5000 deposited for 10 years at 6% compounded continuously.

Solution

$$\begin{aligned} Pe^{in} &= \$5000(e^{0.06})^{10} \\ &= \$5000e^{0.6} \\ &= \$5000(1.82212) \\ &= \$9110.60. \end{aligned}$$

PRACTICE PROBLEM Find the amount resulting from a principal of \$4000 deposited for 4 years at 8% compounded continuously.

Answer: \$5508.51. ■

Before turning to the exercises, let us consider the amounts obtained from \$100 deposited at an interest rate of 12% for 1 year compounded at different time intervals.

$100 at 12% Interest Rate

Compounded	Periods per year	Amount	Effective rate
Annually	1	$112.00	12%
Quarterly	4	$112.55	12.55%
Monthly	12	$112.68	12.68%
Daily	360	$112.74	12.747%
Continuously	~	$112.75	12.749%

EXERCISES

Set A–Effective Interest Rate

For the given interest rate in Exercises 1 through 5, find the effective interest rate with continuous compounding.

1. 5% 2. 9% 3. 10% 4. 14% 5. 15%

Set B–Problems on Continuous Compounding of Interest

Future value
6. Find the amount resulting from a principal of $2000 deposited for 4 years at 6% interest compounded continuously.

Future value
7. Find the amount resulting from a principal of $5000 deposited for 4 years at 7% interest compounded continuously.

Future value
8. Find the amount resulting from a principal of $5000 deposited for 4 years at 8% compounded continuously.

Difference in interest
9. How much difference is there between the interest on $5000 deposited at 6% interest compounded continuously for 10 years and at 6% interest compounded quarterly for 10 years?

Future value
10. At the birth of a child suppose $2000 were placed on deposit at 6% interest compounded continuously. Assuming no money is added or withdrawn from the account, how much money will be in the account when the person reaches age 60?

Present value
11. What is the present value of a dcposit invested at 8% interest compounded continuously for 1 year when the amount in the account at the end of the year is $6000?

Present value
12. What is the present value of a deposit invested at 10% interest compounded continuously for 1 year when the amount in the account at the end of the year is $8000?

Present value
13. What is the present value of a deposit invested at 8% interest compounded continuously for 4 years if the amount in the account at the end of the 4 years is $10,000?

Present value

14. What is the present value of a deposit invested at 7.5% interest compounded continuously for 4 years if the amount in the account at the end of the 4-year period is \$20,000?

Interest rate

15. Suppose the interest earned in one year on \$2000 deposited at an interest rate compounded continuously is \$188.34. What is the annual interest rate?

Interest rate

16. Suppose the interest earned in one year on \$3000 deposited at an interest rate compounded continuously is \$153.81. What is the annual interest rate?

Interest rate

17. Suppose the interest earned in 6 years on \$4000 deposited at an interest rate compounded continuously is \$1399.43. What is the annual interest rate?

Interest rate

18. Suppose the interest earned in 8 years on \$6000 deposited at an interest rate compounded continuously is \$4932.71. What is the annual interest rate?

8.6 INTEREST ON PERIODIC SAVINGS

In the preceding two sections our attention was focused on interest obtained from a given principal. Although such savings as money market certificates and certificates of deposit are of this nature, many types of savings are of a periodic nature. Let us look at some examples.

Quarterly savings

EXAMPLE 1 Suppose at the beginning of a given year a person deposits \$300 at 6% interest compounded quarterly. Suppose further that at the beginning of each quarter thereafter another \$300 is deposited and that the interest is left on deposit. How much would be on deposit at the end of 2 years? (*Note:* We do not assume that there is a final deposit at the end of 2 years. More about this later.)

Solution: The first deposit of \$300 will be on deposit for 8 quarters (2 years) and the amount A_8 of this deposit at that time will be

$$A_8 = \$300\left(1 + \frac{0.06}{4}\right)^8 = \$300(1.015)^8.$$

The second deposit of \$300 will be on deposit for 7 quarters and the amount A_7 for this deposit will be

$$A_7 = \$300(1.015)^7.$$

Continuing, we see that the last deposit of \$300 will draw interest for one quarter, and its amount A_1 will be

$$A_1 = \$300(1.015).$$

The total amount A of all the deposits is

$$A = A_1 + A_2 + A_3 + \cdots + A_7 + A_8;$$

that is,

$$A = \$300(1.015) + \$300(1.015)^2 + \cdots + \$300(1.015)^8.$$

This is the sum of a geometric sequence with $a_1 = 300(1.015)$, $r = 1.015$, and $n = 8$. Using the formula for the sum of a geometric sequence, we obtain

$$\begin{aligned} A &= 300(1.015)\frac{1.015^8 - 1}{1.015 - 1} \\ &= \frac{300(1.015)}{0.015}(1.015^8 - 1) \\ &= (20{,}300)(0.12649) \\ &= \$2567.80. \end{aligned}$$

Note: Since the total deposits amount to 8(\$300) = \$2400, the interest accrued is \$167.80.

In general, let i be the annual interest rate, q be the number of deposits per year, and m be the total number of deposits. It should be clear from examining Example 1 that the **amount** A at the end of m periods resulting from m regular deposits of D dollars at the *beginning of each period* where interest is compounded at $j = i/q$ percent per period is given by

Periodic deposits

$$A = \frac{D}{j}[(1 + j)^{m+1} - (1 + j)].$$

Monthly savings

EXAMPLE 2 Suppose deposits of \$100 drawing 6% interest compounded monthly are made on the first of each month. What is the amount on deposit at the end of 2 years?

Solution: In this case, $D = \$100$, $j = 0.06/12 = 0.005$, and $m = 2(12) = 24$. Therefore,

$$\begin{aligned} A &= \frac{\$100}{0.005}[(1.005)^{25} - 1.005] \\ &= \$20{,}000[0.127\,795\,6] \\ &= \$2555.91 \end{aligned}$$

PRACTICE PROBLEM Suppose deposits of \$50 earning 6% interest compounded monthly are made on the first of each month. (a) What is the amount on deposit at the end of 4 years? (b) How much interest was earned?

Answer: (a) \$2718.41. (b) \$318.41. ■

Suppose a person can obtain 6% interest compounded quarterly on periodic savings for an extended period of time. How much should the person deposit at the beginning of each quarter over a 10-year period in order to save \$30,000? In this case, $A = \$30{,}000$, $j = 0.06/4 = 0.015$, $m = 40$, and D is to be determined.

$$\begin{aligned} \$30{,}000 &= \frac{D}{0.015}[(1.015)^{41} - 1.015], \\ \$450 &= D(0.826\,228\,7), \\ D &= \$544.64 \end{aligned}$$

Therefore \$544.64 should be deposited at the beginning of each quarter.

Let us return to Example 1. Suppose that for some reason the person making the \$300 quarterly deposits had to miss the third deposit. What would be the total savings under these circumstances? To answer the question, we can still use the general formula so long as we realize that the solution to Example 1 will be decreased by just A_6, the amount and interest resulting from the *third* deposit. Therefore, the total amount at the end of 2 years would be

$$\begin{aligned} \$2567.80 - A_6 &= \$2567.80 - \$300(1.015)^6 \\ &= \$2567.80 - \$328.03 \\ &= \$2239.77 \end{aligned}$$

instead of the \$2567.80 obtained in Example 1.

annuity

A periodic savings plan is an example of what is called an *annuity.* An **annuity** involves equal payments (deposits) made at equal intervals. Apart from equal deposits at regular intervals, installment loan payments on automobiles, house payments, life insurance premiums, etc., are examples of annuities. An annuity is called **simple** if the conversion period for interest coincides with the payment periods. (We consider only simple annuities.) A simple annuity where the payments are made at the *beginning of each period* is called an **annuity due.** If the payments are made at the *end of each period,* it is called an **ordinary annuity.** The savings problems we have been considering are examples of annuity due problems; life insurance premiums are often in the same category. Car payments are usually ordinary annuities.

simple annuity

annuity due

ordinary annuity

Let us reconsider Example 1 but as an ordinary annuity. That is, suppose \$300 is deposited for 8 quarters *at the end of each quarter* where $j = 0.06/4 = 0.015$ is the interest rate per quarter. The total amount $\overline{A}$ at the end of 8 quarters from the deposits would be

$$\begin{aligned} \overline{A} &= \$300(1.015)^7 + \$300(1.015)^6 + \cdots + \$300(1.015) + \$300 \\ &= \$300[1 + 1.015 + (1.015)^2 + \cdots + (1.015)^6 + (1.015)^7]. \end{aligned}$$

Observe that the last deposit would earn no interest and the first deposit would earn interest for only 7 quarters. Using the formula for the sum of a geometric sequence, we get

$$\begin{aligned} \overline{A} &= \$300\frac{(1.015)^8 - 1}{1.015 - 1} \\ &= \frac{\$300}{0.015}[(1.015)^8 - 1] \\ &= \$20{,}000(0.126\,492\,6) \\ &= \$2529.85. \end{aligned}$$

Note: The annuity due yielded \$2567.80.

The general formula for determining the amount for an **ordinary annuity** is

ordinary annuity

$$\bar{A} = \frac{D}{j}[(1 + j)^m - 1],$$

where j is the interest rate per period, m is the number of periods, and D is the amount of each of the equal deposits (payments).

Future value of an annuity

EXAMPLE 3 Find the future value of an annuity where \$1000 is invested at the end of each quarter for 5 years if the interest rate is 8% compounded quarterly.

Solution: This is an ordinary annuity where $D = \$1000$, $j = 0.08/4 = 0.02$, and $m = 5(4) = 20$. Hence,

$$\begin{aligned}\bar{A} &= \frac{\$1000}{0.02}[(1.02)^{20} - 1] \\ &= \$50{,}000(0.485\,947) \\ &= \$24{,}297.35\end{aligned}$$

The following is a table listing the amount of \$1.00 in m periods of an ordinary annuity with j as interest rate per period.

Amount of \$1.00 for an Ordinary Annuity

Periods	2%	5%	6%
1	1.000 000	1.000 000	1.000 000
2	2.020 000	2.050 000	2.060 000
3	3.060 400	3.152 500	3.183 600
4	4.121 608	4.310 125	4.373 616
5	5.204 040	5.525 631	5.637 093
6	6.308 121	6.801 913	6.975 319
7	7.434 283	8.142 009	8.393 838
8	8.582 969	9.549 108	9.897 468
9	9.754 628	11.026 564	11.491 316
10	10.949 721	12.577 893	13.180 795

Ordinary annuity

EXAMPLE 4 Verify the entries in the preceding table for $m = 8$.

Solution

(a) If $j = 0.02$ and $m = 8$, then

$$\bar{A} = \frac{\$1.00}{0.02}[(1.02)^8 - 1]$$
$$= 50(0.171\,659\,38)$$
$$= 8.582\,969.$$

(b) If $j = 0.05$ and $m = 8$, then

$$\bar{A} = \frac{\$1.00}{0.05}[(1.05)^8 - 1]$$
$$= 20(0.477\,455\,44)$$
$$= 9.549\,108.$$

(c) If $j = 0.06$ and $m = 8$, then

$$\bar{A} = \frac{\$1.00}{0.06}[(1.06)^8 - 1]$$
$$= \frac{50}{3}(0.593\,848\,1)$$
$$= 9.897\,468.$$

EXERCISES

Set A–Periodic Savings Problems

In Exercises 1 through 14, we assume that all deposits are made at the beginning of the interest period.

Total savings

1. If \$100 a month is deposited at 6% compounded monthly, how much will be on deposit after 5 years?

Total savings

2. If \$100 a month is deposited at 12% compounded monthly, how much will be on deposit after 5 years?

Total savings

3. If \$150 is deposited every 3 months at 8% compounded quarterly, how much will be on deposit after 10 years?

Total savings

4. If $200 is deposited every 3 months at 12% compounded quarterly, how much will be on deposit after 10 years?

Total savings

5. If $200 is deposited every month at 9% compounded monthly, how much will be on deposit after 5 years?

Total savings

6. If $100 is deposited every month at 9% compounded monthly, how much will be on deposit after 5 years?

Omitted deposit

7. In Example 2, what would the final amount be if the third deposit had been missed?

Total savings

8. Suppose $50 a month is deposited each month at 8% compounded monthly. What will be the amount on deposit after 40 years?

Monthly deposit

9. How much would a person need to deposit each month at 6% compounded monthly for 2 years in order to save $800 for the purchase of a television set?

Monthly deposit

10. How much would a person need to save each month at 6% compounded monthly in order to save $4000 in 4 years?

Monthly deposit

11. How much would a person need to save each month at 6% compounded monthly in order to save $200,000 in 40 years?

Monthly deposit

12. How much would a person need to save each month at 6% compounded monthly in order to save $100,000 in 30 years?

Savings and interest

13. At the beginning of each month, $200 is deposited in a savings account paying interest of 6% compounded monthly.
    (a) How much will be in the account after 5 years?
    (b) How many months will it take to save $24,000?
    (c) How much of the $24,000 is interest?

Savings and interest

14. At the beginning of each month $300 is deposited in a savings account paying interest of 6% compounded monthly.
    (a) How much will be in the account after 5 years?
    (b) How many months will it take to save $30,000?
    (c) How much of the $30,000 is interest?

15. Suppose an individual 30 years old can invest money at 12% compounded monthly for an indefinite period of time. How much must he/she invest each month to be a millionaire by the age of 60?

Set B—Ordinary Annuity and Annuity Due Problems

16. Find the amount for each of the indicated annuities:
    (a) $1000 invested annually at the end of the year at 6% compounded annually for 4 years.
    (b) $1000 invested annually at the beginning of the year at 6% compounded annually for 4 years.

17. Find the amount for each of the indicated annuities:
    (a) \$2000 invested annually at the end of the year at 8% compounded annually for 4 years.
    (b) \$2000 invested annually at the beginning of the year at 8% compounded annually for 4 years.
18. Determine the amount at the end of 6 years of an annuity of \$4000 per year at 6% compounded annually provided it is
    (a) an ordinary annuity.
    (b) an annuity due.
19. Determine the amount at the end of 10 years of an annuity of \$2000 per year at 5% compounded annually provided it is
    (a) an ordinary annuity.
    (b) an annuity due.
20. Verify the entries in the ordinary annuity table for $m = 4$.
21. Verify the entries in the ordinary annuity table for $m = 20$.

8.7 INSTALLMENT PAYMENTS AND RETIREMENT INCOME

Let us now investigate another type of savings. Suppose two parents want to set up a trust fund for their child. If they are able to obtain 6% interest compounded quarterly and want the child to receive \$1500 quarterly for the next 20 years, how much is needed to set up the fund? The problem here is to find what is called the **present value** *of the given annuity.*

Let us keep this example in mind while deriving the general formula where V is the present value of the annuity, $j = 0.06/4 = 0.015$ is the interest rate for each of the $m = (4)(20) = 80$ periods, and $P = \$1500$ is the periodic payment made to the child. At the end of the first period before the first payment P is withdrawn, the amount on deposit is the present value V (amount deposited to set up the fund) plus Vj, the interest accumulated on V during the first period; that is, it is

$$V + Vj = V(1 + j).$$

After withdrawal of the first payment P, the amount on deposit *at the end of the first period is*

$$V(1 + j) - P.$$

Now, this amount will earn interest during the next period.

Hence, the amount on deposit at the end of the second period before P is withdrawn is

$$[V(1 + j) - P](1 + j) = V(1 + j)^2 - P(1 + j).$$

Therefore, the amount on deposit *at the end of the second period* (after withdrawal of P) *is*

$$V(1 + j)^2 - P(1 + j) - P.$$

Similarly, we find that the amount on deposit after *three periods* is

$$V(1 + j)^3 - P(1 + j)^2 - P(1 + j) - P = V(1 + j)^3 - P[(1 + j)^2 + (1 + j) + 1].$$

Finally, after m periods it is

$$V(1 + j)^m - P[1 + (1 + j) + (1 + j)^2 + \cdots + (1 + j)^{m-1}].$$

Using the formula for the sum of a geometric sequence, we find that the amount after m periods is given by

$$V(1 + j)^m - P\frac{(1 + j)^m - 1}{j}.$$

Since the withdrawals are to exhaust the funds,

$$V(1 + j)^m - P\frac{(1 + j)^m - 1}{j} = 0$$

and

$$V(1 + j)^m = \frac{P}{j}[(1 + j)^m - 1].$$

By dividing each side of the preceding equation by $(1 + j)^m$, we obtain the following formula for the **present value** V of an annuity for m periodic payments P at the end of each period where the interest rate is j percent for each period. The formula is

present value

$$V = \frac{P}{j}[1 - (1 + j)^{-m}].$$

In our original problem, $P = \$1500$, $j = 0.015$, and $m = 80$. Therefore,

$$\begin{aligned} V &= \frac{\$1500}{0.015}[1 - (1.015)^{-80}] \\ &= \$100{,}000(0.696\,109\,85) \\ &= \$69{,}610.99. \end{aligned}$$

It is worthwhile to observe that the child will eventually receive

$$(\$1500)(80) = \$120{,}000$$

from the approximately \$69,611 set aside in trust.

30-year annuity

EXAMPLE 1 At the age of 40 a person inherits \$150,000 (after taxes) and decides to buy a 30-year annuity earning 8% annually.

(a) How much can the person expect to receive each year from annual payments?

(b) What is the total amount the person will receive during the 30-year period?

Solution

(a) The present value V is known; it is \$150,000. The problem is to find P where $j = 0.08$ and $m = 30$. Thus

$$\begin{aligned} \$150{,}000 &= \frac{P}{0.08}[1 - (1.08)^{-30}], \\ \$12{,}000 &= P(0.900\,622\,67), \\ P &= \$13{,}324.12. \end{aligned}$$

(b) The total amount received is $(30)(\$13{,}324.12) = \$399{,}723.60$. (Note that the total amount received is about $2\frac{2}{3}$ times the cost of the annuity.)

PRACTICE PROBLEM A person inherits \$100,000 (after taxes) and decides to buy a 20-year annuity earning 6% annually. How much can the person expect to receive each year from annual payments?

Answer: \$8718.45. ■

Suppose a person borrows \$4000 to buy a car; the annual interest rate is 12%. How much should the equal monthly payments be if the loan is to be paid off in 36 months? Essentially, this is an annuity problem from the standpoint of the *lender* where the present value V is \$4000. Since $j = 0.012/12 = 0.01$ and $m = 36$,

$$\$4000 = \frac{P}{0.01}[1 - (1.01)^{-36}],$$

$$\$40 = P(0.301\,075),$$

$$P = \$132.86.$$

Note that the total amount paid is $(36)(132.86) = \$4782.96$; the amount of interest paid is $\$4782.96 - \$4000 = \$782.96$.

PRACTICE PROBLEM A person borrows \$3000 to buy a car; the annual interest rate is 9%. How much should the equal monthly payments be if the loan is to be paid off in 24 months?

Answer: \$137.06. ■

Mortgage payments

EXAMPLE 2 The figures given in Section 8.1 from a newspaper clipping indicated that a 30-year mortgage of \$50,000 at 15% required monthly payments of \$623.23. Verify this amount.

Solution: Essentially, $V = \$50{,}000$ is the present value of an annuity on the part of the lender. The problem is to find P where $j = 0.15/12 = 0.0125$ and $m = 360$. Hence

$$\$50{,}000 = \frac{P}{0.0125}[1 - (1.0125)^{-360}],$$

$$\$625 = P(0.988\,576\,78),$$

$$P = \$632.22^{+}.$$

The lender will "round up" to \$623.23.

Savings and retirement payments

EXAMPLE 3 A couple decides at age 20 that they would like to be assured of an income of \$4000 every 3 months for 15 years after they reach the age of 55. If they are able to arrange for an 8% annual interest rate compounded quarterly, how much per quarter do they need to save to meet their goal?

Solution: First we need to find the present value (at age 55) of their annuity where $P = \$4000$, $j = 0.08/4 = 0.02$, and $m = (4)(15) = 60$.

$$V = \frac{\$4000}{0.02}[1 - (1.02)^{-60}]$$
$$= \$200{,}000(0.695\,217\,7)$$
$$= \$139{,}043.54.$$

Now we resort to the savings plan formula using $A = \$139{,}043.54$, $j = 0.02$, and $m = (3)(35) = 140$ to find out how much they need to deposit at the beginning of each quarter for 35 years in order to obtain \$139,043.54. Since

$$A = \frac{D}{j}[(1 + j)^{n+1} - (1 + j)],$$

it follows that

$$\$139{,}043.54 = \frac{D}{0.02}[(1.02)^{141} - 1.02],$$
$$\$2780.87 = D(15.296\,395),$$
$$D = \$181.80.$$

They would need to save \$181.80 every three months to meet their goal.

---

add-on interest

Suppose a person borrows \$6000 to buy a car and plans to pay off the loan in 48 equal monthly payments. The lender states that there is an **add-on interest rate** of 8.5% per year. The payments would be figured as follows. The interest for 1 year is

$$0.085 \times 6000 = \$510.$$

Therefore for 4 years (48 months) the interest is

$$4 \times \$510 = \$2040.$$

Since the loan plus the interest is $\$6000 + \$2040 = \$8040$, the monthly payments are

$$\frac{\$8040}{48} = \$167.60.$$

APR

Federal law requires that the **annual percentage rate,** denoted by APR, must be disclosed. Although solving the equation necessary to find the APR for the 4-year 8.5% annual add-on interest rate is not easy algebraically, it is easy to verify that it is 15.17%. To do this, note that from the lender's viewpoint \$6000 is the present value of an annuity where $m = 48$, $j = 0.1517/12 = 0.012\ 641\ 6\overline{6}$, and P is the monthly payments. Using the formula

$$V = \frac{P}{j}[1 - (1 + j)^{-m}],$$

we find that

$$\begin{aligned} \$6000(0.012\ 641\ 667) &= P(0.452\ 830\ 4), \\ P &= \$167.50. \end{aligned}$$

Since \$167.50 was the computed monthly payments, the APR is 15.17%.

Add-on interest and APR

EXAMPLE 4

(a) Suppose a \$1200 loan is obtained for 1 year with a 6% add-on interest rate. What would be the monthly payments?

(b) Verify that the APR for this loan is approximately 10.9%.

Solution

(a) Since $0.06 \times \$1200 = \$72$ is the interest, the monthly payments are $(1200 + \$72)/12 = \$106$.

(b) We use the present-value formula for an annuity with $V = \$1200$, $j = 0.109/12 = 0.009\ 083\ 3\overline{3}$, and $m = 12$:

$$\begin{aligned} \$1200 &= \frac{P}{j}[1 - (1 + j)^{-12}], \\ \$10.90 &= P(0.102\ 828), \\ P &= \$106. \end{aligned}$$

Before turning to the exercises, consider the following problem. Suppose you could deposit \$100 at the beginning of each month for 20 years at an interest rate of 6% compounded monthly. For how many years thereafter could you withdraw \$300 a month from the account? (Make a guess.)

First we need to find the amount on deposit after 20 years of making monthly deposits where $D = \$100$, $j = 0.06/12 = 0.005$, and $m = (12)(20) = 240$.

$$\begin{aligned} A &= \frac{\$100}{0.005}[(1.005)^{241} - 1.005] \\ &= \$20{,}000(2.321\ 755\ 5) \\ &= \$46{,}435.20. \end{aligned}$$

Now, $46,435.20 should be considered as the present value V of an annuity where $P = \$300$, $j = 0.005$, and m is unknown. From

$$V = \frac{P}{j}[1 - (1 + j)^{-m}],$$

we get

$$\begin{aligned} \$46{,}435.20 &= \frac{\$300}{0.005}[1 - (1.005)^{-m}], \\ 0.077\ 392 &= 1 - (1.005)^{-m}, \\ -0.226\ 08 &= -(1.005)^{-m} \\ 0.226\ 08 &= (1.005)^{-m} \\ &= \frac{1}{(1.005)^{m}}, \\ (1.005)^{m} &= \frac{1}{0.226\ 08} \\ &= 4.423\ 213, \\ m[\log(1.005)] &= \log(4.423\ 213) \\ (0.002\ 166)m &= 0.645\ 737\ 85 \\ m &= 298^{+}. \end{aligned}$$

Therefore, you could get $300 a month for 298 months, that is, for 24 years and 10 months. (Note that you can withdraw $300 a month for 298 months as a result of only 240 monthly deposits of $100.)

EXERCISES

Set A–Installment Payments and Retirement Income Problems

Trust fund

1. Two parents want to set up a trust fund for their child. They can obtain 8% interest compounded quarterly and want the

child to receive \$300 quarterly for the next 10 years. How much is needed to set up the fund?

Trust fund

2. Two parents want to set up a trust fund for their child. They can obtain 8% interest compounded quarterly and want the child to receive \$1000 quarterly for the next 10 years. How much is needed to set up the fund?

20-year annuity

3. Upon graduation, a college graduate is given a 20-year annuity at a cost of \$20,000 paying 8% interest compounded annually.
   (a) How much will she receive each year?
   (b) What is the total amount that she will receive?

20-year annuity

4. Upon graduation, a college graduate is given a 20-year annuity at a cost of \$30,000 paying 6% compounded annually.
   (a) How much will he receive each year?
   (b) What is the total amount he will receive?

Car payments

5. Suppose a person borrows \$5000 to buy a car. The interest rate is 1% per month. How much should the equal monthly payments be if the loan is to be paid off in 36 months?

Car payments

6. Suppose a person borrows \$6000 to buy a car. The interest rate is $1\frac{1}{2}$% per month. How much should the equal monthly payments be if the loan is to be paid off in 36 months?

Mortgage payments

In Exercises 7 through 10, verify the monthly payment given in Section 8.1 for the 30-year mortgage of \$50,000 at:

7. 12%   8. 8%   9. 10%   10. 18%

Add-on interest

11. Suppose a person borrows \$4000 for 3 years with an annual add-on interest rate of 9%. What are the monthly payments?

Add-on interest

12. Suppose a person borrows \$4000 for 4 years with an annual add-on interest rate of 9%. What are the monthly payments?

APR

13. A 3-year \$4000 loan is secured from a bank. The monthly payments to repay the loan are \$141.11. Verify that the APR is 16.24%. (See Exercise 11.)

APR

14. A 4-year \$4000 loan is secured from a bank. The monthly payments to repay the loan are \$113.33. Verify that the APR is 15.98%. (See Exercise 12.)

Savings and retirement payments

15. Suppose a person were to deposit \$200 a month for 20 years at 6% compounded monthly. For how many years thereafter could the person withdraw \$1000 a month from the account?

Savings and retirement payments

16. Suppose a person were to deposit \$100 a month for 30 years at 8% compounded monthly. For how many years thereafter could the person withdraw \$500 a month from the account?

Savings and retirement payments

17. A couple decide that in 30 years they would like to receive an income of \$3000 every three months for 10 years. If they can

obtain 6% compounded quarterly, how much per quarter must they save to meet their goal?

Savings and retirement payments

18. A couple decide that in 30 years they would like to receive an income of \$5000 every three months for 10 years. If they can obtain 8% compounded quarterly, how much per quarter do they need to save to meet their goal?

8.8 SUMMARY AND REVIEW EXERCISES

1. ARITHMETIC SEQUENCES

(a) If a_1 is the first term of an arithmetic sequence then $a_1 + d$ is the second term where d is the **common difference.** The nth term is

$$a_n = a_1 + (n - 1)d.$$

(b) The **sum** s_n of the first n terms of an arithmetic sequence is given by

$$s_n = \frac{n}{2}[2a_1 + (n - 1)d] = \frac{n}{2}(a_1 + a_n).$$

2. GEOMETRIC SEQUENCES

(a) If a is the first term of a geometric sequence then ar is the second term where r is the **common ratio.** The nth term is

$$a_n = ar^{n-1}.$$

(b) The **sum** s_n of the first n terms of a geometric sequence is given by

$$s_n = \frac{ar^n - a}{r - 1} = \frac{a - ar^n}{1 - r} = a\left(\frac{1 - r^n}{1 - r}\right)$$

provided $r \neq 1$.

3. COMPOUND INTEREST

(a) If i is the annual interest rate, q is the number of compoundings per year for n years, then $j = i/q$ is the interest rate per period, $m = nq$ is the number of periods, and the **amount** A_m (future value) of an initial principal P after m periods is given by

$$A_m = P(1 + j)^m.$$

(b) If i is the annual interest rate and q is the number of compoundings per year, then the **effective interest rate** E_q is given by

$$E_q = \left[\left(1 + \frac{i}{q}\right)^q - 1\right] \times 100.$$

(c) If i is the annual interest rate and P is the initial principal, then the amount in n years resulting from **continuous compounding** is

$$A_C = Pe^{in}.$$

(d) The **effective interest rate** for an annual interest rate of i percent compounded continuously is given by

$$E_C = (e^i - 1)(100).$$

4. INTEREST ON PERIODIC SAVINGS

(a) The **amount** A at the end of m periods resulting from m regular deposits of D dollars *at the beginning of each period* at $j = i/q$ percent interest rate per period is given by

$$A = \frac{D}{j}[(1 + j)^{m+1} - (1 + j)].$$

(b) The **amount** $\bar{A}$ at the end of m periods resulting from m regular deposits of D dollars *at the end of each period* at $j = i/q$ percent interest rate per period is given by

$$\bar{A} = \frac{D}{j}[(1 + j)^m - 1].$$

5. INSTALLMENT PAYMENTS AND RETIREMENT INCOME

(a) The **present value** V of an annuity for m periodic payments P at the end of each period where the interest rate is j percent for each period is given by

$$V = \frac{P}{j}[(1 - (1 + j)^{-m}].$$

6. REVIEW EXERCISES

For each arithmetic sequence in Exercises 1 through 4, do each of the following.
(a) Find the common difference. (b) List the next 3 terms.

(c) Find the sum of the first 30 terms of the sequence.

1. 40, 32, 24, . . .

2. 1.5, 2.7, 3.9, . . .

3. $\frac{2}{3}, \frac{7}{6}, \frac{5}{3}, \ldots$

4. $-20, -17, -14, \ldots$

For each geometric sequence in Exercises 5 and 6, do each of the following.
(a) Find the common ratio. (b) List the next 3 terms.

5. $64, -16, 4, \ldots$

6. 9, 6, 4, . . .

For each geometric sequence in Exercises 7 and 8, do each of the following.
(a) Find the common ratio. (b) Find x and y.

7. $2.5, x, y, 6.86, \ldots$

8. $5, x, y, 8.64, \ldots$

9. The sum of an arithmetic sequence with 20 terms is 130 and the 20th term is 54.
(a) Find the first term.
(b) What is the common difference?

10. The sum of an arithmetic sequence with 30 terms is 180 and the 30th term is 62.
(a) Find the first term.
(b) What is the common difference?

11. Suppose that a person obtains a job with a starting salary of \$18,000 and receives a \$1200 raise each year thereafter.
(a) How much would the person's salary be during the 10th year?
(b) What would the person's total gross income be for the 10-year period?

12. Suppose that a person obtains a job with a starting salary of \$18,000 and that he receives a 6% raise each year thereafter.
(a) How much would the person's salary be during the 10th year?
(b) What would the person's total gross income be for the 10-year period?

13. What is the total interest from a fixed deposit of \$300 for 4 years earning an annual interest rate of 12% compounded quarterly?

14. What is the present value of \$8000 if the interest rate is 8% compounded quarterly for 4 years?

15. How much should an investor put in a 4-year savings certificate earning 8% interest compounded quarterly in order to have \$10,000 on deposit at maturity of the certificate?

16. What is the total interest from a fixed deposit of \$1500 for 1 year at 6% interest compounded continuously?

17. Suppose \$155.76 is the interest earned in 1 year from a \$2000 deposit with interest compounded continuously. What is the annual interest rate?

18. Suppose \$100 is deposited each month and earns 8% compounded monthly. What will be the amount on deposit after 20 years?

19. How much would a person need to save each month earning 9% compounded monthly in order to save \$100,000 in 20 years?

20. What should be the monthly payment for a 30-year home mortgage of \$60,000 at 12% annual interest?

CHAPTER

9

SOME STATISTICAL CONCEPTS

9

9.1 INTRODUCTION

This chapter presents a short introduction to the subject of statistics. Our discussion will include a brief look at descriptive statistics and a study of some normal populations. A mastery of this material will enable you to deal with problems such as the following.

1. A pharmaceutical company estimates that a certain medicine for headaches is 75% effective. If a random sample of 225 headache sufferers are given the medicine, what is the probability that at least 180 individuals are cured of their headaches?
2. The failure times of the computer chip ZUK/0095 are normally distributed with mean 25.25 months and standard deviation 0.75 month. The Zukumuku Company (which makes the ZUK/0095) will replace, free of charge, any chip which fails within 2 years after purchase. What proportion of chips will the company have to replace? ■

9.2 GRAPHICAL REPRESENTATION OF DATA

The word *data* usually refers to a collection of numbers; it is the object of *statistics* to analyze data. This means that if data are given, a successful statistical analysis will obtain information from the data. A familiar example of this is the use of opinion polls. At election time polls are conducted in order to determine, as accurately as possible, the outcome of the election. As another example, a manufacturer wishes to have information about the use of 10 brands of soap. In a group of 1000 persons it would be possible to determine how many use each of the 10 brands. Once these data are collected, the manufacturer wishes to draw conclusions or make inferences. For example, is it appropriate for him to conclude that the most popular brand among the 1000 persons interviewed is the most popular brand among all soap users? In an example such as this the collection of all soap users is referred to as the *population*, and the 1000 persons interviewed are said to be a *sample*.

population
sample

In this section we are interested in that branch of statistics called *descriptive statistics*. The other branch, called *inferential statistics*, will be considered in later sections. The object of descriptive statistics is to represent data in such a way that they

descriptive statistics
inferential statistics

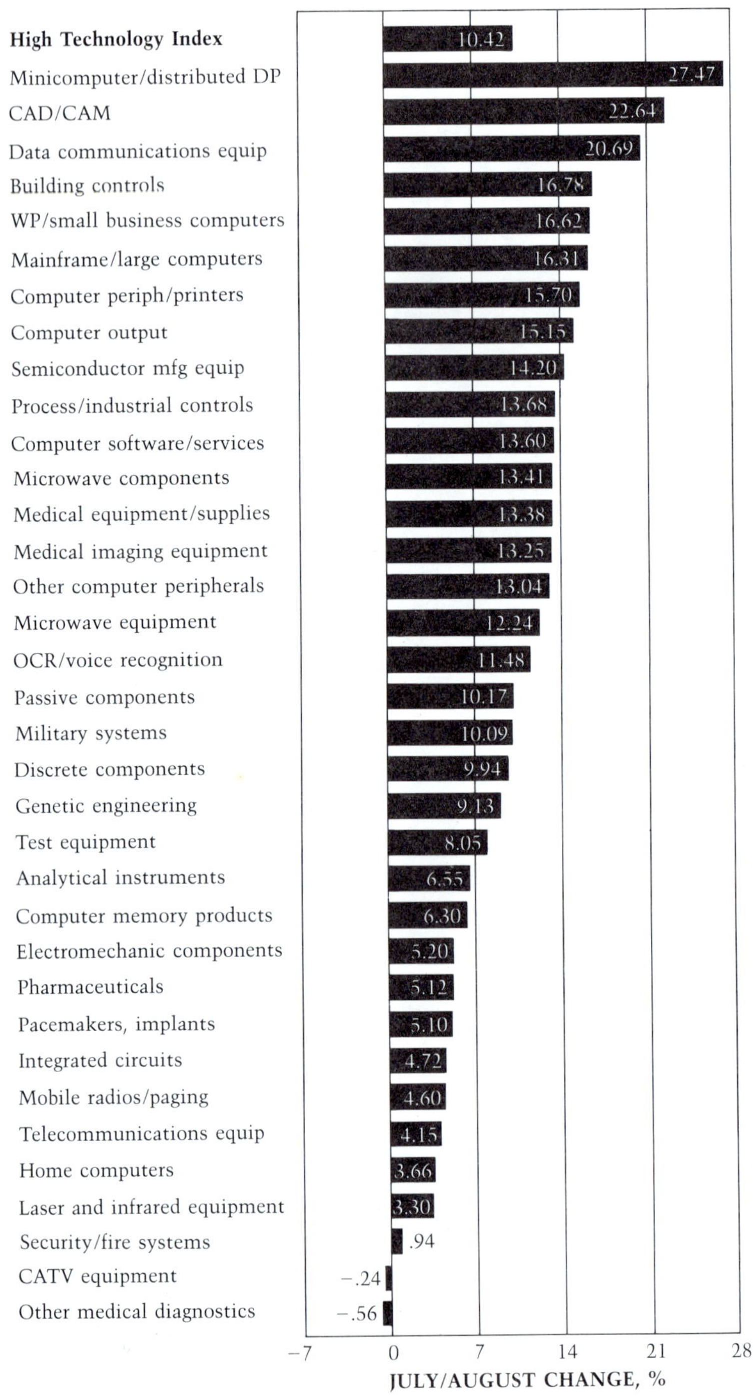

FIG. 9.1. Bar graph: high-technology index.

may be more easily comprehended. A large part of this aspect of the subject deals with *graphical* representations which will be the main focus here. Figure 9.1 shows a *bar graph* which gives information about sectors of industry related to high technology. This is only one of many different kinds of graphs which are used in statistics. The graph used most often here is the *histogram.*

To illustrate these new ideas, consider the set of data in Table 9.1 which represents the sulphur dioxide count in London over a 30-day period (given in micrograms/cubic meter). The table seems to be a jumble of numbers without any clear structure—it lacks organization. As an aid to better understanding, we list the numbers in order from smallest to largest in Table 9.2.

TABLE 9.1

69	101	77	103	72	77
83	78	86	87	90	79
60	75	77	77	77	87
64	79	102	94	89	78
70	83	101	78	76	82

TABLE 9.2

60				
64				
69				
70				
72				
75				
76				
77	77	77	77	77
78	78	78		
79	79			
82				
83	83			
86				
87	87			
89				
90				
94				
101	101			
102				
103				

Table 9.2 helps us to grasp the given data more easily, but it is not entirely satisfactory because of the large spread of these numbers. In order to make the table more compact, we may group

classes

numbers into *classes.* The classes used in this case are

Class 1: 56–60
Class 2: 61–65
Class 3: 66–70
Class 4: 71–75
Class 5: 76–80
Class 6: 81–85
Class 7: 86–90
Class 8: 91–95
Class 9: 96–100
Class 10: 101–105

frequency distribution

Table 9.3, which is called a *frequency distribution,* is the result of counting the numbers in each of these classes.

TABLE 9.3

Class	Tally	Frequency
56–60	I	1
61–65	I	1
66–70	II	2
71–75	II	2
76–80	~~IIII~~ ~~IIII~~ I	11
81–85	III	3
86–90	~~IIII~~	5
91–95	I	1
96–100		0
101–105	IIII	4

Frequency tables may be converted into graphs by listing the given data along the x-axis and the frequencies on the y-axis. If the data have been grouped into classes as in Table 9.3, the midpoint of each class is used to represent that class. For example, the number 88 represents the class 86–90. Using this technique, we have graphed the data in Table 9.3 in Figure 9.2.

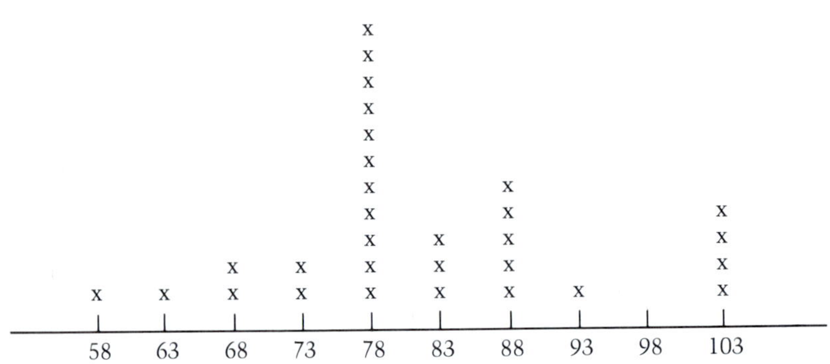

FIG. 9.2.

Figure 9.2 supplies us with much more structure in the original data than was obtained simply by looking at Table 9.1. It is possible, of course, to arrive at other representations of these data. We could have chosen, for example, another way of grouping numbers into classes. To some extent the groupings are arbitrary, but some guidelines for doing this should be observed.

RULES FOR CONSTRUCTING DATA CLASSES

1. Be reasonable in choosing the number of classes. Try not to make the number too small or too large.
2. Make all classes the same length.
3. Make sure that each number from the data set falls into one and only one class.

Frequency distribution

EXAMPLE 1 The following data represent the amount of water (gal/day) used by a small rural community over a 60-day period. Construct a frequency distribution based on the class intervals

39–44, 45–50, 51–56, 57–62, 63–68, 69–74, 75–80

Water used, gal/day

43	54	58	39	51	52	71	53	50	53
48	50	61	39	60	59	40	49	41	60
66	56	53	47	54	46	40	69	60	54
52	46	39	75	61	55	62	72	45	50
48	51	50	75	42	40	54	39	51	50
55	45	44	40	67	59	63	67	46	49

Solution: Counting the frequency for each class produces the distribution given in the accompanying table.

Class	Tally	Frequency
39–44	卌 卌 ||	12
45–50	卌 卌 卌	15
51–56	卌 卌 卌	15
57–62	卌 ||||	9
63–68	||||	4
69–74	|||	3
75–80	||	2

PRACTICE PROBLEM The following data represent the number of new applications/day for Workman's Compensation during the month of November, 1983, in a large city (there were 20 working days in this month). Construct a frequency distribution based on the classes 21–25, 26–30, 31–35, 36–40, 41–45.

Applications per day

40	23	33	42	39	29	39	31	34	30
29	29	34	30	29	29	39	36	38	29

Answer

Class	Tally	Frequency
21–25	|	1
26–30	~~||||~~ |||	8
31–35	||||	4
36–40	~~||||~~ |	6
41–45	|	1 ■

class marks

Figure 9.2 is a graph of the data given in Table 9.1. That graph was constructed by finding the midpoints of the classes. These midpoints are called the *class marks.* Recall that midpoints are obtained by the formula

$$\frac{x_1 + x_2}{2}.$$

Using this formula, we calculate the class marks in the accompanying table for the classes of Example 1.

Class	Class mark	Frequency
39–44	$(39 + 44)/2 = 41.5$	12
45–50	$(45 + 50)/2 = 47.5$	15
51–56	$(51 + 56)/2 = 53.5$	15
57–62	$(57 + 62)/2 = 59.5$	9
63–68	$(63 + 68)/2 = 65.5$	4
69–74	$(69 + 74)/2 = 71.5$	3
75–80	$(75 + 80)/2 = 77.5$	2

A graph of the data from Example 1 is constructed in Figure 9.3 by using the class marks and frequencies.

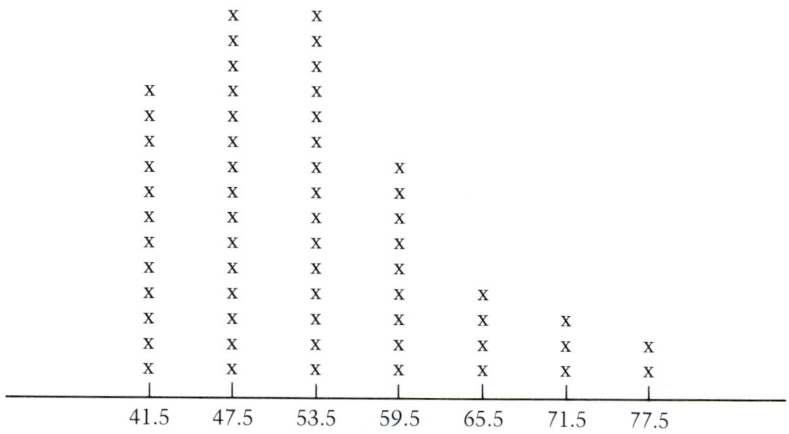

FIG. 9.3. Graph of data from Example 1.

histogram

At this point a *histogram* associated with these data is constructed in the following way. Each class is extended to the halfway point of the next class. For example, consider the classes 39–44 and 45–50:

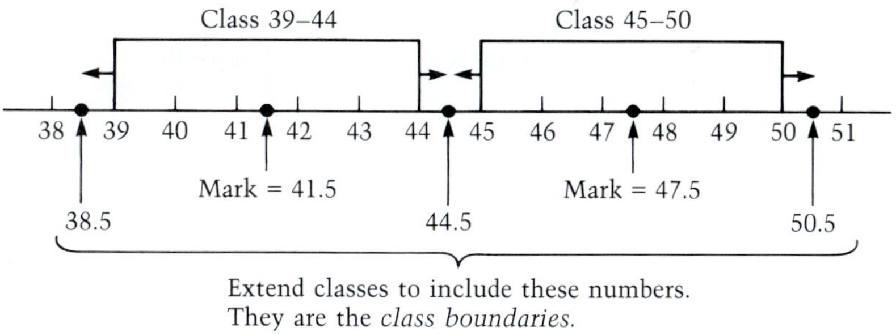

class boundaries

The class 39–44 is replaced by the class 38.5–44.5 and class 45–50 by 44.5–50.5. The numbers at the endpoints of these intervals are called the *class boundaries*. The histogram is completed by erecting a rectangle on each new class so that the base has the length of the class (from halfway point to halfway point, in our case from 38.5 to 44.5, etc.) and the height is the frequency of the class. This is done in Figure 9.4 for the data of Example 1.

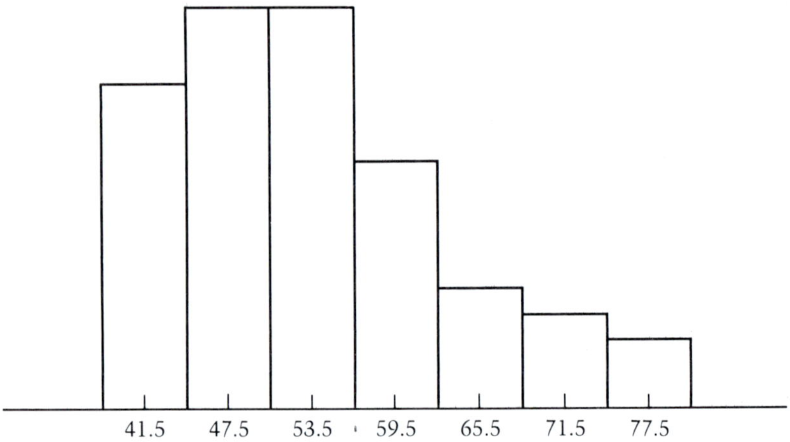

FIG. 9.4.

Histogram

EXAMPLE 2 The following data represent the number of titles of books on applied sciences published annually in Ireland over the period 1963–1982.

Number of titles in applied sciences published in Ireland

31	14	24	33	30	20	30	22	25	21
20	20	25	21	20	20	30	27	29	20

(a) Construct a frequency distribution based on the intervals 14–16, 17–19, 20–22, 23–25, 26–28, 29–31, 32–34.

(b) Construct a histogram from the frequency distribution.

Solution

(a)

Class	Tally	Frequency
14–16	I	1
17–19		0
20–22	~~IIII~~ IIII	9
23–25	III	3
26–28	I	1
29–31	~~IIII~~	5
32–34	I	1

(b) To construct the histogram we first determine the class marks and boundaries.

Class	Class mark	Class boundaries	Frequency
14–16	15	13.5–16.5	1
17–19	18	16.5–19.5	0
20–22	21	19.5–22.5	9
23–25	24	22.5–25.5	3
26–28	27	25.5–28.5	1
29–31	30	28.5–31.5	5
32–34	33	31.5–34.5	1

The graph is completed in Figure 9.5.

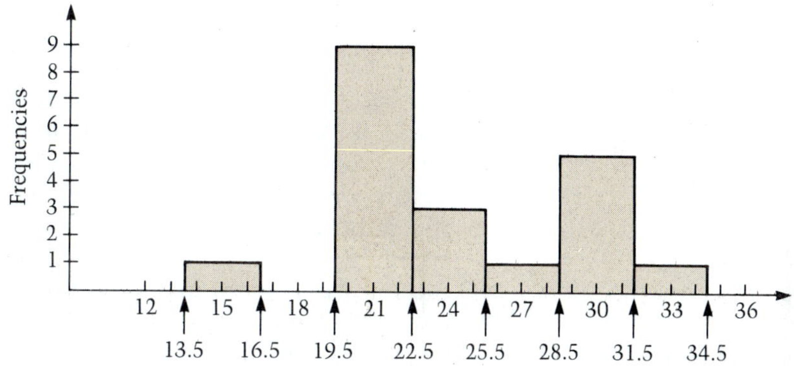

FIG. 9.5. Histogram for Example 2.

PRACTICE PROBLEM The following data represent the number of deaths/year in auto accidents per 1000 of population in Canada during the years 1963–1982:

Deaths in auto accidents/1000 in Canada, 1963–1982

29	23	32	22	25	19	27	26	20	20
26	27	25	30	25	20	28	26	31	23

Construct a frequency distribution based on the classes 19–21, 22–24, 25–27, 28–30, 31–33 and then construct the corresponding histogram.

Answer

Class	Class mark	Boundaries	Tally	Frequencies
19–21	20	18.5–21.5	||||	4
22–24	23	21.5–24.5	|||	3
25–27	26	24.5–27.5	~~||||~~ ||||	9
28–30	29	27.5–30.5	|||	3
31–33	32	30.5–33.5	||	2

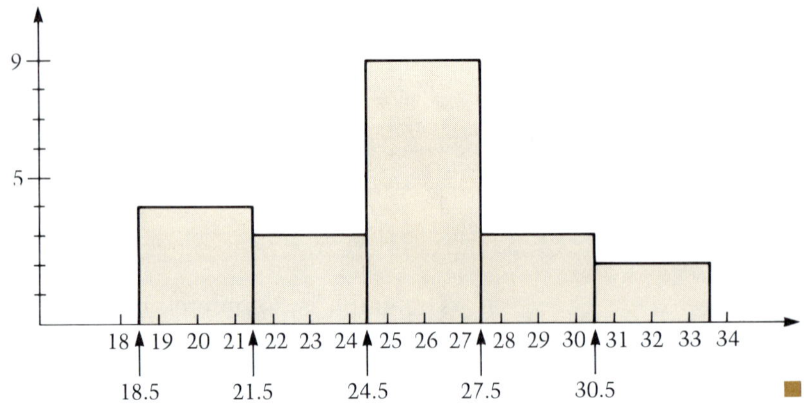

Suppose we toss a coin 30 times and count the number of heads. If this experiment is repeated 100 times, information of the type contained in Table 9.4 will be obtained. For example, the first row of the table indicates that 9 heads were obtained on 4 of the 100 trials of the experiment. The fourth column of the table contains the *relative frequencies.* These are obtained by dividing the numbers in column 3 by the total number of trials (in this case 100). For example, the relative frequency in row number 1 is $\frac{4}{100} = 0.04$. We obtain a *relative frequency histogram* just as we do an ordinary histogram except that the relative frequencies are plotted on the vertical axis rather than the actual frequencies. This has been done for Table 9.4 in Figure 9.6. It should be noted that in this case (and with all our binomial histograms) the class marks are the possible number of successes and the class boundaries extend one-half unit on each side of these.

relative frequency

relative frequency histogram

TABLE 9.4

Number of heads	Class boundaries	Number of experiments	Relative frequency
9	8.5–9.5	4	0.04
10	9.5–10.5	2	0.02
11	10.5–11.5	4	0.04
12	11.5–12.5	8	0.08
13	12.5–13.5	13	0.13
14	13.5–14.5	12	0.12
15	14.5–15.5	15	0.15
16	15.5–16.5	10	0.10
17	16.5–17.5	14	0.14
18	17.5–18.5	10	0.10
19	18.5–19.5	5	0.05
20	19.5–20.5	3	0.03

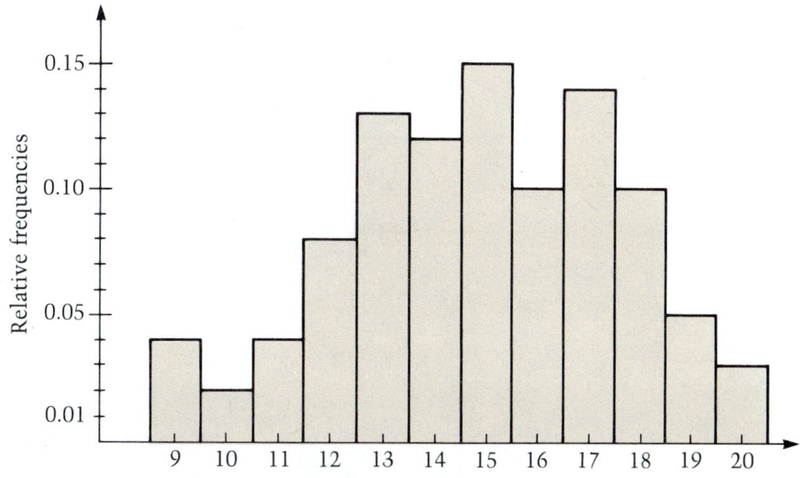

FIG. 9.6. Histogram for Table 4.

As we shall see in Section 9.4, the relative frequency histogram is especially useful in studying binomial distributions. Example 3 demonstrates the standard method for constructing a histogram for a binomial experiment.

Histogram for a binomial experiment

EXAMPLE 3 Construct a relative frequency histogram for the binomial experiment having parameters $n = 10$ and $p = 0.5$. (Recall that n is the number of trials and p is the probability of success.)

Solution: For this example the number of successes k can range only from 0 to 10, and these are the class marks. We first plot the values of k on the k-axis, along with their class boundaries.

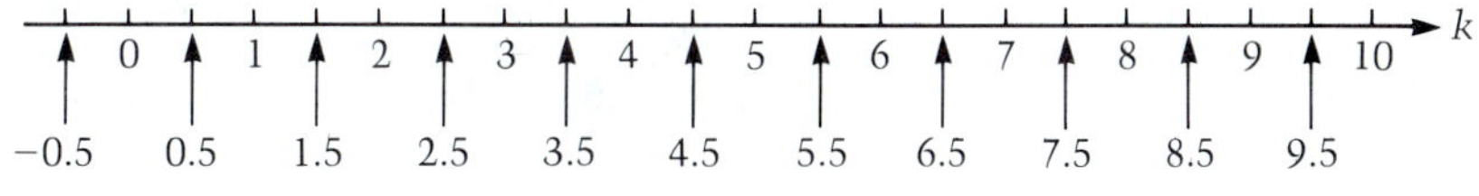

For each value of k we determine $b(k; 0.5, 10)$; Table 5 of Appendix A gives each of these values which are shown in the table below. These probabilities are now plotted on the vertical axis. The histogram is shown in Figure 9.7.

k	0	1	2	3	4	5	6	7	8	9	10
$b(k; 0.5, 10)$	0.001	0.010	0.044	0.117	0.205	0.246	0.205	0.117	0.044	0.010	0.001

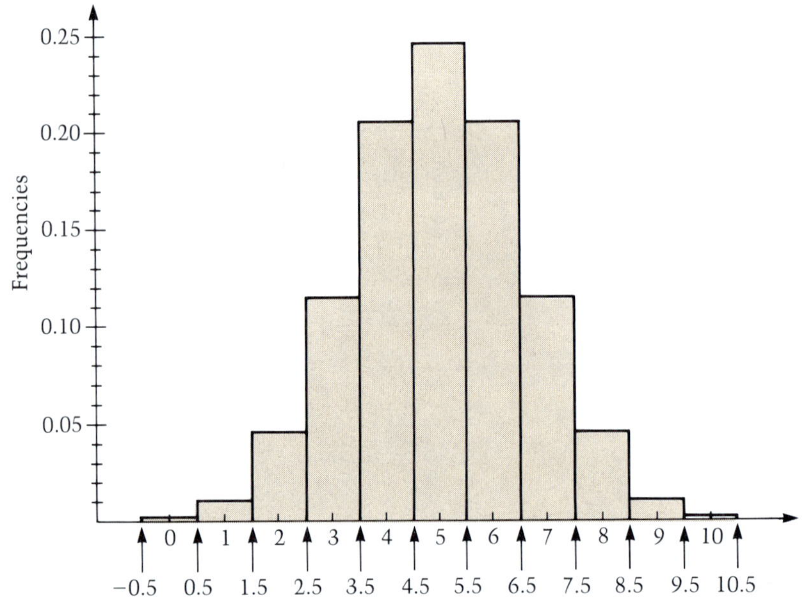

FIG. 9.7. Histogram for Example 3.

EXERCISES

Set A–Frequency Distributions and Histograms

Births/month in a large hospital

1. The following data represent the number of births/month in a large hospital over a two-year period. Construct a frequency distribution based on the intervals 63–65, 66–68, 69–71, 72–74, 75–77.

Number of births/ month

68	63	71	70	64	64	70	71	68	74	69	63
72	70	76	66	66	72	65	75	76	74	72	76

Energy consumption in the USA

2. The following data represent the amount of energy consumed per person/year in the USA in equivalent tons of oil over the period 1968–1982. Construct a frequency distribution.

Amount of oil consumed/person

9	8	8	10	9	10	9	9	9	9	9	9	10	9	8

Labor disputes in Belgium

3. The data given below represent the number of labor disputes/year in Belgium over a period of 28 years. Construct a frequency distribution based on the intervals 231–233, 234–236, 237–239, etc.

Number of labor disputes in Belgium

246	243	243	244	240	245	246	244	251	243	248	246	243	239
245	245	234	234	255	244	248	231	241	249	245	246	235	250

Steel production in the USA

4. The data here represent the steel production (in millions of tons) in the USA over a 25-year period. Construct a frequency distribution based on the intervals 116–118, 119–121, etc.

Steel production in the USA

132	122	132	116	126	127	132	122	128	136	131	132	124
122	126	130	129	121	132	124	122	128	127	133	129	

Legal abortions in Australia

5. The data given below represent the number of legal abortions/year (in thousands) in Australia over a 23-year period. Construct a frequency distribution.

Legal abortions in Australia (thousands)

61	61	57	56	58	60	60	60	61	59	59	61
60	60	62	59	63	62	60	59	61	59	60	

Female deaths

6. The data of this problem give the number of female deaths due to breast cancer per 100,000 of total population in West Germany over a 20-year period. Construct a frequency distribution with one number in each class and from this construct a histogram.

Number of deaths due to breast cancer/100,000 of population

29	31	33	30	29	29	33	30	32	29
30	29	29	31	25	30	27	31	31	30

Live births

7. The data here represent the number of live births per 100,000 of population in Norway over a period of 20 years. Construct a frequency distribution with one number in each class and from this construct a histogram.

Number of live births/100,000 of population in Norway

13	13	14	16	15	13	13	14	14	14
15	14	15	13	13	14	14	13	16	14

Political prisoners

8. The following data represent the number of political prisoners/year in the USSR "adopted" by Amnesty International over a 30-year period. Construct a frequency distribution based on the intervals 150–199, 200–249, 250–299, 300–349, etc. From the frequency distribution construct the corresponding histogram.

Number of political prisoners in USSR "adopted" by Amnesty International

274	274	228	276	369	296	320	281	423	169
425	309	471	396	277	224	261	230	379	284
252	315	341	153	319	272	154	166	206	299

Police in Denmark

9. The following data represent the number of police/year per 100,000 inhabitants in Denmark over a period of 25 years. Construct a frequency distribution based on the intervals 5–7, 8–10, 11–13, 14–16 and then construct a histogram based on the frequency distribution.

Number of police in Denmark

6	12	6	8	8	11	5	8	15	10	9	7	7
6	9	6	7	10	8	9	14	7	6	7	11	

Sex offenses in USA

10. The following data represent the yearly average number of sex offenses/hour in the USA over a period of 30 years. Construct a frequency distribution based on the intervals 0–2, 3–5, 6–8, 9–11, 12–14 and from this construct the corresponding histogram.

Number of sex offenses/hour in the USA

9	9	5	3	3	4	3	4	0	8	3	5	7	7	2
15	1	5	4	4	1	9	6	9	9	6	6	3	5	2

Set B–Binomial Histograms

In each of Exercises 11–20 construct a histogram for the given binomial experiment.

Binomial histogram

11. Construct a histogram for the binomial distribution having parameters $n = 5$, $p = 0.5$.
12. Construct a histogram for the binomial distribution having parameters $n = 5$, $p = 0.6$.
13. Construct a histogram for the binomial distribution having parameters $n = 6$, $p = 0.4$.
14. Construct a histogram for the binomial distribution having parameters $n = 6$, $p = 0.7$.
15. Construct a histogram for the binomial experiment having parameters $n = 10$, $p = 0.6$.
16. Construct a histogram for the binomial experiment having parameters $n = 10$, $p = 0.4$.
17. Construct a histogram for the binomial experiment having parameters $n = 15$, $p = 0.7$.
18. Construct a histogram for the binomial experiment having parameters $n = 20$, $p = 0.4$.
19. Construct a histogram for the binomial experiment having parameters $n = 20$, $p = 0.3$.
20. Construct a histogram for the binomial experiment having parameters $n = 20$, $p = 0.8$.

9.3 CENTRAL TENDENCIES, VARIANCE AND CHEBYSHEV'S THEOREM

mean
median

When data are analyzed it is helpful to obtain a "typical number" which can, in some sense, represent the entire data collection. It is possible to do this in several ways, but we will discuss only two: the *mean* and the *median.* These are called *measures of central tendency.*

The **median** of a set of data is a number which is located in the middle if the numbers are arranged in their proper order. This implies that *half of the data are less than the median and half are larger.* To illustrate this consider the data

20 8 11 7 18

To determine the median these numbers are first arranged in their natural order:

7 8 11 18 20

We see that 11 is the number in the middle, so the median is 11. Note that in this case the size of the set of numbers is $n = 5$. Now suppose the set of numbers is

7 8 18 20

This time we have only $n = 4$ numbers, and there is no number in our set which is in the middle. The middle is between 8 and 18 but there are many numbers between these two. In this case, we define the median to be the average of the two numbers closest to the middle:

$$\frac{8 + 18}{2} = 13$$

RULE FOR DETERMINING THE MEDIAN OF A SET OF NUMBERS Suppose a set of data contains n numbers, and these numbers are arranged in their natural order. Then

1. If n is odd, there is a middle number in the set and it is the median of the collection.
2. If n is even, the average of the two numbers nearest the middle is the median.

Median of data

EXAMPLE 1 Find the median of each set of data given below.

(a) 9 9 8 11 10 10 9 8 10

(b) 43 41 39 40 40 43 41 39 38 40

Solution

(a) The first step is to place the numbers in their natural order:

8 8 9 9 9 10 10 10 11

In this case $n = 9$, an odd number, and this indicates that there is a middle number in the set. Since $n/2 = 4.5$, the middle number is the fifth one. Therefore, the median of the set is 9.

(b) Placing the numbers in order we have

38 39 39 40 40 40 41 41 43 43

This set of data contains an even number: $n = 10$. Since $n/2 = 5$, the numbers nearest the middle are the fifth and sixth ones, namely, 40 and 40:

38	39	39	40	40	40	41	41	43	43
				5th	6th				

Then, by definition, the median is $(40 + 40)/2 = 40$.

PRACTICE PROBLEM Determine the median of each set of data.
(a) 70 71 68 74 69 (b) 0 8 3 5 7 7 2 15

Answer: (a) 70. (b) 6. ■

Suppose that we have a data sample:

$$x_1, x_2, x_3, x_4, \ldots, x_n.$$

The subscripts here indicate the order in which the numbers were obtained in the sample. For example, x_4 was the fourth number obtained in the sample.

It is often necessary to indicate the sum of the numbers of a sample such as the x's given above. One way to do this is to write

$$x_1 + x_2 + x_3 + x_4 + \cdots + x_n,$$

sigma notation

where the dots $(+ \cdots +)$ indicate that we continue with all the terms until we arrive at the last one. Another way to do this is to use the so-called *sigma notation*. This notation makes use of a variation on the Greek letter sigma Σ (hence the name) to indicate a sum. Thus we write

$$\sum x_i \quad \text{or} \quad \sum x$$

as a way of indicating that the x-values should be added.

Sigma notation

EXAMPLE 2 For the x-values given in the sample below, compute Σx.

Sample x: 5 6 5 2 5 3 2 3 5 5 3 3

Solution: We have

$$\sum x = 5 + 6 + 5 + 2 + 5 + 3 + 2 + 3 + 5 + 5 + 3 + 3 = 47.$$

PRACTICE PROBLEM Compute Σy for the y-values given below:

Sample y: 18 10 10 25 11 11 17 34 25

Answer: 161. ■

DEFINITION OF THE SAMPLE MEAN Suppose that a sample consists of the numbers $x_1, x_2, x_3, \ldots, x_n$. Then the *mean of the sample,* denoted by $\overline{X}$, is

$$\overline{X} = \left(\frac{1}{n}\right) \sum x.$$

We may describe the sample mean as the sum of all the numbers in the sample divided by the number of data in the sample.

Mean and median of a sample

EXAMPLE 3 A dentist obtained a random sample of his 10-year old patients and determined the number of instances of caries each had.

Sample x: 6 5 2 5 3 2 3 5 5 3 3 6

Determine the median and mean of this sample.

Solution: To determine the median we arrange the numbers in order.

2 2 3 3 3 3 5 5 5 5 6 6

Since $n = 12$, $n/2 = 6$, and the median is the average of the 6th and 7th numbers in the list:

$$\frac{3 + 5}{2} = 4.$$

By definition of the mean, we have

$$\overline{X} = \frac{6 + 5 + 2 + 5 + 3 + 2 + 3 + 5 + 5 + 3 + 3 + 6}{12}.$$

Therefore, $\overline{X} = \frac{48}{12} = 4$.

PRACTICE PROBLEM The dentist of Example 3 obtained another random sample which is listed below. Determine the median and mean.

Sample y: 5 4 5 6 5 4 7 4 6 6

Answer: Median = 5, $\overline{Y} = 5.2$. ■

A geometric interpretation of the median is the observation (loosely speaking) that half of the sample is on the left of the median and the other half on the right. A geometric explanation of the mean is not so obvious. If we imagine that each number in a sample has one unit of weight and these weights are properly distributed along the number line, then the mean is the position on the line where a fulcrum may be placed to ensure that the line will balance. To illustrate this situation consider the sample of Example 3:

6 5 2 5 3 2 3 5 5 3 3 6

Place these numbers on the number line:

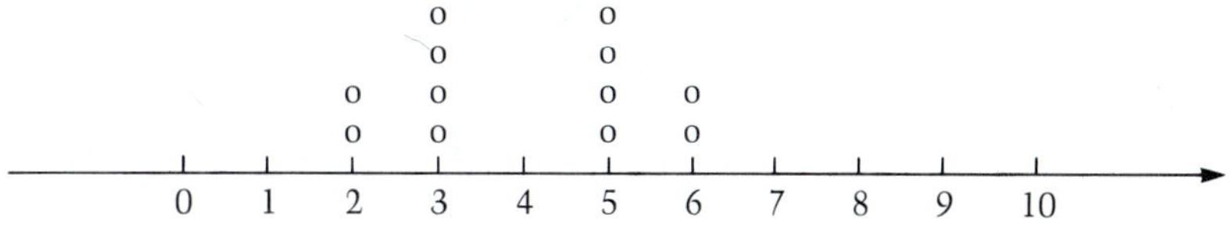

In Example 3 it was determined that $\overline{X} = 4$ and 4 is the balancing point on the line:

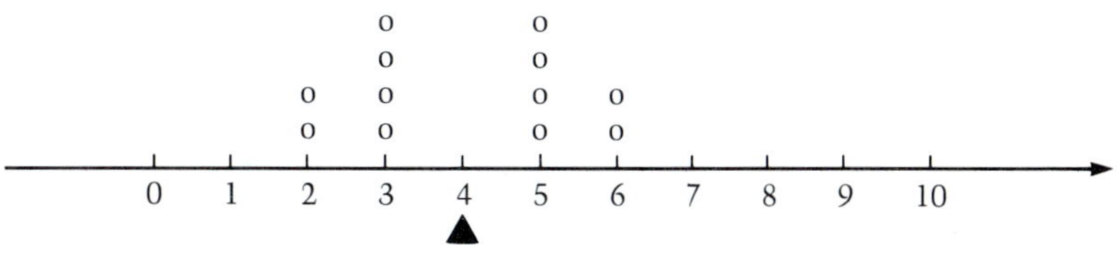

In Figure 9.8 we show diagrams of three different sets of data each having mean $\overline{X} = 4$. This figure shows that the mean alone does not give us information about how the data are spread out relative to the mean. What is needed is a measure of the spread. For such a measure it is desirable that it be largest in case (b) of Figure 9.8 and smallest for case (c). Two such measures are the *variance* and *standard deviation.*

variance
standard deviation

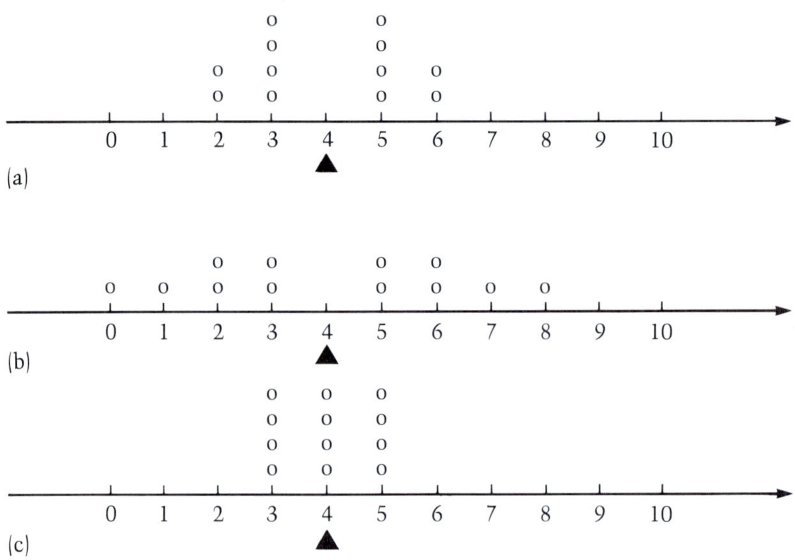

FIG. 9.8. Three different sets of data each having mean $\overline{X} = 4$.

DEFINITION OF VARIANCE AND STANDARD DEVIATION
Suppose that $x_1, x_2, x_3, \ldots, x_n$ is a population sample and that $\overline{X}$ is the mean. Then the *variance* V is defined as follows:

$$V = \frac{\sum (x - \overline{X})^2}{n},$$

where the sum is taken for all values of x. The *standard deviation* s is the positive square root of the variance V.

These definitions will be illustrated by computing V and s for the data given in Example 3. This is easiest to do by arranging the work in a table as has been done in Table 9.5.

TABLE 9.5

	x	$x - \overline{X}$	$(x - \overline{X})^2$
	2	−2	4
	2	−2	4
	3	−1	1
	3	−1	1
	3	−1	1
	3	−1	1
	5	1	1
	5	1	1
	5	1	1
	5	1	1
	6	2	4
	6	2	4
Totals	48	0	24

Table 9.5 is constructed by placing each value of x in column 1. From the sum in column 1 the mean $\overline{X}$ is easily determined. Column 2 is obtained by subtracting $\overline{X}$ from each value of x in column 1, and column 3 is simply the square of each value in column 2. Finally, the variance V is the sum in column 3 divided by the size of the sample (in this case 12). Thus from Table 9.5, we have

$$\overline{X} = \frac{48}{12} = 4, \qquad V = \frac{24}{12} = 2, \qquad s = \sqrt{2}.$$

We may now compare the variance and standard deviation for the data given in the three graphs of Figure 9.8. Table 9.6 provides this comparison.

TABLE 9.6

	V	s
(a)	2	$\sqrt{2}$ (1.414)
(b)	5.833	2.415
(c)	0.666	0.816

We see from Table 9.6 that the variance and standard deviation are largest for (b) and smallest for (c) as they should be. Note that the numbers in this table have been rounded to three places. Calculations in this chapter will always be rounded to three places of accuracy. It should be noted that some authors define the variance and standard deviation by dividing by $n - 1$ rather than n as we have done. For large values of n the results of these two calculations are very close.

Variance and standard deviation

EXAMPLE 4 Calculate the variance and standard deviation of the data given in Figure 9.8(b).

Solution: The data given in this figure are

0 1 2 2 3 3 5 5 6 6 7 8

For the appropriate calculations these data are arranged in the accompanying table.

	x	$x - \overline{X}$	$(x - \overline{X})^2$
	0	−4	16
	1	−3	9
	2	−2	4
	2	−2	4
	3	−1	1
	3	−1	1
	5	1	1
	5	1	1
	6	2	4
	6	2	4
	7	3	9
	8	4	16
Totals	48	0	70

From the table, we have

$$\overline{X} = \frac{48}{12} = 4, \qquad V = \frac{70}{12} = 5.833, \qquad s = 2.415.$$

PRACTICE PROBLEM Verify the variance and standard deviation given in Table 9.6 for the data of Figure 9.8(c). ■

The geometric meaning of standard deviation is expressed by the following result which is due to the Russian mathematician Chebyshev.

Chebyshev's Theorem

CHEBYSHEV'S THEOREM Let the sample $x_1, x_2, x_3, \ldots, x_n$ be given and have mean $\overline{X}$ and standard deviation s. Then the fraction of the numbers of the sample which lie within k standard deviations of the mean is at least

$$1 - \frac{1}{k^2}.$$

Chebyshev's Theorem is illustrated in Figure 9.9.

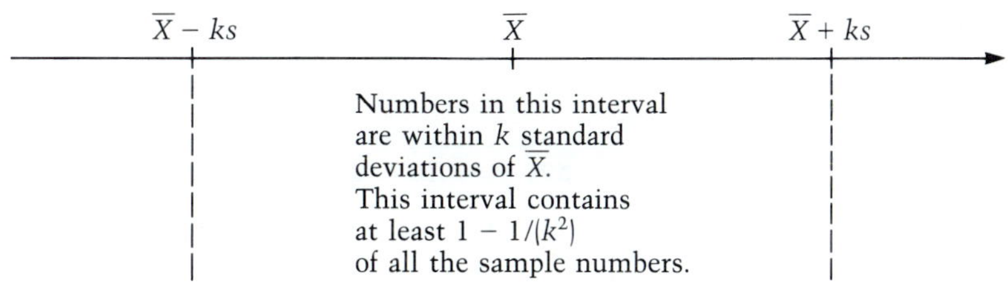

FIG. 9.9. Chebyshev's Theorem.

To illustrate this result, suppose $k = 2$. Then

$$1 - \frac{1}{k^2} = 1 - \frac{1}{4} = \frac{3}{4}.$$

Therefore, at least 75% of all numbers in any sample lie within 2 standard deviations of the mean $\overline{X}$ and no more than 25% can lie outside this range.

Eye surgery

EXAMPLE 5 Suppose we know that the mean recovery time from eye surgery is 20 days with a standard deviation of 2 days. What percent of patients can be expected to take longer than 26 days to recover from this surgery?

Solution: It is given that $\overline{X} = 20$ and $s = 2$. To use the Chebyshev result it is necessary to find the distance that 26 is from 20 *in standard deviation units.* Since 26 is 6 units from the mean 20 and $s = 2$, 26 is 3 standard deviations $(\frac{6}{2})$ from the mean. We know from Chebyshev's Theorem that at least

$$1 - \frac{1}{k^2} = 1 - \frac{1}{3^2} = 1 - \frac{1}{9} = \frac{8}{9} = 0.888 = 89\%$$

of a sample will lie within 3 standard deviations of the mean, and hence, no more than 11% will be outside this range. Therefore, we can expect no more than about 11% of the patients to take longer than 26 days to recover.

PRACTICE PROBLEM Using the information of Example 5 determine the percent of patients that can be expected to recover within 24 days.

Answer: 75%. ■

EXERCISES

Set A–Finding Medians of Samples

Waiting times

1. A sample of customers in a supermarket gave the following waiting times (in minutes) in the checkout line:

   7.7 6.7 5.1 5.4 6 6.7 4.6 1.7 4.7

   Find the median of the sample.

Routine aircraft maintenance

2. A sample of airplanes of a particular type produced the following hours of required routine maintenance (in a month):

   4.6 4.6 5.5 6.8 5.8 5.9 5.9 7.4 4.7

   Determine the median.

Typos in a manuscript

3. The typos/page were counted in a 25-page manuscript. The following numbers of typos/page were found:

   3 3 3 0 2 4 3 0 3 3 2 3 1 3 2 2 3 3 3 3 0
   2 1 2 2

   Find the median.

Auto accidents

4. The reported number of accidents/year over a 20-year period at a particular intersection is given below

   11 5 8 12 11 7 11 8 9 7 7 7 9 7 7 7 11 10 11 7

   Find the median.

Number of instances of caries

5. A dentist took a random sample of 25 of his patients and determined the number of caries/patient to be as follows:

12 8 14 7 9 6 11 10 6 6 10 11 9 13 10 6 11
10 14 8 7 11 7 13 14

Find the median.

Withdrawals from a statistics class

6. An instructor kept records over a 16-term period and found that in an introductory statistics course the following percentages of the class did not finish the course:

45.7 44.1 42 36.5 34.7 34.3 50.6 38.7 51.6 36 38.5
41.5 38.9 43 43.6 45.1

Find the median.

Unemployment

7. In one unemployment office the following number of new applications/week for Workman's Compensation were accepted over a 30-week period:

20 10 18 19 18 19 23 12 19 31 27 25 19 15 22
30 19 18 21 20 24 24 14 20 17 23 23 31 30 19

Find the median.

Cases of rubella

8. The following is the number of new cases/month of rubella reported in the state of Florida during 1983:

6 5 6 7 6 6 6 7 5 5 5 6

Find the median.

16-ounce cereal boxes

9. A sample of 50 boxes of cereal (each marked 16 ounces) were weighed to obtain the following data:

16.2 16.1 16.6 16 16.9 16.5 15.9 15.6 15.8 15.7
16.4 15.9 15.5 15.3 15.9 15.3 15.8 15.6 16.7 15.7
16 16.1 16.5 15.5 16.1 17.2 16.3 16.1 15.9 15.9
17.7 16.1 15.7 15.9 16.3 15.9 16.1 17.1 15.8 15.6
15.8 16.6 16.6 16.7 16 15.6 15.7 15.9 15.7 15.9

Find the median.

Salesman's contacts

10. The number of contacts/week made by a salesman over a 25-week period are given below:

12 17 14 15 16 16 13 21 13 15 14 14 19 17 16
18 17 16 16 14 15 13 20 10 16

Find the median.

Set B–Finding Means and Standard Deviations

Waiting times

11. Find the mean and standard deviation of the data given in Exercise 1.

Routine aircraft maintenance

12. Find the mean and standard deviation of the data given in Exercise 2.

Typos in a manuscript

13. The number of typos/page of a 10-page manuscript is given below:

3 3 3 0 2 0 4 3 0 1

Find the mean and standard deviation of these data.

Auto accidents

14. The number of reported accidents/year at a particular intersection over a 12-year period is given as follows:

11 8 11 11 9 7 7 9 7 11 10

Find the mean and standard deviation of these data.

Number of instances of caries

15. A dentist finds the following number of caries cases/patient in 15 patients:

12 14 9 11 6 10 9 10 11 14 7 7 14 9 8

Find the mean and standard deviation.

Withdrawals from a statistics class

16. Find the mean and standard deviation of the data given in Exercise 6.

Cases of rubella

17. Find the mean and standard deviation of the data given in Exercise 8.

Energy consumption

18. The following data represent the amount of energy consumed per person/year in the USA in equivalent tons of oil over the period 1968–1982. Find the mean and standard deviation.

9 8 8 10 9 10 9 9 9 9 9 9 10 9 8

Legal abortions

19. The following data represent the number of legal abortions/year (in thousands) in Australia over a 23-year period. Find the mean and standard deviation.

61 61 57 56 58 60 60 60 61 59 59 61 60 60 62
59 63 62 60 59 61 59 60

Live births

20. The data given below represent the number of live births per 100,000 of population in Norway over a period of 20 years. Find the mean and standard deviation.

29 31 33 30 29 29 33 30 32 29 30 29 29 31 25
30 27 31 31 30

Set C–Applying Chebyshev's Theorem

21. What is the smallest percentage of a sample that will be within 1.5 standard deviations of the mean?

22. What is the smallest percentage of a sample that will be within 1.8 standard deviations of the mean?

23. What is the largest percentage of a sample that will be farther than 2.25 standard deviations from the mean?

24. What is the largest percentage of a sample that will be farther than 2.36 standard deviations from the mean?

Number of caries cases

25. The mean number of caries for 10-year olds is 5 with a standard deviation of 0.5. What is the largest percentage of 10-year olds that will have more than 7 caries?

Loan approval times

26. A credit union will either approve or deny a loan application from a member. The mean number of days for a certain credit union to make a decision is 5 with a standard deviation of 0.8 day. What is the smallest percentage of applications that will be decided within 6 days?

Selling times

27. The mean number of days to sell a used car which is advertised in the *Atlanta Constitution* is 12 with a standard deviation of 2.25 days. What is the largest percentage of used cars that will require more than 17 days to sell?

Sulfur dioxide count in London

28. The mean sulfur dioxide count/day in London is 80 $\mu g/m^3$ with a standard deviation of 4.38. What is the largest percentage of days this count will be more than 100 $\mu g/m^3$?

Unemployment

29. The mean number of applications/week for Workmen's Compensation at an unemployment office is 33 with a standard deviation of 5.75. What is the smallest percentage of weeks the number of applications will be more than 40?

Number of political prisoners in USSR

30. The mean number of political prisoners/year in the USSR adopted by Amnesty International is 300 with standard deviation 15. What is the largest percentage of years Amnesty International will adopt more than 325 political prisoners?

9.4 NORMAL CURVES

In order to introduce the normal curve, we will consider again histograms of binomial experiments. Recall that this means that we construct graphs of binomial probabilities. To illustrate this in detail consider the binomial distribution having parameters $n = 5$ and $p = 0.5$. From Table 5 of Appendix A we see that

$$b(0; 5, 0.5) = 0.031,$$
$$b(1; 5, 0.5) = 0.156,$$
$$b(2; 5, 0.5) = 0.312,$$
$$b(3; 5, 0.5) = 0.312,$$
$$b(4; 5, 0.5) = 0.156,$$
$$b(5; 5, 0.5) = 0.031.$$

A graph of these probabilities is constructed as follows: The possible values for k (which are 0, 1, 2, 3, 4, 5) are plotted on the horizontal axis, and the probabilities are plotted on the vertical axis as shown in Figure 9.10. The next step is to construct a vertical line at each value of k and this line will have a length equal to $b(k; 5, 0.5)$. (See Figure 9.11.) The histogram is completed by choosing the midpoints of the intervals as shown in Figure 9.11 and drawing rectangles as shown in Figure 9.12. Note that the area of each rectangle is exactly equal to the corresponding probability since the base of each rectangle has width 1 and the height is equal to the probability. This is convenient because it allows us to translate questions about probabilities into questions about areas of rectangles. For example, the probability of 2, 3, or 4 successes in 5 trials is the sum of the areas of the rectangles whose bases are centered at 2, 3, and 4.

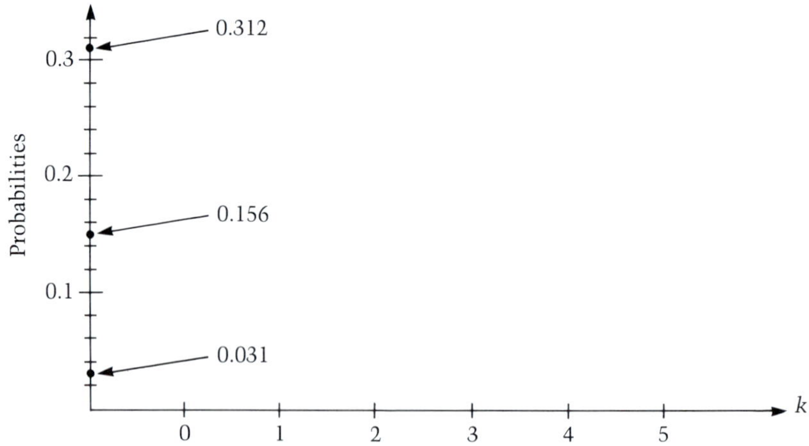

FIG. 9.10.

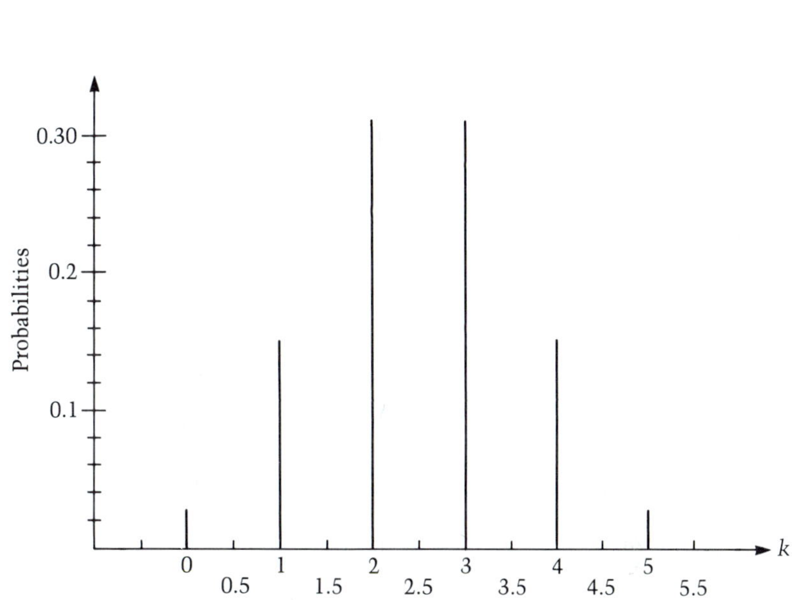

FIG. 9.11.

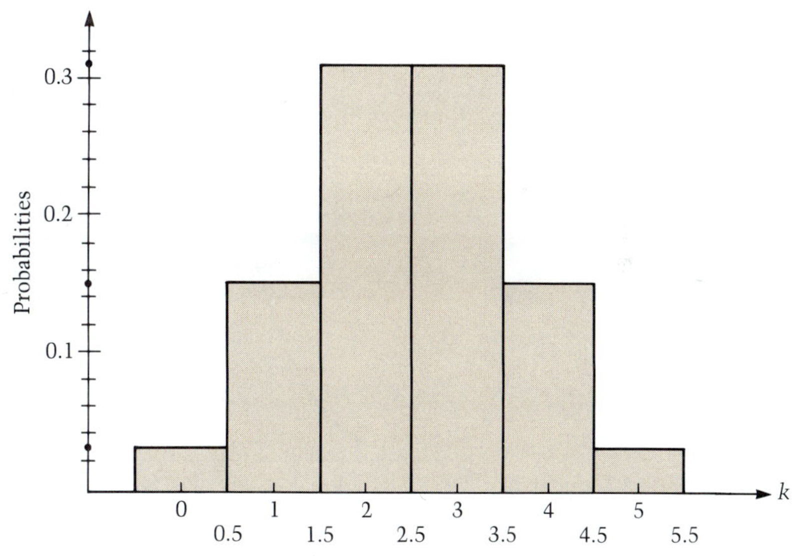

FIG. 9.12.

We now construct histograms for binomial experiments having $n = 6, 10, 20, 25$, and, for each of these, $p = 0.5$. Figures 9.13 through 9.16 show these histograms. Note in these figures that the graphs seem to have a "bell shape" as n increases. The bell shape has been emphasized by the smooth curve which has been imposed on the histograms. The smooth graph is called a *normal curve*, and the pictures suggest that the binomial probabilities

normal curve

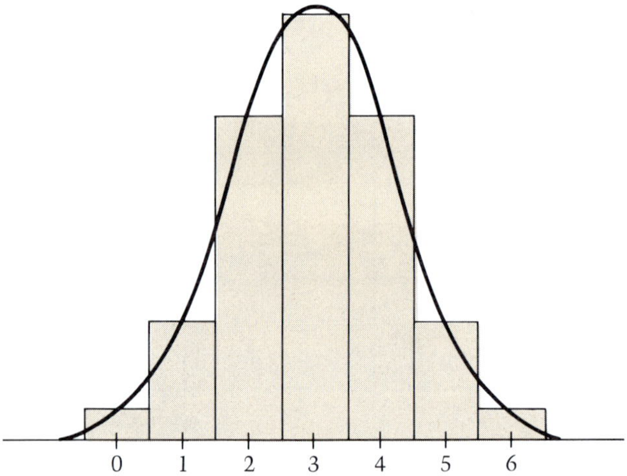

FIG. 9.13. $n = 5$, $p = 0.5$.

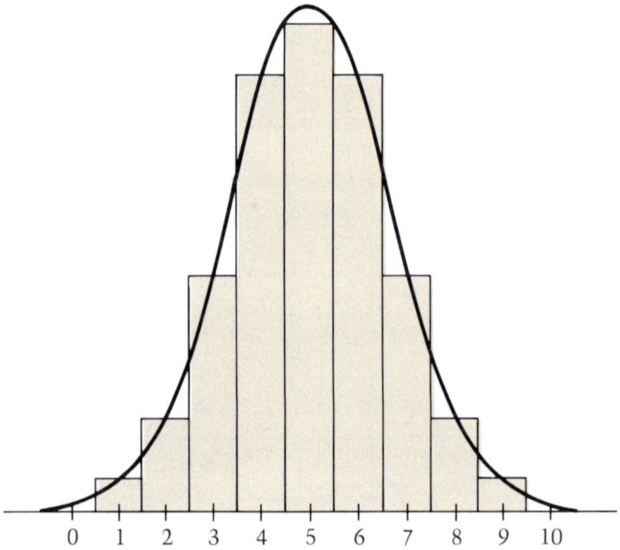

FIG. 9.14. $n = 10$, $p = 0.5$.

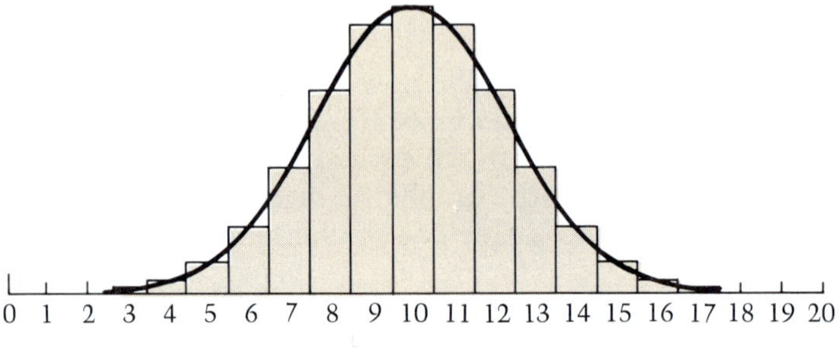

FIG. 9.15. $n = 20$, $p = 0.5$.

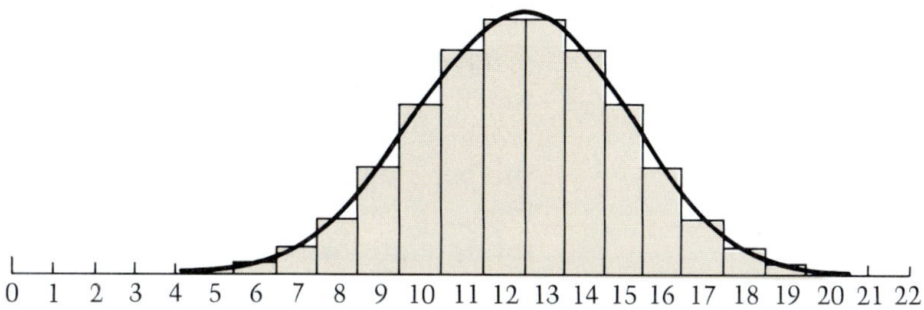

FIG. 9.16. $n = 25, p = 0.5$.

which give rise to the histograms are closely related to normal curves. In Section 9.5 the relation will be investigated more carefully.

Before applications are discussed we must consider some of the details which are fundamental for understanding normal curves. First of all, each normal curve is associated with two numbers, called *parameters*. These are the *mean* μ (mu) and the *standard deviation* σ (sigma). We dealt with the mean $\overline{X}$ and the standard deviation s before, but $\overline{X}$ and s refer to *samples*, whereas μ and σ refer to the *population* mean and standard deviation. If a sample is given, we can compute $\overline{X}$ and s. In actual practice, a population is not often *given* in the sense that we know μ and σ. This is why we need $\overline{X}$ and s; they are used to estimate population parameters. In the following discussions it is frequently assumed that we know these parameters.

parameters

samples

population

In Figure 9.17 the mean is illustrated for a typical normal curve. The mean is a number which locates the center of the curve; in Figure 9.17 the mean is 0, that is, $\mu = 0$. The mean of a normal curve can be any number: positive, negative, or zero.

The standard deviation, which can be a positive number only, has the following property: If we move from the mean μ

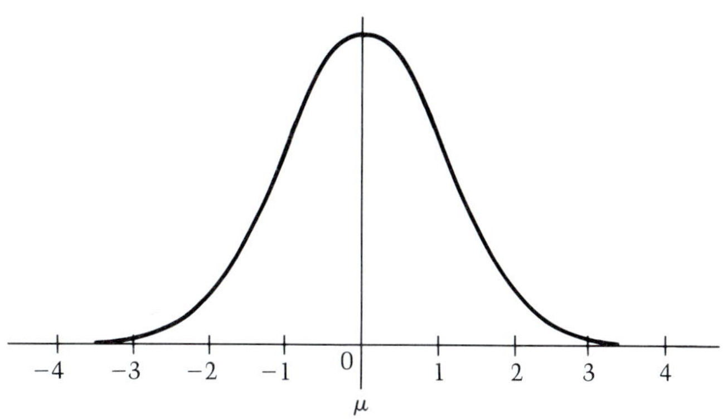

FIG. 9.17. A typical normal curve; $\mu = 0$.

along the horizontal axis to the right 1 standard deviation σ, we reach a point on the curve known as an *inflection point*. From the top of the curve to the inflection point the curve is bending *downward* and after that it is bending *upward;* at the inflection point, the bending changes from downward to upward as illustrated in Figure 9.18. The population parameters μ and σ also have the interpretations of their counterparts $\overline{X}$ and s.

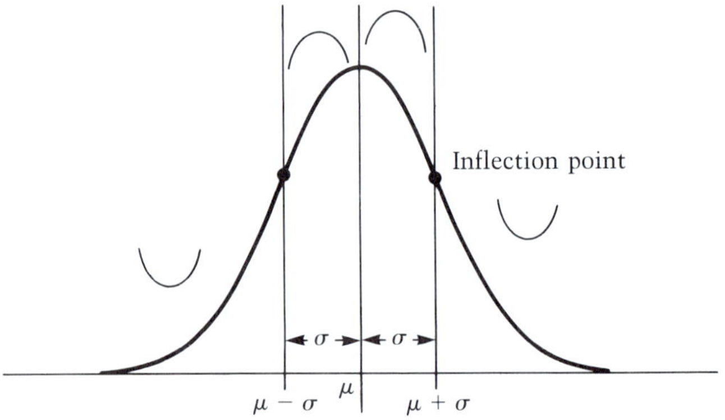

FIG. 9.18. Inflection points.

Normal curves have the convenient property that they are *symmetric with respect to their mean.* This means that if we fold the paper along the mean line shown in Figure 9.18, the two parts of the curve on opposite sides of the mean will fit each other exactly. This fact is important because it allows us the convenience of working with only half of the curve. We will see the advantage of this fact in some of the examples below.

Two convenient values of μ and σ are $\mu = 0$ and $\sigma = 1$. When these particular parameters are chosen, we obtain the *standard normal curve.*

standard normal

DEFINITION OF THE STANDARD NORMAL CURVE The *standard normal* curve (or distribution) is the normal curve having parameters

$$\mu = 0 \qquad \text{and} \qquad \sigma = 1.$$

All our computations with normal curves are referred to this particular one. If a number on the horizontal axis is referred to the standard normal curve, then it is called a *z-number;* otherwise it is called an *x-number.*

We discussed previously how areas in histograms are related to probabilities. In this connection, it is necessary to compute areas under normal curves. Table 6 of Appendix A lists readings which give areas associated with the standard normal curve, and Figure 9.19 illustrates a typical area given by this table. Note in Figure 9.19 that the number z is on the right-hand side of the mean; Table 6 gives only readings for z-numbers which are nonnegative, and the resulting areas are those under the curve and to the left of the number z. It may seem at first sight that the information given in this table is too restrictive to allow us to handle all area problems. However, this arrangement is sufficient for every problem which we will encounter.

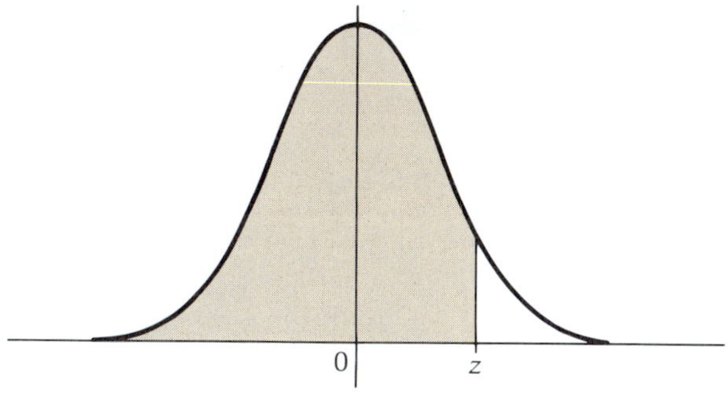

FIG. 9.19.

Table 6 of Appendix A may be used to solve the following types of area problems:

(a) Areas under the standard normal curve to the left of z-numbers (see Figure 9.19).

(b) Areas under the standard normal curve to the right of z-numbers (see Figure 9.20).

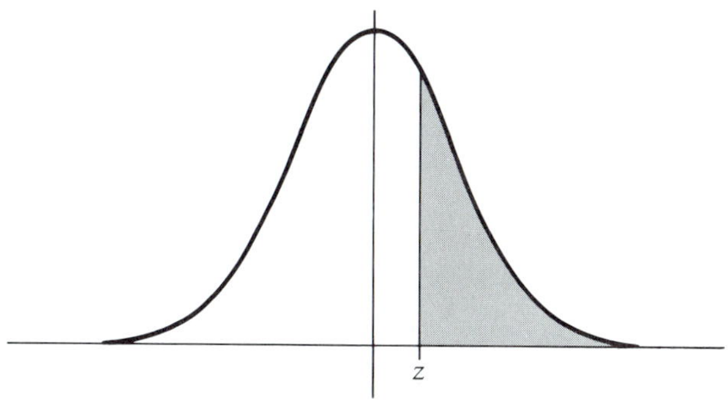

FIG. 9.20.

(c) Areas under the standard normal curve between two z-numbers (see Figure 9.21).

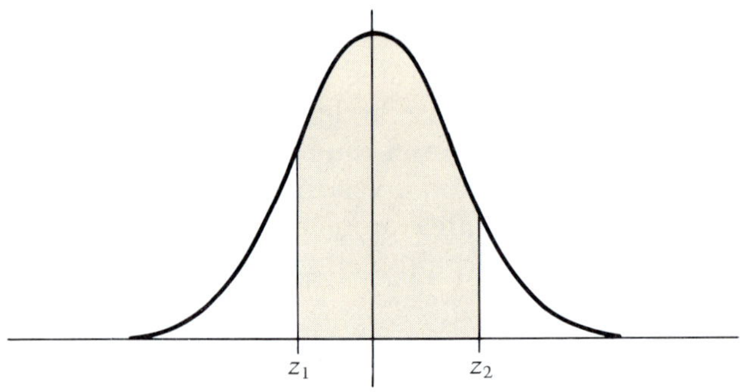

FIG. 9.21.

Area under the standard normal curve

EXAMPLE 1 Find the area under the standard normal curve to the left of $z = 1.27$.

Solution: In Table 6 of Appendix A we locate 1.2 in the left-hand column headed by z and in the first row opposite to z we locate 0.07 $(1.27 = 1.20 + 0.07)$. Now the area is read from the table, as shown in Figure 9.22. Therefore, we conclude that the area under the standard normal curve to the left of 1.27 is 0.8980 (to four places of accuracy).

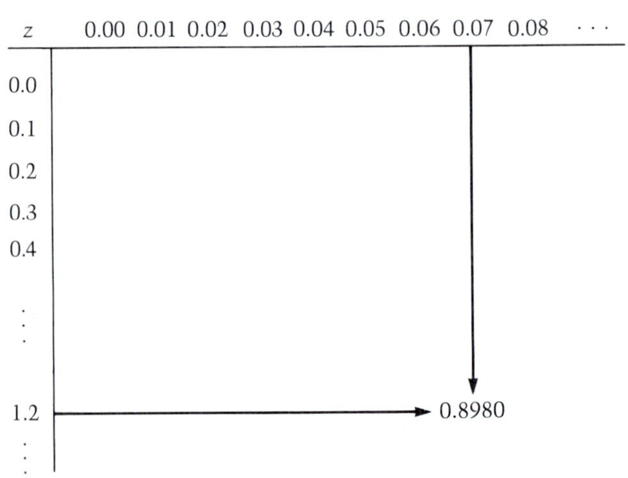

FIG. 9.22.

PRACTICE PROBLEM Find the area under the standard normal curve to the left of $z = 2.38$.

Answer: 0.9913. ■

To compute the areas of type (b) as shown in Figure 9.20, we cannot use Table 6 directly. However, it is easy to solve this kind of problem by using the fact that *the area under the entire curve is* 1. If we desire the area under the standard normal curve to the right of $z = 1.27$, we first find the area to the left of 1.27; this was found in Example 1 to be 0.8980. Then

$$1 - 0.8980 = 0.1020$$

is the area under the curve to the *right* of 1.27 (see Figure 9.23).

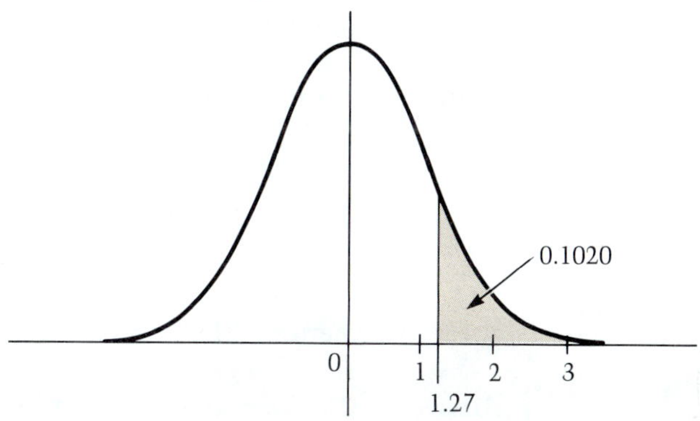

FIG. 9.23.

Standard normal curve area

EXAMPLE 2 Find the area under the standard normal curve to the right of 0.43.

Solution: From Table 6 of Appendix A we find a reading of 0.6664 corresponding to the number $z = 0.43$. Therefore, the area under the standard normal curve to the left of 0.43 is 0.6664 and the area to the right of this number is

$$1 - 0.6664 = 0.3336.$$

PRACTICE PROBLEM Find the area under the standard normal curve to the right of the number $z = 0.76$.

Answer: $1 - 0.7764 = 0.2236$. ■

Standard normal curve area

EXAMPLE 3 Find the area under the standard normal curve between the numbers $z = 0.43$ and $z = 1.27$ (see Figure 9.24).

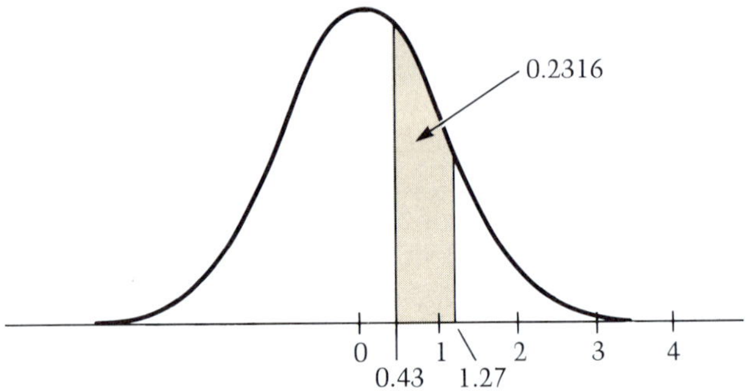

FIG. 9.24.

Solution: We may solve this problem by observing that the desired area may be expressed as the area to the left of 1.27 less the area to the left of 0.43. These two areas have been obtained in Examples 1 and 2. Hence, the area under the standard normal curve between 0.43 and 1.27 is

$$0.8980 - 0.6664 = 0.2316.$$

PRACTICE PROBLEM Find the area under the standard normal curve between $z = 0.98$ and $z = 2.15$.

Answer: $0.9842 - 0.8365 = 0.1477$. ■

The problem of finding areas associated with negative numbers can be easily handled by using the symmetry of the curve. Consider Figure 9.25: The area to the left of $-z$ is the same as the area to the right of z. Hence, if we desire to know the area under the standard normal curve to the left of $-z$, we calculate the area under the curve to the right of z.

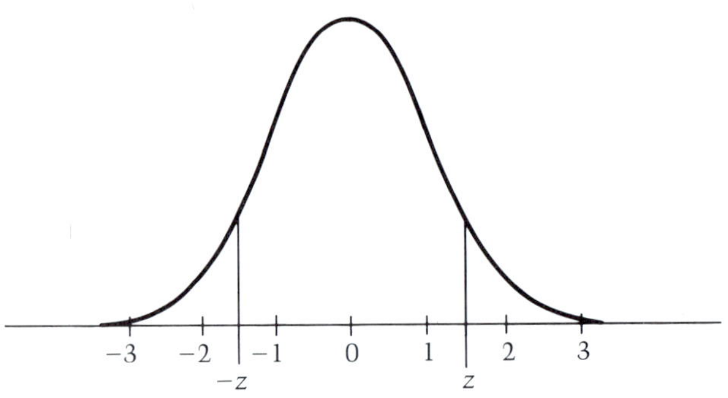

FIG. 9.25.

Standard normal curve area

EXAMPLE 4 Find the area under the standard normal curve to the left of −2.23.

Solution: The area needed here is the same as the area to the right of 2.23. From Table 6, the area under the curve to the left of 2.23 is 0.9871. Therefore, the area under the curve to the right of 2.23 is

$$1 - 0.9871 = 0.0129,$$

which implies that the area to the left of −2.23 is 0.0129.

PRACTICE PROBLEM Find the area under the standard normal curve to the left of −1.71.

Answer: 0.0436. ■

Standard normal curve area

EXAMPLE 5 Find the area under the standard normal curve to the right of −1.10.

Solution: The area desired in this case is the same as the area to the left of 1.10 (see Figure 9.26). Therefore, the area to the right of −1.10 is 0.8643.

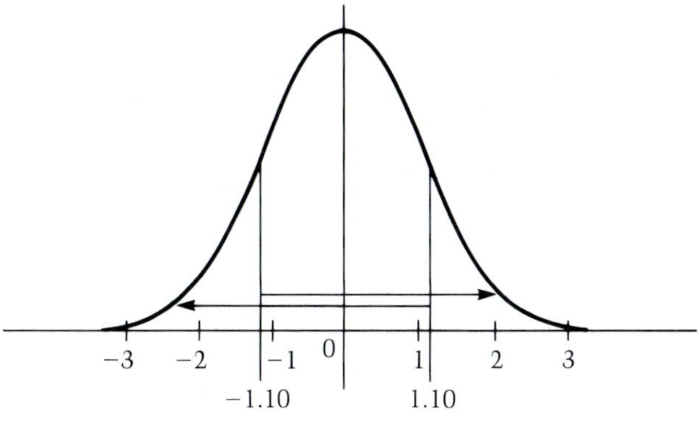

FIG. 9.26.

PRACTICE PROBLEM Find the area under the standard normal curve to the right of −2.26.

Answer: 0.9881. ■

Standard normal curve area

EXAMPLE 6 Find the number z so that the area under the standard normal curve to the left of z is 0.9906.

Solution: In the body of Table 6 we look for the number 0.9906. This number is found opposite the number $z = 2.35$. Therefore, the area to the left of $z = 2.35$ is 0.9906, and 2.35 is the number we seek.

PRACTICE PROBLEM Find the number z so that the area under the standard normal curve to the left of z is 0.9948.

Answer: 2.56. ■

Standard normal curve area

EXAMPLE 7 Find the number z so that the area under the standard normal curve between $-z$ and z is 0.8788.

Solution: The type of area in this case is shown in Figure 9.27. Because of the symmetry, the area of region A is the same as the area of region B. Therefore, the area of A is $\frac{1}{2}$ of the given area of 0.8788:

$$\text{Area of region } A = \left(\frac{1}{2}\right)(0.8788) = 0.4394.$$

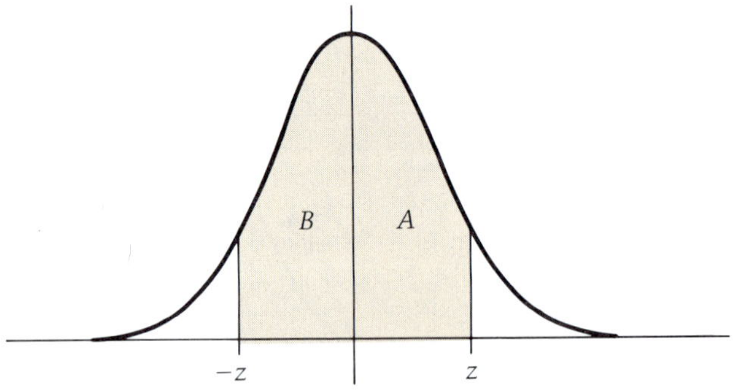

FIG. 9.27.

Since the area on the left of the mean is 0.5000, the number

$$(\text{Area of region } A) + 0.500 = 0.4394 + 0.5000 = 0.9394$$

is the area shown in Figure 9.28. From Table 6, we find that the area to the left of 1.55 is 0.9394. In this case we see that $z = 1.55$ is the answer to the problem.

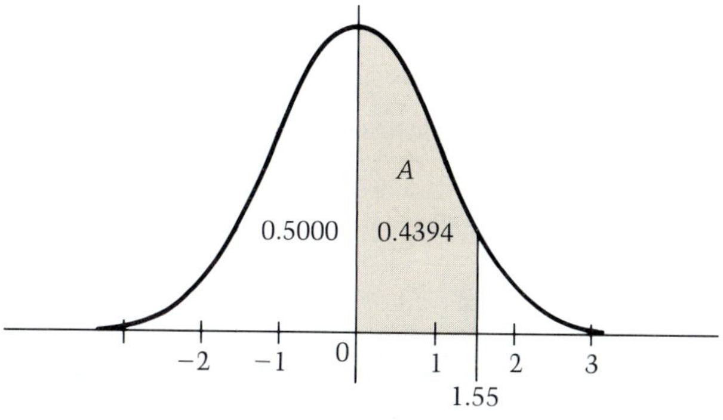

FIG. 9.28.

PRACTICE PROBLEM Find the number z so that the area under the standard normal curve between $-z$ and z is 0.6970.

Answer: 1.03. ■

Table 6 is limited in some respects. For example, the number $z = 1.553$ cannot be read from the table. In this case, we round 1.553 to 1.55 and use 1.55 instead of 1.553. If we are asked to find a z-number such that the area to the left of z is 0.9050, we will not find this in the table either. In cases of this type, we will follow the practice of taking the number in the table which is closest to the number we are looking for. The number in Table 6 closest to 0.9050 is 0.9049. Therefore, we would take $z = 1.31$ as the solution to this problem.

Suppose we wish to determine areas associated with normal curves other than the standard normal curve. How do we proceed? Fortunately, this problem is not difficult to solve. The basic idea behind the solution is to refer all such questions to the standard normal curve. This kind of solution is made possible by the formula

$$z = \frac{x - \mu}{\sigma} \tag{1}$$

where μ and σ are the mean and standard deviation of a given normal curve. The x in formula (1) refers to a number associated with the given normal curve and as usual z refers to the corresponding number for the standard normal curve. These corresponding numbers are illustrated in Figure 9.29.

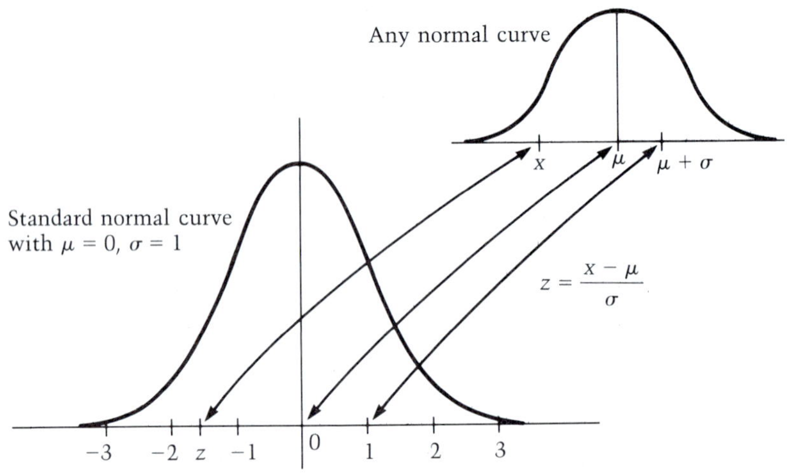

FIG. 9.29.

Convert x to z

EXAMPLE 8 Given a normal curve with $\mu = 100$ and $\sigma = 10$, find the z-number of the standard normal curve which corresponds to the x-number 125.

Solution: Using formula (1), we have

$$z = \frac{x - \mu}{\sigma} = \frac{125 - 100}{10} = 2.5.$$

Thus $z = 2.5$ corresponds to $x = 125$.

PRACTICE PROBLEM For the normal curve of Example 8 find the z-number which corresponds to $x = 98$.

Answer: −0.2. ■

Convert z to x

EXAMPLE 9 Given a normal curve with $\mu = 100$ and $\sigma = 10$, find the x-number for this normal curve which corresponds to the z-number 1.5 of the standard normal curve.

Solution: In this case, $z = 1.5$ and x is unknown. Using formula (1), we have

$$z = \frac{x - \mu}{\sigma}$$

$$1.5 = \frac{x - 100}{10}.$$

Solving this equation for x gives

$$x - 100 = 15$$
$$x = 115.$$

PRACTICE PROBLEM For the normal curve of Example 9, find the x-number which corresponds to $z = 1$.

Answer: $x = 110$. ■

We may now explain how formula (1) is used in the computation of areas. Suppose we are given the normal curve $\mu = 100$ and $\sigma = 10$ and we wish to determine the area under this curve to the left of $x = 125$. Using formula (1), we find the z-number corresponding to $x = 125$; in Example 8 we found this to be 2.5. Now using Table 6, we determine the area under the standard normal curve to the left of 2.5, and this is the area under the normal curve to the left of 125.

Area under a normal curve

EXAMPLE 10 Find the area to the left of 105 under the curve which has mean 100 and standard deviation 10.

Solution: In formula (1) we have

$$z = \frac{105 - 100}{10} = \frac{5}{10} = 0.5.$$

From Table 6 we find the area to the left of 0.5 under the standard normal curve to be 0.6915. Therefore, the area to the left of 105 under the given normal curve is 0.6915.

PRACTICE PROBLEM For the normal curve of Example 10, find the area to the left of 110.

Answer: 0.8413. ■

Area under a normal curve

EXAMPLE 11 Given the normal curve with parameters $\mu = 75$ and $\sigma = 20$, find the area under the curve between $x_1 = 60$ and $x_2 = 80$.

Solution: Using formula (1), we find z-numbers which correspond to x_1 and x_2:

$$z_1 = \frac{x_1 - \mu}{\sigma} = \frac{60 - 75}{20} = \frac{-15}{20} = -0.75,$$

$$z_2 = \frac{x_2 - \mu}{\sigma} = \frac{80 - 75}{20} = \frac{5}{20} = 0.25.$$

Now the problem is to find the area under the standard normal curve between −0.75 and 0.25. Using Table 6, we find the area to the left of 0.25 to be 0.5987. Also the area to the left of −0.75 is

$$1 - 0.7734 = 0.2266.$$

From this we find that the area under the standard normal curve between −0.75 and 0.25 is

$$0.5987 - 0.2266 = 0.3721.$$

Therefore, the area between 60 and 80 under the given normal curve is 0.3721 (see Figure 9.30).

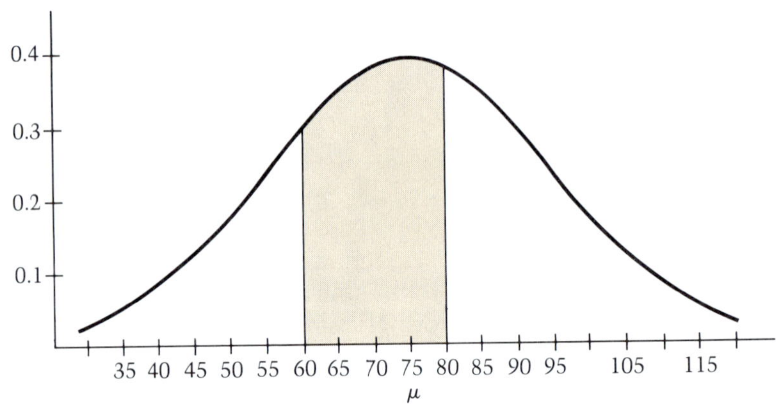

FIG. 9.30.

PRACTICE PROBLEM For the normal curve of Example 11, find the area from 65 to 85.

Answer: 0.3830. ■

Area under a normal curve

EXAMPLE 12 Given the normal curve with mean 200 and standard deviation 40, find the x-number to the right of the mean 200 so that the area under this curve between 200 and x is 21.9% of the total area under the curve.

Solution: Since the area under the curve to the left of the mean is $\frac{1}{2}$ the total area, we wish to find the number x such that the area under the curve to the left of x is

$$0.219 + 0.5000 = 0.7190.$$

Thus we wish to find the z-number such that the area under the standard normal curve to the left of z is 0.7190. From Table 6, we find $z = 0.58$. Therefore, the number x which we want is determined by (1):

$$0.58 = \frac{x - 200}{40}.$$

Solving for x gives

$$x - 200 = 23.2$$
$$x = 223.2.$$

PRACTICE PROBLEM For the normal curve of Example 12, find the x-number to the right of the mean 200 such that the area under this curve between 200 and x is 17% of the total area under the curve.

Answer: 217.6. ■

central region

It is customary in certain problems to refer to a *central region* of a normal curve. This is illustrated in Figure 9.31. The central

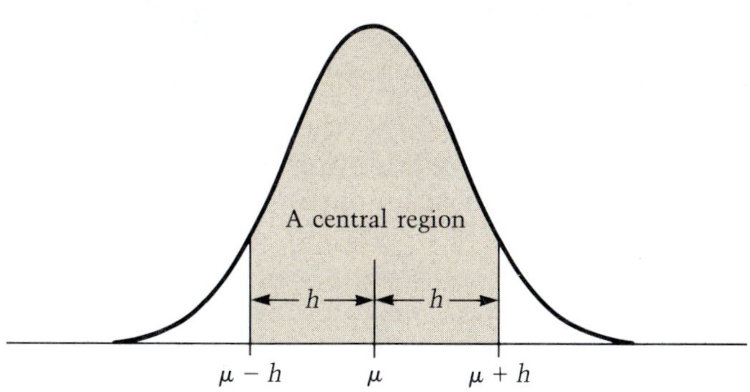

FIG. 9.31.

region shown in that figure is the region under the curve from $\mu - h$ to $\mu + h$, where h is some positive number. In Example 12, we have a normal curve with mean 200 and standard deviation 40. It was determined that for this normal curve the area between 200 and 223.2 is 21.9% of the entire area. Since 223.2 = 200 + 23.2 and 176.8 = 200 − 23.2, we can talk about the central region between 176.8 and 223.2. In this case the central area from 176.8 to 223.2 is 2(21.9) = 43.8% of all the area under the curve.

We frequently refer to the central region of a normal curve *within 1 or 2 standard deviations of the mean;* that is the region between $\mu - \sigma$ and $\mu + \sigma$ (within 1 standard deviation σ) and $\mu - 2\sigma$ and $\mu + 2\sigma$ (within 2 standard deviations σ), respectively. For the standard normal curve, the numbers from -1 $(= \mu - \sigma)$ to 1 $(= \mu + \sigma)$ are within 1 standard deviation of the mean, and Table 6 indicates that this central region has area 0.6826. This tells us that more than 68% of the area of the standard normal curve is contained within 1 standard deviation of the mean.

Central area within 1 standard deviation of the mean

EXAMPLE 13 Determine the central area contained within 1 standard deviation of the mean for any normal curve.

Solution: Suppose we have a normal curve with parameters μ and σ. We must determine the area under the curve between $\mu - \sigma$ and $\mu + \sigma$. Using formula (1) with

$$x_1 = \mu - \sigma \qquad \text{and} \qquad x_2 = \mu + \sigma,$$

we have

$$z_1 = \frac{x_1 - \mu}{\sigma} = \frac{(\mu - \sigma) - \mu}{\sigma} = \frac{-\sigma}{\sigma} = -1,$$

$$z_2 = \frac{x_2 - \mu}{\sigma} = \frac{(\mu + \sigma) - \mu}{\sigma} = \frac{\sigma}{\sigma} = 1.$$

This shows that the z-numbers -1 and 1 correspond to $\mu - \sigma$ and $\mu + \sigma$ for *every* normal curve. Thus the area within 1 standard deviation of the mean for each normal curve is the same as the area within 1 standard deviation for the standard normal curve. From this it may be concluded that about 68% of all the area under each normal curve is contained within 1 standard deviation of the mean.

PRACTICE PROBLEM Find the area contained under any normal curve within 2 standard deviations of the mean.

Answer: 0.9544. ■

The table below shows some common central areas for normal curves.

k	Area within k standard deviations of the mean
1	0.6826
2	0.9544
3	0.9974

EXERCISES

Set A–Left-hand Areas Under the Standard Normal Curve

In Exercises 1 through 12, use Table 6 of Appendix A to find the area under the standard normal curve to the left of the given z-numbers.

1. 1.19 **2.** 2.64 **3.** 1.92 **4.** 3.37
5. 2.96 **6.** 3.94 **7.** −3.42 **8.** −2.58
9. −0.52 **10.** −1.54 **11.** −0.79 **12.** −0.11

Set B–Determining z-numbers from Given Areas

In Exercises 13 through 20, each of the numbers given is an area to the left of a z-number on the standard normal curve. Use Table 6 to determine the correct z-number in each case.

13. 0.8577 **14.** 0.9345 **15.** 0.9973 **16.** 0.5040
17. 0.6141 **18.** 0.8790 **19.** 0.9699 **20.** 0.7764

Set C–Using Formula (1)

In Exercises 21 through 28, use formula (1) to convert each of the given x-numbers to corresponding z-numbers.

21. $x = 90, \mu = 100, \sigma = 10$ **22.** $x = 120, \mu = 50, \sigma = 10$
23. $x = 120, \mu = 200, \sigma = 50$ **24.** $x = 22, \mu = 30, \sigma = 10$
25. $x = 105, \mu = 100, \sigma = 12$ **26.** $x = 60, \mu = 50, \sigma = 15$
27. $x = 225, \mu = 200, \sigma = 40$ **28.** $x = 33, \mu = 30, \sigma = 12$

Set D–Right-hand Areas Under the Standard Normal Curve

In Exercises 29 through 40, use Table 6 to find the area under the standard normal curve to the right of each given z-number.

29. 3.09 **30.** 0.94 **31.** 3.17 **32.** 1.73

33. 2.60
34. 0.07
35. -1.12
36. -2.14
37. -1.29
38. -1.04
39. -2.34
40. -2.67

Set E–Areas Under the Standard Normal Curve Between Given Numbers

In Exercises 41 through 48, use Table 6 to determine the areas under the standard normal curve between the two given z-numbers.

41. 0.55, 1.40
42. 1.05, 1.95
43. 0.49, 2.82
44. 2.06, 3.37
45. -2.0, 1.15
46. -1.14, 1.42
47. -2.42, 1.73
48. -2.00, 2.14

Set F–Finding z-numbers Corresponding to Given Areas

In Exercises 49 through 58, each of the given numbers is an area to the right of some z-number on the standard normal curve. Determine the correct z-number, using Table 6.

49. 0.5871
50. 0.8749
51. 0.9726
52. 0.9898
53. 0.7486
54. 0.2451
55. 0.1056
56. 0.0409
57. 0.3936
58. 0.0055

Set G–Central Areas Under the Standard Normal Curve

In Exercises 59 through 64, use Table 6 to determine the area of the central region determined by the given z-numbers.

59. -1.52, 1.52
60. -0.97, 0.97
61. -2.85, 2.85
62. -3.12, 3.12
63. -0.25, 0.25
64. -1.32, 1.32

Set H–Finding x-numbers from Given z-numbers

In Exercises 65 through 69, use formula (1) to find x-numbers which correspond to the given z-numbers.

65. $z = 1.42, \mu = 250, \sigma = 25$
66. $z = 2.85, \mu = 300, \sigma = 15$
67. $z = 1.18, \mu = 75, \sigma = 5$
68. $z = -1.14, \mu = 80, \sigma = 7$
69. $z = -2.47, \mu = 62.5, \sigma = 8.2$

Set I–Areas Under a Normal Curve

In Exercises 70 through 81, you are given a normal curve with parameters $\mu = 250$ and $\sigma = 25$. Find the requested area:

70. area to the left of 265.
71. area to the left of 255.
72. area to the left of 241.
73. area to the left of 223.

74. area to the right of 257.
75. area to the right of 277.
76. area to the right of 232.
77. area to the right of 221.
78. area between 230 and 260.
79. area between 240 and 250.
80. area between 240 and 280.
81. area between 200 and 300.

Set J–Finding Central Regions

In Exercises 82 through 89, you are given a normal curve with $\mu = 250$ and $\sigma = 25$. Find the central region of this curve which contains the area listed in the exercise.

82. 0.6102 83. 0.8472 84. 0.5222 85. 0.0638
86. 0.9016 87. 0.9624 88. 0.2398 89. 0.7646

9.5 NORMAL CURVES AND BINOMIAL PROBABILITIES

It has been suggested in the previous section that normal curves are closely related to binomial probabilities. Roughly speaking, we can say that if the number of trials in a binomial experiment is large enough, then there is a normal curve which approximates the binomial histogram with reasonable accuracy. This implies right away that in some instances normal curves may take the place of histograms. How can we determine if the number of trials is large enough? A fairly reliable rule is this: If $p \geq 0.5$, then np should be at least 5 and if $p < 0.5$, then $n(1 - p)$ should be at least 5.

In this section we consider the details of approximating binomial probabilities by a normal curve. As we have seen, the histogram of a binomial experiment consists of a number of rectangles, and the areas of these rectangles are the appropriate probabilities. In the normal approximation we replace the histogram by a normal curve. This means that each rectangle of the histogram is replaced by a certain region under the normal curve (see Figure 9.32). Hence, questions relating to binomial probabilities are converted to questions about areas under the normal curve. A very important part of this conversion may be seen in Figure 9.33: the rectangle centered at the number k is replaced by the region of the curve between $k - 0.5$ and $k + 0.5$. The base of each rectangle in the histogram is 1 unit in length and since k is centered on the base, the endpoints are $k - 0.5$ and $k + 0.5$.

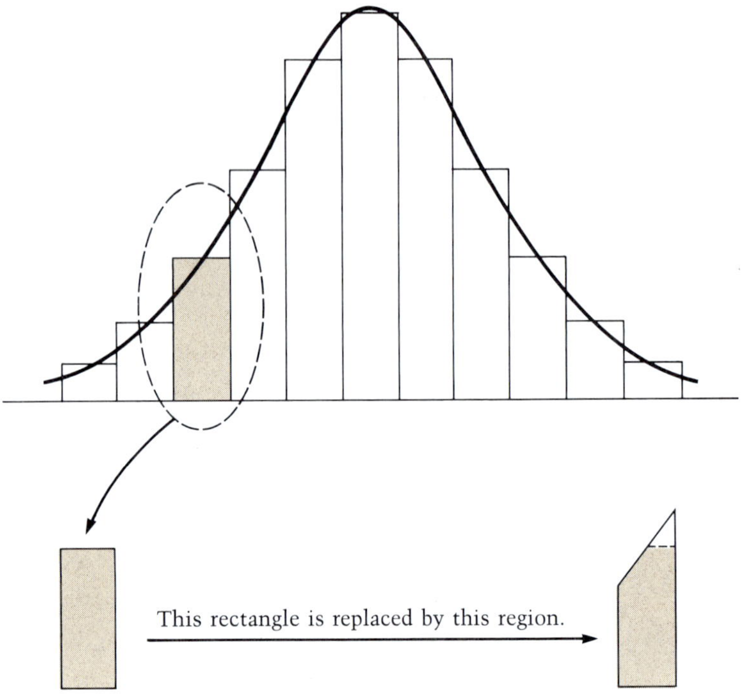

FIG. 9.32.

Each normal curve is uniquely determined by its parameters, and this implies that the problem of imposing the "best-fitting" normal curve on a histogram of binomial probabilities is the same as the problem of determining the appropriate parameters. It is

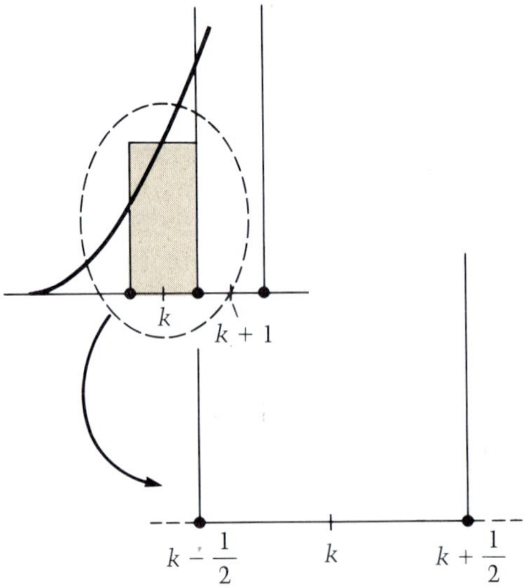

FIG. 9.33.

not difficult to show that the appropriate parameters are given by the formulas below.

MEAN AND STANDARD DEVIATION OF A BINOMIAL EXPERIMENT For a binomial experiment of n trials and success probability p, the mean and standard deviation are given as follows:

$$\mu = np, \qquad \sigma = \sqrt{np(1-p)}.$$

The parameters μ and σ tell us how to "fit" the proper normal curve to the histogram. As an example let $n = 8$ and $p = 0.5$. In Figure 9.34 we have drawn a line graph for the probabilities and also indicated the corresponding points on the normal curve by the small circles. Note that the small circles are not always placed at the end of the lines; this indicates the discrepancy between the numbers $b(k; 8, 0.5)$ and the normal curve.

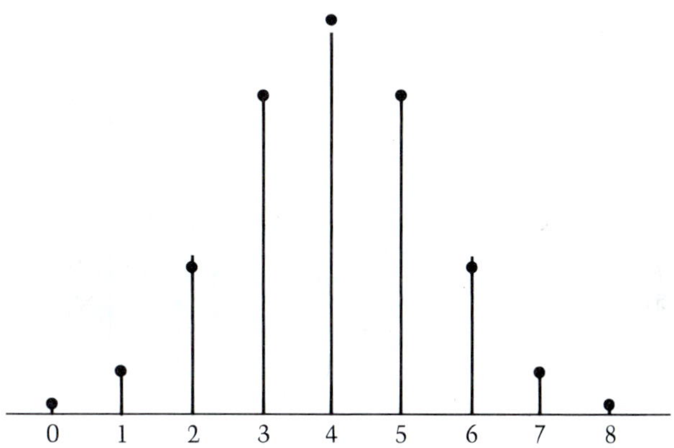

FIG. 9.34.

In Figure 9.35 we have the line graph for a binomial experiment in which $n = 25$ and $p = 0.3$, and the position of the normal curve is indicated by the small circles. In Figures 9.36 and 9.37 we have the histograms obtained from the line graphs of Figures 9.34 and 9.35.

To combine these notions into a specific problem, suppose we toss a fair coin 400 times and ask for the probability of obtaining exactly 200 heads. This is a binomial experiment in which $n = 400$ and $p = 0.5$, and if we solve it in the usual way, we must compute

$$b(200; 400, 0.5) = \frac{400!}{200!200!}(0.5)^{400}.$$

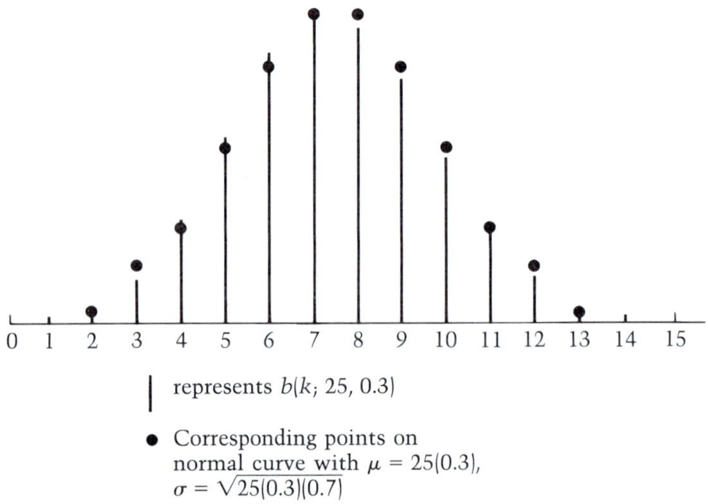

FIG. 9.35.

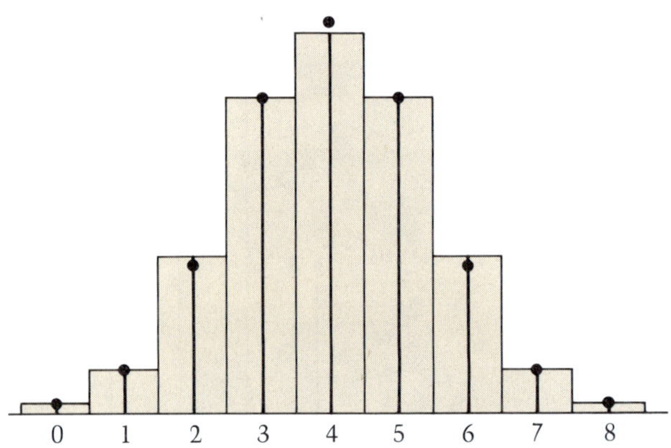

FIG. 9.36.

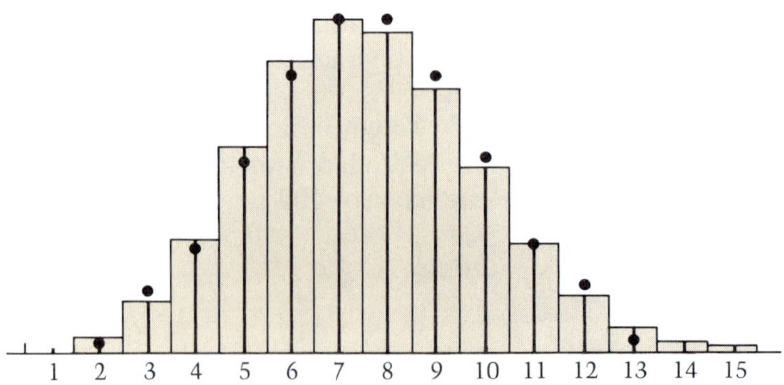

FIG. 9.37.

Such a calculation is not inviting, even on a computer. It is for problems of this type that the normal curve is so useful. The best fitting normal curve in this instance is the one having parameters

$$\mu = np = 400(0.5) = 200,$$
$$\sigma = \sqrt{np(1-p)} = \sqrt{400(0.5)(0.5)} = 20(0.5) = 10.$$

In Figure 9.38 we have indicated the relation between that part of the histogram for $b(k; 400, 0.5)$ which shows $b(200; 400, 0.5)$ and the normal curve with $\mu = 200$ and $\sigma = 10$. The probability of obtaining exactly 200 heads is the area of the rectangle shown in Figure 9.38, and from this figure we can clearly see that to obtain an approximation by the normal curve requires that we calculate the area under the curve from 199.5 to 200.5. To accomplish this, we convert the x-numbers to the corresponding z-numbers:

$$z_1 = \frac{199.5 - 200}{10} = \frac{-0.5}{10} = -0.05,$$
$$z_2 = \frac{200.5 - 200}{10} = \frac{0.5}{10} = 0.05.$$

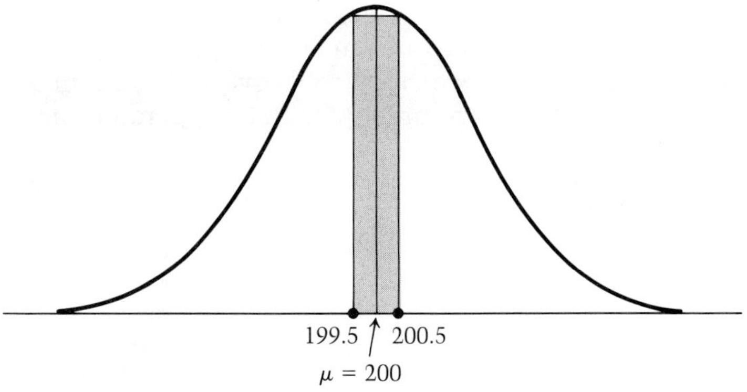

FIG. 9.38.

Using Table 6, we find that the area under the standard normal curve from −0.05 to 0.05 is 0.0398. Therefore, we conclude that the probability of obtaining exactly 200 heads in 400 tosses of a fair coin is approximately 0.04. This indicates that it is very unlikely that an even split will be obtained when a fair coin is tossed 400 times. This result is certainly not surprising; we do not expect to obtain an exactly even split—only something in the neighborhood of such a result. For example, the probability is very high that we obtain between 180 and 220 heads; this is true because 180 and 220 are within 2 standard deviations of the mean.

Approximating a probability by a normal curve

EXAMPLE 1 A fair die is rolled 100 times. What is the probability of obtaining a 4 at least 25 times?

Solution: We interpret this problem as a binomial experiment in which obtaining a 4 is "success." Then $p = \frac{1}{6} = 0.167$ (rounded) and $q = \frac{5}{6} = 0.833$. Since $p < 0.5$, we should not use a normal approximation unless $nq > 5$. But $nq = 100(0.833) = 83.3 > 5$, and here a normal curve may be used to give an accurate approximation in this case. First, we calculate the parameters:

$$\mu = np = 100(0.167) = 16.7,$$
$$\sigma = \sqrt{npq} = \sqrt{100(0.167)(0.833)} = 3.73.$$

In Figure 9.39 the relation between the histogram and the approximating normal curve is indicated. We see that the actual probability is the sum of the areas under the rectangles marked 25, 26, 27, . . . , 99, 100. This area we may approximate as the area under the normal curve to the right of 24.5 (note that it is not the area under the curve to the right of 25 that we desire). Thus we need to convert the x-number 24.5 to the corresponding z-number for the standard normal curve:

$$z = \frac{24.5 - 16.7}{3.73} = 2.10.$$

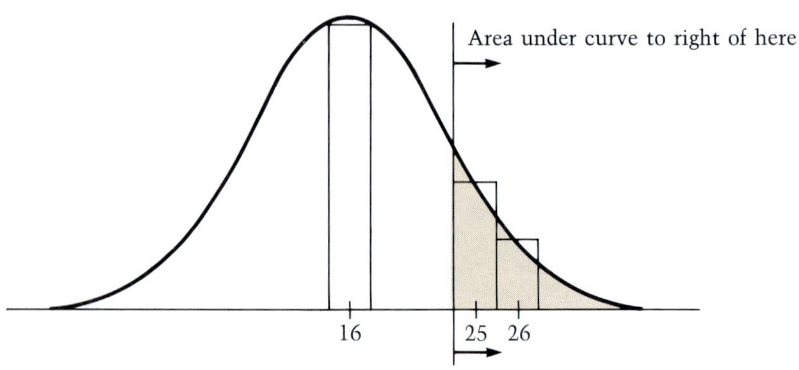

FIG. 9.39.

We therefore wish to find the area under the standard normal curve to the right of 2.10 and by Table 6 this area is $1 - 0.9821 = 0.0179$. Thus we conclude that the probability of getting a 4 at least 25 times on 100 rolls of a fair die is approximately 0.018.

PRACTICE PROBLEM Use a normal approximation to find the probability of obtaining a 6 no more than 20 times on 100 rolls of a fair die.

Answer: 0.8461. ■

Approximating a probability by a normal curve

EXAMPLE 2 A student takes a true-false test on a subject he knows nothing about. He answers each question by tossing a coin; he chooses answer T if heads turns up and F if tails turns up. There are 100 questions on the test, and a passing grade requires at least 60 correct answers. What is the probability that the student passes?

Solution: We may look upon this as a binomial experiment with $n = 100$ and $p = 0.5$. The problem is to compute the probability that the student achieves at least 60 successes in 100 trials. The exact probability is given by the sum

$$\sum b(k; 100, 0.5)$$

where k ranges from 60 to 100, and this would involve considerable computation. Since $p \geq 0.5$ and $np = 100(0.5) = 50 > 5$, it is appropriate to use a normal approximation. The relation between the exact probability (the rectangles) and the normal approximation is shown in Figure 9.40. From the figure it is clear that we

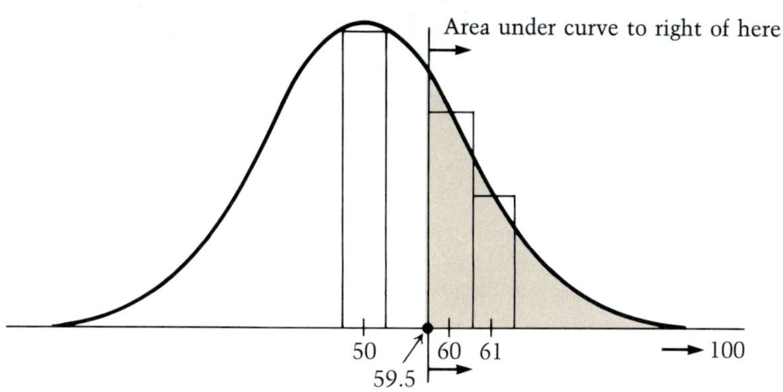

FIG. 9.40.

need to calculate the area under the normal curve to the right of 59.5. The mean and standard deviation are

$$\mu = np = 100(0.5) = 50,$$
$$\sigma = \sqrt{npq}$$
$$= \sqrt{100(0.5)(0.5)} = 5.$$

Then

$$z = \frac{59.5 - 50}{5} = 1.9.$$

Hence, we need the area under the standard normal curve to the right of 1.9. Using Table 6 we find that this area is $1 - 0.9713 = 0.1287$. From this we conclude that the probability of the student's receiving a passing grade is less than 0.13.

PRACTICE PROBLEM In Example 2 find the probability that the student gets at least half of the questions correct.

Answer: 0.5398. ■

Medical problem

EXAMPLE 3 A pharmaceutical company estimates that a certain medicine for headaches is 75% effective. If a random sample of 225 headache sufferers are given the medicine, what is the probability that at least 180 individuals are cured of their headaches?

Solution: Consider this as a binomial experiment in which "success" is a cured headache. Then $n = 225$ and $p = 0.75$. A normal curve may be used to solve this problem, since $np = 225(0.75) = 168.75 > 5$. The parameters for the normal curve are

$$\mu = np = 168.75,$$
$$\sigma = \sqrt{npq}$$
$$= \sqrt{225(0.75)(0.25)}$$
$$= 6.495.$$

We must determine the probability of at least 180 cures, and this implies that we need to find the area under the normal curve to the right of 179.5:

$$z = \frac{179.5 - 168.75}{6.495} = 1.655.$$

From Table 6, the area to the right of 1.655 under the standard normal curve is

$$1 - 0.9515 = 0.0485.$$

Therefore, we conclude that the probability is about 0.05% that at least 180 of the 225 individuals will be cured of their headaches.

PRACTICE PROBLEM In Example 3 determine the probability that no more than 165 headaches are cured.

Answer: 0.3085. ■

EXERCISES

Set A–The Normal Approximation

Each of Exercises 1 through 10 presents a binomial experiment. In each case, use the normal approximation to determine the probability requested.

1. If $n = 50$ and $p = 0.5$, what is the probability of obtaining no more than 40 successes?
2. If $n = 50$ and $p = 0.5$, what is the probability of obtaining at least 30 successes?
3. If $n = 60$ and $p = 0.45$, what is the probability of at least 30 but no more than 40 successes?
4. If $n = 100$ and $p = 0.62$, what is the probability of obtaining exactly 55 successes?
5. If $n = 125$ and $p = 0.48$, what is the probability of obtaining exactly 62 successes?
6. If $n = 185$ and $p = 0.66$, what is the probability of obtaining fewer than 120 successes?
7. If $n = 190$ and $p = 0.37$, what is the probability of obtaining more than 75 successes?
8. If $n = 260$ and $p = 0.25$, what is the probability of obtaining at least 60 but no more than 70 successes?
9. If $n = 635$ and $p = 0.475$, what is the probability of obtaining at least 300 but no more than 325 successes?

10. If $n = 945$ and $p = 0.835$, what is the probability of obtaining at least 775 but no more than 810 successes?

Set B–Applications

In Exercises 11 through 20, use the normal approximation to solve the problems given.

11. A fair coin is flipped 100 times. What is the probability of getting an even split between heads and tails?

12. A fair die is rolled 100 times. What is the probability of obtaining a number larger than 4 at least 40 times?

13. A box contains 35 white and 65 red marbles. An experiment consists of drawing 1 marble randomly and replacing it. If this experiment is repeated 400 times, what is the probability of drawing at least 95 but no more than 120 white marbles?

14. A student takes a test of 100 questions. At least 60 correct answers are necessary for a passing grade. If the student knows 60% of the questions of the type asked, what is the probability that she passes?

Quality control

15. A manufacturer of radiator caps packs the caps in cartons of 1000 caps each. It is known from experience that about 1% of the caps are defective. What is the probability that a randomly chosen box of caps contains more than 10 defective caps?

Agriculture

16. A research worker in agriculture knows that about 80% of a certain kind of seed will germinate when planted. If 250 of these seeds are randomly chosen and planted, what is the probability that at least 215 will germinate?

Insurance

17. An insurance company is interested in the mortality rate of men in the age range from 50 to 65. The company knows that about 15% of 50-year olds die before they reach age 65. If the company writes insurance policies on 450 men aged 50, what is the probability that the company will have more than 74 claims from the beneficiaries of this group within the next 15 years?

Health care

18. Under certain conditions it is known that the infection rate of mononucleosis is about 12%. If 300 individuals are exposed, what is the probability that at least 40 will contract the disease?

Political science

19. A political analyst estimates that 60% of the members of the U.S. House of Representatives favor congressional reform. If reform legislation is introduced in the House, what is the probability it will pass? Assume that the bill can be passed by a simple majority (the House has 435 members).

Restaurant planning

20. Alice's Restaurant estimates that 65% of its lunch customers order the Blue Plate Special. If Alice is expecting 500 customers for lunch on Tuesday, what is the smallest number of specials that should be prepared to make sure that the probability of the restaurant's not being able to fill all orders is no more than 5%.

9.6 SOME NORMAL POPULATIONS

In this section we are working with a very important class of populations and shall see how to answer some statistical questions about them. These populations are called *normal* because their histograms resemble normal curves. In the previous section we saw how histograms of binomial experiments resemble normal curves and how this resemblance can be exploited. The relation between histograms and normal curves is by no means restricted to binomial experiments. Many populations exhibit this characteristic. When we speak of such a population, we frequently say that the population is *normally distributed* or is *approximately normally distributed.*

normally distributed

As an example of the kind of population under discussion here, consider the heights of American males. This population is approximately normally distributed. To illustrate this statement we will construct a histogram resulting from a random sample of male heights. Table 9.7 shows a sample of size $n = 50$, where heights have been measured to the nearest 0.5 inch.

TABLE 9.7

Heights	Heights	Heights	Heights	Heights
69.0	70.5	64.0	65.0	67.0
60.5	73.0	62.0	68.0	67.5
65.0	66.5	65.0	73.0	68.5
68.5	63.5	61.5	63.5	69.0
66.5	64.0	63.0	70.0	68.0
75.0	70.0	63.0	71.0	67.5
73.0	68.5	66.0	66.5	61.5
66.0	74.5	68.5	65.0	70.0
70.0	64.0	71.0	74.0	66.0
71.5	64.5	66.5	64.0	59.0

The distribution of the data given in Table 9.7 is shown in Table 9.8.

TABLE 9.8

Interval	Tally	Frequency
57.75–59.75	I	1
59.75–61.75	III	3
61.75–63.75	~~IIII~~	5
63.75–65.75	~~IIII~~ IIII	9
65.75–67.75	~~IIII~~ ~~IIII~~	10
67.75–69.75	~~IIII~~ III	8
69.75–71.75	~~IIII~~ III	8
71.75–73.75	III	3
73.75–75.75	III	3

The histogram corresponding to the distribution of Table 9.8 has been constructed in Figure 9.41. The histogram is suggestive of a normal shape, but not strongly so. The normal shape can be made more pronounced by increasing the size of the sample and decreasing the length of the intervals.

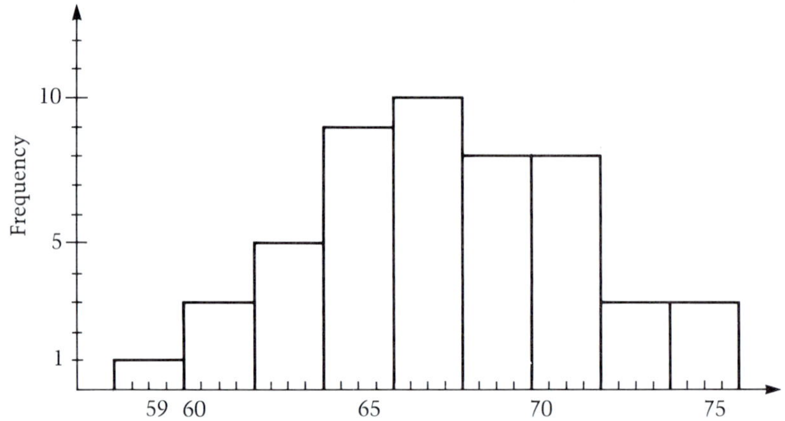

FIG. 9.41.

In this example, it is possible to measure heights to a much greater accuracy than has been done in Table 9.7. Theoretically it is possible for such measurements to be any numbers (within an appropriate range). These measurements can change continuously from one to the other. For this reason, the population is said to be *continuous*. On the other hand, the population consisting of the number of children per family, for example, can consist only of the numbers 0, 1, 2, 3, 4, . . . A family obviously cannot have 2.5 children. Note that these numbers cannot change continu-

continuous

discrete

continuity correction

ously from one to the other; instead, they jump from one to the next. A population of this kind is said to be *discrete.* Our concern here is to answer questions about populations which are approximately normally distributed by using appropriate normal curves. In order to make the procedure most effective we will need to take into account whether populations are discrete or continuous. Discrete populations require that we make a *continuity correction* while no such correction is needed for a continuous population. The continuity correction will be illustrated in Example 2. Example 1 deals with a continuous population for which no correction is necessary.

Continuous normal population

EXAMPLE 1 It is known that the heights of adult American males are normally distributed with mean 68 inches and standard deviation 4 inches. What is the probability that a randomly chosen male will be taller than 6 feet?

Solution: This population is continuous and we must find the area under the normal curve to the right of 72 (since 6 feet = 72 inches). This is shown in Figure 9.42. To make this determination we use the formula

$$z = \frac{x - \mu}{\sigma},$$

as before. In this case, $x = 72$, $\mu = 68$, and $\sigma = 4$. Then

$$z = \frac{72 - 68}{4} = \frac{4}{4} = 1,$$

and the area to the right of 1 under the standard normal curve is 0.1587. Therefore, the probability of randomly choosing a male

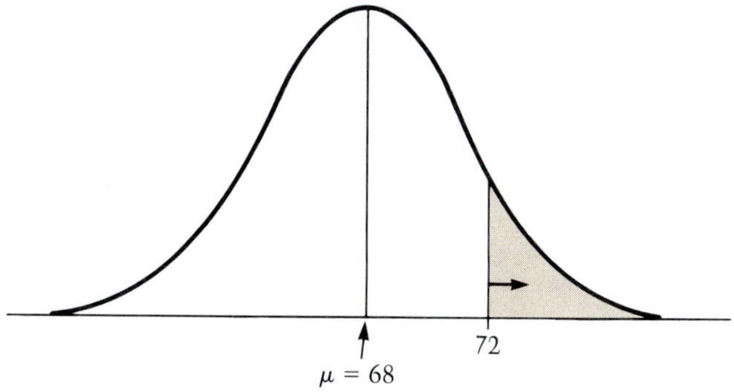

FIG. 9.42.

whose height is greater than 6 feet is approximately 0.16. Note that we may also conclude that about 16% of the population is taller than 6 feet.

PRACTICE PROBLEM For the population of Example 1 determine the proportion of the population that is less than 5 feet tall.

Answer: 0.0228 or slightly more than 2%. ■

Discrete normal population

EXAMPLE 2 Certain electronic parts are shipped in 100-lot cartons. We consider the number of defective parts per carton. This population is known to be approximately normally distributed with mean 10 and standard deviation 1. What is the probability that a randomly chosen carton will contain no more than 7 defective items?

Solution: This population is discrete, and therefore, a continuity correction is appropriate. The terminology has not been used before, but we have seen the continuity correction in connection with binomial distributions, and all discrete populations are similar to those distributions in the way normal curves are related to binomial distributions. In this example, we wish to determine the probability of 7 or fewer defective parts. Thus, we wish to compute the area to the left of 7.5 as shown in Figure 9.43. The probability at 7 must be "spread" over the entire interval from 6.5 to 7.5. Then

$$z = \frac{7.5 - 10}{1} = -2.5.$$

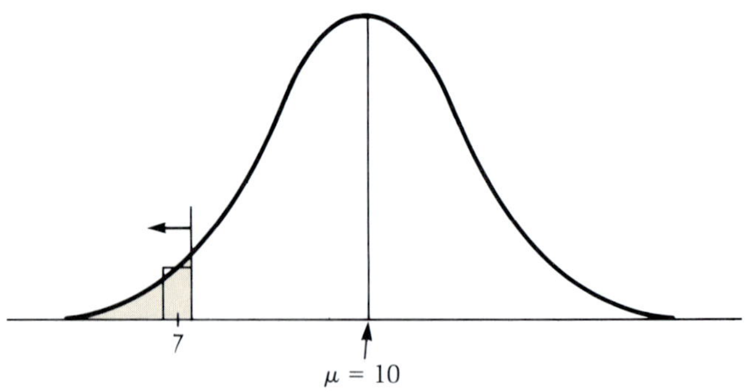

FIG. 9.43.

From Table 6 of Appendix A, the area to the left of -2.5 is 0.0062. Thus the probability is 0.0062 that a randomly chosen carton will contain no more than 7 defective items.

PRACTICE PROBLEM In Example 2 determine the probability that a randomly chosen carton will contain at least 11 defective parts.

Answer: 0.3085. ■

Suppose a department store is concerned about the sizes of men's suits to keep in stock. The store will not want to stock sizes that are not likely to sell. Since there are very few men who are 8 feet tall, for example, the store will most likely not want to stock clothes for men this tall. And similarly the store would find little or no profit in stocking suits for men who are unusually small. It is clear that the store must decide upon a smallest- and a largest-size suit to keep in stock. A decision of this kind can be made in a routine way from a knowledge of the "average" height of adult males. We have assumed in Example 1 that male heights are normally distributed with mean 68 inches and standard deviation 4 inches. If this information is correct, it is easy to determine a range of heights that will include, for example, 95% of all males. We have already seen that 95% of the area under the standard normal curve is contained between $z = -1.960$ and $z = 1.960$. Converting these z-numbers to x-numbers in this example gives $x = 60.61$ and $x = 75.84$:

$$z = \frac{x - \mu}{\sigma}, \qquad z = \frac{x - \mu}{\sigma},$$

$$-1.960 = \frac{x - 68}{4}, \qquad 1.960 = \frac{x - 68}{4},$$

$$x = 60.61; \qquad x = 75.84.$$

We now have the information that about 95% of all adult males are between 60.61 and 75.84 inches (or between about 5.05 feet and 6.32 feet) tall. If the department store decides to stock suit sizes to fit men in this range of heights, they should be able to fit about 95% of their customers.

confidence interval

The interval [5.05, 6.32] which we have just found is said to be a *95% confidence interval* for adult male heights; that is we can be 95% confident that the height of a randomly chosen male will be in this range.

Confidence interval for male heights

EXAMPLE 3 Using the information of Example 2, find a 97% confidence interval for male heights.

Solution: Using Table 6, we find that 97% of the area under the standard normal curve is contained between $z = -2.17$ and $z = 2.17$. Converting these to x-numbers, we get

$$z = \frac{x - \mu}{\sigma}, \qquad z = \frac{x - \mu}{\sigma},$$

$$-2.17 = \frac{x - 68}{4}, \qquad 2.17 = \frac{x - 68}{4},$$

$$-8.68 = x - 68, \qquad 8.68 = x - 68,$$

$$x = 59.32; \qquad x = 76.68.$$

Therefore, the 97% confidence interval is [59.32, 76.86] in inches or [4.9, 6.4] in feet.

PRACTICE PROBLEM Find a 90% confidence interval for male heights.

Answer: [61.4, 74.6] in inches; [5.1, 6.2] in feet. ■

Confidence interval for number of defective parts

EXAMPLE 4 For the population of Example 2 find a 95% confidence interval for the number of defective parts per carton.

Solution: We already know that 95% of the area under the standard normal curve is contained between −1.960 and 1.960. Therefore, we need to convert these two z-numbers into x-numbers for this population. But we must keep in mind that this is a discrete population and that a continuity correction is needed for this conversion. For $z = 1.960$, we have

$$1.960 = \frac{(x + 0.5) - 10}{1}$$

$$= x - 9.5,$$

$$x = 11.460.$$

Since x must be a whole number, we take $x = 11$. This shows that the larger x-number is 1 unit to the right of the mean of this population (remember that the mean is given to be 10), and therefore, the other x-number must be 1 unit to the left of 10. Thus a 95% confidence interval for the number of defective parts is

$$[9, 11].$$

The number of defective parts, 95% of the time, is between 9 and 11.

PRACTICE PROBLEM For the population of Example 2, find a 98% confidence interval for the number of defective parts per carton.

Answer: [8, 12]. ■

EXERCISES

Set A–Probability Questions About Normal Populations

Grades on a quiz

1. Assume that grades on a quiz are normally distributed with mean 70 and standard deviation 12. What is the probability that a grade chosen at random will be at least 80? no more than 65?

Waiting times

2. The waiting times in a checkout line in a supermarket are known to be normally distributed with mean 5 minutes and standard deviation 3 minutes. What is the probability that a randomly selected shopper will have to wait no longer than 6 minutes? at least 4 minutes?

Weights of cats

3. The weights of cats are normally distributed with mean 9.6 pounds and standard deviation 1.5 pounds. What is the probability that a randomly chosen cat weighs over 10 pounds?

IQ scores

4. Assume that IQ scores are normally distributed with mean 100 and standard deviation 10. What is the probability that a randomly chosen person has an IQ of at least 115? 95 or less?

Gestation periods

5. Gestation periods for human babies are normally distributed with mean 268 days and standard deviation 15 days. What is the probability that a randomly chosen birth has a gestation period of more than 300 days?

Sizes of washers

6. A machine stamps out washers which are supposed to measure 5 centimeters in diameter. The machine, when properly adjusted, is known to produce washers having measurements which are normally distributed with mean 5 centimeters and standard deviation 0.05 centimeter. If the machine is in

proper working order, what is the probability that a randomly chosen washer will measure more than 5.5 centimeters?

Number of absences

7. The number of absences per student in a statistics class is normally distributed with a mean of 7 and a standard deviation of 2. What is the probability that a randomly chosen student will be absent no more than 6 times?

Women's shoe sizes

8. Women's shoe sizes are normally distributed with mean 8.5 and standard deviation 1.5. What proportion of women have shoe sizes greater than 9?

GRE

9. Verbal scores on the GRE (Graduate Record Examination) are normally distributed with mean 500 and standard deviation 100. What proportion of those taking the test have scores greater than 675?

Test times

10. The length of time required to finish a standard mathematics test is normally distributed with mean 75 minutes and standard deviation 10 minutes. What proportion of students will finish in 90 minutes?

Set B–Miscellaneous Problems About Normal Populations

Test times

11. Use the information given in Exercise 10 to determine how much time should be allowed for this test so that 90% of the students who take it will finish on time.

Women's shoe sizes

12. Use the information given in Exercise 8. If the manager of a shoe store wishes to stock shoe sizes which will fit 95% of his female customers, what should his range of sizes be?

GRE

13. Use the information given in Exercise 9 to determine the lowest score a student can receive to be in the upper 5% of test takers.

Gestation periods

14. Use the information given in Exercise 5 to determine a 95% confidence interval for gestation periods of human babies.

Failure times of computer chips

15. The failure times of the computer chip ZUK/0095 are normally distributed with mean 25.25 months and standard deviation 0.75 month. The Zukumuku Company (which manufactures the ZUK/0095) will replace, free of charge, any chip which fails within 2 years after purchase. What proportion of the chips will the company have to replace?

Annual incomes

16. The annual incomes of families living in Clarkston, GA, are normally distributed with mean \$19,500 and standard deviation \$600. A bank will not open a branch in Clarkston unless at least 65% of the population earns \$19,300 or more. Will the bank open a branch in Clarkston?

Quiz design

17. An instructor wishes to design a quiz so that no more than 6% of the class fails (achieves a score of less than 60). Assume that grades will be normally distributed and have standard deviation 10. What will the mean of the quiz have to be?

Quiz design

18. An instructor assigns grades so that 5% of his students receive A's, 20% B's, 50% C's, 20% D's, and 5% F's. The scores on a final exam were normally distributed with mean 65 and standard deviation 18. What were the cutoff scores for the letter grades?

Weights of cereal cartons

19. A manufacturer of breakfast foods wants to fill cereal cartons with 16 ounces of cereal. The weights of the cartons are normally distributed and the machine which fills the boxes operates with a standard deviation of 0.5 ounces (when in proper working order). If the manufacturer wishes that no more than 5% of the boxes contain less than 16 ounces, what should be the mean of the weights? How should the machine be regulated so that no more than 5% of the boxes have less than 16 ounces?

Number of requests for brochures

20. The State Park Service of Georgia keeps brochures on hand to distribute to persons interested in the state's parks. The number of requests for these brochures per day is normally distributed with mean 50 and standard deviation 10. Using 95% confidence, find a confidence interval for the number of requests per day. How many should be kept on hand each day so that supplies are exhausted no more than 5% of the time?

9.7 SUMMARY AND REVIEW EXERCISES

REVIEW OF CONCEPTS

1. **Mean of a sample.** If $x_1, x_2, \ldots x_n$ is a sample of a population, then the mean $\overline{X}$ of the sample is

$$\overline{X} = \left(\frac{1}{n}\right) \sum x$$

where x takes all the values of the sample.

2. **Median of a sample.** Suppose a set of data contains n numbers and these are arranged in their natural order. Then
(i) If n is odd, there is a middle number in the set and it is the median of the sample.
(ii) If n is even, the average of the two numbers nearest the middle is the median.

3. **Standard deviation of sample.** Suppose that $x_1, x_2, \ldots, x_n$ is a sample of some population and $\overline{X}$ is the mean of the sample. Then the *variance* V is defined as follows:

$$V = \frac{\sum (x - \overline{X})^2}{n}$$

where the sum is taken over all values of the sample. The *standard deviation* s of the sample is the square root of the variance V.

4. **Chebyshev's Theorem.** Let the sample $x_1, x_2, \ldots, x_n$ be given and have mean $\overline{X}$ and standard deviation s. Then the fraction of the numbers of the sample which lie within k standard deviations of the mean is at least

$$1 - \frac{1}{k^2}.$$

5. **Formula connecting x- and z-numbers.** If μ and σ are the mean and standard deviation of a normal distribution, then

$$z = \frac{x - \mu}{\sigma}.$$

6. **Formulas for mean and standard deviation of a binomial distribution.** If n is the number of trials and p the probability of success on a single trial of a binomial experiment, then

$$\mu = np \qquad \text{and} \qquad \sigma = \sqrt{np(1 - p)}.$$

REVIEW EXERCISES

1. Construct a histogram of the binomial distribution with $n = 10$ and $p = 0.5$.
2. Use the following data for this exercise.

4	5	1	9	4	6	8	8	3	10	6	4	4
10	7	10	10	7	7	4	6	3	2	6	5	5

(a) Construct a frequency distribution.
(b) Construct a histogram from the frequency distribution.

3. Find the mean of the sample

4 5 1 9 4 6 8 8 3 10 6 4 4

4. (a) Find the median of the sample in Exercise 3.
(b) Find the median of

4 6 4 5 10 3 1 9 8 3 4 6

5. Find the standard deviation of the sample in Exercise 3.

6. Suppose the mean germinating time for seed Z is 3.5 days with standard deviation 0.5 day. What is the largest percentage of these seeds which will require more than 5 days to germinate?

7. What is the area under the standard normal curve
(a) to the left of -1.2? (b) to the right of 0.32?
(c) between 0.5 and 1.5?

8. If a normal distribution has mean 125 and standard deviation 3, what is the area under this curve between 120 and 130?

9. A fair coin is tossed 400 times. What is the probability that the number of heads obtained is between 198 and 215?

10. The mean length of time required to finish an English test is 90 minutes with standard deviation 10 minutes. If 500 students take this test, how many will finish within 75 minutes?

APPENDIX

A

TABLES

TABLE 1 **Compound Interest $(1 + i)^n$**

n	1%	$1\frac{1}{2}$%	2%	$2\frac{1}{2}$%	3%	4%	5%	6%
1	1.0100	1.0150	1.0200	1.0250	1.0300	1.0400	1.0500	1.0600
2	1.0201	1.0302	1.0404	1.0506	1.0609	1.0816	1.1025	1.1236
3	1.0303	1.0457	1.0612	1.0769	1.0927	1.1249	1.1576	1.1910
4	1.0406	1.0614	1.0824	1.1038	1.1255	1.1699	1.2155	1.2625
5	1.0510	1.0773	1.1041	1.1314	1.1593	1.2167	1.2763	1.3382
6	1.0615	1.0934	1.1262	1.1597	1.1941	1.2653	1.3401	1.4185
7	1.0721	1.1098	1.1487	1.1887	1.2299	1.3159	1.4071	1.5036
8	1.0829	1.1265	1.1717	1.2184	1.2668	1.3686	1.4775	1.5938
9	1.0937	1.1434	1.1951	1.2489	1.3048	1.4233	1.5513	1.6895
10	1.1046	1.1605	1.2190	1.2801	1.3439	1.4802	1.6289	1.7908
11	1.1157	1.1779	1.2434	1.3121	1.3842	1.5395	1.7103	1.8983
12	1.1268	1.1956	1.2682	1.3449	1.4258	1.6010	1.7959	2.0122
13	1.1381	1.2136	1.2936	1.3785	1.4685	1.6651	1.8856	2.1329
14	1.1495	1.2318	1.3195	1.4130	1.5126	1.7317	1.9799	2.2609
15	1.1610	1.2502	1.3459	1.4483	1.5580	1.8009	2.0789	2.3966
16	1.1726	1.2690	1.3728	1.4845	1.6047	1.8730	2.1829	2.5404
17	1.1843	1.2880	1.4002	1.5216	1.6528	1.9479	2.2920	2.6928
18	1.1961	1.3073	1.4282	1.5597	1.7024	2.0258	2.4066	2.8543
19	1.2081	1.3270	1.4568	1.5987	1.7535	2.1068	2.5270	3.0256
20	1.2202	1.3469	1.4859	1.6386	1.8061	2.1911	2.6533	3.2071
21	1.2324	1.3671	1.5157	1.6796	1.8603	2.2788	2.7860	3.3996
22	1.2447	1.3876	1.5460	1.7216	1.9161	2.3699	2.9253	3.6035
23	1.2572	1.4084	1.5769	1.7646	1.9736	2.4647	3.0715	3.8197
24	1.2697	1.4295	1.6084	1.8087	2.0328	2.5633	3.2251	4.0489
25	1.2824	1.4509	1.6406	1.8539	2.0938	2.6658	3.3864	4.2919
26	1.2953	1.4727	1.6734	1.9003	2.1566	2.7725	3.5557	4.5494
27	1.3082	1.4948	1.7069	1.9478	2.2213	2.8834	3.7335	4.8223
28	1.3213	1.5172	1.7410	1.9965	2.2879	2.9987	3.9201	5.1117
29	1.3345	1.5400	1.7758	2.0464	2.3566	3.1187	4.1161	5.4184
30	1.3478	1.5631	1.8114	2.0976	2.4273	3.2434	4.3219	5.7435
31	1.3613	1.5865	1.8476	2.1500	2.5001	3.3731	4.5380	6.0881
32	1.3749	1.6103	1.8845	2.2038	2.5751	3.5081	4.7649	6.4534
33	1.3887	1.6345	1.9222	2.2589	2.6523	3.6484	5.0032	6.8406
34	1.4026	1.6590	1.9607	2.3153	2.7319	3.7943	5.2533	7.2510
35	1.4166	1.6839	1.9999	2.3732	2.8139	3.9461	5.5160	7.6861
36	1.4308	1.7091	2.0399	2.4325	2.8983	4.1039	5.7918	8.1473
37	1.4451	1.7348	2.0807	2.4933	2.9852	4.2681	6.0814	8.6361
38	1.4595	1.7608	2.1223	2.5557	3.0748	4.4388	6.3855	9.1543
39	1.4741	1.7872	2.1647	2.6196	3.1670	4.6164	6.7048	9.7035
40	1.4889	1.8140	2.2080	2.6851	3.2620	4.8010	7.0400	10.2857

TABLE 1 Compound Interest $(1 + i)^n$ (*cont.*)

n	7%	$7\frac{1}{2}$%	8%	$8\frac{1}{2}$%	9%	10%	11%	12%
1	1.0700	1.0750	1.0800	1.0850	1.0900	1.1000	1.1100	1.1200
2	1.1449	1.1556	1.1664	1.1772	1.1881	1.2100	1.2321	1.2544
3	1.2250	1.2423	1.2597	1.2773	1.2950	1.3310	1.3676	1.4049
4	1.3108	1.3355	1.3605	1.3859	1.4116	1.4641	1.5181	1.5735
5	1.4026	1.4356	1.4693	1.5037	1.5386	1.6105	1.6851	1.7623
6	1.5007	1.5433	1.5869	1.6315	1.6771	1.7716	1.8704	1.9738
7	1.6058	1.6590	1.7138	1.7701	1.8280	1.9487	2.0762	2.2107
8	1.7182	1.7835	1.8509	1.9206	1.9926	2.1436	2.3045	2.4760
9	1.8385	1.9172	1.9990	2.0839	2.1719	2.3579	2.5580	2.7731
10	1.9672	2.0610	2.1589	2.2610	2.3674	2.5937	2.8394	3.1058
11	2.1049	2.2156	2.3316	2.4532	2.5804	2.8531	3.1518	3.4785
12	2.2522	2.3818	2.5182	2.6617	2.8127	3.1384	3.4985	3.8960
13	2.4098	2.5604	2.7196	2.8879	3.0658	3.4523	3.8833	4.3635
14	2.5785	2.7524	2.9372	3.1334	3.3417	3.7975	4.3104	4.8871
15	2.7590	2.9589	3.1722	3.3997	3.6425	4.1772	4.7846	5.4736
16	2.9522	3.1808	3.4259	3.6887	3.9703	4.5950	5.3109	6.1304
17	3.1588	3.4194	3.7000	4.0023	4.3276	5.0545	5.8951	6.8660
18	3.3799	3.6758	3.9960	4.3425	4.7171	5.5599	6.5436	7.6900
19	3.6165	3.9515	4.3157	4.7116	5.1417	6.1159	7.2633	8.6128
20	3.8697	4.2479	4.6610	5.1120	5.6044	6.7275	8.0623	9.6463
21	4.1406	4.5664	5.0338	5.5466	6.1088	7.4002	8.9492	10.8038
22	4.4304	4.9089	5.4365	6.0180	6.6586	8.1403	9.9336	12.1003
23	4.7405	5.2771	5.8715	6.5296	7.2579	8.9543	11.0263	13.5523
24	5.0724	5.6729	6.3412	7.0846	7.9111	9.8497	12.2392	15.1786
25	5.4274	6.0983	6.8485	7.6868	8.6231	10.8347	13.5855	17.0001
26	5.8074	6.5557	7.3964	8.3401	9.3992	11.9182	15.0799	19.0401
27	6.2139	7.0474	7.9881	9.0490	10.2451	13.1100	16.7386	21.3249
28	6.6488	7.5759	8.6271	9.8182	11.1671	14.4210	18.5799	23.8839
29	7.1143	8.1441	9.3173	10.6528	12.1722	15.8631	20.6237	26.7499
30	7.6123	8.7550	10.0627	11.5583	13.2677	17.4494	22.8923	29.9599
31	8.1451	9.4116	10.8677	12.5407	14.4618	19.1943	25.4104	33.5551
32	8.7153	10.1174	11.7371	13.6067	15.7633	21.1138	28.2056	37.5817
33	9.3253	10.8763	12.6760	14.7632	17.1820	23.2252	31.3082	42.0915
34	9.9781	11.6920	13.6901	16.0181	18.7284	25.5477	34.7521	47.1425
35	10.6766	12.5689	14.7853	17.3796	20.4140	28.1024	38.5749	52.7996
36	11.4239	13.5115	15.9682	18.8569	22.2512	30.9127	42.8181	59.1356
37	12.2236	14.5249	17.2456	20.4597	24.2538	34.0039	47.5281	66.2318
38	13.0793	15.6143	18.6253	22.1988	26.4367	37.4043	52.7562	74.1797
39	13.9948	16.7853	20.1153	24.0857	28.8160	41.1448	58.5593	83.0812
40	14.9745	18.0442	21.7245	26.1330	31.4094	45.2593	65.0009	93.0510

TABLE 2 Present Value of a Dollar $(1 + i)^{-n}$

n	1%	$1\frac{1}{2}$%	2%	$2\frac{1}{2}$%	3%	4%	5%	6%
1	.99010	.98522	.98039	.97561	.97087	.96154	.95238	.94340
2	.98030	.97066	.96117	.95181	.94260	.92456	.90703	.89000
3	.97059	.95632	.94232	.92860	.91514	.88900	.86384	.83962
4	.96098	.94218	.92385	.90595	.88849	.85480	.82270	.79209
5	.95147	.92826	.90573	.88385	.86261	.82193	.78353	.74726
6	.94205	.91454	.88797	.86230	.83748	.79031	.74622	.70496
7	.92372	.90103	.87056	.84127	.81309	.75992	.71068	.66506
8	.92348	.88771	.85349	.82075	.78941	.73069	.67684	.62741
9	.91434	.87459	.83676	.80073	.76642	.70259	.64461	.59190
10	.90529	.86167	.82035	.78120	.74409	.67556	.61391	.55839
11	.89632	.84893	.80426	.76214	.72242	.64958	.58468	.52679
12	.88745	.83639	.78849	.74356	.70138	.62460	.55684	.49697
13	.87866	.82403	.77303	.72542	.68095	.60057	.53032	.46884
14	.86996	.81185	.75788	.70773	.66112	.57748	.50507	.44230
15	.86135	.79985	.74301	.69047	.64186	.55526	.48102	.41727
16	.85282	.78803	.72845	.67362	.62317	.53391	.45811	.39365
17	.84438	.77639	.71416	.65720	.60502	.51337	.43630	.37136
18	.83602	.76491	.70016	.64117	.58739	.49363	.41552	.35034
19	.82774	.75361	.68643	.62553	.57029	.47464	.39573	.33051
20	.81954	.74247	.67297	.61027	.55368	.45639	.37689	.31180
21	.81143	.73150	.65978	.59539	.53755	.43883	.35894	.29416
22	.80340	.72069	.64684	.58086	.52189	.42196	.34185	.27751
23	.79544	.71004	.63416	.56670	.50669	.40573	.32557	.26180
24	.78757	.69954	.62172	.55288	.49193	.39012	.31007	.24698
25	.77977	.68921	.60953	.53939	.47761	.37512	.29530	.23300
26	.77205	.67902	.59758	.52623	.46369	.36069	.28124	.21981
27	.76440	.66899	.58586	.51340	.45019	.34682	.26785	.20737
28	.75684	.65910	.57437	.50088	.43708	.33348	.25509	.19563
29	.74934	.64936	.56311	.48866	.42435	.32065	.24295	.18456
30	.74192	.63976	.55207	.47674	.41199	.30832	.23138	.17411
31	.73458	.63031	.54125	.46511	.39999	.29646	.22036	.16425
32	.72730	.62099	.53063	.45377	.38834	.28506	.20987	.15496
33	.72010	.61182	.52023	.44270	.37703	.27409	.19987	.14619
34	.71297	.60277	.51003	.43191	.36604	.26355	.19035	.13791
35	.70591	.59387	.50003	.42137	.35538	.25342	.18129	.13011
36	.69892	.58509	.49022	.41109	.34503	.24367	.17266	.12274
37	.69200	.57644	.48061	.40107	.33498	.23430	.16444	.11579
38	.68515	.56792	.47119	.39128	.32523	.22529	.15661	.10924
39	.67837	.55953	.46195	.38174	.31575	.21662	.14915	.10306
40	.67165	.55126	.45289	.37243	.30656	.20829	.14205	.09722

TABLE 2 **Present Value of a Dollar $(1 + i)^{-n}$ (*cont.*)**

n	7%	$7\frac{1}{2}$%	8%	$8\frac{1}{2}$%	9%	10%	11%	12%
1	0.9346	0.9302	0.9259	0.9217	0.9174	0.9091	0.9009	0.8929
2	0.8734	0.8653	0.8573	0.8495	0.8417	0.8264	0.8116	0.7972
3	0.8163	0.8050	0.7938	0.7829	0.7722	0.7513	0.7312	0.7118
4	0.7629	0.7488	0.7350	0.7216	0.7084	0.6830	0.6587	0.6355
5	0.7130	0.6966	0.6806	0.6650	0.6499	0.6209	0.5935	0.5674
6	0.6663	0.6480	0.6302	0.6129	0.5963	0.5645	0.5346	0.5066
7	0.6227	0.6028	0.5835	0.5649	0.5470	0.5132	0.4817	0.4523
8	0.5820	0.5607	0.5403	0.5207	0.5019	0.4665	0.4339	0.4039
9	0.5439	0.5216	0.5002	0.4799	0.4604	0.4241	0.3909	0.3606
10	0.5083	0.4852	0.4632	0.4423	0.4224	0.3855	0.3522	0.3220
11	0.4751	0.4513	0.4289	0.4076	0.3875	0.3505	0.3173	0.2875
12	0.4440	0.4199	0.3971	0.3757	0.3555	0.3186	0.2858	0.2567
13	0.4150	0.3906	0.3677	0.3463	0.3262	0.2897	0.2575	0.2292
14	0.3878	0.3633	0.3405	0.3191	0.2992	0.2633	0.2320	0.2046
15	0.3624	0.3380	0.3152	0.2941	0.2745	0.2394	0.2090	0.1827
16	0.3387	0.3144	0.2919	0.2711	0.2519	0.2176	0.1883	0.1631
17	0.3166	0.2925	0.2703	0.2499	0.2311	0.1978	0.1696	0.1456
18	0.2959	0.2720	0.2502	0.2303	0.2120	0.1799	0.1528	0.1300
19	0.2765	0.2531	0.2317	0.2122	0.1945	0.1635	0.1377	0.1161
20	0.2584	0.2354	0.2145	0.1956	0.1784	0.1486	0.1240	0.1037
21	0.2415	0.2190	0.1987	0.1803	0.1637	0.1351	0.1117	0.0926
22	0.2257	0.2037	0.1839	0.1662	0.1502	0.1228	0.1007	0.0826
23	0.2109	0.1895	0.1703	0.1531	0.1378	0.1117	0.0907	0.0738
24	0.1971	0.1763	0.1577	0.1412	0.1264	0.1015	0.0817	0.0659
25	0.1842	0.1640	0.1460	0.1301	0.1160	0.0923	0.0736	0.0588
26	0.1722	0.1525	0.1352	0.1199	0.1064	0.0839	0.0663	0.0525
27	0.1609	0.1419	0.1252	0.1105	0.0976	0.0763	0.0597	0.0469
28	0.1504	0.1320	0.1159	0.1019	0.0895	0.0693	0.0538	0.0419
29	0.1406	0.1228	0.1073	0.0939	0.0822	0.0630	0.0485	0.0374
30	0.1314	0.1142	0.0994	0.0865	0.0754	0.0573	0.0437	0.0334
31	0.1228	0.1063	0.0920	0.0797	0.0691	0.0521	0.0394	0.0298
32	0.1147	0.0988	0.0852	0.0735	0.0634	0.0474	0.0355	0.0266
33	0.1072	0.0919	0.0789	0.0677	0.0582	0.0431	0.0319	0.0238
34	0.1002	0.0855	0.0730	0.0624	0.0534	0.0391	0.0288	0.0212
35	0.0937	0.0796	0.0676	0.0575	0.0490	0.0356	0.0259	0.0189
36	0.0875	0.0740	0.0626	0.0530	0.0449	0.0323	0.0234	0.0169
37	0.0818	0.0688	0.0580	0.0489	0.0412	0.0294	0.0210	0.0151
38	0.0765	0.0640	0.0537	0.0450	0.0378	0.0267	0.0190	0.0135
39	0.0715	0.0596	0.0497	0.0415	0.0347	0.0243	0.0171	0.0120
40	0.0668	0.0554	0.0460	0.0383	0.0318	0.0221	0.0154	0.0107

TABLE 3 **Exponential Functions (e^x)**

x	e^x	x	e^x	x	e^x	x	e^x
0.00	1.0000	**0.50**	1.6487	**1.00**	2.7183	**1.50**	4.4817
0.01	1.0101	0.51	1.6653	1.01	2.7456	1.51	4.5267
0.02	1.0202	0.52	1.6820	1.02	2.7732	1.52	4.5722
0.03	1.0305	0.53	1.6989	1.03	2.8011	1.53	4.6182
0.04	1.0408	0.54	1.7160	1.04	2.8292	1.54	4.6646
0.05	1.0513	**0.55**	1.7333	**1.05**	2.8577	**1.55**	4.7115
0.06	1.0618	0.56	1.7507	1.06	2.8864	1.56	4.7588
0.07	1.0725	0.57	1.7683	1.07	2.9154	1.57	4.8066
0.08	1.0833	0.58	1.7860	1.08	2.9447	1.58	4.8550
0.09	1.0942	0.59	1.8040	1.09	2.9743	1.59	4.9037
0.10	1.1052	**0.60**	1.8221	**1.10**	3.0042	**1.60**	4.9530
0.11	1.1163	0.61	1.8404	1.11	3.0344	1.61	5.0028
0.12	1.1275	0.62	1.8589	1.12	3.0649	1.62	5.0531
0.13	1.1388	0.63	1.8776	1.13	3.0957	1.63	5.1039
0.14	1.1503	0.64	1.8965	1.14	3.1268	1.64	5.1552
0.15	1.1618	**0.65**	1.9155	**1.15**	3.1582	**1.65**	5.2070
0.16	1.1735	0.66	1.9348	1.16	3.1899	1.66	5.2593
0.17	1.1853	0.67	1.9542	1.17	3.2220	1.67	5.3122
0.18	1.1972	0.68	1.9739	1.18	3.2544	1.68	5.3656
0.19	1.2092	0.69	1.9937	1.19	3.2871	1.69	5.4195
0.20	1.2214	**0.70**	2.0138	**1.20**	3.3201	**1.70**	5.4739
0.21	1.2337	0.71	2.0340	1.21	3.3535	1.71	5.5290
0.22	1.2461	0.72	2.0544	1.22	3.3872	1.72	5.5845
0.23	1.2586	0.73	2.0751	1.23	3.4212	1.73	5.6407
0.24	1.2712	0.74	2.0959	1.24	3.4556	1.74	5.6973
0.25	1.2840	**0.75**	2.1170	**1.25**	3.4903	**1.75**	5.7546
0.26	1.2969	0.76	2.1383	1.26	3.5254	1.76	5.8124
0.27	1.3100	0.77	2.1598	1.27	3.5609	1.77	5.8709
0.28	1.3231	0.78	2.1815	1.28	3.5966	1.78	5.9299
0.29	1.3364	0.79	2.2034	1.29	3.6328	1.79	5.9895
0.30	1.3499	**0.80**	2.2255	**1.30**	3.6693	**1.80**	6.0496
0.31	1.3634	0.81	2.2479	1.31	3.7062	1.81	6.1104
0.32	1.3771	0.82	2.2705	1.32	3.7434	1.82	6.1719
0.33	1.3910	0.83	2.2933	1.33	3.7810	1.83	6.2339
0.34	1.4049	0.84	2.3164	1.34	3.8190	1.84	6.2965
0.35	1.4191	**0.85**	2.3396	**1.35**	3.8574	**1.85**	6.3598
0.36	1.4333	0.86	2.3632	1.36	3.8962	1.86	6.4237
0.37	1.4477	0.87	2.3869	1.37	3.9354	1.87	6.4883
0.38	1.4623	0.88	2.4109	1.38	3.9749	1.88	6.5535
0.39	1.4770	0.89	2.4351	1.39	4.0149	1.89	6.6194
0.40	1.4918	**0.90**	2.4596	**1.40**	4.0552	**1.90**	6.6859
0.41	1.5068	0.91	2.4843	1.41	4.0960	1.91	6.7531
0.42	1.5220	0.92	2.5093	1.42	4.1371	1.92	6.8210
0.43	1.5373	0.93	2.5345	1.43	4.1787	1.93	6.8895
0.44	1.5527	0.94	2.5600	1.44	4.2207	1.94	6.9588
0.45	1.5683	**0.95**	2.5857	**1.45**	4.2631	**1.95**	7.0287
0.46	1.5841	0.96	2.6117	1.46	4.3060	1.96	7.0993
0.47	1.6000	0.97	2.6379	1.47	4.3492	1.97	7.1707
0.48	1.6161	0.98	2.6645	1.48	4.3929	1.98	7.2427
0.49	1.6323	0.99	2.6912	1.49	4.4371	1.99	7.3155
0.50	1.6487	**1.00**	2.7183	**1.50**	4.4817	**2.00**	7.3891

TABLE 3 Exponential Functions (e^x) (*cont.*)

x	e^x	x	e^x	x	e^x	x	e^x
2.00	7.3891	**2.50**	12.182	**3.00**	20.086	**3.50**	33.115
2.01	7.4633	2.51	12.305	3.01	20.287	3.51	33.448
2.02	7.5383	2.52	12.429	3.02	20.491	3.52	33.784
2.03	7.6141	2.53	12.554	3.03	20.697	3.53	34.124
2.04	7.6906	2.54	12.680	3.04	20.905	3.54	34.467
2.05	7.7679	**2.55**	12.807	**3.05**	21.115	**3.55**	34.813
2.06	7.8460	2.56	12.936	3.06	21.328	3.56	35.163
2.07	7.9248	2.57	13.066	3.07	21.542	3.57	35.517
2.08	8.0045	2.58	13.197	3.08	21.758	3.58	35.874
2.09	8.0849	2.59	13.330	3.09	21.977	3.59	36.234
2.10	8.1662	**2.60**	13.464	**3.10**	22.198	**3.60**	36.598
2.11	8.2482	2.61	13.599	3.11	22.421	3.61	36.966
2.12	8.3311	2.62	13.736	3.12	22.646	3.62	37.338
2.13	8.4149	2.63	13.874	3.13	22.874	3.63	37.713
2.14	8.4994	2.64	14.013	3.14	23.104	3.64	38.092
2.15	8.5849	**2.65**	14.154	**3.15**	23.336	**3.65**	38.475
2.16	8.6711	2.66	14.296	3.16	23.571	3.66	38.861
2.17	8.7583	2.67	14.440	3.17	23.807	3.67	39.252
2.18	8.8463	2.68	14.585	3.18	24.047	3.68	39.646
2.19	8.9352	2.69	14.732	3.19	24.288	3.69	40.045
2.20	9.0250	**2.70**	14.880	**3.20**	24.533	**3.70**	40.447
2.21	9.1157	2.71	15.029	3.21	24.779	3.71	40.854
2.22	9.2073	2.72	15.180	3.22	25.028	3.72	41.264
2.23	9.2999	2.73	15.333	3.23	25.280	3.73	41.679
2.24	9.3933	2.74	15.487	3.24	25.534	3.74	42.098
2.25	9.4877	**2.75**	15.643	**3.25**	25.790	**3.75**	42.521
2.26	9.5831	2.76	15.800	3.26	26.050	3.76	42.948
2.27	9.6794	2.77	15.959	3.27	26.311	3.77	43.380
2.28	9.7767	2.78	16.119	3.28	26.576	3.78	43.816
2.29	9.8749	2.79	16.281	3.29	26.843	3.79	44.256
2.30	9.9742	**2.80**	16.445	**3.30**	27.113	**3.80**	44.701
2.31	10.074	2.81	16.610	3.31	27.385	3.81	45.150
2.32	10.176	2.82	16.777	3.32	27.660	3.82	45.604
2.33	10.278	2.83	16.945	3.33	27.938	3.83	46.063
2.34	10.381	2.84	17.116	3.34	28.219	3.84	46.525
2.35	10.486	**2.85**	17.288	**3.35**	28.503	**3.85**	46.993
2.36	10.591	2.86	17.462	3.36	28.789	3.86	47.465
2.37	10.697	2.87	17.637	3.37	29.079	3.87	47.942
2.38	10.805	2.88	17.814	3.38	29.371	3.88	48.424
2.39	10.913	2.89	17.993	3.39	29.666	3.89	48.911
2.40	11.023	**2.90**	18.174	**3.40**	29.964	**3.90**	49.402
2.41	11.134	2.91	18.357	3.41	30.265	3.91	49.899
2.42	11.246	2.92	18.541	3.42	30.569	3.92	50.400
2.43	11.359	2.93	18.728	3.43	30.877	3.93	50.907
2.44	11.473	2.94	18.916	3.44	31.187	3.94	51.419
2.45	11.588	**2.95**	19.106	**3.45**	31.500	**3.95**	51.935
2.46	11.705	2.96	19.298	3.46	31.817	3.96	52.457
2.47	11.822	2.97	19.492	3.47	32.137	3.97	52.985
2.48	11.941	2.98	19.688	3.48	32.460	3.98	53.517
2.49	12.061	2.99	19.886	3.49	32.786	3.99	54.055
2.50	12.182	**3.00**	20.086	**3.50**	33.115	**4.00**	54.598

TABLE 3 Exponential Functions (e^x) (*cont.*)

x	e^x	x	e^x	x	e^x	x	e^x
4.00	54.598	**4.50**	90.017	**5.00**	148.41	**5.50**	244.69
4.01	55.147	4.51	90.922	5.01	149.90	5.55	257.24
4.02	55.701	4.52	91.836	5.02	151.41	5.60	270.43
4.03	56.261	4.53	92.759	5.03	152.93	5.65	284.29
4.04	56.826	4.54	93.691	5.04	154.47	5.70	298.87
4.05	57.397	**4.55**	94.632	**5.05**	156.02	**5.75**	314.19
4.06	57.974	4.56	95.583	5.06	157.59	5.80	330.30
4.07	58.557	4.57	96.544	5.07	159.17	5.85	347.23
4.08	59.145	4.58	97.514	5.08	160.77	5.90	365.04
4.09	59.740	4.59	98.494	5.09	162.39	5.95	383.75
4.10	60.340	**4.60**	99.484	**5.10**	164.02	**6.00**	403.43
4.11	60.947	4.61	100.48	5.11	165.67	6.05	424.11
4.12	61.559	4.62	101.49	5.12	167.34	6.10	445.86
4.13	62.178	4.63	102.51	5.13	169.02	6.15	468.72
4.14	62.803	4.64	103.54	5.14	170.72	6.20	492.75
4.15	63.434	**4.65**	104.58	**5.15**	172.43	**6.25**	518.01
4.16	64.072	4.66	105.64	5.16	174.16	6.30	544.57
4.17	64.715	4.67	106.70	5.17	175.91	6.35	572.49
4.18	65.366	4.68	107.77	5.18	177.68	6.40	601.85
4.19	66.023	4.69	108.85	5.19	179.47	6.45	632.70
4.20	66.686	**4.70**	109.95	**5.20**	181.27	**6.50**	655.14
4.21	67.357	4.71	111.05	5.21	183.09	6.55	699.24
4.22	68.033	4.72	112.17	5.22	184.93	6.60	735.10
4.23	68.717	4.73	113.30	5.23	186.79	6.65	772.78
4.24	69.408	4.74	114.43	5.24	188.67	6.70	812.41
4.25	70.105	**4.75**	115.58	**5.25**	190.57	**6.75**	854.06
4.26	70.810	4.76	116.75	5.26	192.48	6.80	897.85
4.27	71.522	4.77	117.92	5.27	194.42	6.85	943.88
4.28	72.240	4.78	119.10	5.28	196.37	6.90	992.27
4.29	72.966	4.79	120.30	5.29	198.34	6.95	1043.1
4.30	73.700	**4.80**	121.51	**5.30**	200.34	**7.00**	1096.6
4.31	74.440	4.81	122.73	5.31	202.35	7.05	1152.9
4.32	75.189	4.82	123.97	5.32	204.38	7.10	1212.0
4.33	75.944	4.83	125.21	5.33	206.44	7.15	1274.1
4.34	76.708	4.84	126.47	5.34	208.51	7.20	1339.4
4.35	77.478	**4.85**	127.74	**5.35**	210.61	**7.25**	1408.1
4.36	78.257	4.86	129.02	5.36	212.72	7.30	1480.3
4.37	79.044	4.87	130.32	5.37	214.86	7.35	1556.2
4.38	79.838	4.88	131.63	5.38	217.02	7.40	1636.0
4.39	80.640	4.89	132.95	5.39	219.20	7.45	1719.9
4.40	81.451	**4.90**	134.29	**5.40**	221.41	**7.50**	1808.0
4.41	82.269	4.91	135.64	5.41	223.63	7.55	1900.7
4.42	83.096	4.92	137.00	5.42	225.88	7.60	1998.2
4.43	83.931	4.93	138.38	5.43	228.15	7.65	2100.6
4.44	84.775	4.94	139.77	5.44	230.44	7.70	2208.3
4.45	85.627	**4.95**	141.17	**5.45**	232.76	**7.75**	2321.6
4.46	86.488	4.96	142.59	5.46	235.10	7.80	2440.6
4.47	87.357	4.97	144.03	5.47	237.46	7.85	2565.7
4.48	88.235	4.98	145.47	5.48	239.85	7.90	2697.3
4.49	89.121	4.99	146.94	5.49	242.26	7.95	2835.6
4.50	90.017	**5.00**	148.41	**5.50**	244.69	**8.00**	2981.0

TABLE 4 Four-Place Logarithms (Base Ten)

t	0	1	2	3	4	5	6	7	8	9
1.0	0.0000	0.0043	0.0086	0.0128	0.0170	0.0212	0.0253	0.0294	0.0334	0.0374
1.1	0.0414	0.0453	0.0492	0.0531	0.0569	0.0607	0.0645	0.0682	0.0719	0.0755
1.2	0.0792	0.0828	0.0864	0.0899	0.0934	0.0969	0.1004	0.1038	0.1072	0.1106
1.3	0.1139	0.1173	0.1206	0.1239	0.1271	0.1303	0.1335	0.1367	0.1399	0.1430
1.4	0.1461	0.1492	0.1523	0.1553	0.1584	0.1614	0.1644	0.1673	0.1703	0.1732
1.5	0.1761	0.1790	0.1818	0.1847	0.1875	0.1903	0.1931	0.1959	0.1987	0.2014
1.6	0.2041	0.2068	0.2095	0.2122	0.2148	0.2175	0.2201	0.2227	0.2253	0.2279
1.7	0.2304	0.2330	0.2355	0.2380	0.2405	0.2430	0.2455	0.2480	0.2504	0.2529
1.8	0.2553	0.2577	0.2601	0.2625	0.2648	0.2672	0.2695	0.2718	0.2742	0.2765
1.9	0.2788	0.2810	0.2833	0.2856	0.2878	0.2900	0.2923	0.2945	0.2967	0.2989
2.0	0.3010	0.3032	0.3054	0.3075	0.3096	0.3118	0.3139	0.3160	0.3181	0.3201
2.1	0.3222	0.3243	0.3263	0.3284	0.3304	0.3324	0.3345	0.3365	0.3385	0.3404
2.2	0.3424	0.3444	0.3464	0.3483	0.3502	0.3522	0.3541	0.3560	0.3579	0.3598
2.3	0.3617	0.3636	0.3655	0.3674	0.3692	0.3711	0.3729	0.3747	0.3766	0.3784
2.4	0.3802	0.3820	0.3838	0.3856	0.3874	0.3892	0.3909	0.3927	0.3945	0.3962
2.5	0.3979	0.3997	0.4014	0.4031	0.4048	0.4065	0.4082	0.4099	0.4116	0.4133
2.6	0.4150	0.4166	0.4183	0.4200	0.4216	0.4232	0.4249	0.4265	0.4281	0.4298
2.7	0.4314	0.4330	0.4346	0.4362	0.4378	0.4393	0.4409	0.4425	0.4440	0.4456
2.8	0.4472	0.4487	0.4502	0.4518	0.4533	0.4548	0.4564	0.4579	0.4594	0.4609
2.9	0.4624	0.4639	0.4654	0.4669	0.4683	0.4698	0.4713	0.4728	0.4742	0.4757
3.0	0.4771	0.4786	0.4800	0.4814	0.4829	0.4843	0.4857	0.4871	0.4886	0.4900
3.1	0.4914	0.4928	0.4942	0.4955	0.4969	0.4983	0.4997	0.5011	0.5024	0.5038
3.2	0.5051	0.5065	0.5079	0.5092	0.5105	0.5119	0.5132	0.5145	0.5159	0.5172
3.3	0.5185	0.5198	0.5211	0.5224	0.5237	0.5250	0.5263	0.5276	0.5289	0.5302
3.4	0.5315	0.5328	0.5340	0.5353	0.5366	0.5378	0.5391	0.5403	0.5416	0.5428
3.5	0.5441	0.5453	0.5465	0.5478	0.5490	0.5502	0.5514	0.5527	0.5539	0.5551
3.6	0.5563	0.5575	0.5587	0.5599	0.5611	0.5623	0.5635	0.5647	0.5658	0.5670
3.7	0.5682	0.5694	0.5705	0.5717	0.5729	0.5740	0.5752	0.5763	0.5775	0.5786
3.8	0.5798	0.5809	0.5821	0.5832	0.5843	0.5855	0.5866	0.5877	0.5888	0.5899
3.9	0.5911	0.5922	0.5933	0.5944	0.5955	0.5966	0.5977	0.5988	0.5999	0.6010
4.0	0.6021	0.6031	0.6042	0.6053	0.6064	0.6075	0.6085	0.6096	0.6107	0.6117
4.1	0.6128	0.6138	0.6149	0.6160	0.6170	0.6180	0.6191	0.6201	0.6212	0.6222
4.2	0.6232	0.6243	0.6253	0.6263	0.6274	0.6284	0.6294	0.6304	0.6314	0.6325
4.3	0.6335	0.6345	0.6355	0.6365	0.6375	0.6385	0.6395	0.6405	0.6415	0.6425
4.4	0.6435	0.6444	0.6454	0.6464	0.6474	0.6484	0.6493	0.6503	0.6513	0.6522
4.5	0.6532	0.6542	0.6551	0.6561	0.6571	0.6580	0.6590	0.6599	0.6609	0.6618
4.6	0.6628	0.6637	0.6646	0.6656	0.6665	0.6675	0.6684	0.6693	0.6702	0.6712
4.7	0.6721	0.6730	0.6739	0.6749	0.6758	0.6767	0.6776	0.6785	0.6794	0.6803
4.8	0.6812	0.6821	0.6830	0.6839	0.6848	0.6857	0.6866	0.6875	0.6884	0.6893
4.9	0.6902	0.6911	0.6920	0.6928	0.6937	0.6946	0.6955	0.6964	0.6972	0.6981
5.0	0.6990	0.6998	0.7007	0.7016	0.7024	0.7033	0.7042	0.7050	0.7059	0.7067
5.1	0.7076	0.7084	0.7093	0.7101	0.7110	0.7118	0.7126	0.7135	0.7143	0.7152
5.2	0.7160	0.7168	0.7177	0.7185	0.7193	0.7202	0.7210	0.7218	0.7226	0.7235
5.3	0.7243	0.7251	0.7259	0.7267	0.7275	0.7284	0.7292	0.7300	0.7308	0.7316
5.4	0.7324	0.7332	0.7340	0.7348	0.7356	0.7364	0.7372	0.7380	0.7388	0.7396
t	**0**	**1**	**2**	**3**	**4**	**5**	**6**	**7**	**8**	**9**

TABLE 4 **Four-Place Logarithms (Base Ten) (*cont.*)**

t	0	1	2	3	4	5	6	7	8	9
5.5	0.7404	0.7412	0.7419	0.7427	0.7435	0.7443	0.7451	0.7459	0.7466	0.7474
5.6	0.7482	0.7490	0.7497	0.7505	0.7513	0.7520	0.7528	0.7536	0.7543	0.7551
5.7	0.7559	0.7566	0.7574	0.7582	0.7589	0.7597	0.7604	0.7612	0.7619	0.7627
5.8	0.7634	0.7642	0.7649	0.7657	0.7664	0.7672	0.7679	0.7686	0.7694	0.7701
5.9	0.7709	0.7716	0.7723	0.7731	0.7738	0.7745	0.7752	0.7760	0.7767	0.7774
6.0	0.7782	0.7789	0.7796	0.7803	0.7810	0.7818	0.7825	0.7832	0.7839	0.7846
6.1	0.7853	0.7860	0.7868	0.7875	0.7882	0.7889	0.7896	0.7903	0.7910	0.7917
6.2	0.7924	0.7931	0.7938	0.7945	0.7952	0.7959	0.7966	0.7973	0.7980	0.7987
6.3	0.7993	0.8000	0.8007	0.8014	0.8021	0.8028	0.8035	0.8041	0.8048	0.8055
6.4	0.8062	0.8069	0.8075	0.8082	0.8089	0.8096	0.8102	0.8109	0.8116	0.8122
6.5	0.8129	0.8136	0.8142	0.8149	0.8156	0.8162	0.8169	0.8176	0.8182	0.8189
6.6	0.8195	0.8202	0.8209	0.8215	0.8222	0.8228	0.8235	0.8241	0.8248	0.8254
6.7	0.8261	0.8267	0.8274	0.8280	0.8287	0.8293	0.8299	0.8306	0.8312	0.8319
6.8	0.8325	0.8331	0.8338	0.8344	0.8351	0.8357	0.8363	0.8370	0.8376	0.8382
6.9	0.8388	0.8395	0.8401	0.8407	0.8414	0.8420	0.8426	0.8432	0.8439	0.8445
7.0	0.8451	0.8457	0.8463	0.8470	0.8476	0.8482	0.8488	0.8494	0.8500	0.8506
7.1	0.8513	0.8519	0.8525	0.8531	0.8537	0.8543	0.8549	0.8555	0.8561	0.8567
7.2	0.8573	0.8579	0.8585	0.8591	0.8597	0.8603	0.8609	0.8615	0.8621	0.8627
7.3	0.8633	0.8639	0.8645	0.8651	0.8657	0.8663	0.8669	0.8675	0.8681	0.8686
7.4	0.8692	0.8698	0.8704	0.8710	0.8716	0.8722	0.8727	0.8733	0.8739	0.8745
7.5	0.8751	0.8756	0.8762	0.8768	0.8774	0.8779	0.8785	0.8791	0.8797	0.8802
7.6	0.8808	0.8814	0.8820	0.8825	0.8831	0.8837	0.8842	0.8848	0.8854	0.8859
7.7	0.8865	0.8871	0.8876	0.8882	0.8887	0.8893	0.8899	0.8904	0.8910	0.8915
7.8	0.8921	0.8927	0.8932	0.8938	0.8943	0.8949	0.8954	0.8960	0.8965	0.8971
7.9	0.8976	0.8982	0.8987	0.8993	0.8998	0.9004	0.9009	0.9015	0.9020	0.9025
8.0	0.9031	0.9036	0.9042	0.9047	0.9053	0.9058	0.9063	0.9069	0.9074	0.9079
8.1	0.9085	0.9090	0.9096	0.9101	0.9106	0.9112	0.9117	0.9122	0.9128	0.9133
8.2	0.9138	0.9143	0.9149	0.9154	0.9159	0.9165	0.9170	0.9175	0.9180	0.9186
8.3	0.9191	0.9196	0.9201	0.9206	0.9212	0.9217	0.9222	0.9227	0.9232	0.9238
8.4	0.9243	0.9248	0.9253	0.9258	0.9263	0.9269	0.9274	0.9279	0.9284	0.9289
8.5	0.9294	0.9299	0.9304	0.9309	0.9315	0.9320	0.9325	0.9330	0.9335	0.9340
8.6	0.9345	0.9350	0.9355	0.9360	0.9365	0.9370	0.9375	0.9380	0.9385	0.9390
8.7	0.9395	0.9400	0.9405	0.9410	0.9415	0.9420	0.9425	0.9430	0.9435	0.9440
8.8	0.9445	0.9450	0.9455	0.9460	0.9465	0.9469	0.9474	0.9479	0.9484	0.9489
8.9	0.9494	0.9499	0.9504	0.9509	0.9513	0.9518	0.9523	0.9528	0.9533	0.9538
9.0	0.9542	0.9547	0.9552	0.9557	0.9562	0.9566	0.9571	0.9576	0.9581	0.9586
9.1	0.9590	0.9595	0.9600	0.9605	0.9609	0.9614	0.9619	0.9624	0.9628	0.9633
9.2	0.9638	0.9643	0.9647	0.9652	0.9657	0.9661	0.9666	0.9671	0.9675	0.9680
9.3	0.9685	0.9689	0.9694	0.9699	0.9703	0.9708	0.9713	0.9717	0.9722	0.9727
9.4	0.9731	0.9736	0.9741	0.9745	0.9750	0.9754	0.9759	0.9763	0.9768	0.9773
9.5	0.9777	0.9782	0.9786	0.9791	0.9795	0.9800	0.9805	0.9809	0.9814	0.9818
9.6	0.9823	0.9827	0.9832	0.9836	0.9841	0.9845	0.9850	0.9854	0.9859	0.9863
9.7	0.9868	0.9872	0.9877	0.9881	0.9886	0.9890	0.9894	0.9899	0.9903	0.9908
9.8	0.9912	0.9917	0.9921	0.9926	0.9930	0.9934	0.9939	0.9943	0.9948	0.9952
9.9	0.9956	0.9961	0.9965	0.9969	0.9974	0.9978	0.9983	0.9987	0.9991	0.9996
t	0	1	2	3	4	5	6	7	8	9

TABLE 5 Binomial Probabilities

		p												
	k	**0.01**	**0.1**	**0.2**	**0.3**	**0.33**	**0.4**	**0.5**	**0.6**	**0.67**	**0.7**	**0.8**	**0.9**	**0.99**
$N = 1$	0	0.990	0.900	0.800	0.700	0.670	0.600	0.500	0.400	0.330	0.300	0.200	0.100	0.010
	1	0.010	0.100	0.200	0.300	0.330	0.400	0.500	0.600	0.670	0.700	0.800	0.900	0.990
$N = 2$	0	0.980	0.810	0.640	0.490	0.445	0.360	0.250	0.160	0.111	0.090	0.040	0.010	
	1	0.020	0.180	0.320	0.420	0.444	0.480	0.500	0.480	0.444	0.420	0.320	0.180	0.020
	2		0.010	0.040	0.090	0.111	0.160	0.250	0.360	0.445	0.490	0.640	0.810	0.980
$N = 3$	0	0.970	0.729	0.512	0.343	0.301	0.216	0.125	0.064	0.036	0.027	0.008	0.001	
	1	0.030	0.243	0.384	0.441	0.444	0.432	0.375	0.288	0.219	0.189	0.096	0.027	
	2		0.027	0.096	0.189	0.219	0.288	0.375	0.432	0.444	0.441	0.384	0.243	0.030
	3		0.001	0.008	0.027	0.036	0.064	0.125	0.216	0.301	0.343	0.512	0.729	0.970
$N = 4$	0	0.961	0.656	0.410	0.240	0.202	0.130	0.063	0.026	0.012	0.008	0.002		
	1	0.039	0.292	0.410	0.412	0.397	0.346	0.250	0.154	0.096	0.076	0.026	0.004	
	2		0.048	0.154	0.265	0.293	0.346	0.375	0.346	0.293	0.265	0.154	0.048	
	3		0.004	0.026	0.076	0.096	0.154	0.250	0.346	0.397	0.412	0.410	0.292	0.039
	4			0.002	0.008	0.012	0.026	0.063	0.130	0.202	0.240	0.410	0.656	0.961
$N = 5$	0	0.951	0.590	0.328	0.168	0.135	0.078	0.031	0.010	0.004	0.002			
	1	0.048	0.328	0.410	0.360	0.332	0.259	0.156	0.077	0.040	0.028	0.006		
	2	0.001	0.073	0.205	0.309	0.328	0.346	0.313	0.230	0.161	0.132	0.051	0.008	
	3		0.008	0.051	0.132	0.161	0.230	0.313	0.346	0.328	0.309	0.205	0.073	0.001
	4			0.006	0.028	0.040	0.077	0.156	0.259	0.332	0.360	0.410	0.328	0.048
	5				0.002	0.004	0.010	0.031	0.078	0.135	0.168	0.328	0.590	0.951
$N = 6$	0	0.941	0.531	0.262	0.118	0.090	0.047	0.016	0.004	0.001	0.001			
	1	0.057	0.354	0.393	0.303	0.267	0.187	0.094	0.037	0.016	0.010	0.002		
	2	0.002	0.098	0.246	0.324	0.330	0.311	0.234	0.138	0.080	0.060	0.015	0.001	
	3		0.015	0.082	0.185	0.216	0.276	0.313	0.276	0.216	0.185	0.082	0.015	
	4		0.001	0.015	0.060	0.080	0.138	0.234	0.311	0.330	0.324	0.246	0.098	0.002
	5			0.002	0.010	0.016	0.037	0.094	0.187	0.267	0.303	0.393	0.354	0.057
	6				0.001	0.001	0.004	0.016	0.047	0.090	0.118	0.262	0.531	0.941
$N = 7$	0	0.932	0.478	0.210	0.083	0.061	0.028	0.008	0.002					
	1	0.066	0.372	0.367	0.247	0.209	0.131	0.055	0.017	0.006	0.004			
	2	0.002	0.124	0.275	0.318	0.309	0.261	0.164	0.077	0.037	0.025	0.004		
	3		0.023	0.115	0.227	0.253	0.290	0.273	0.194	0.125	0.097	0.029	0.003	
	4		0.003	0.029	0.097	0.125	0.194	0.273	0.290	0.253	0.227	0.115	0.023	
	5			0.004	0.025	0.037	0.077	0.164	0.261	0.309	0.318	0.275	0.124	0.002
	6				0.004	0.006	0.017	0.055	0.131	0.209	0.247	0.367	0.372	0.066
	7						0.002	0.008	0.028	0.061	0.083	0.210	0.478	0.932
$N = 8$	0	0.923	0.430	0.168	0.058	0.041	0.017	0.004	0.001					
	1	0.075	0.383	0.336	0.198	0.160	0.090	0.031	0.008	0.002	0.001			
	2	0.002	0.149	0.294	0.296	0.276	0.209	0.109	0.041	0.016	0.010	0.001		
	3		0.033	0.147	0.254	0.272	0.279	0.219	0.124	0.066	0.047	0.009		
	4		0.005	0.046	0.136	0.167	0.232	0.273	0.232	0.167	0.136	0.046	0.005	
	5			0.009	0.047	0.066	0.124	0.219	0.279	0.272	0.254	0.147	0.033	
	6			0.001	0.010	0.016	0.041	0.109	0.209	0.276	0.296	0.294	0.149	0.002
	7				0.001	0.002	0.008	0.031	0.090	0.160	0.198	0.336	0.383	0.075
	8						0.001	0.004	0.017	0.041	0.058	0.168	0.430	0.923

TABLE 5 Binomial Probabilities (*cont.*)

		p												
	k	**0.01**	**0.1**	**0.2**	**0.3**	**0.33**	**0.4**	**0.5**	**0.6**	**0.67**	**0.7**	**0.8**	**0.9**	**0.99**
N = 9	0	0.914	0.387	0.134	0.040	0.027	0.010	0.002						
	1	0.083	0.387	0.302	0.156	0.121	0.060	0.018	0.004	0.001				
	2	0.003	0.172	0.302	0.267	0.238	0.161	0.070	0.021	0.007	0.004			
	3		0.045	0.176	0.267	0.273	0.251	0.164	0.074	0.032	0.021	0.003		
	4		0.007	0.066	0.172	0.202	0.251	0.246	0.167	0.099	0.074	0.017	0.001	
	5		0.001	0.017	0.074	0.099	0.167	0.246	0.251	0.202	0.172	0.066	0.007	
	6			0.003	0.021	0.032	0.074	0.164	0.251	0.273	0.267	0.176	0.045	
	7				0.004	0.007	0.021	0.070	0.161	0.238	0.267	0.302	0.172	0.003
	8					0.001	0.004	0.018	0.060	0.121	0.156	0.302	0.387	0.083
	9							0.002	0.010	0.027	0.040	0.134	0.387	0.914
N = 10	0	0.904	0.349	0.107	0.028	0.018	0.006	0.001						
	1	0.092	0.387	0.268	0.121	0.090	0.040	0.010	0.002					
	2	0.004	0.194	0.302	0.233	0.199	0.121	0.044	0.011	0.003	0.001			
	3		0.057	0.201	0.267	0.262	0.215	0.117	0.042	0.015	0.009	0.001		
	4		0.011	0.088	0.200	0.225	0.251	0.205	0.111	0.055	0.037	0.006		
	5		0.001	0.026	0.103	0.133	0.201	0.246	0.201	0.133	0.103	0.026	0.001	
	6			0.006	0.037	0.055	0.111	0.205	0.251	0.225	0.200	0.088	0.011	
	7			0.001	0.009	0.015	0.042	0.117	0.215	0.262	0.267	0.201	0.057	
	8				0.001	0.003	0.011	0.044	0.121	0.199	0.233	0.302	0.194	0.004
	9						0.002	0.010	0.040	0.090	0.121	0.268	0.387	0.092
	10							0.001	0.006	0.018	0.028	0.107	0.349	0.904
N = 15	0	0.860	0.206	0.035	0.005	0.002								
	1	0.130	0.343	0.132	0.031	0.018	0.005							
	2	0.009	0.267	0.231	0.092	0.063	0.022	0.003						
	3	0.001	0.129	0.250	0.170	0.134	0.063	0.014	0.002					
	4		0.043	0.188	0.219	0.198	0.127	0.042	0.007	0.001	0.001			
	5		0.010	0.103	0.206	0.215	0.186	0.092	0.024	0.006	0.003			
	6		0.002	0.043	0.147	0.176	0.207	0.153	0.061	0.021	0.012	0.001		
	7			0.014	0.081	0.111	0.177	0.196	0.118	0.055	0.035	0.003		
	8			0.003	0.035	0.055	0.118	0.196	0.177	0.111	0.081	0.014		
	9			0.001	0.012	0.021	0.061	0.153	0.207	0.176	0.147	0.043	0.002	
	10				0.003	0.006	0.024	0.092	0.186	0.215	0.206	0.103	0.010	
	11				0.001	0.001	0.007	0.042	0.127	0.198	0.219	0.188	0.043	
	12						0.002	0.014	0.063	0.134	0.170	0.250	0.129	0.001
	13							0.003	0.022	0.063	0.092	0.231	0.267	0.009
	14								0.005	0.018	0.031	0.132	0.343	0.130
	15									0.002	0.005	0.035	0.206	0.860

TABLE 5 Binomial Probabilities (*cont.*)

		p												
	k	**0.01**	**0.1**	**0.2**	**0.3**	**0.33**	**0.4**	**0.5**	**0.6**	**0.67**	**0.7**	**0.8**	**0.9**	**0.99**
N = *20*	0	0.818	0.122	0.012	0.001									
	1	0.165	0.270	0.058	0.007	0.003								
	2	0.016	0.285	0.137	0.028	0.015	0.003							
	3	0.001	0.190	0.205	0.072	0.045	0.012	0.001						
	4		0.090	0.218	0.130	0.095	0.035	0.005						
	5		0.032	0.175	0.179	0.149	0.075	0.015	0.001					
	6		0.009	0.109	0.192	0.184	0.124	0.037	0.005	0.001				
	7		0.002	0.055	0.164	0.181	0.166	0.074	0.015	0.003	0.001			
	8			0.022	0.114	0.145	0.180	0.120	0.036	0.009	0.004			
	9			0.007	0.065	0.095	0.160	0.160	0.071	0.023	0.012			
	10			0.002	0.031	0.052	0.117	0.176	0.117	0.052	0.031	0.002		
	11				0.012	0.023	0.071	0.160	0.160	0.095	0.065	0.007		
	12				0.004	0.009	0.036	0.120	0.180	0.145	0.114	0.022		
	13				0.001	0.003	0.015	0.074	0.166	0.181	0.164	0.055	0.002	
	14					0.001	0.005	0.037	0.124	0.184	0.192	0.109	0.009	
	15						0.001	0.015	0.075	0.149	0.179	0.175	0.032	
	16							0.005	0.035	0.095	0.130	0.218	0.090	
	17							0.001	0.012	0.045	0.072	0.205	0.190	0.001
	18								0.003	0.015	0.028	0.137	0.285	0.016
	19									0.003	0.007	0.058	0.270	0.165
	20										0.001	0.012	0.122	0.818

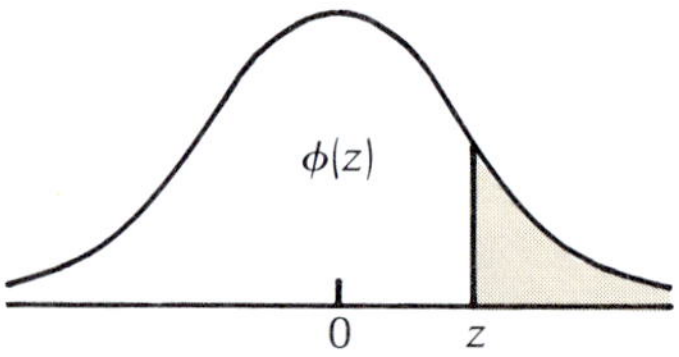

TABLE 6 The Normal Distribution

z	.00	.01	.02	.03	.04	.05	.06	.07	.08	.09
.0	.5000	.5040	.5080	.5120	.5160	.5199	.5239	.5279	.5319	.5359
.1	.5398	.5438	.5478	.5517	.5557	.5596	.5636	.5675	.5714	.5753
.2	.5793	.5832	.5871	.5910	.5948	.5987	.6026	.6064	.6103	.6141
.3	.6179	.6217	.6255	.6293	.6331	.6368	.6406	.6443	.6480	.6517
.4	.6554	.6591	.6628	.6664	.6700	.6736	.6772	.6808	.6844	.6879
.5	.6915	.6950	.6985	.7019	.7054	.7088	.7123	.7157	.7190	.7224
.6	.7257	.7291	.7324	.7357	.7389	.7422	.7454	.7486	.7517	.7549
.7	.7580	.7611	.7642	.7673	.7704	.7734	.7764	.7794	.7823	.7852
.8	.7881	.7910	.7939	.7967	.7995	.8023	.8051	.8078	.8106	.8133
.9	.8159	.8186	.8212	.8238	.8264	.8289	.8315	.8340	.8365	.8389
1.0	.8413	.8438	.8461	.8485	.8508	.8531	.8554	.8577	.8599	.8621
1.1	.8643	.8665	.8686	.8708	.8729	.8749	.8770	.8790	.8810	.8830
1.2	.8849	.8869	.8888	.8907	.8925	.8944	.8962	.8980	.8997	.9015
1.3	.9032	.9049	.9066	.9082	.9099	.9115	.9131	.9147	.9162	.9177
1.4	.9192	.9207	.9222	.9236	.9251	.9265	.9279	.9292	.9306	.9319
1.5	.9332	.9345	.9357	.9370	.9382	.9394	.9406	.9418	.9429	.9441
1.6	.9452	.9463	.9474	.9484	.9495	.9505	.9515	.9525	.9535	.9545
1.7	.9554	.9564	.9573	.9582	.9591	.9599	.9608	.9616	.9625	.9633
1.8	.9641	.9649	.9656	.9664	.9671	.9678	.9686	.9693	.9699	.9706
1.9	.9713	.9719	.9726	.9732	.9738	.9744	.9750	.9756	.9761	.9767
2.0	.9772	.9778	.9783	.9788	.9793	.9798	.9803	.9808	.9812	.9817
2.1	.9821	.9826	.9830	.9834	.9838	.9842	.9846	.9850	.9854	.9857
2.2	.9861	.9864	.9868	.9871	.9875	.9878	.9881	.9884	.9887	.9890
2.3	.9893	.9896	.9898	.9901	.9904	.9906	.9909	.9911	.9913	.9916
2.4	.9918	.9920	.9920	.9925	.9927	.9929	.9931	.9932	.9934	.9936
2.5	.9938	.9940	.9941	.9943	.9945	.9946	.9948	.9949	.9951	.9952
2.6	.9953	.9955	.9956	.9957	.9959	.9960	.9961	.9962	.9963	.9964
2.7	.9965	.9966	.9967	.9968	.9969	.9970	.9971	.9972	.9973	.9974
2.8	.9974	.9975	.9976	.9977	.9977	.9978	.9979	.9979	.9980	.9981
2.9	.9981	.9982	.9982	.9983	.9984	.9984	.9985	.9985	.9986	.9986
3.0	.9987	.9987	.9987	.9988	.9988	.9989	.9989	.9989	.9990	.9990
3.1	.9990	.9991	.9991	.9991	.9992	.9992	.9992	.9992	.9993	.9993
3.2	.9993	.9993	.9994	.9994	.9994	.9994	.9994	.9995	.9995	.9995
3.3	.9995	.9995	.9996	.9996	.9996	.9996	.9996	.9996	.9996	.9997
3.4	.9997	.9997	.9997	.9997	.9997	.9997	.9997	.9997	.9997	.9998

APPENDIX

B

ANSWERS TO ODD-NUMBERED EXERCISES AND SOLUTIONS TO SELECTED EXERCISES

B

CHAPTER 1. SETS, NUMBERS, AND COORDINATE SYSTEMS

1.2 THE LANGUAGE OF SETS

1. 1, 2 **3.** 6, 7 **5.** 1, 2, 3 **7.** 1, 3

9. 1, 2, 3, 4, 6, 8, 12, 24 **11.** There are no elements to list

13. {2, 3, 4, 5, 6, 7} **15.** {1, 4, 9, 16}

17. {1, 3, 5, 7, 9} **19.** {1, 4, 9, 10, 16, 20, 25, 30}

21. (a) **23.** (a), (b), (c) **25.** (b)

27. (a) $\{a\}, \{b\}, \{c\}, \{d\}$
(b) $\{a, b\}, \{a, c\}, \{a, d\}, \{b, c\}, \{b, d\}, \{c, d\}$
(c) $\{a, b, c\}, \{a, b, d\}, \{b, c, d\}, \{a, c, d\}$

29. 1, 2, 3, 4, 5 **31.** 1, 2, 3, 4, 5, 6, 8, 10, 12

33. 1, 2, 3, 4, 5, 7, 9, 11 **35.** 1, 2, 3, 4, 5, 6, 7, 8, 9, 10, 11, 12

37. $A' = \{2, 4, 6, 8, 10, 12\}$

39. $A' = \{1, 3, 4, 5, 6, 7, 8, 9, 11, 12\}$

41. $A' = \varnothing$ **43.** True **45.** True **47.** False **49.** False

1.3 COUNTING ELEMENTS IN SETS

1. 8 **3.** 19 **5.** 3 **7.** 15 **9.** Not consistent

11. 136 **13.** (a) 40 (b) 30 (c) 60 **15.** 119

1.4 THE REAL NUMBERS

1. $\frac{11}{10}$ **3.** $\frac{31}{5}$ **5.** $\frac{62}{9}$ **7.** $\frac{281}{100}$ **9.** $\frac{829}{99}$ **11.** $\frac{529}{99}$

13. $-\frac{2450}{999}$ **15.** $-\frac{15989}{198}$ **17.** $\frac{1123}{198}$ **19.** 1.5

21. 0.4 **23.** 0.7 **25.** 1.1 **27.** 0.08 **29.** 0.03

31. 0.19 **33.** −0.625 **35.** $1.1\overline{6}$ **37.** $-0.1\overline{6}$

39. $-0.2\overline{6}$ **41.** $0.\overline{1}$ **43.** $-0.\overline{037}$

45.

47.

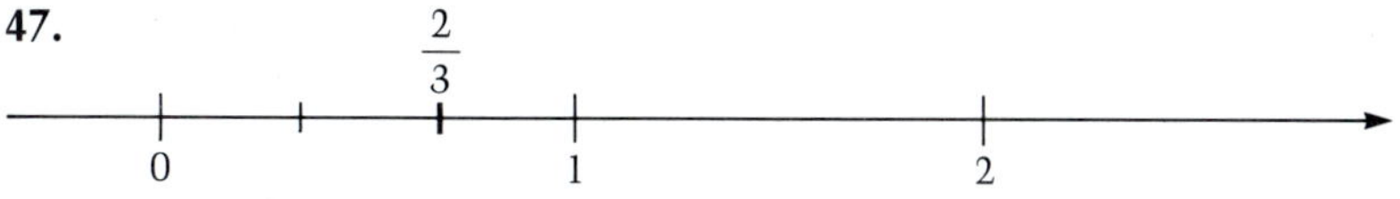

49.

51.

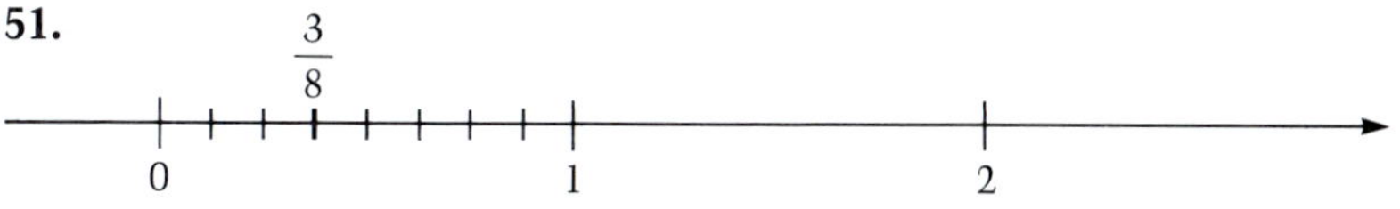

53. Since $\sqrt{10} \doteq 3.16$, plot 3.2 for one decimal place.

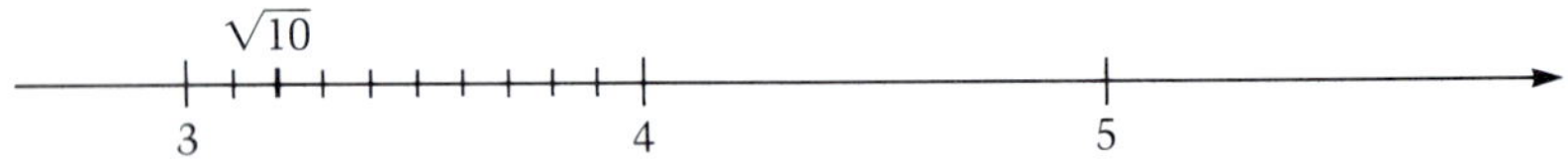

55. $\sqrt{24} \doteq 4.898$ (to three places). Take 4.90 as a two-place approximation: $-\sqrt{24} \doteq -4.90$.

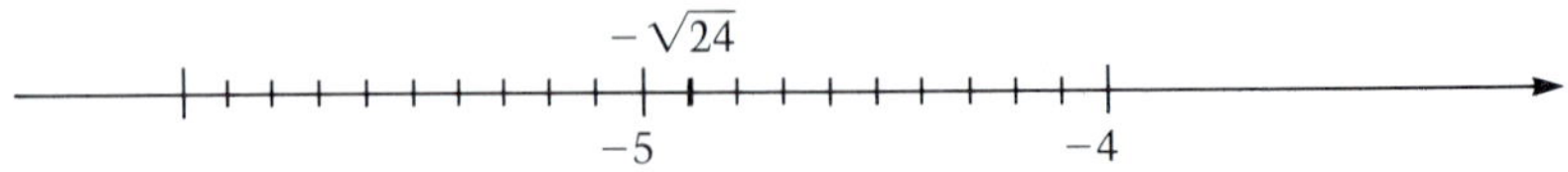

57. $\sqrt{41} \doteq 6.403$ (to three places). Take $\sqrt{41} \doteq 6.40$ to two places.

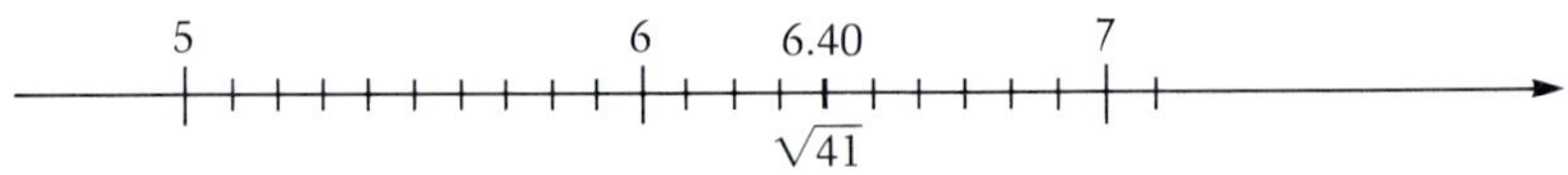

1.5 RECTANGULAR COORDINATE SYSTEMS

Answers to Exercises 1, 3, 5, 7, 9 are on one coordinate system.

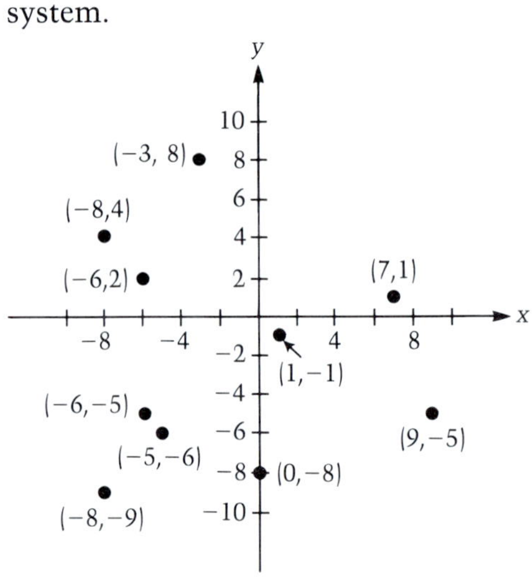

11. $(-5, -3.5)$ **13.** $(4, -2)$ **15.** $(-5, -5)$ **17.** $(-3, 2)$
19. $(9, -3.5)$ **21.** $5\sqrt{2}$ **23.** $\sqrt{2}$ **25.** $\sqrt{193}$
27. $\sqrt{113}$ **29.** $5\sqrt{10}$ **31.** $-\frac{1}{15}$ **33.** $-\frac{12}{13}$
35. $-\frac{5}{6}$ **37.** -8 **39.** $-\frac{1}{11}$ **41.** PQ
43. RS **45.** RS **47.** RS
49. PQ, RS have same steepness.

1.6 REVIEW EXERCISES

1. (a) $\{1, 2, 3, 6, 9\}$ (b) $\{-1, 1, 3\}$
2. $\{a, b, c\}$ $\{a, b, d\}$ $\{a, b, e\}$ $\{a, c, d\}$ $\{a, c, e\}$ $\{a, d, e\}$
3. $\{2, 4, 6, 8\}$
4. all positive odd integers
5. (a) 14 (b) 70
6. (a) 0.56 (b) $2.5\overline{3}$
7. (a) $\frac{43}{20}$ (b) $\frac{5229}{99}$
8. $\sqrt{35} = 5.92$

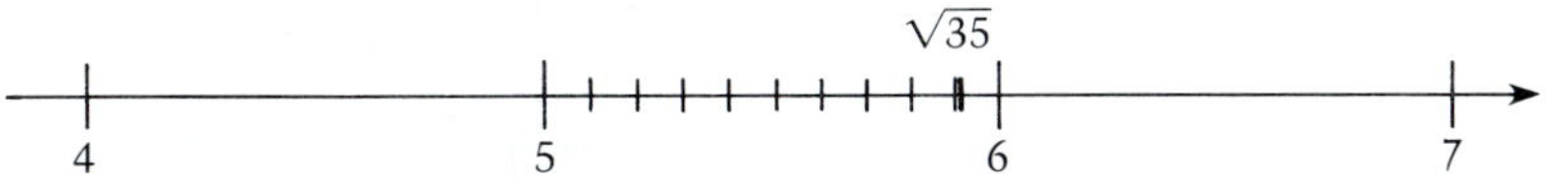

9.

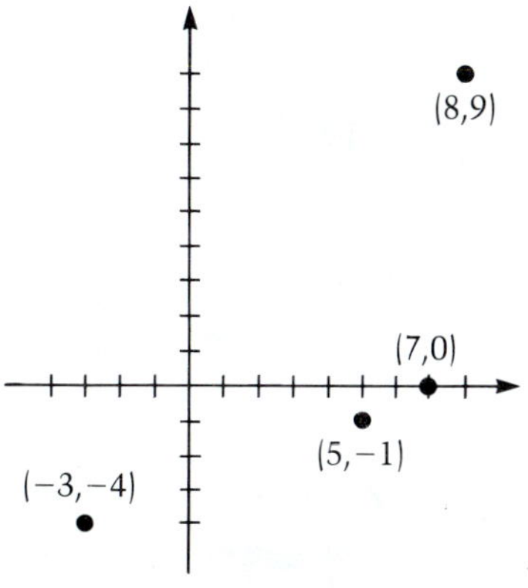

10. (a) $(-3, 7)$ (b) $(-2, 0.5)$
11. (a) 8 (b) $\sqrt{197}$
12. (a) 0 (b) $-\frac{1}{14}$
13. (a) (b)

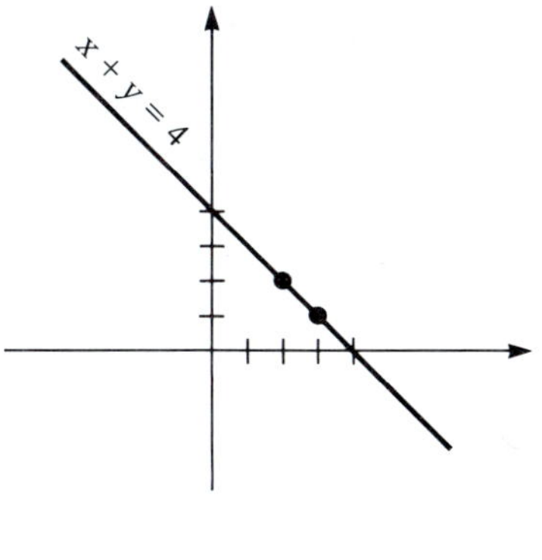

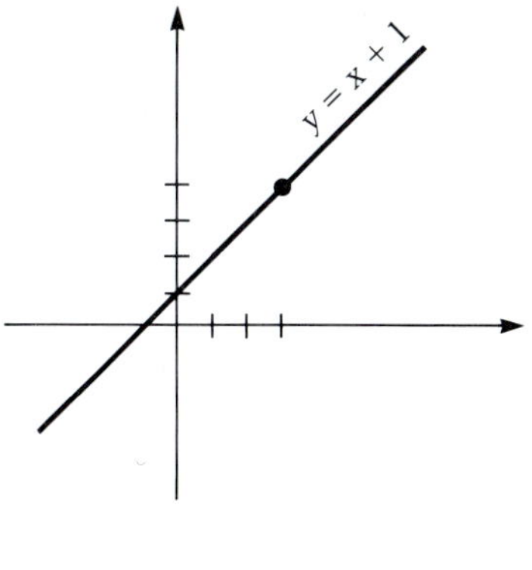

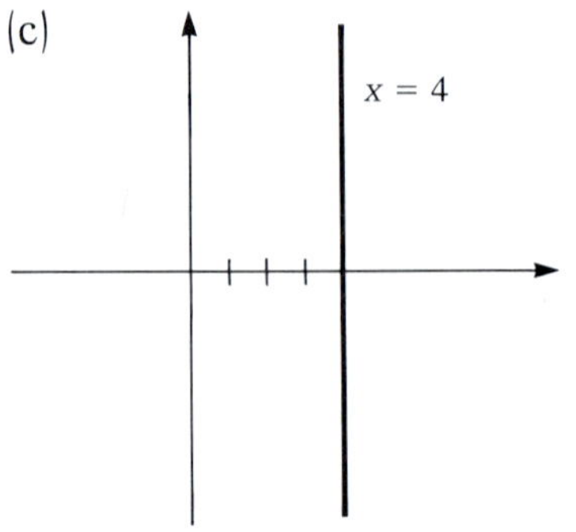

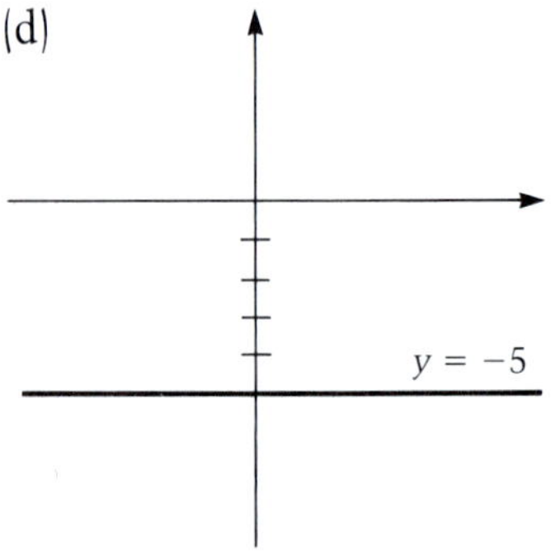

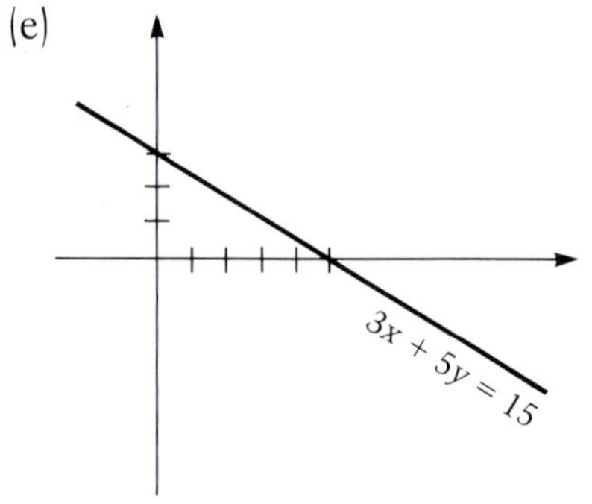

CHAPTER 2. LINEAR MODELS

2.2 LINEAR EQUATIONS IN ONE VARIABLE

1. $\frac{11}{3}$ **3.** 6.4 **5.** $\frac{1}{5}$
7. $\frac{31}{12}$ **9.** 204 **11.** 150
13. 1 **15.** $-\frac{23}{4}$ **17.** 37; 23
19. $3.60 **21.** 128; 16 **23.** 96; 71
25. 100 miles **27.** 115 **29.** 48 mph
31. 5 hours 37.5 minutes

2.3 LINEAR INEQUALITIES IN ONE VARIABLE

1. $x > -\frac{9}{2}$ **3.** $x \le -\frac{6}{13}$ **5.** $x \ge 7$ **7.** $x < -\frac{9}{26}$
9. $x \le \frac{2}{3}$ **11.** $x < \frac{27}{13}$ **13.** $x \le \frac{29}{21}$ **15.** $x > -40$
17. $x < 41$ **19.** $x > -34.8$

21. Let x be the cost of 19-inch set. Then $912 - x$ is cost of 21-inch set. $7x + 4(912 - x) \le 4620$; $7x + 3648 - 4x \le 4620$; $3x \le 972$; $x \le 324$. 324 dollars.

23. Let x be the distance the pickup travels. Then $240 - x$ is the distance the panel truck travels. $\frac{x}{12} + \frac{240 - x}{10} \le 22$; $5x + 1440 - 6x \le 1320$; $-x \le 120$; $x \ge 120$. 120 miles.

25. (a) Let x be the cost of a reel mower. Then $576 - x$ is the cost of a rotary mower. $5x + 9(576 - x) \le 4336$; $5x + 5184 - 9x \le 4336$; $-4x \le 848$; $x \ge 212$. \$212.
(b) $-x \le 212$ and $576 - x \le 364$. \$364.

27. Let x be units sold. Hence the profit per unit is $3x$, and x must be such that $3x \ge 150{,}000$, $x > 50{,}000$. Over 50,000 units.

2.4 LINEAR EQUATIONS IN TWO VARIABLES

1. $\sqrt{34}$; $(\frac{11}{2}, \frac{17}{2})$ **3.** $\sqrt{193}$; $(-1, -\frac{1}{2})$ **5.** $6\sqrt{5}$; $(\frac{1}{2}, 1)$

7. $\frac{1}{2}$ **9.** $-\frac{2}{9}$ **11.** $\frac{4}{7}$

13. $y = -7$ **15.** $x = 2$ **17.** $2x + y = -2$

19. $4x - 3y = 34$ **21.** $11x - 4y = 46$ **23.** $x + 3y = 20$

25. $x = -3$ **27.** $10x + 7y = 42$ **29.** $3x - 5y = -23$

31. $4x + 2y = 3$ (See below.)

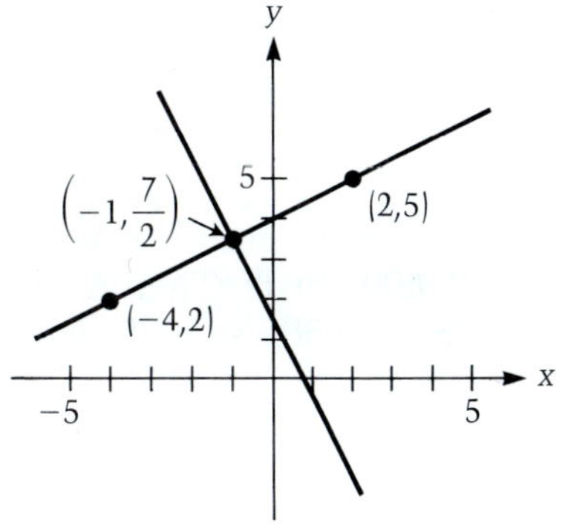

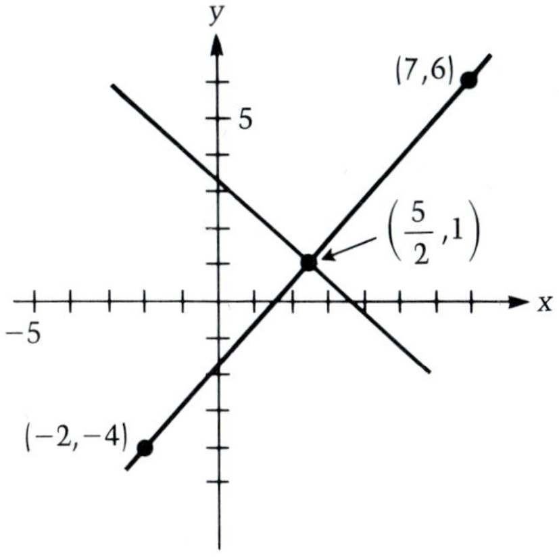

33.

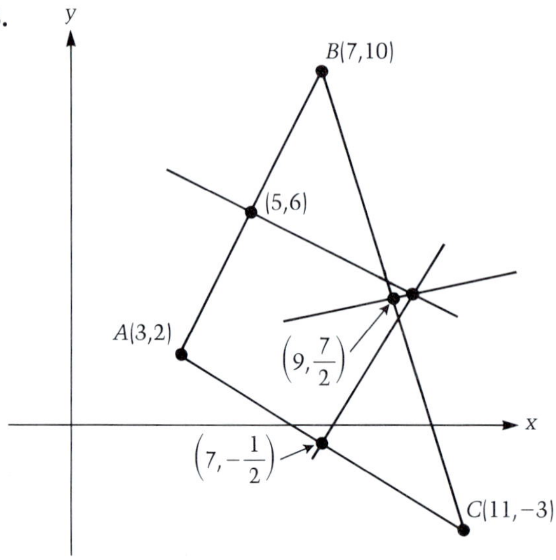

35. Warehouse A.

2.5 LINEAR SYSTEMS AND THEIR APPLICATIONS

1. $x = 2, y = -3$ **3.** $x = \frac{1}{2}, y = 1$

5. $x = \frac{8}{41}, y = \frac{76}{41}$ **7.** $x = 8, y = -12$

9. $x = \frac{4}{3}, y = -\frac{5}{4}$

11. Let x be the number of \$6.00 shirts and y be the number of \$9.00 shirts. Then $6x + 9y = 7680$ and $6.25x + 9.50y = 8050$, which are equivalent to $2x + 3y = 2560$ and $x + 1.52y = 1288$; $x = 680$ and $y = 400$.

13. The perpendicular bisector of AB is $x + 2y = 17$. The perpendicular bisector of BC is $8x - 26y = -19$. The solution is $x = \frac{202}{21}$ and $y = \frac{155}{42}$; that is, approximately 9.6 miles east and 3.7 miles north of City Hall.

15. Let x be the number of shares at \$20 a share and let y be the number of shares at \$15 a share. Then $20x + 15y = 16{,}250$ and $1.60x + 1.00y = 1300$. $x = 750$ and $y = 100$.

17. Let x be the number of gallons of 9 percent alcohol and let y be the number of gallons of 12 percent alcohol. Then $x + y = 2000$ and $.09x + .13y = 200$. $x = 1500$ gallons and $y = 500$ gallons.

19. Let x be the number of Chewey bars in the shipment and let y be the number of Crispy bars. Then

$.75x + .35y = 171.25$ and $.50x + .22y = 111.50$. $x = 135$ and $y = 200$.

21. Let x be the number of gallons of 25 percent solution and let y be the number of gallons of 14 percent solution. Then $x + y = 500$ and $.25x + .14y = .18(500)$. $x = 181.82$ gallons and $y = 318.18$ gallons.

23. Let x be the amount invested in stocks and let y be the amount invested in bonds. Then $x + y = 85{,}000$ and $.12x + .08y = 8{,}680$. $x = \$47{,}000$.

2.6 FUNCTIONS AND THEIR GRAPHS

1. (a) 19 (b) 27 (c) No (d) No

3. (a) 12 (b) 27 (c) No (d) No

5. (a) 8 (b) 13 (c) No (d) No

7. (a) 8 (b) 27 (c) No (d) Yes

9. (a)

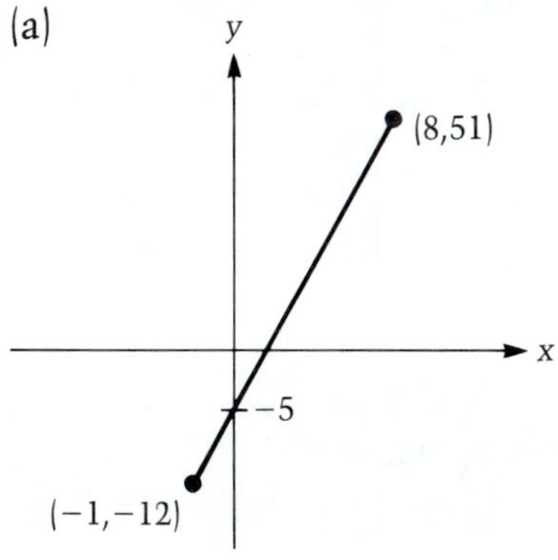

(b) 51 (c) −12

11. (a)

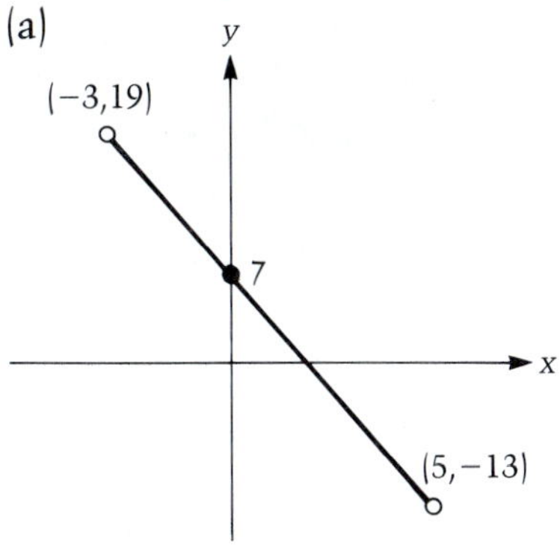

(b) None (c) None

13. (a)

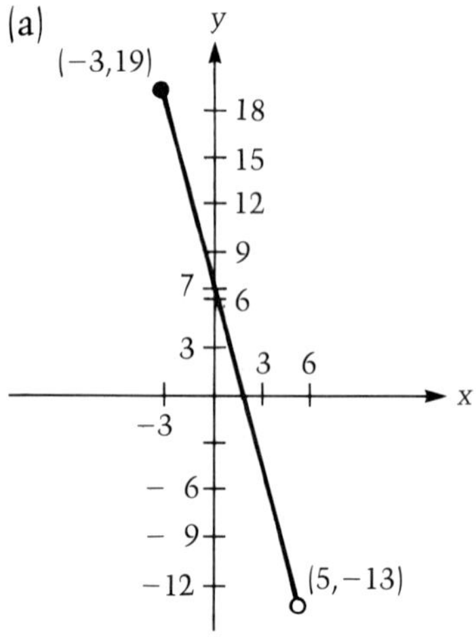

(b) 19 (c) None

15. (a)

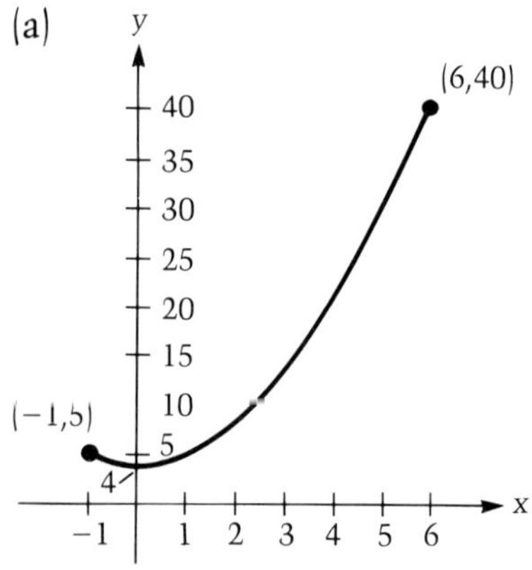

(b) 40 (c) 4

17. (a)

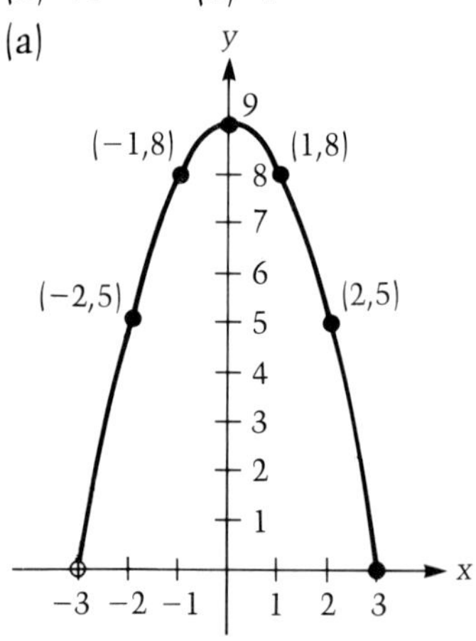

(b) 9 (c) 0

19. (a)

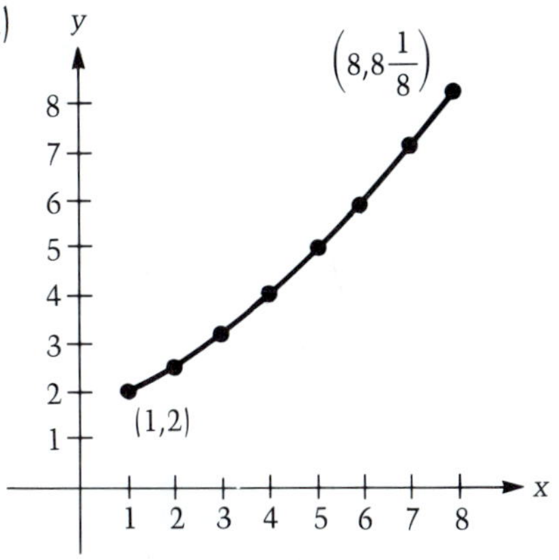

(b) $\frac{65}{8}$ (c) 2

21. (a)

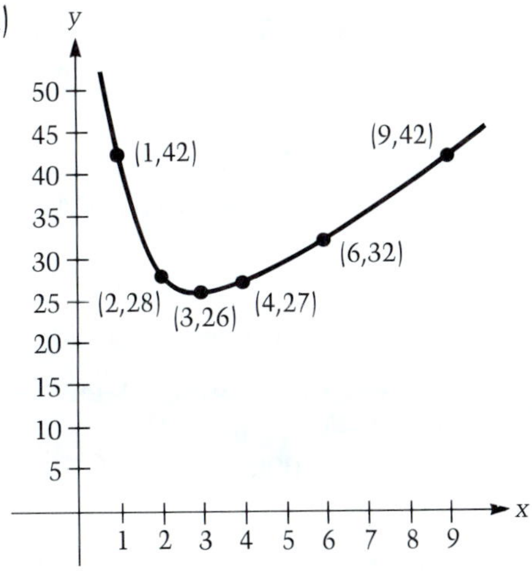

(b) None (c) $f(3) = 26$

2.7 APPLICATIONS OF LINEAR FUNCTIONS

1. (a) $V(t) = 8000 - 1200t$ (b) $V(2) = \$5600$
(c) $8000 - 1200t = 0$ when $t = 6\frac{2}{3}$ years = 6 years 8 months
(d) $V(2.5) = \$5000$

3. (a) $D(3.50) = 48{,}000$ units (b) $D(6) = 28{,}000$ units
(c) $100(76 - 8p) = 0$ when $p = \$9.50$ (d) The change (increase or decrease) in demand when the price is increased from p to $p + 1$ dollars. It is decreased by \$8000.

5. (a) $S(3.50) = 28{,}000$ units (b) $S(6) = 48{,}000$ units
(c) $1000(76 - 8p) = 8000p$ when $p = \$4.75$.
(d) $S(4.75) = 38{,}000$ units

7. (a) $C(t) = kt$, $43.50 = k(\frac{3}{2})$, and $k = 29$. (b) The hourly charge. (c) $C(t) = 29t$ (d) $C(4.75) = \$137.75$
9. (a) $3.50 + 20(.15) = \$6.50$ (b) $C(x) = .15x + 3.50$ (c) $C(35) = \$8.75$ (d) $\overline{C}(x) = .15 + \dfrac{3.50}{x}$ (e) $\overline{C}(35) = \$0.25$

(f)
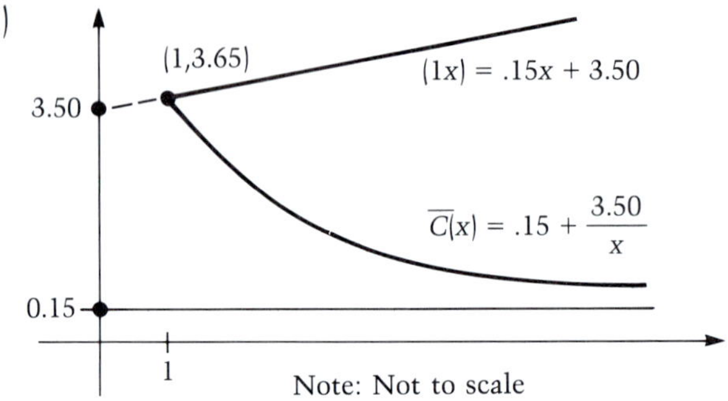

11. $.15x + 3.50 = .10x + 5.00$ when $x = 30$ checks.
13. (a) $R(x) = 600N(x) = 24{,}000x + 7{,}200{,}000$ (b) $R(90) = \$9{,}360{,}000$
15. (a) $P(x) = R(x) - C(x) = 5000x - 1{,}700{,}000$ (b) $P(90) = -\$1{,}250{,}000$, a loss of \$1,250,000.
17. \$3,300,000
19. (a) $R(x) = 500N(x) = 15{,}000x + 4{,}000{,}000$ (b) $R(100) = \$5{,}500{,}000$
21. (a) $P(x) = R(x) - C(x) = 2{,}000x - 1{,}700{,}000$ (b) \$1,500,000 loss.
23. \$300,000

2.8 SUMMARY AND REVIEW EXERCISES

1. $x \le -\frac{52}{29}$
2. $x > 3$
3. $2x - 3y = -13$
4. $5x + 6y = -13$
5. $6x - y = 6$
6. $2x + 3y = 8$
7. $x = \frac{3}{2}$, $y = -2$.
8. $x = -3$, $y = 7$
9. Maximum is -4 and minimum is -28.
10. Maximum is 53 and minimum is 4.
11. 106 of the \$35 radios.
12. 146 dimes.
13. 34 gallons
14. 384 miles
15. Approximately 77 miles west and 69 miles south.
16. 70 nut bars and 120 chocolate bars.
17. 30 of the \$328 motors and 23 of the \$450 motors.
18. 15 units
19. $C(t) = 37.50t$
20. $\overline{C}(x) = 3x + 2 + \dfrac{27}{x}$, $x = 3$, and $\overline{C}(3) = 20$.

CHAPTER 3. MATRICES AND LINEAR SYSTEMS

3.2 SOLVING LINEAR SYSTEMS

1. $x = 3, y = 3$ **3.** $x = -1, y = 1$ **5.** $x = 1, y = 1$

7. $x = -5, y = 1$ **9.** $x = 1, y = 1$ **11.** $x = -2, y = 5$

13. $x = -4, y = 7$ **15.** $x = 1, y = -1$ **17.** $x = -1, y = -1$

19. $x = -3, y = 1$

21. Dependent. Two solutions: $x = 1$, $y = 1$ and $x = 4$, $y = 8$

23. Unique solution: $x = 1$, $y = 1$

25. Inconsistent **27.** Inconsistent **29.** Inconsistent

31. Dependent. Two solutions: $x = -\frac{7}{5}$, $y = 0$ and $x = 0$, $y = -\frac{7}{2}$

3.3 LINEAR SYSTEMS IN MORE THAN TWO UNKNOWNS

1. (a) $x = 0, y = 0, z = 6$ (b) $x = 5, y = 1, z = 1$

3. (a) $x = 2, y = 2, z = 3, w = 6$
(b) $x = 13, y = \frac{1}{2}, z = 7, w = -4$

5. If $x = 1$, $y = 1$, $z = 1$, then

$$2x - 3y + z = 0 \qquad \text{and} \qquad 3x + y - z = 3$$

but

$$x - y + 2z \neq 1.$$

7. If $x_1 = x_4$, $x_2 = 1 - x_4$, $x_3 = 2 + 2x_4$, then

$$\begin{aligned} 6x_1 + 2x_2 + x_3 - 6x_4 &= 6x_4 + 2(1 - x_4) + (2 + 2x_4) - 6x_4 \\ &= 6x_4 + 2 - 2x_4 + 2 + 2x_4 - 6x_4 \\ &= 4 \end{aligned}$$

$$\begin{aligned} 5x_1 + 2x_2 + x_3 - 5x_4 &= 5x_4 + 2(1 - x_4) + (2 + 2x_4) - 5x_4 \\ &= 5x_4 + 2 - 2x_4 + 2 + 2x_4 - 5x_4 \\ &= 4 \end{aligned}$$

$$\begin{aligned} 3x_1 + x_2 + x_3 - 4x_4 &= 3x_4 + (1 - x_4) + (2 + 2x_4) - 4x_4 \\ &= 3x_4 + 1 - x_4 + 2 + 2x_4 - 4x_4 \\ &= 3 \end{aligned}$$

9. $x = 1 - y$, $z = -4 + 3y$

11. $x = 1 + 7y$, $z = -1 - 10y$

13. (a) $x = (\frac{1}{23})(36 - 13y)$, $z = (\frac{1}{23})(61 - 15y)$

15. $x = (\frac{1}{3})(11y + 3)$, $z = (\frac{1}{6})(y + 12)$

17. $x = (\frac{1}{2})(8 - y)$, $z = (\frac{1}{3})(y - 11)$

19. $x = -(\frac{1}{19})(4y + 11)$, $z = (\frac{1}{19})(44y - 69)$

21. $x = 0, y = 8, z = 0$ **23.** $x = 6, y = -5, z = -3$

25. $x = -9, y = -2, z = 4$ **27.** $x = -6, y = 9, z = -4$

29. $x = -7, y = 2, z = 0$ **31.** $x = -4, y = 8, z = 0$

33. $y = -3 - 5x,\ z = 3,\ w = -7 - 5x$

35. $x = -1 + 8z,\ y = -2 + 5z,\ w = 4 - z$

37. $x = 4 - 3w - 6v,\ y = 4 - 9w + 3v,\ z = 3w - 2v$

3.4 MATRICES AND LINEAR SYSTEMS

1. Matrix of coefficients: $\begin{bmatrix} 1 & 1 \\ 2 & -3 \end{bmatrix}$ Augmented matrix: $\left[\begin{array}{cc|c} 1 & 1 & 1 \\ 2 & -3 & 2 \end{array}\right]$

3. Matrix of coefficients: $\begin{bmatrix} 3 & -5 & -6 \\ -7 & 9 & -6 \end{bmatrix}$ Augmented matrix: $\left[\begin{array}{ccc|c} 3 & -5 & -6 & 4 \\ -7 & 9 & -6 & -8 \end{array}\right]$

5. Matrix of coefficients: $\begin{bmatrix} 9 & 0 & 4 & 0 \\ 1 & 6 & 6 & -3 \\ 4 & 3 & -7 & -9 \\ 2 & 4 & -1 & 9 \end{bmatrix}$ Augmented matrix: $\left[\begin{array}{cccc|c} 9 & 0 & 4 & 0 & 0 \\ 1 & 6 & 6 & -3 & -6 \\ 4 & 3 & -7 & -9 & -8 \\ 2 & 4 & -1 & 9 & 8 \end{array}\right]$

7. $x = 0, y = 1$ **9.** $x = 1, y = 0$

11.
$$x = 1 - y,\ z = -4 + 3y,$$
or
$$y = 1 - x,\ z = -1 - 3x,$$
or
$$x = -(\tfrac{1}{3})(1 + z),\ y = (\tfrac{1}{3})(4 + z)$$

13.
$$x = (\tfrac{1}{15})(-11 + 13z),\ y = (\tfrac{1}{15})(61 - 23z),$$
or
$$z = (\tfrac{1}{13})(11 + 15x),\ y = (\tfrac{1}{13})(36 - 23x),$$
or
$$x = (\tfrac{1}{23})(36 - 13y),\ z = (\tfrac{1}{23})(61 - 15y)$$

15.
$$y = 8 - 2x,\ z = -(\tfrac{1}{3})(3 + 2x),$$
or
$$x = (\tfrac{1}{2})(8 - y),\ z = (\tfrac{1}{3})(-11 + y),$$
or
$$x = (-\tfrac{3}{2})(1 + z),\ y = 11 + 3z$$

17. $x = 0, y = 8, z = 0$ **19.** $x = 6, y = -5, z = -3$

21. $x = -9, y = -2, z = 4$ **23.** $x = -6, y = 9, z = -4$

25. $x = -7, y = 2, z = 0$ **27.** $x = -4, y = 8, z = 0$

29. $y = -3 - 5x,\ z = 3,\ w = -7 - 5x$

31. $x = -1 + 8z,\ y = -2 + 5z,\ w = 4 - z$

3.5 SOME MATRIX COMPUTATIONS

1. $\begin{bmatrix} 1 & 2 \\ 2 & 4 \end{bmatrix}$ **3.** $\begin{bmatrix} 3 & 4 \\ 6 & 7 \\ 11 & 12 \\ 18 & 19 \end{bmatrix}$

5. $[0 \quad -3 \quad -8 \quad -15 \quad -24]$

7. $\begin{bmatrix} 1 & 0 & 0 \\ 0 & 1 & 0 \\ 0 & 0 & 1 \end{bmatrix}$

9. $\begin{bmatrix} 2 & 0 & 0 & 0 \\ 3 & 3 & 1 & 1 \\ 4 & 4 & 4 & 2 \\ 5 & 5 & 5 & 5 \end{bmatrix}$

11. $\begin{bmatrix} 1 & -1 & 1 \\ 2 & 6 & -1 \\ 1 & 8 & 3 \end{bmatrix}$ **13.** $\begin{bmatrix} 2 & 1 & -4 \\ 7 & -3 & -5 \\ 8 & -5 & -3 \end{bmatrix}$

15. $\begin{bmatrix} -2 & 0 & 2 \\ -6 & -2 & 4 \\ -6 & -2 & 0 \end{bmatrix}$ **17.** $\begin{bmatrix} 1 & 1 & -3 \\ 4 & -4 & -3 \\ 5 & -6 & -3 \end{bmatrix}$

19. $\begin{bmatrix} -6 & -4 & 14 \\ -22 & 14 & 16 \\ -26 & 22 & 12 \end{bmatrix}$

21. $x = -1,\ y = 7,\ z = 2,\ w = 1$

23. $x = 0.5,\ y = -1.5,\ z = 2.5,\ w = \frac{1}{3}$

25. No solution

27. $x = \begin{bmatrix} 1 & 1 & 5 \\ -2 & 6 & 5 \\ -3 & 8 & 5 \end{bmatrix}$ **29.** $x = \begin{bmatrix} \frac{1}{2} & -\frac{3}{2} & -\frac{3}{2} \\ 3 & 3 & -\frac{7}{2} \\ \frac{5}{2} & 4 & \frac{1}{2} \end{bmatrix}$

31. $x = \begin{bmatrix} 3 & 2 & 13 \\ -4 & 18 & 12 \\ -7 & 24 & 14 \end{bmatrix}$

3.6 MATRIX MULTIPLICATION

1. -3 **3.** 12 **5.** 18 **7.** 16

9. (a) $y = 2$ (b) $y = 7$ **11.** (a) $y = -11$ (b) $y = -\frac{17}{2}$

13. A is a row vector, B is a column vector, and the number of entries in A and B is the same.

15. A has 5 rows and B has 7 columns.

17. A is a square matrix.

19. 8 **21.** -3 **23.** 6 **25.** 11

27. $\begin{bmatrix} 10 & 11 \\ 25 & -11 \end{bmatrix}$ **29.** $\begin{bmatrix} -10 \\ -5 \end{bmatrix}$ **31.** $\begin{bmatrix} -30 & -35 & 40 \\ 24 & 37 & -5 \end{bmatrix}$

33. $\begin{bmatrix} -15 & -16 \\ 0 & 55 \\ -2 & 35 \end{bmatrix}$ **35.** $\begin{bmatrix} 2 & 1 & 1 \\ 1 & 2 & 1 \\ 1 & 1 & 2 \end{bmatrix}$ **37.** $\begin{bmatrix} 0 & -1 & 4 \\ -2 & 5 & 3 \\ 2 & 4 & 5 \end{bmatrix}$

39. $\begin{bmatrix} 3 & 2 & 3 \\ 2 & 3 & 3 \\ 3 & 3 & 2 \end{bmatrix}$ **41.** $\begin{bmatrix} 4 & 3 & -1 \\ 1 & 8 & 3 \\ 7 & 9 & 6 \end{bmatrix}$

43. $x = 1, y = 2$ **45.** $x = 1, y = 1$

47. $\begin{aligned} 2x - 3y &= 10, \\ x + 4y &= 8 \end{aligned}$ **49.** $\begin{aligned} -8x + 6y - 6z &= -4, \\ 3x + 4y + 6z &= 3, \\ -7y - 2z &= 1 \end{aligned}$

51. $\begin{aligned} 3x \qquad + z - 2w &= 1, \\ -y \qquad + 5w &= -3, \\ 5x + 3y + 2z - 4w &= 4 \end{aligned}$

3.7 INVERSES OF SQUARE MATRICES

1. $\begin{bmatrix} 1 & 0 \\ -1 & 1 \end{bmatrix}$ **3.** $\begin{bmatrix} 1 & -5 \\ 0 & 1 \end{bmatrix}$ **5.** $\begin{bmatrix} 1 & 4 \\ 0 & -1 \end{bmatrix}$

7. $\begin{bmatrix} 36 & -5 \\ -7 & 1 \end{bmatrix}$ **9.** $\begin{bmatrix} -11 & 4 \\ 3 & -1 \end{bmatrix}$ **11.** $\begin{bmatrix} 17 & 2 \\ 8 & 1 \end{bmatrix}$

13. No inverse **15.** $\frac{1}{3}\begin{bmatrix} -1 & -2 \\ 2 & 1 \end{bmatrix}$ **17.** $\begin{bmatrix} 1 & 0 & 0 \\ 0 & 1 & 0 \\ 0 & 0 & 1 \end{bmatrix}$

19. $\begin{bmatrix} 1 & 1 & -1 \\ 0 & 1 & 0 \\ 0 & -1 & 1 \end{bmatrix}$ **21.** $\begin{bmatrix} -1 & -1 & -3 \\ 0 & -1 & -3 \\ 0 & 0 & 1 \end{bmatrix}$ **23.** $\begin{bmatrix} 30 & -1 & -3 \\ 31 & -1 & -3 \\ -8 & 0 & 1 \end{bmatrix}$

25. $\begin{bmatrix} 12 & -1 & -3 \\ 13 & -1 & -3 \\ -6 & 0 & 1 \end{bmatrix}$ **27.** $\begin{bmatrix} 12 & -1 & -14 \\ 13 & -1 & -15 \\ -6 & 0 & 7 \end{bmatrix}$

29. No inverse **31.** No inverse

33. $\frac{1}{8}\begin{bmatrix} 3 & -2 & -4 & 1 \\ 2 & 4 & 0 & -2 \\ -1 & -2 & 4 & -3 \\ 4 & 0 & 0 & 4 \end{bmatrix}$

35. $A = \begin{bmatrix} 1 & -1 \\ -7 & 8 \end{bmatrix}$, $A^{-1} = \begin{bmatrix} 8 & 1 \\ 7 & 1 \end{bmatrix}$, $B = \begin{bmatrix} -9 \\ 68 \end{bmatrix}$,

$$X = \begin{bmatrix} x \\ y \end{bmatrix} = \begin{bmatrix} -4 \\ 5 \end{bmatrix}$$

37. $A = \begin{bmatrix} 1 & 2 \\ 8 & 15 \end{bmatrix}$, $A^{-1} = \begin{bmatrix} -15 & 2 \\ 8 & -1 \end{bmatrix}$, $B = \begin{bmatrix} -1 \\ -12 \end{bmatrix}$,

$$X = \begin{bmatrix} x \\ y \end{bmatrix} = \begin{bmatrix} -9 \\ 4 \end{bmatrix}$$

39. $A = \begin{bmatrix} 1 & -1 \\ 7 & -6 \end{bmatrix}$, $A^{-1} = \begin{bmatrix} -6 & 1 \\ -7 & 1 \end{bmatrix}$, $B = \begin{bmatrix} 7 \\ 44 \end{bmatrix}$,

$$X = \begin{bmatrix} x \\ y \end{bmatrix} = \begin{bmatrix} 2 \\ -5 \end{bmatrix}$$

41. $A = \begin{bmatrix} -1 & 4 \\ 8 & -33 \end{bmatrix}$, $A^{-1} = \begin{bmatrix} -33 & -4 \\ -8 & -1 \end{bmatrix}$, $B = \begin{bmatrix} 23 \\ -190 \end{bmatrix}$,

$$X = \begin{bmatrix} x \\ y \end{bmatrix} = \begin{bmatrix} 1 \\ 6 \end{bmatrix}$$

43. $A = \begin{bmatrix} 1 & 0 & 9 \\ 5 & -1 & 53 \\ 4 & 0 & 36 \end{bmatrix}$

A^{-1} does not exist; the system is dependent. Equation (3) is redundant and may be deleted. We may then write the system as

$$\begin{aligned} x \quad + 9z &= -9, \\ 5x - y + 53z &= -12. \end{aligned}$$

If z is used as a parameter, the system may be expressed as

$$\begin{aligned} x \quad &= -9 - 9z, \\ 5x - y &= -12 - 53z. \end{aligned}$$

Then

$$A = \begin{bmatrix} 1 & 0 \\ 5 & -1 \end{bmatrix}, \quad A^{-1} = \begin{bmatrix} 1 & 0 \\ 5 & -1 \end{bmatrix}, \quad B = \begin{bmatrix} -9 - 9z \\ -12 - 53z \end{bmatrix},$$

$$X = \begin{bmatrix} x \\ y \end{bmatrix} = \begin{bmatrix} -9 - 9z \\ -33 + 8z \end{bmatrix}$$

45. $A = \begin{bmatrix} 1 & 2 & 3 \\ 2 & 5 & 6 \\ 5 & 10 & 16 \end{bmatrix}, \quad A^{-1} = \begin{bmatrix} 20 & -2 & -3 \\ -2 & 1 & 0 \\ -5 & 0 & 1 \end{bmatrix},$

$$B = \begin{bmatrix} -12 \\ -23 \\ -63 \end{bmatrix}, \quad X = \begin{bmatrix} x \\ y \\ z \end{bmatrix} = \begin{bmatrix} -5 \\ 1 \\ -3 \end{bmatrix}$$

47. $A = \begin{bmatrix} -1 & -9 & 7 \\ -4 & -35 & 32 \\ 8 & 72 & -57 \end{bmatrix}, \quad A^{-1} = \begin{bmatrix} -309 & -9 & -43 \\ 28 & 1 & 4 \\ -8 & 0 & -1 \end{bmatrix},$

$$B = \begin{bmatrix} -22 \\ -52 \\ 169 \end{bmatrix}, \quad X = \begin{bmatrix} x \\ y \\ z \end{bmatrix} = \begin{bmatrix} -1 \\ 8 \\ 7 \end{bmatrix}$$

49. $A = \begin{bmatrix} 5 & 3 & 3 & 3 \\ 5 & 3 & 2 & 2 \\ 3 & 1 & 0 & 1 \\ 2 & 1 & 1 & 1 \end{bmatrix}, \quad A^{-1} = \begin{bmatrix} -1 & 0 & 0 & 3 \\ 1 & 1 & 0 & -5 \\ -1 & 0 & -1 & 4 \\ 2 & -1 & 1 & -4 \end{bmatrix},$

$$B = \begin{bmatrix} 2 \\ 2 \\ 1 \\ 1 \end{bmatrix}, \quad X = \begin{bmatrix} x \\ y \\ z \\ w \end{bmatrix} = \begin{bmatrix} 1 \\ -1 \\ 1 \\ -1 \end{bmatrix}$$

3.8 SOME APPLICATIONS OF MATRICES

1. (a) 9 (b) 0 (c) 33 (d) 5 (e) 32 (f) 66 (g) 6 (h) 42

3. (a) $T - S = [116 \quad 192 \quad 276 \quad 93 \quad 120]$
(b) $[404 \quad 502 \quad 708 \quad 283 \quad 267]$

5. (a) \$212.13 (b) \$26.75 (c) 3 by 2
(d) a_{12} = labor costs of producing an executive desk at the Warrenton plant
a_{21} = labor costs of producing a secretary desk at the Filmore plant

a_{22} = labor costs of producing a secretary desk at the Warrenton plant

a_{13} = labor costs of producing a student desk at the Filmore plant

a_{23} = labor costs of producing a student desk at the Warrenton plant

(e) $\begin{bmatrix} 13.50 & 12.69 \\ 8.10 & 6.75 \\ 4.59 & 4.05 \end{bmatrix}$

(f) $wc' = \begin{bmatrix} 229.10 & 212.63 \\ 157.55 & 145.80 \\ 31.16 & 28.89 \end{bmatrix}$

7. $\begin{bmatrix} 3.95 & 4.05 \\ 0.23 & 0.18 \\ 0.33 & 0.31 \end{bmatrix}$

9. (a) Let $T = [1.5 \quad 1 \quad 0.5 \quad 2]$. Then

$$Tx = [0.87 \quad 0.10 \quad 0.58 \quad 0.55],$$

which shows the amount of each vitamin A, B, C, D (in this order) obtained.

(b) Cost of A = (0.20)(0.87) = 0.17
Cost of B = (0.10)(0.5) = 0.05
Cost of C = (0.58)(0.30) = 0.17
Cost of D = (0.55)(0.40) = 0.22

Total cost of vitamins = 0.61

3.9 SOME FURTHER APPLICATIONS OF LINEAR SYSTEMS—MATRICES

1.

	Junior	Competition	Pro
Cutting	1	1	1.5
Laminating	1	1.5	2
Finishing	1.5	2	2.5

3. Let

x = Number of juniors

y = Number of competition

z = Number of pro

If z is used as a parameter, then

$$x = 80 - z \qquad \text{and} \qquad y = 70 - 0.5z.$$

All possible schedules can be found by choosing z as any of the numbers 0, 1, 2, 3, . . . , 80. The maximum number of Pro models that can be made is 80.

5. Using the notation of (3) and z as a parameter gives

$$x = 22 - 0.5z,$$
$$y = 54 - z.$$

All possible schedules can be found by choosing z as any of the numbers 0, 1, 2, 3, . . . , 44.

7. Invest \$40,000 in I and II, \$20,000 in III.

3.10 FURTHER APPLICATIONS—INPUT/OUTPUT ANALYSIS

1. (a) $\begin{bmatrix} 0.026 & 0.038 \\ 0.053 & 0.087 \end{bmatrix}$ (b) $\begin{bmatrix} 700 \\ 910 \end{bmatrix}$ (c) $\begin{bmatrix} 760 \\ 1040 \end{bmatrix}$

3. (a) $\begin{bmatrix} 0.063 & 0.066 \\ 0.094 & 0.011 \end{bmatrix}$ (b) $\begin{bmatrix} 240 \\ 870 \end{bmatrix}$ (c) $\begin{bmatrix} 320 \\ 910 \end{bmatrix}$

5. (a) $\begin{bmatrix} 0.170 & 0.013 \\ 0.038 & 0.104 \end{bmatrix}$ (b) $\begin{bmatrix} 430 \\ 670 \end{bmatrix}$ (c) $\begin{bmatrix} 530 \\ 770 \end{bmatrix}$

7. (a) $\begin{bmatrix} 0.130 & 0.018 & 0.117 \\ 0.039 & 0.046 & 0.117 \\ 0.049 & 0.041 & 0.054 \end{bmatrix}$ (b) $\begin{bmatrix} 498 \\ 845 \\ 470 \end{bmatrix}$ (c) $\begin{bmatrix} 670 \\ 983 \\ 574 \end{bmatrix}$

9. $$I - A = \begin{bmatrix} 0.965 & -0.120 \\ -0.088 & 0.947 \end{bmatrix}$$

$$(I - A)^{-1} = \begin{bmatrix} 1.048 & 0.133 \\ 0.097 & 1.068 \end{bmatrix} \begin{bmatrix} 460 \\ 660 \end{bmatrix} = \begin{bmatrix} 569.860 \\ 749.500 \end{bmatrix}$$

11. $$I - A = \begin{bmatrix} 0.890 & -0.110 \\ -0.020 & 0.890 \end{bmatrix}$$

$$(I - A)^{-1}D = \begin{bmatrix} 1.127 & 0.139 \\ 0.025 & 1.127 \end{bmatrix} \begin{bmatrix} 360 \\ 230 \end{bmatrix} = \begin{bmatrix} 437.591 \\ 268.210 \end{bmatrix}$$

13. $$I - A = \begin{bmatrix} 0.800 & -0.160 \\ -0.120 & 0.930 \end{bmatrix}$$

$$(I - A)^{-1}D = \begin{bmatrix} 1.283 & 0.221 \\ 0.166 & 1.104 \end{bmatrix}\begin{bmatrix} 110 \\ 490 \end{bmatrix} = \begin{bmatrix} 249.42 \\ 559.220 \end{bmatrix}$$

15. $$I - A = \begin{bmatrix} 0.933 & -0.048 & -0.144 \\ -0.047 & 0.952 & -0.158 \\ -0.069 & -0.043 & 0.884 \end{bmatrix}$$

$$(I - A)^{-1}D = \begin{bmatrix} 1.089 & 0.063 & 0.189 \\ 0.068 & 1.063 & 0.201 \\ 0.088 & 0.057 & 1.156 \end{bmatrix}\begin{bmatrix} 384 \\ 472 \\ 469 \end{bmatrix} = \begin{bmatrix} 536.553 \\ 622.117 \\ 602.860 \end{bmatrix}$$

3.11 REVIEW EXERCISES

1. (a) $x = 5,\ y = -6$ (b) $x = 8,\ y = -3$

2. (a) $x = 3,\ y = -3,\ z = 3$ (b) $x = \frac{1}{29}(10 - 29y)$, $z = \frac{1}{29}(181 + 29y)$

3. (a) $x = -5,\ y = 4$ (b) $x = 2,\ y = -3,\ z = 1$ (c) $y = (\frac{1}{17})(23x - 251),\ z = (\frac{1}{17})(73 + 16x)$

4. (a) $\begin{bmatrix} 0 & -5 & -5 \\ 1 & 9 & 3 \\ -10 & 11 & -14 \end{bmatrix}$ (b) $\begin{bmatrix} 2 & -5 & -9 \\ 11 & -9 & 1 \\ 0 & 7 & 2 \end{bmatrix}$

(c) $\begin{bmatrix} 5 & -10 & -20 \\ 27 & -27 & 1 \\ 5 & 12 & 12 \end{bmatrix}$ (d) $\begin{bmatrix} -11 & 23 & -5 \\ 44 & 34 & 47 \\ 47 & -47 & 87 \end{bmatrix}$

(e) $\begin{bmatrix} 59 & -59 & 39 \\ -18 & 4 & 4 \\ -10 & -4 & 47 \end{bmatrix}$

5. (a) $\frac{1}{36}\begin{bmatrix} -1 & -6 \\ 5 & -6 \end{bmatrix}$ (b) No inverse

6. $\frac{1}{3}\begin{bmatrix} -3 & 5 & 11 \\ -16 & 18 & 0 \\ -5 & -5 & -10 \end{bmatrix}$

7. (a) $x = -1,\ y = -4$ (b) $x = 8,\ y = -2,\ z = 4$

CHAPTER 4. LINEAR PROGRAMMING IN THE PLANE

4.2 GRAPHS OF LINEAR INEQUALITIES

1.

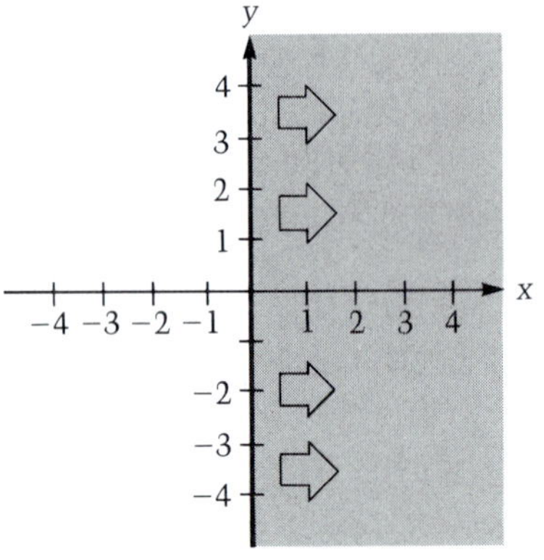

3.

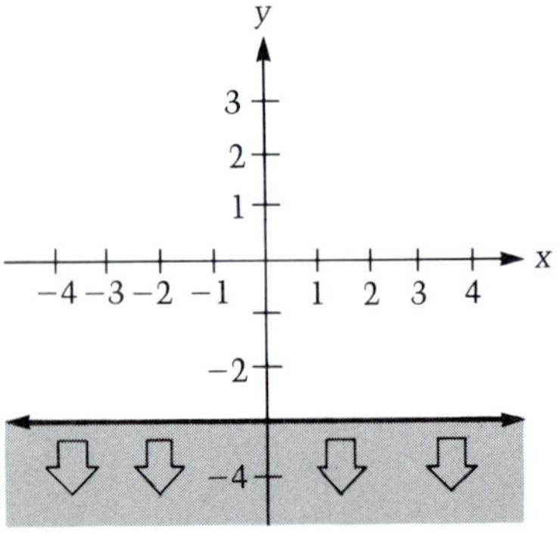

5.

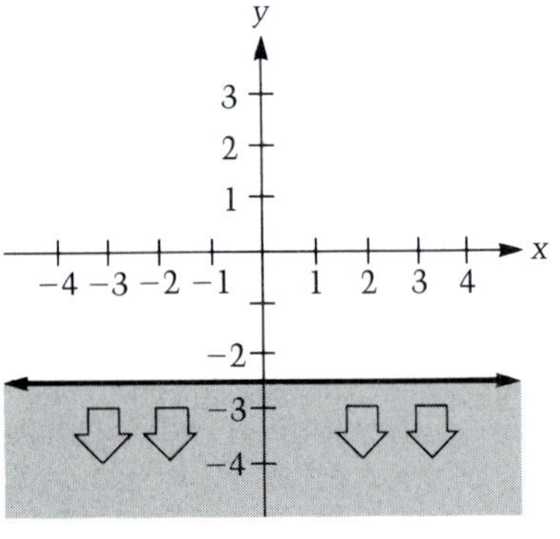

7.

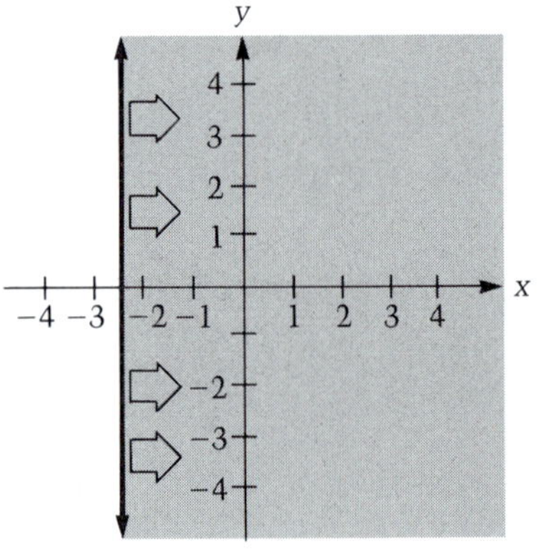

9.

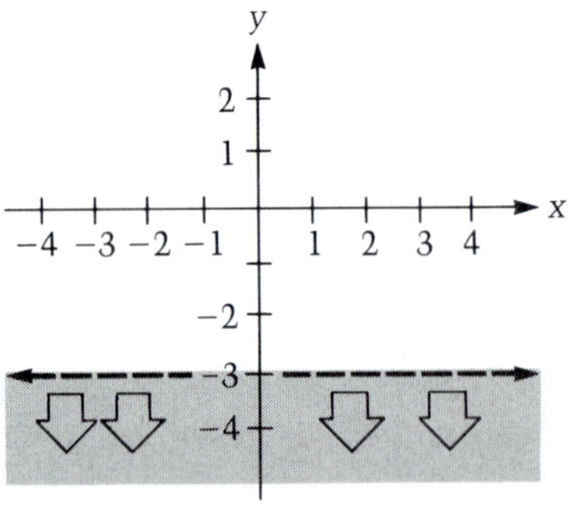

11.

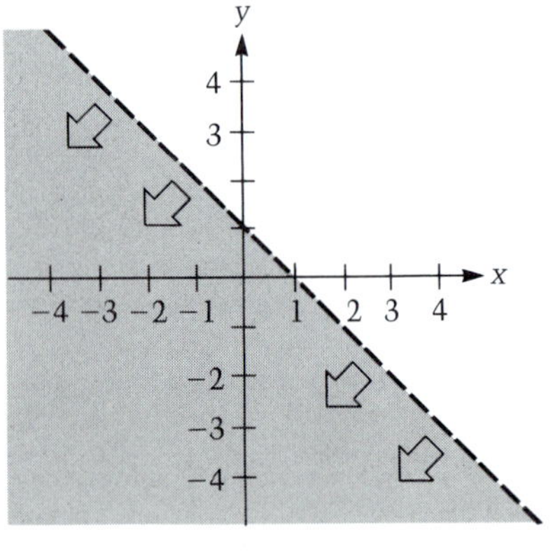

13.

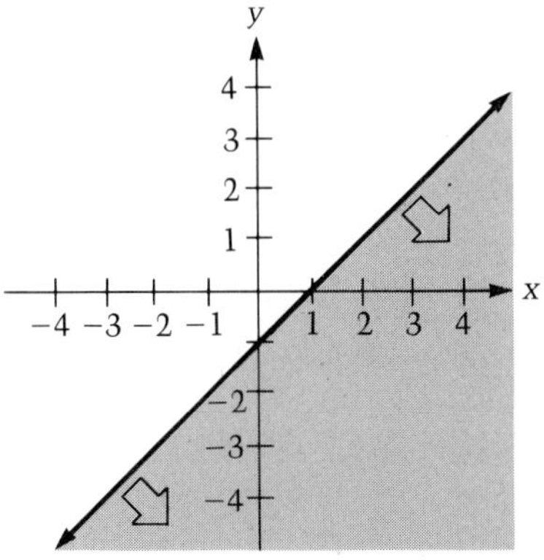

15.

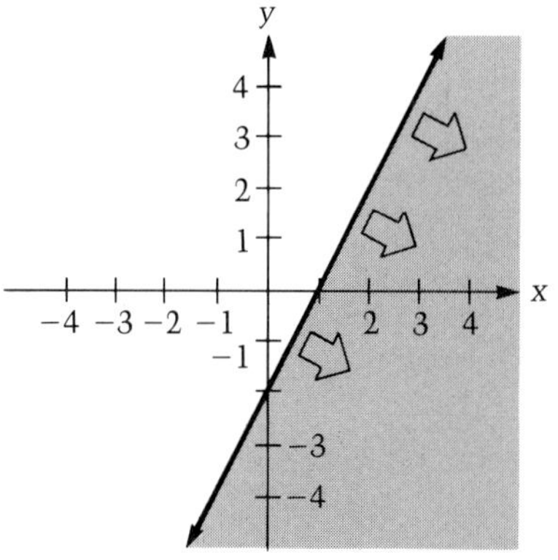

17.

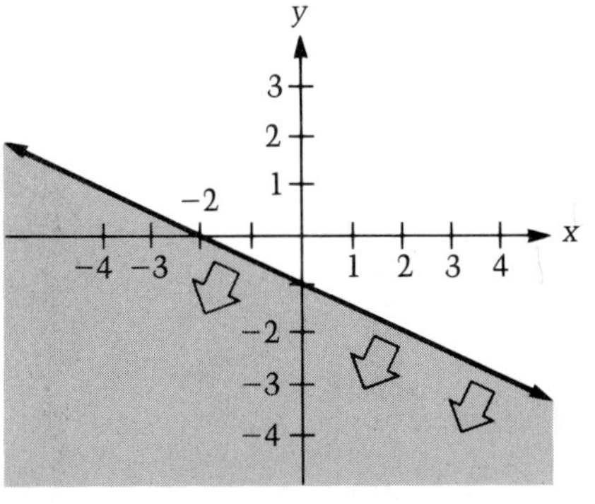

19.

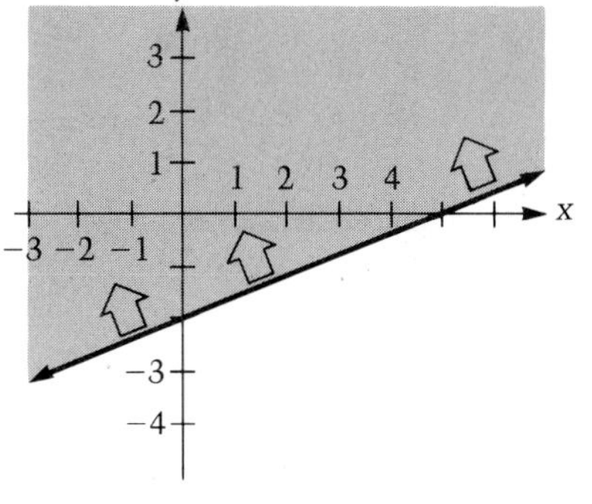

21.

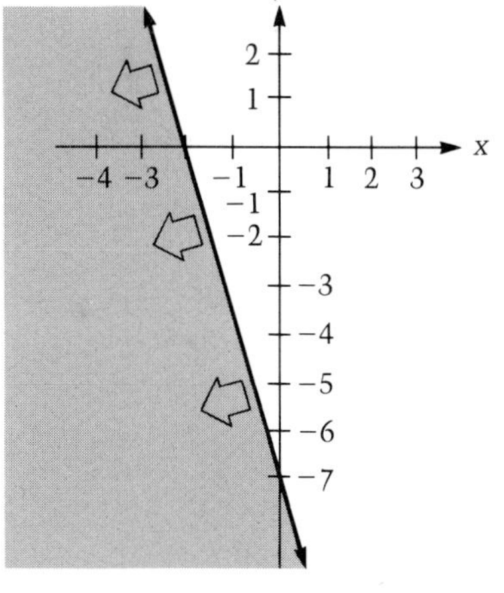

23.

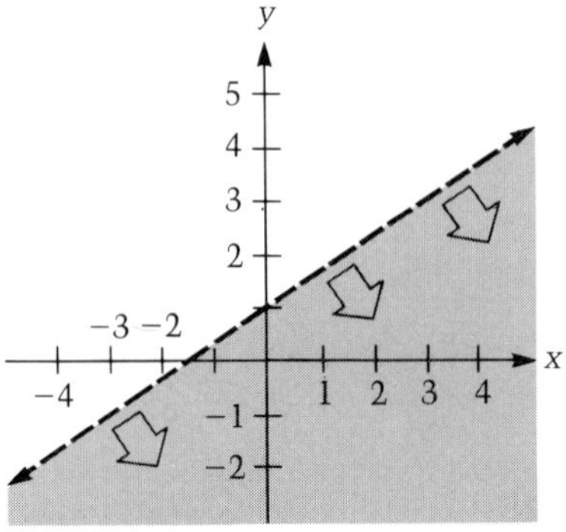

25.

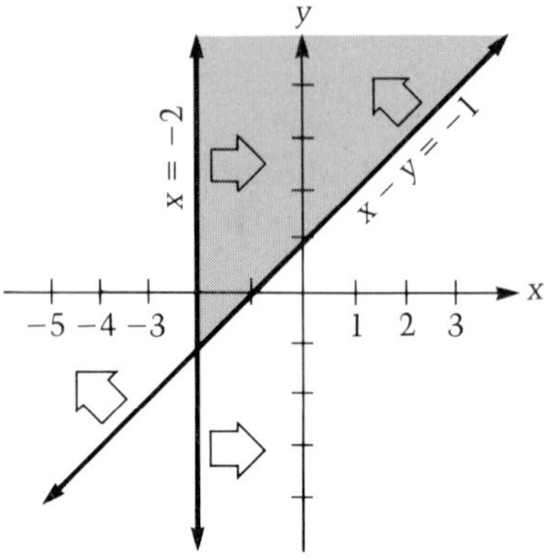

27.

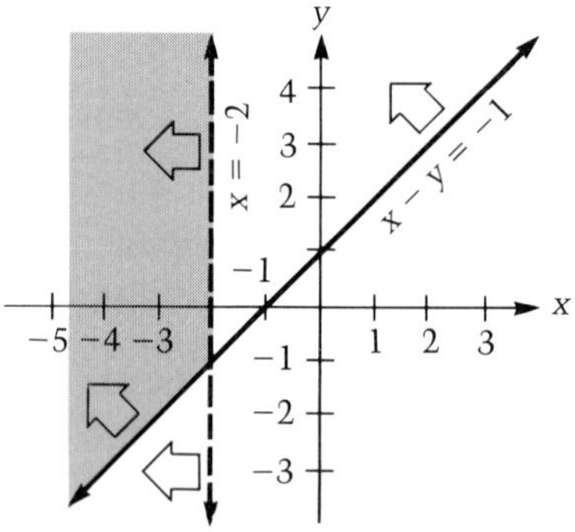

29.

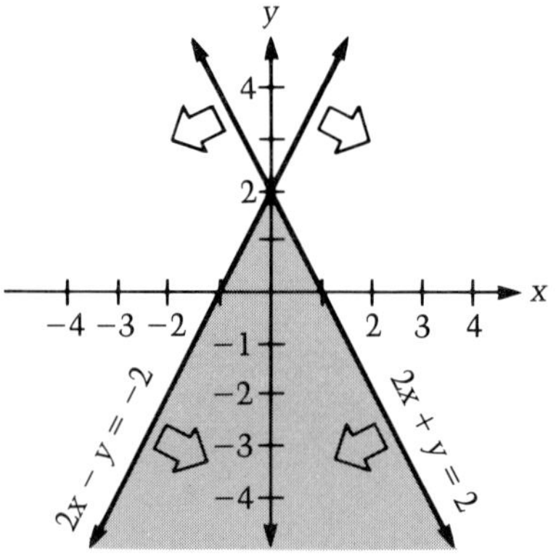

31.

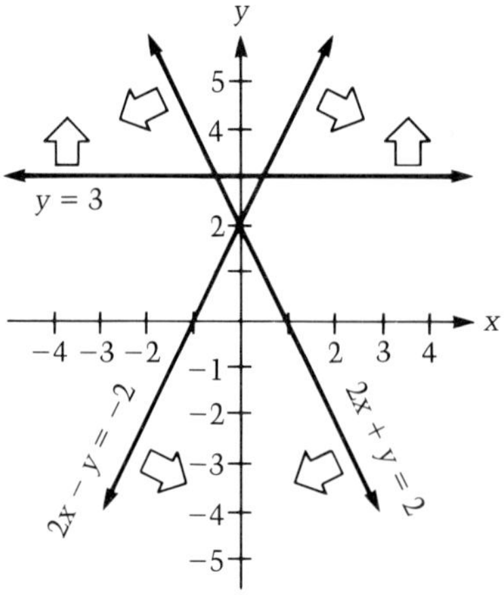

33.

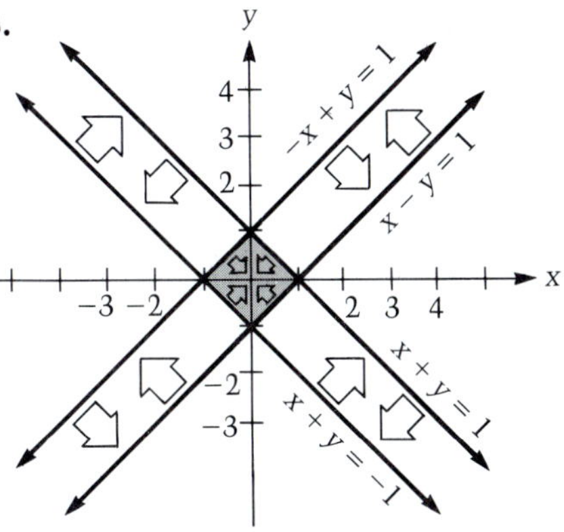

4.3 LINEAR PROGRAMMING PROBLEMS IN THE PLANE

1. (a) Max: 3; min: 1
   (b) Max: 1; min: −1
   (c) Max: 9; min: −1
   (d) Max: 5; min: −3

3. (a) Max: 12; min: 1
   (b) Max: −1; min: −12
   (c) Max: 2; min: −8
   (d) Max: 8, min: −2

5. Max $P = 60$ at $x = 0$, $y = 12$
7. Max $P = 150$ at $x = 6$, $y = 0$
9. Min $C = -35$ at $x = 7$, $y = 0$
11. Max $P = 42$ at $x = 0$, $y = 6$
13. Min $C = -132$ at $x = 11$, $y = 0$

4.4 LINEAR PROGRAMMING MODELS—APPLICATIONS

1. (a) $y \geq x$ (b) $7y \geq 3x$
3. \$112,500,000 in loans, \$37,500,000 in securities for a profit of \$28,125,000
5. No pine, 6 acres of fir for a cost of \$6000
7. Operate A 8 days and B 6 days. The minimum cost will be \$152,000
9. Operate each mine 2 days. The minimum cost will be \$1120

4.5 REVIEW EXERCISES

1. (a)

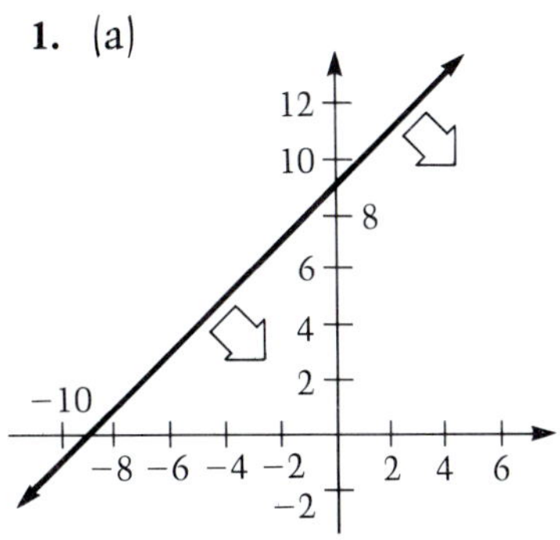

(b)

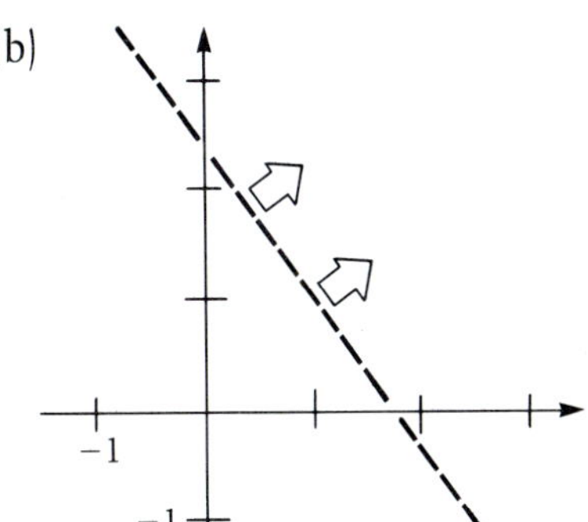

(c)

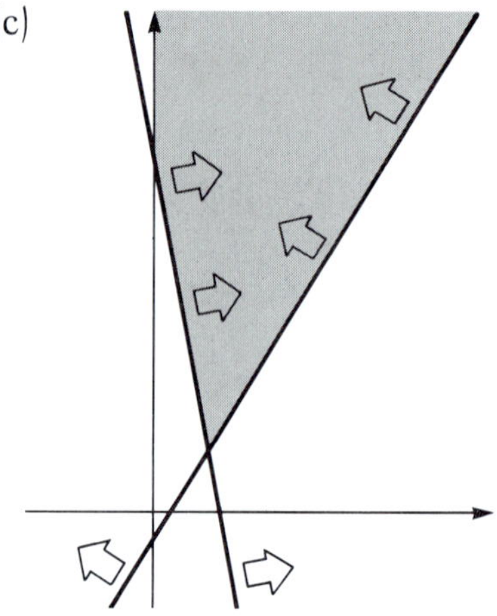

(d)

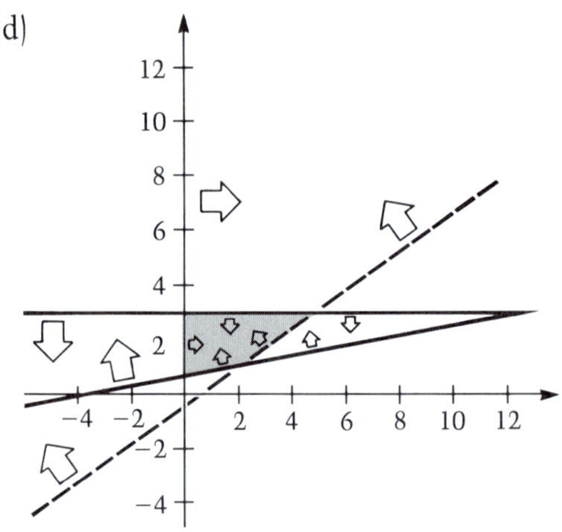

2. (a) max = 38 min = 9
   (b) max = 29 min = 6
   (c) max = −6 min = −28
   (d) max = 25 min = −1
3. (a) $P = 18$ at (4, 2) (b) $P = 12$ at (2, 1)
4. \$5000 stocks, \$5000 bonds

CHAPTER 5. THE SIMPLEX ALGORITHM

5.2 THE SIMPLEX ALGORITHM—PART I

1. Using slack variables u, v, the linear system is

$$\begin{aligned} 2x + y + u &= 12, \\ x + y + v &= 9, \\ -25x + 13y + P &= 0, \end{aligned}$$

and the initial simplex tableau is

$$\left[\begin{array}{rr|rrr|r} 2 & 1 & 1 & 0 & 0 & 12 \\ 1 & 1 & 0 & 1 & 0 & 9 \\ \hline -25 & 13 & 0 & 0 & 1 & 0 \end{array}\right]$$

3. Using slack variables u, v, the linear system is

$$\begin{aligned} 2x + y + u &= 6, \\ -x + 2y + v &= 2, \\ -2x + y + P &= 0, \end{aligned}$$

and the initial simplex tableau is

$$\left[\begin{array}{rr|rrr|r} 2 & 1 & 1 & 0 & 0 & 6 \\ -1 & 2 & 0 & 1 & 0 & 2 \\ \hline -2 & 1 & 0 & 0 & 1 & 0 \end{array}\right]$$

5. Using slack variables u, v, w, the linear system is

$$\begin{aligned} 5x - y + 2z + u &= 20, \\ 2x + y + 3z + v &= 15, \\ x + 4y - z + w &= 18, \\ -2x - 3y - 4z + P &= 0, \end{aligned}$$

and the initial simplex tableau is

$$\left[\begin{array}{rrr|rrrr|r} 5 & -1 & 2 & 1 & 0 & 0 & 0 & 20 \\ 2 & 1 & 3 & 0 & 1 & 0 & 0 & 15 \\ 1 & 4 & -1 & 0 & 0 & 1 & 0 & 18 \\ \hline -2 & -3 & -4 & 0 & 0 & 0 & 1 & 0 \end{array}\right]$$

7.
$$\begin{array}{c} \begin{array}{rrrrr} x & y & u & v & P \end{array} \\ \left[\begin{array}{rr|rrr|r} 2 & 3 & 1 & 0 & 0 & 12 \\ 2 & 5 & 0 & 1 & 0 & 9 \\ \hline -11 & -6 & 0 & 0 & 1 & 0 \end{array}\right] \end{array}$$
$x = 0$, $y = 0$, $u = 12$, $v = 9$, $P = 0$

9.
$$\begin{array}{c} \begin{array}{rrrrrr} x & y & u & v & w & P \end{array} \\ \left[\begin{array}{rr|rrrr|r} 4 & 3 & 1 & 0 & 0 & 0 & 24 \\ 2 & 1 & 0 & 1 & 0 & 0 & 9 \\ 2 & 4 & 0 & 0 & 1 & 0 & 24 \\ \hline 4 & -5 & 0 & 0 & 0 & 1 & 0 \end{array}\right] \end{array}$$
$x = 0$, $y = 0$, $u = 24$, $v = 9$, $w = 24$, $P = 0$

11. $x = 10$, $y = 0$, max $P = 80$
13. $x = 8$, $y = 0$, max $P = 160$
15. $x = 10$, $y = \frac{17}{3}$, max $P = 105\frac{1}{3}$
17. $x = 8$, $y = 1$, $z = 0$, max $P = 52$
19. $x = 0$, $y = 1$, $z = 3$, max $P = 22$
21. Invest \$51,000 in stocks, nothing in bonds, and \$49,000 in certificates. Maximum profit is \$18,040.
23. (a) Make 4 of each model. Profit is \$1420. (b) \$75

5.3 THE SIMPLEX ALGORITHM—PART II

1. (a) $x = 1$, $y = 1$, max $P = 9$ (b) $x = 2$, $y = 0$, max $P = 12$
3. $x = 4$, $y = 0$, max $P = 4$
5. (a) $x = 0$, $y = 2$, max $P = 4$ (b) $x = 2$, $y = 0$, max $P = 4$
7. $x = 2$, $y = 1$, max $P = 7$
9. $x = 0$, $y = 0$, $z = 10$, max $P = 40$
11. Produce 6,000,000 barrels of diesel fuel 4,000,000 barrels of gasoline for a maximum profit of \$16,000,000.
13. Invest \$10,000 in stocks, \$5000 in bonds for a maximum profit of \$2550.

5.4 THE SIMPLEX ALGORITHM—PART III

1. $x = 3$, $y = 0$, min $C = 3$
3. $x = 5$, $y = 10$, min $C = 100$
5. $x = \frac{3}{4}$, $y = \frac{3}{4}$, min $C = \frac{3}{2}$ **7.** $x = 5$, $y = 10$, min $C = 35$
9. (a) $x = 1$, $y = 1$, min $C = 17$ (b) $x = 1$, $y = 2$, min $C = -7$
11. (a) $x = 0$, $y = 0$, $z = 15$, min $C = 15$ (b) $x = 0$, $y = 5$, $z = 15$, min $C = 10$
13. (a) $x = 4$, $y = 0$, $z = 0$, min $C = 4$ (b) $x = 0$, $y = 0$, $z = 4$, min $C = 4$
15. Do not operate Abe I; operate Abe II $1\frac{1}{3}$ days; operate Abe III $7\frac{1}{3}$ days. The minimum cost is \$2980.
17. Ship 50 heaters from I to north, 25 heaters from I to south, no heaters from II to north, 15 heaters from II to south. Minimum cost is \$415.

5.5 REVIEW PROBLEMS

1.
$$\begin{aligned} 3x + y + z + s &= 15 \\ x + 2y + 3z + t &= 10 \\ 2x + 3y + 5z + u &= 12 \\ -5x - 9y - z + P &= 0 \end{aligned}$$

$$\left[\begin{array}{ccc|cccc|c} 3 & 1 & 1 & 1 & 0 & 0 & 0 & 15 \\ 1 & 2 & 3 & 0 & 1 & 0 & 0 & 10 \\ 2 & 3 & 5 & 0 & 0 & 1 & 0 & 12 \\ \hline -5 & -9 & -1 & 0 & 0 & 0 & 1 & 0 \end{array}\right]$$

2. $x = 0$, $y = 0$, $z = 0$, $s = 15$, $t = 10$, $u = 12$, $P = 0$
3. $P = 52$ when $x = 8$, $y = 0$, $z = 2$
4. $P = 5$ when $x = 2$, $y = 3$
5. $P = -20$ when $x = 0$, $y = 20$, $z = 0$

CHAPTER 6. COMBINATORICS

6.2 EXPERIMENTS AND THE COUNTING PRINCIPLE

1. 5 **3.** 6; $\{M, T, R, L, F, D\}$ where M is for Meatloaf, T for Turkey, etc.
5. (a) 20; $\{1, 2, 3, \ldots, 19, 20\}$ (b) 2; $\{B, W\}$
7. Outcomes: (H, H), (H, T), (T, H), (T, T)
   Sample space: $\{(H, H), (H, T), (T, H), (T, T)\}$

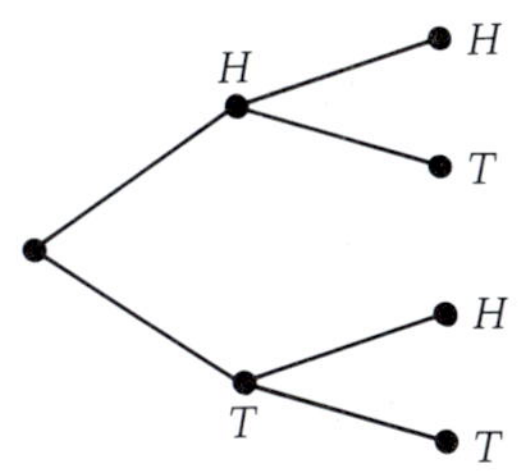

9. Outcomes: $(0, a)$, $(0, b)$, $(0, c)$, $(1, a)$, $(1, b)$, $(1, c)$
   Sample space: $\{(0, a), (0, b), (0, c), (1, a), (1, b), (1, c)\}$

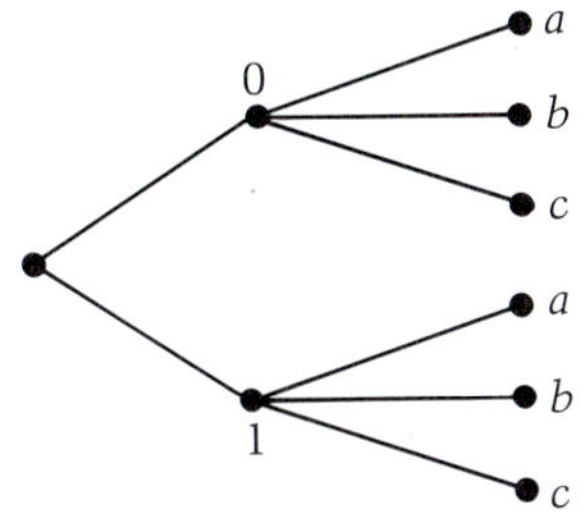

11. Outcomes: $(0, 1)$, $(0, 2)$, $(0, 3)$, $(1, 0)$, $(1, 2)$, $(1, 3)$, $(2, 0)$, $(2, 1)$, $(2, 3)$, $(3, 0)$, $(3, 1)$, $(3, 2)$

Sample space: {(0, 1), (0, 2), (0, 3), (1, 0), (1, 2), (1, 3), (2, 0), (2, 1), (2, 3), (3, 0), (3, 1), (3, 2)}

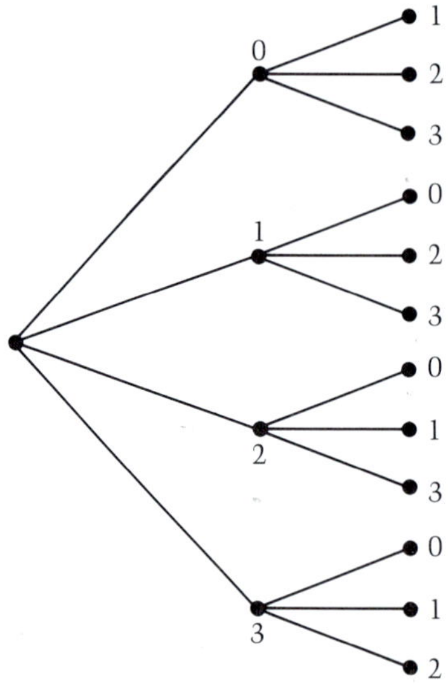

13. 20 **15.** 12 **17.** 216 **19.** 2^{10}, or 1024
21. 4^{10}, or 1,048,576 **23.** 120 **25.** (a) 50 (b) 600
27. (a) 10^7, or 10,000,000 (b) $8 \times 10^6 = 8{,}000{,}000$ **29.** 20
31. (a) $2^6 = 64$ (b) $12^2 = 144$ (c) $3^8 = 6561$ (d) 8

6.3. PERMUTATIONS

1. (a)

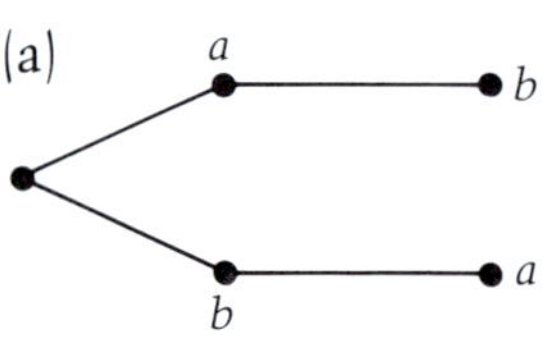

(b) {(a, b), (b, a)}
(c) 2

3. (a)

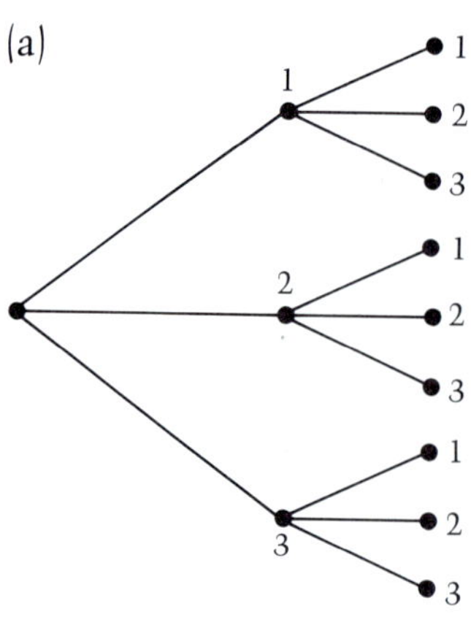

(b) {(1, 1), (1, 2), (1, 3), (2, 1), (2, 2), (2, 3), (3, 1), (3, 2), (3, 3)}
(c) 8 (d) 6

5. (a)

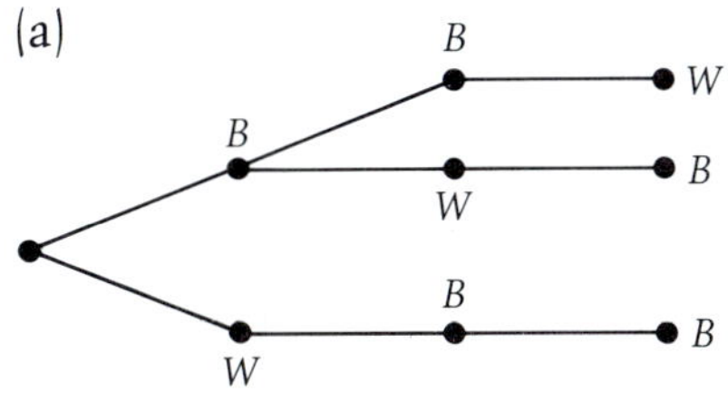

(b) {(*B*, *B*, *W*), (*B*, *W*, *B*), (*W*, *B*, *B*)}

7. (a) 24 (b) 6 (c) 12 (d) 4 (e) 24

9. (a) 125 (b) 60 **11.** (a) 900 (b) 450 (c) 180

13. 9900 **15.** *aabb* *baab*
abab *baba*
abba *bbaa*

17. (a) *abcd* (b) *bacd* (c) *cabd* (d) *dabc*
abdc *badc* *cadb* *dacb*
acbd *bcad* *cbad* *dbac*
acdb *bcda* *cbda* *dbca*
adbc *bdac* *cdab* *dcab*
adcb *bdca* *cdba* *dcba*

19. 24 **21.** 5 · 4 · 3 · 2 · 1 (= 120) **23.** 24 **25.** 256

27. 120 **29.** 30,240

6.4 PERMUTATIONS (*CONTINUED*)

1. 12 **3.** 5 **5.** 1680 **7.** 70 **9.** 220

11. (a) 26 (b) 720 **13.** (a) 12 (b) 720

15. (a) 240 (b) 16 **17.** 28 **19.** 4060

21. (a) 6 (b) 30 (c) 120 (d) 360 (e) 720 (f) 720

23. (a) 840 (b) 24 (c) 120 (d) 240 (e) 240 (f) 20 (g) 360

25. 720 **27.** 210 **29.** 5040

6.5 COMBINATIONS

1. 28 **3.** 35 **5.** 4950 **7.** 56 **9.** 462

11. 10 **13.** 4845

15. (a) 1 (b) 9 (c) 36 (d) 84 (e) 126 (f) 126 (g) 84
(h) 36 (i) 9 (j) 1

17. 560 **19.** 2300 **21.** (a) 6 (b) 3 (c) 3

23. (a) 5 (b) 4 (c) 3 (d) 2 (e) 1 **25.** 35

27. (a) 30,240 (b) 100,000 **29.** (a) 414 (b) 38,709

31. 792 **33.** 362,880 **35.** 15

37. (a) 220 (b) 7,187,400

39. (a) 2016 (b) 3136 (c) 1536 **41.** 6 **43.** 12
45. 1260 **47.** 6435 **49.** 293,930 **51.** 1,562,275

6.6 REVIEW PROBLEMS

1. 12 **2.** 59,049 **3.** 1024
4. (a) 90 (b) 5040 (c) 286 **5.** 1680 **6.** 120
7. 792 **8.** 1848 **9.** 630 **10.** 18 **11.** 36

CHAPTER 7. PROBABILITY

7.2 WHAT IS PROBABILITY?

1. 39 **3.** 481 **5.** 161 **7.** 14 per 10,000 hands
9. 15 per 10,000,000 hands
11. (a) 0.1 (b) 0.5 (c) 0.4 (d) 0.4 (e) 0.2
13. (a) 0.5 (b) 0.4 (c) 0.7 (d) 0.5 (e) 0.3
15. (a) $\frac{1}{3}$ (b) $\frac{2}{3}$
17. (a) $\frac{1}{12}$ (b) $\frac{1}{12}$ (c) $\frac{1}{6}$ (d) $\frac{1}{6}$ (e) $\frac{1}{6}$ (f) $\frac{1}{6}$ (g) $\frac{1}{12}$ (h) $\frac{5}{12}$ (i) $\frac{7}{12}$
19. {1, 2}, {1, 3}, {1, 4}, {2, 3}, {2, 4}, {3, 4}; probability is $\frac{1}{6}$ for each. (a) $\frac{1}{6}$ (b) $\frac{5}{6}$ (c) $\frac{1}{6}$ (d) $\frac{1}{6}$ (e) $\frac{1}{3}$
21. Outcomes: 12, 13, 14, 21, 23, 24, 31, 32, 34, 41, 42, 43.
(a) $\frac{1}{12}$ (b) $\frac{1}{4}$ (c) $\frac{2}{3}$ (d) $\frac{5}{6}$ (e) $\frac{5}{6}$

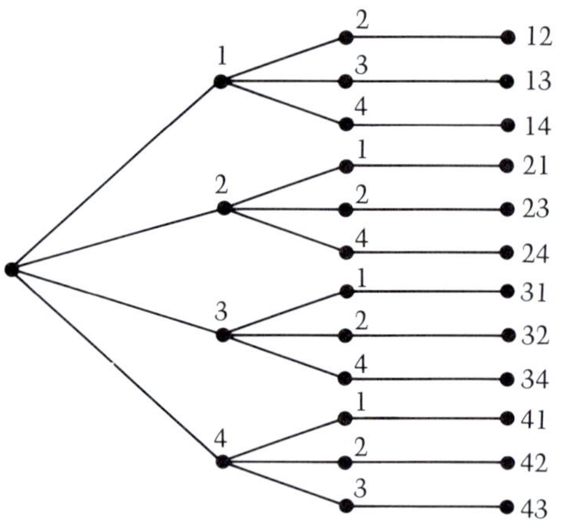

23. *zxxyy*	*xzxyy*	*xxzyy*	*xxyzy*	*xxyyz*
zxyxy	*xzyxy*	*xyzxy*	*xyxzy*	*xyxyz*
zxyyx	*xzyyx*	*xyzyx*	*xyyzx*	*xyyxz*
zyxxy	*yzxxy*	*yxzxy*	*yxxzy*	*yxxyz*
zyxyx	*yzxyx*	*yxzyx*	*yxyzx*	*yxyxz*
zyyxx	*yzyxx*	*yyzxx*	*yyxzx*	*yyxxz*

$\frac{1}{30}$ is the probability of each simple event. (a) $\frac{12}{30}$ (b) $\frac{10}{30}$ (c) $\frac{12}{30}$ (d) $\frac{25}{30}$ (e) $\frac{18}{30}$

25. $\frac{1}{2}$

7.3 SOME PRINCIPLES OF PROBABILITY

1. 0.49 **3.** 0.82 **5.** (a) 0 (b) 0.90

7. (a) 0 (b) 0.93 **9.** (a) 0 (b) 0.85 0 (c) 0.65 (d) 0.50

11. Yes. A single fair coin toss; M = "heads," N = "tails."

13. $P(A) = \frac{2}{3}$, $P(B) = \frac{1}{3}$ **15.** 0.20

17. $P(A) = 0.05$, $P(B) = 0.35$, $P(C) = 0.6$

19. (a) $\frac{1}{6}$ (b) $\frac{5}{36}$ (c) $\frac{7}{12}$ (d) $\frac{1}{2}$ (e) $\frac{1}{2}$ (f) $\frac{5}{6}$

21. (a) $\frac{16}{49}$ (b) $\frac{9}{49}$ (c) $\frac{24}{49}$ **23.** (a) $\frac{7}{12}$ (b) $\frac{5}{12}$

25. (a) 0.62 (b) 0.73 (c) 0.44 (d) 0.56

7.4 CONDITIONAL PROBABILITY AND INDEPENDENT EVENTS

1.

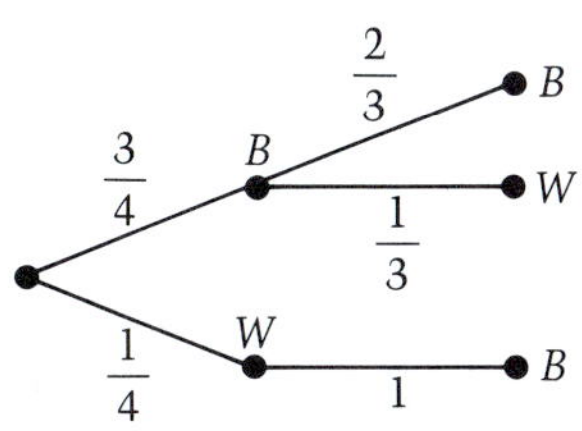

Outcomes	*Probabilities*
BB	$\frac{1}{2}$
BW	$\frac{1}{4}$
WB	$\frac{1}{4}$

(a) 0 (b) $\frac{1}{2}$ (c) $\frac{1}{2}$ (d) $\frac{1}{2}$

3.

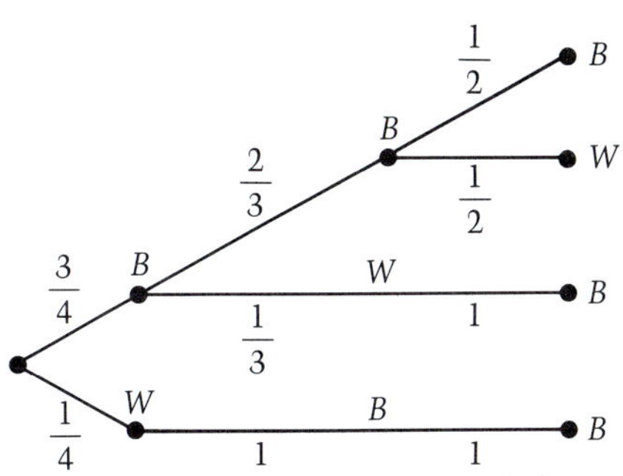

Outcomes	*Probabilities*
BBB	$\frac{1}{4}$
BBW	$\frac{1}{4}$
BWB	$\frac{1}{4}$
WBB	$\frac{1}{4}$

(a) 0 (b) $\frac{1}{4}$ (c) $\frac{1}{4}$ (d) 0 (e) 0

5.

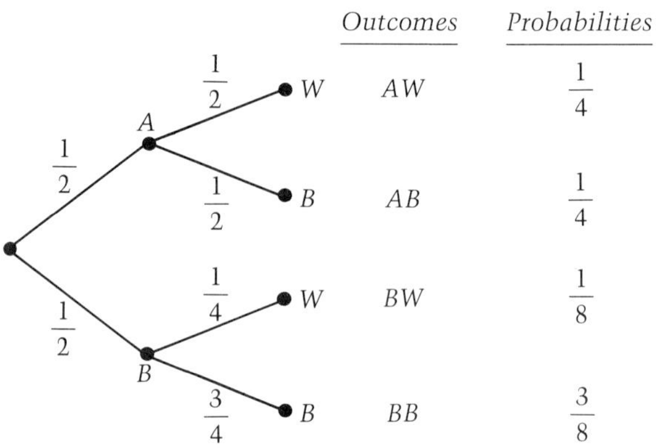

(a) $\frac{1}{4}$ (b) $\frac{1}{4}$ (c) $\frac{1}{8}$ (d) $\frac{3}{8}$

7. $\frac{1}{3}$ **9.** $\frac{1}{2}$

11. (a) 10 (b) {1, 2}, {1, 3}, {1, 4}, {1, 5}, {2, 3}, {2, 4}, {2, 5}, {3, 4}, {3, 5}, {4, 5} (c) $\frac{2}{5}$ (d) $\frac{1}{4}$ (e) $\frac{1}{4}$ (f) $\frac{2}{7}$

13. (a) $\frac{1}{2}$ (b) $\frac{1}{4}$ (c) $\frac{1}{7}$ **15.** (a) $\frac{1}{17}$ (b) $\frac{11}{51}$ (c) $\frac{3}{13}$

17. (a) $\frac{167}{330}$ or $0.5\overline{06}$ (b) $\frac{163}{330}$ or $0.4\overline{93}$

19. (a) $0.0\overline{21}$ (b) $0.\overline{14}$ (c) 0.076

21. (a) $\frac{744}{1566}$ or 0.475 (b) $\frac{822}{1566}$ or 0.525
(c) $\frac{451}{1566}$ or 0.288 (d) $\frac{355}{1566}$ or 0.227
(e) $\frac{215}{451}$ or 0.477 (f) $\frac{201}{822}$ or 0.245
(g) $\frac{385}{728}$ or 0.529

23. (a) $\frac{1}{3}$ (b) $\frac{55}{176}$ or 0.313 (c) F and P are not independent (d) $\frac{2}{25}$ (e) $\frac{3}{20}$ (f) S and F are not independent

25.

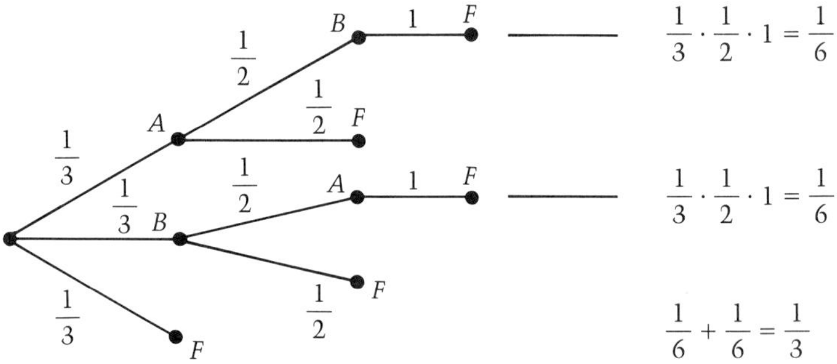

The probability that both A and B are tried before the food is reached is $\frac{1}{3}$.

27. $\frac{1}{4}$, $\frac{1}{2}$ **29.** (a) 0.45 (b) 0.5 (c) 0.35

31. (a) 0.25 (b) 0.375 (c) 0.375

33. (a) 0.42 (b) 0.348 (c) 0.232

7.5 BINOMIAL EXPERIMENTS

1. 0.5 **3.** 0.3456 **5.** 0.1536
7. 0.32768 **9.** 0.200 658 125 **11.** 0.167 790 298
13. 0.051 **15.** 0.234 **17.** 0.005
19. 0.003 **21.** 0.012 **23.** 0.000
25. 0.012 **27.** 0.150 **29.** 0.000
31. 0.938 **33.** 1.000 **35.** 0.962
37. 0.222 **39.** 0.264 **41.** 0.723

7.6 BAYES' THEOREM

1. (a) $\frac{1}{3}$ (b) $\frac{1}{4}$ (c) $\frac{17}{48}$ (d) $\frac{31}{48}$
3. (a) 0.775 (b) 0.4 (c) 0.3 (d) 0.51 (e) 0.18 (f) 0.31
5. (a) $\frac{8}{17}$ (b) $\frac{9}{17}$ (c) $\frac{4}{31}$ (d) $\frac{27}{31}$

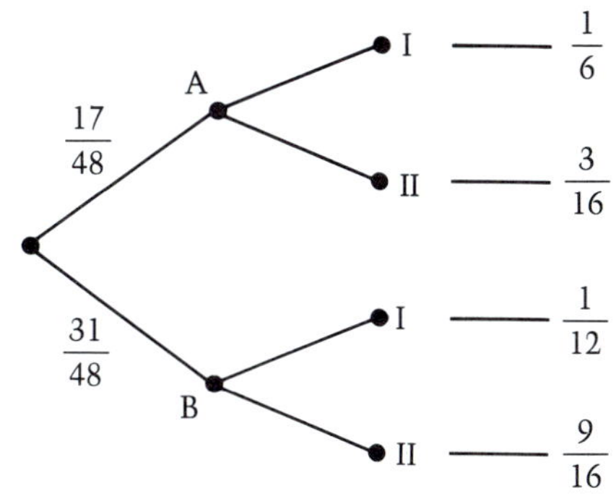

7. (a) 0.608 (b) $0.\overline{4}$ (c) 0.032 (d) 0.157 (e) $0.\overline{1}$ (f) 0.323 (g) 0.235 (h) $0.\overline{4}$ (i) 0.645

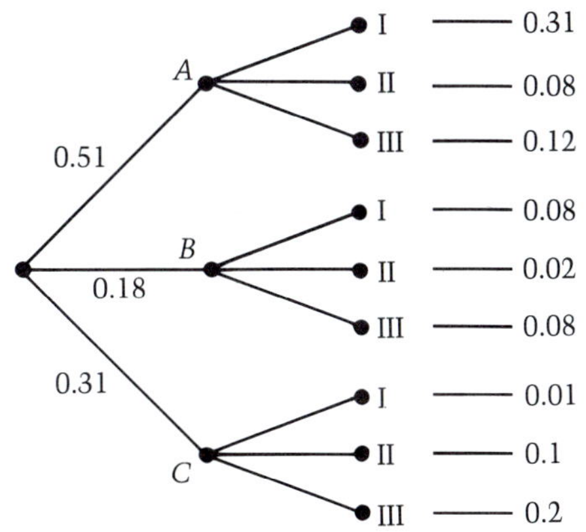

9. 0.3 **11.** (a) 0.678 (b) 0.683
13. (a) 0 (b) $\frac{1}{3}$ (c) $\frac{2}{3}$

15. (a) $\frac{9}{14}$ (b) $\frac{9}{19}$ (c) $\frac{3}{8}$ (d) $\frac{9}{104}$

17. 0.499 **19.** (a) 0.4372 (b) 0.3443 (c) 0.2186

21. 0.1099

7.7 MARKOV CHAINS

1. (a) 0.5 (b) 0.7 (c) 0.36 (d) 0.6

3. (a) 0.30 (b) 0.10 (c) 0.70 (d) 0.31 (e) 0.35 (f) 0.41

5. (a) 0.472 (b) 0.528 (c) 0.7168 (d) 0.2832

7. (a) 0.527 (b) 0.289 (c) 0.184 (d) 0.5131 (e) 0.2974 (f) 0.1895

9. $\langle 0.5 \quad 0.5 \rangle$ **11.** $\langle \frac{1}{3} \quad \frac{1}{3} \quad \frac{1}{3} \rangle$

13. (a) $\begin{bmatrix} 0.5 & 0.5 \\ 1 & 0 \end{bmatrix}$

(b) $\langle 1 \quad 0 \rangle$ (c) 0.5 (d) 0.75 (e) 0.375 (f) $\langle \frac{2}{3} \quad \frac{1}{3} \rangle$ (g) About $\frac{2}{3}$ (h) Number 1

15. (a) $T = \begin{bmatrix} 0 & \frac{1}{2} & \frac{1}{2} & 0 \\ \frac{1}{3} & 0 & \frac{1}{3} & \frac{1}{3} \\ \frac{1}{3} & \frac{1}{3} & 0 & \frac{1}{3} \\ 0 & \frac{1}{2} & \frac{1}{2} & 0 \end{bmatrix}$

(b) $\frac{1}{6}$ (c) $\langle \frac{1}{5} \quad \frac{3}{10} \quad \frac{3}{10} \quad \frac{1}{5} \rangle$ (d) $\frac{1}{5}$ (e) $\frac{3}{5}$

17. (a) $T = \begin{bmatrix} 0 & \frac{1}{3} & 0 & \frac{1}{3} & \frac{1}{3} \\ \frac{1}{3} & 0 & \frac{1}{3} & 0 & \frac{1}{3} \\ 0 & \frac{1}{3} & 0 & \frac{1}{3} & \frac{1}{3} \\ \frac{1}{3} & 0 & \frac{1}{3} & 0 & \frac{1}{3} \\ \frac{1}{4} & \frac{1}{4} & \frac{1}{4} & \frac{1}{4} & 0 \end{bmatrix}$

(b) $\frac{11}{36}$ (c) $\langle \frac{3}{16} \quad \frac{3}{16} \quad \frac{3}{16} \quad \frac{3}{16} \quad \frac{1}{4} \rangle$ (d) $\frac{3}{16}$ (e) $\frac{3}{8}$

19. (a) $T = \begin{bmatrix} 0 & \frac{1}{2} & 0 & \frac{1}{2} \\ \frac{1}{2} & 0 & \frac{1}{2} & 0 \\ 0 & \frac{1}{2} & 0 & \frac{1}{2} \\ \frac{1}{2} & 0 & \frac{1}{2} & 0 \end{bmatrix}$

(b) $\langle \frac{1}{4} \quad \frac{1}{4} \quad \frac{1}{4} \quad \frac{1}{4} \rangle$ (c) States are equally likely.

21. $\frac{2}{7}$ **23.** $\frac{7}{33}$ for Democrats. $\frac{1}{11}$ for Republicans. $\frac{23}{33}$ for Independents.

7.8 REVIEW EXERCISES

1. 135 **2.** $\frac{5}{26}$ **3.** 0.458 **4.** $\frac{48}{95}$ **5.** $\frac{13}{20}$

6. $\frac{1}{2}$ **7.** $\frac{2}{3}$ **8.** 0.3125 **9.** 0.175104 **10.** $\frac{16}{41}$

11. (a) 0.8 (b) 0.35 (c) 0.5 (d) $\langle 0.4\overline{6} \quad 0.5\overline{3} \rangle$ (e) $0.5\overline{3}$

CHAPTER 8. SEQUENCES AND MATHEMATICS OF FINANCE

8.2 ARITHMETIC SEQUENCES

1. (a) 3 (b) 17, 20, 23 (c) 798
3. (a) 4 (b) 13, 17, 21 (c) 861
5. (a) 1.1 (b) 5.6, 6.7, 7.8 (c) 279.3
7. (a) −5 (b) −12, −17, −22 (c) −987
9. (a) $\frac{1}{6}$ (b) $\frac{7}{6}, \frac{4}{3}, \frac{3}{2}$ (c) 49
11. (a) 249 (b) 6325 **13.** 3.1
15. (a) 3 (b) 21 (c) 735 **17.** 1395
19. (a) \$26,200 (b) \$309,000
21. (a) (i) \$411,000 (ii) \$414,000
23. \$3339.75 (b) Yes. $\frac{n(n+1)}{2} = 300$; $n = 24$
27. \$289.38 **29.** \$987.75

8.3 GEOMETRIC SEQUENCES

1. −1 and 3 **3.** $-\frac{1}{2}$ and $\frac{1}{16}$ **5.** $-\frac{1}{3}$ and $\frac{1}{9}$
7. $\frac{1}{2}$ and $\frac{1}{96}$ **9.** $1.2 = \frac{6}{5}$ and 8.957952
11. $\frac{819}{128}$ **13.** (a) $3\sqrt{2}$ (b) 24 and 36
15. (a) $13{,}200(1.05)^9 = \$20{,}477.53$ (b) \$166,028.18
17. (a) $20{,}000(1.06)^9 = \$33{,}789.58$ (b) \$263,615.90
19. (a) \$19,500 from (1) and \$20,273.75 from (2). (b) \$172,500 from (1) and \$158,169.54 from (2). (c) \$395,000 from (1) and \$441,427.09 from (2).
21. \$29,428.47

8.4 INTEREST: SIMPLE AND COMPOUND

1. $j = 0.005$ and $m = 36$ **3.** $j = 0.025$ and $m = 8$
5. $j = 0.06$ and $m = 10$ **7.** \$1050.94
9. \$3804.72 **11.** 5.0945% **13.** 12.6825%
15. \$618.31 **17.** 47 quarters **19.** 70 months
21. \$3642.23 **23.** \$6231.67 **25.** \$12,405.21

8.5 CONTINUOUS INTEREST

1. 5.127% **3.** 10.517% **5.** 16.183%
7. \$6615.65 **9.** \$40.50 **11.** \$5538.70
13. \$7261.49 **15.** 9% **17.** 5%

8.6 INTEREST ON PERIODIC SAVINGS

1. \$7011.88 **3.** \$9241.50 **5.** \$15,197.96
7. \$2444.31 **9.** \$31.30 **11.** \$99.93
13. (a) \$14,023.77 (b) 94 months (c) \$5200 approximately.
15. \$283.30 **17.** (i) \$9012.22 (ii) \$8733.20
19. (i) \$25,155.78 (ii) \$26,413.57

8.7 INSTALLMENT PAYMENTS AND RETIREMENT INCOME

1. \$8206.64 **3.** (a) \$2037.04 (b) \$40,740.80
5. \$166.07 **7.** \$514.31 **9.** \$438.79
11. \$141.11 **13.** 16.24%

15. $A = \frac{200}{0.005}[(1.005)^{241} - 1.005] = \$92{,}870.22$. Now,

$$92{,}870.22 = \frac{1000}{0.005}[1 - (1.005)^{-m}]$$

$$(1.005)^{-m} = 0.5356489, \qquad (1.005)^{m} = 1.8668945,$$

$$m = 125^{+}\text{months} \doteq 10 \text{ years } 5 \text{ months}.$$

17. $V = \left(\frac{3000}{0.015}\right)[1 - (1.015)^{-40}] = \$89{,}747.53$ is the amount they will need in 30 years to buy the 10-year annuity. Thus, we now need to find D where

$$89{,}747.53 = \frac{D}{0.015}[(1.015)^{121} - 1.015]; \qquad D = \$266.90.$$

8.8 SUMMARY AND REVIEW EXERCISES

1. (a) -8 (b) 16, 8, 0 (c) -2880
3. (a) $\frac{1}{2}$ (b) $\frac{13}{6}, \frac{8}{3}, \frac{19}{6}$ (c) $\frac{475}{2}$
5. (a) $-\frac{1}{4}$ (b) $-1, \frac{1}{4}, -\frac{1}{16}$
7. (a) 1.4 (b) $x = 3.5$ and $y = 4.9$
9. (a) -41 (b) 5 **11.** (a) \$28,800 (b) \$234,000
13. \$181.41 **15.** \$7284.46 **17.** 7.5% **19.** \$148.62

CHAPTER 9. STATISTICS

9.2 GRAPHICAL REPRESENTATION OF DATA

1.

Interval	Tally	Frequency
63–65	𝍸	5
66–68	||||	4
69–71	𝍸 |	6
72–74	𝍸	5
75–77	||||	4

3.

Interval	Tally	Frequency
231–233	|	1
234–236	|||	3
237–239	|	1
240–242	||	2
243–245	~~||||~~ ~~||||~~ |	11
246–248	~~||||~~ |	6
249–251	|||	3
252–254		0
255–257	|	1

5.

Interval	Tally	Frequency
56–57	||	2
58–59	~~||||~~ |	6
60–61	~~||||~~ ~~||||~~ ||	12
62–63	|||	3

7.

Interval	Tally	Frequency
13–13	~~||||~~ ||	7
14–14	~~||||~~ |||	8
15–15	|||	3
16–16	||	2

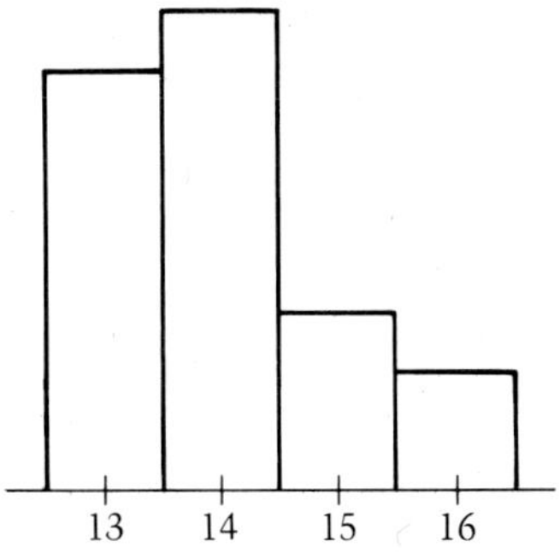

9.

Interval	Tally	Frequency
5–7	~~||||~~ ~~||||~~ |	11
8–10	~~||||~~ ||||	9
11–13	|||	3
14–16	||	2

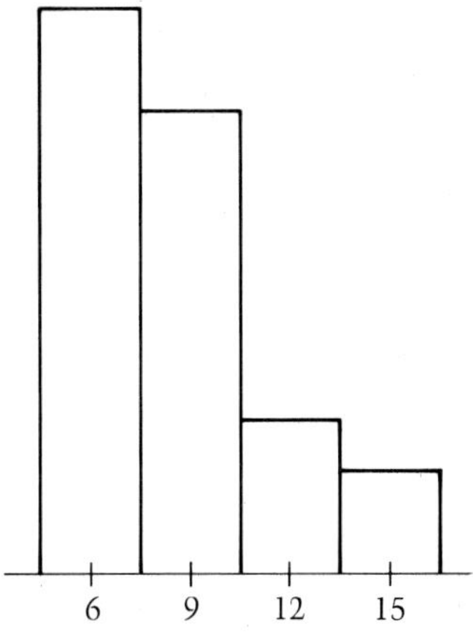

11.

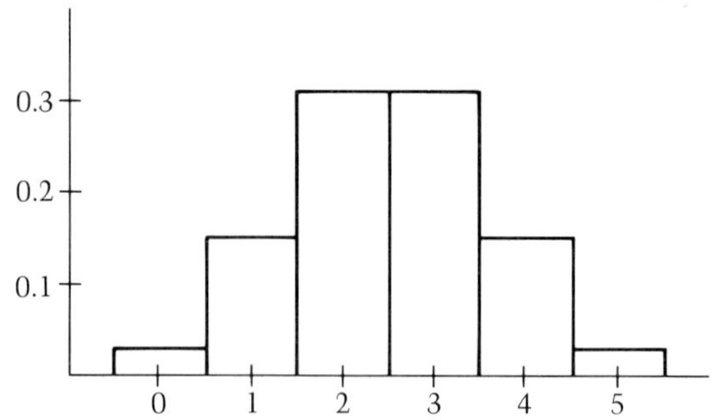
0.3
0.2
0.1
0
1
2
3
4
5

13.

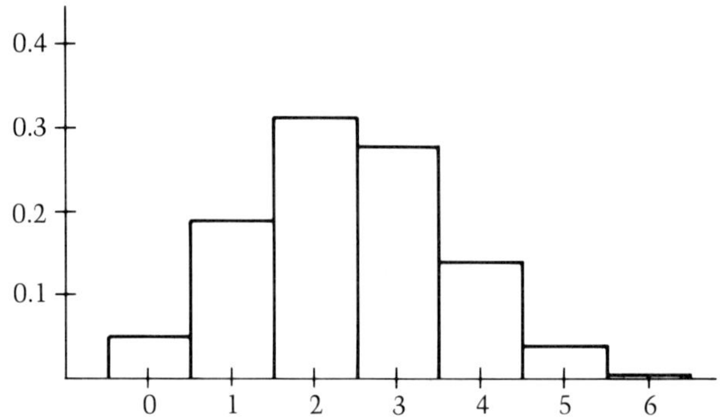
0.4
0.3
0.2
0.1
0
1
2
3
4
5
6

15.

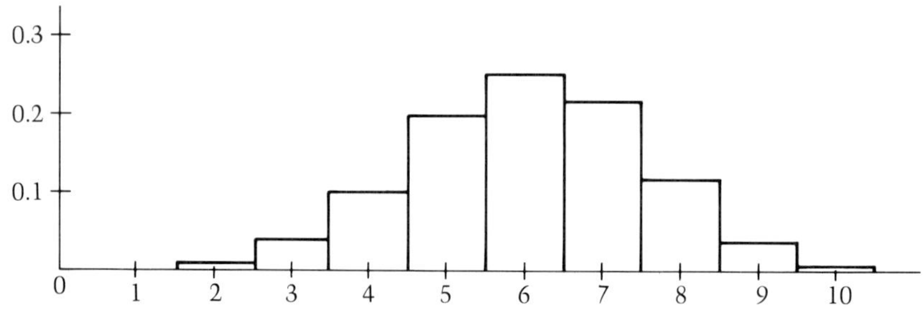
0.3
0.2
0.1
0
1
2
3
4
5
6
7
8
9
10

17.

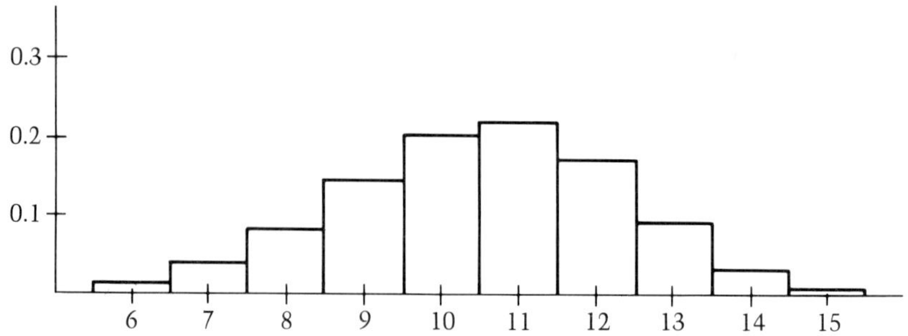
0.3
0.2
0.1
6
7
8
9
10
11
12
13
14
15

19.

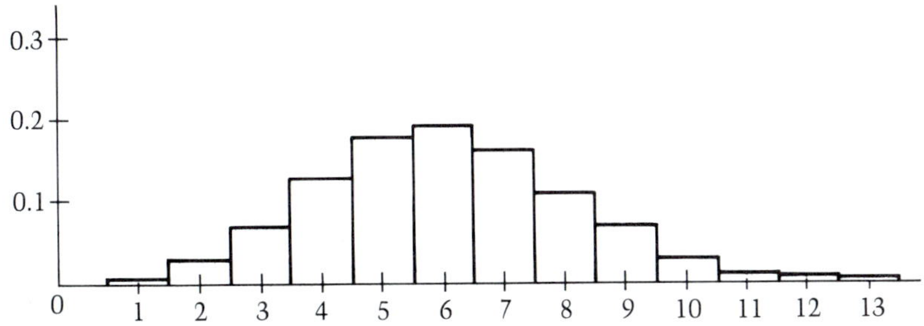

9.3 CENTRAL TENDENCIES, VARIANCE, AND CHEBYCHEV'S THEOREM

1. 5.4 **3.** 3 **5.** 10 **7.** 20 **9.** 15.9
11. $\overline{X} = 5.4$, $s = 1.730$ **13.** $\overline{X} = 1.9$, $s = 1.524$
15. $\overline{X} = 10.067$, $s = 2.604$ **17.** $\overline{X} = 5.833$, $s = 0.718$
19. $\overline{X} = 59.913$, $s = 1.593$ **21.** 55.6% **23.** 19.8%
25. 6.3% **27.** 20.3% **29.** 32.5%

9.4 THE NORMAL CURVE

1. 0.8830 **3.** 0.9726 **5.** 0.9985
7. 0.0003 **9.** 0.3015 **11.** 0.2148
13. 1.07 **15.** 2.78 **17.** 0.29
19. 1.88 **21.** −1 **23.** −1.6
25. 0.417 **27.** 0.625 **29.** 0.0010
31. 0.0008 **33.** 0.0047 **35.** 0.8686
37. 0.9015 **39.** 0.9904 **41.** 0.2104
43. 0.3097 **45.** 0.8521 **47.** 0.9504
49. −0.22 **51.** −1.92 **53.** −0.67
55. 1.25 **57.** 0.27 **59.** 0.8714
61. 0.9956 **63.** 0.1974 **65.** 285.5
67. 80.9 **69.** 42.246 **71.** 0.5793
73. 0.1401 **75.** 0.1401 **77.** 0.8770
79. 0.1554 **81.** 0.9544 **83.** 214.25, 285.75
85. 248, 252 **87.** 198, 302 **89.** 220.25, 279.75

9.5 NORMAL CURVES AND BINOMIAL PROBABILITIES

1. 1.0000 **3.** 0.2576 **5.** 0.0672 **7.** 0.2177
9. 0.5388 **11.** 0.0796 **13.** 0.0207 **15.** 0.4364
17. 0.8868 **19.** 1.000

9.6 REVIEW EXERCISES

1.

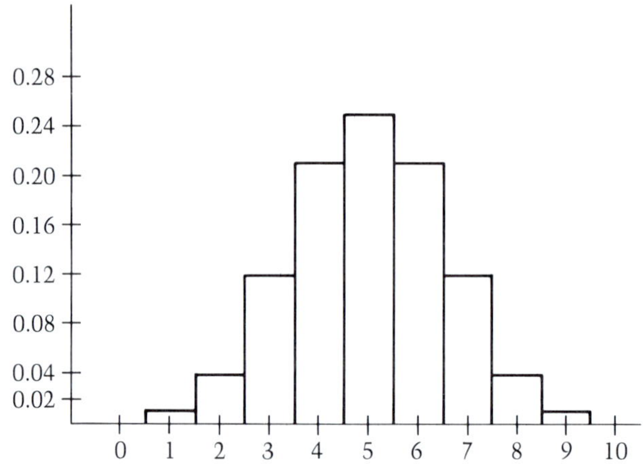

2.

1–2	||	2
3–4	𝍸 ||	7
5–6	𝍸 ||	7
7–8	𝍸	5
9–10	𝍸	5

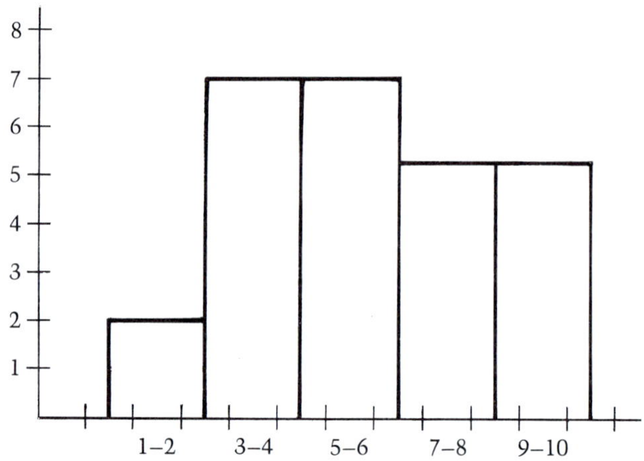

3. $5.\overline{538461}$ **4.** (a) 5 (b) 4.5

5. 2.4997 **6.** 0.13%

7. (a) 0.1151 (b) 0.3745 (c) 0.2417

8. 0.9050 **9.** 0.5125 **10.** 33

INDEX

R

S

T

U

V

W

X

Y

1. $\#(A \cup B) = \#(A) + \#(B) - \#(A \cap B)$ (The number of elements in $A \cup B$. Chapter 1)

2. $m = \dfrac{b - d}{a - c}, \quad a \neq c$ (Slope of line on points (a, b), (c, d). Chapter 1)

3. $y = mx + b$ (Slope-intercept form of a line. Chapter 1)

4. $y - b = m(x - a)$ (Point-slope form of a line. Chapter 1)

5. $d = \sqrt{(a - c)^2 + (b - d)^2}$ (Distance between points (a, b), (c, d). Chapter 1)

6. $P(n, r) = n(n - 1)(n - 2) \cdots (n - r + 1)$ (Number of permutations of n objects taken r at a time. Chapter 6)

7. $\dbinom{n}{r} = \dfrac{n!}{r!(n - r)!}$ (Number of combinations of n objects taken r at a time. Chapter 6)

8. $P(A \text{ or } B) = P(A) + P(B) - P(A \text{ and } B)$ (Probability of event A or B. Chapter 7)

9. $P(A \mid B) = \dfrac{P(A \text{ and } B)}{P(B)}, \quad P(B) \neq 0$ (Probability of A given B. Chapter 7)

10. $P(A \text{ and } B) = P(A) \cdot P(B)$, if A, B are independent. (Probability of A and B if A, B are independent. Chapter 7)

11. $a_n = a_1 + (n - 1)d$ (nth term of an arithmetic sequence. Chapter 8)

12. $s_n = \dfrac{n}{2}[2a_1 + (n - 1)d]$ (Sum of an arithmetic sequence. Chapter 8)

13. $s_n = \dfrac{n}{2}(a_1 + a_n)$ (Sum of an arithmetic sequence. Chapter 8)